Orbital Motion

Orbital Motion

Third Edition

A E Roy, Ph.D., F.R.A.S., F.R.S.E., F.B.I.S.
Professor of Astronomy,
University of Glasgow

Adam Hilger, Bristol, Philadelphia and New York

British Library Cataloguing in Publication Data

Roy, A. E. (Archie Edmiston), *1924*–
    Orbital motion.—3rd ed.
    1. Astronomical bodies. Orbits
    I. Title
    521.3

ISBN 0-85274-228-2 (hbk)
        0-85274-229-0 (pbk)

Library of Congress Cataloging-in-Publication Data are available

First published 1978
Reprinted 1979
Second edition 1982
Third edition 1988
Reprinted 1991

Consultant Editor: Professor A J Meadows

Published under the Adam Hilger imprint by IOP Publishing Ltd
Techno House, Redcliffe Way, Bristol BS1 6NX, England
242 Cherry Street, Philadelphia, PA 19106, USA

Printed in Great Britain by J W Arrowsmith Ltd, Bristol

Dedicated to the Memory of
Michael William Ovenden (1926–1987)

Contents

4 The Two-Body Problem

5 The Many-Body Problem

10   Artificial Satellites

11   Rocket Dynamics and Transfer Orbits

12   Interplanetary and Lunar Trajectories

13   Orbit Determination and Interplanetary Navigation

14   Binary and Other Few-Body Systems

Preface to First Edition

The term 'orbital motion' implies the movement of one body about another. Traditionally this may involve the revolution of a satellite about a planet, or a planet about the Sun, or one star about another or about the centre of the Galaxy. Even galaxies orbit each other. Branches of dynamics have been developed ever since the days of Newton to tackle the problems associated with such matters.

Celestial mechanics, the oldest of these branches, was begun by Newton himself and received its fullest classical development during the eighteenth and nineteenth centuries, culminating in the elegant work of Poincaré and the exhaustive treatment of the Moon's motion by Hill and Brown. Such treatments involved for the most part analytical methods, many important sections of applied mathematics stemming from the development of such methods.

More recently, with the invention of high-speed computers, a branch of celestial mechanics has been created in which numerical experiments are carried out, investigating the orbital evolution of three or more (sometimes many more!) gravitating masses. Valuable insights into problems hitherto unsolved by more conventional methods have resulted from such experiments. Such methods have been applied, not only to the cases of satellites about planets or planetary orbital evolution, but also to binary star creation and the dynamics of star clusters.

Again, the advent of artificial satellites and interplanetary probes has posed interesting and hitherto unencountered classes of orbital motion problems urgently requiring solutions. The term 'astrodynamics' has been applied to the rapidly developing branch of space science that deals with the study of the behaviour of man-made missiles within the Solar System under the action of Newton's law of gravitation and impulsive rocket action. Also within the scope of that subject are powered flight and the effects of atmospheric drag, electromagnetic fields and, where applicable, radiation pressure.

The whole field of orbital motion studies (celestial mechanics, astrodynamics, stellar kinematics, stellar dynamics, binary motion and so on) has therefore seen many significant and exciting developments in recent years and it is hoped that the present book will go some way towards introducing the reader to these branches of astronomy. This text is therefore aimed principally at university and technical college students requiring an introduction to the subject of orbital motion. Such students may have little or no knowledge of astronomy or of space science but, for one reason or another, feel the need to learn the basic concepts, techniques and results achieved in the subject. Much of the material has been used in courses given by the author in the departments of astronomy, aeroengineering and geography of Glasgow University over the past fifteen years or so. Parts of an earlier work by the author† (now out of print) have been extensively rewritten, updated and incorporated in the present text.

The book contains over 200 carefully selected references that will enable the reader to broaden his knowledge of those fields that interest him. Most of these references themselves provide extensive bibliographies. The problems at the end of certain chapters will give the student a means of testing his understanding of the subject matter of such chapters. Many are original or have appeared in examinations set by the Department of Astronomy of Glasgow University in the past; others are versions of problems that have appeared in many places in many forms. Answers and hints towards solutions are provided where necessary at the end of the book.

Throughout the book a knowledge of calculus and elementary vector methods is assumed. Figures and equations are given the number of the chapter in which they appear, followed by a number denoting their order in that chapter.

The Appendices provide lists of relevant astronomical and mathematical data.

The author acknowledges with sincere gratitude his personal debt to those workers, not only in the reference lists who have influenced his thoughts in times past, but also to his former teachers, his colleagues and students past and present for many helpful discussions.

He would also like to thank the Controller of Her Majesty's Stationery Office for permission to reproduce from the *Explanatory Supplement to the Astronomical Ephemeris* the astronomical data on the planets and satellites in Appendices I, III and IV, the Council of the British Astronomical Association for permission to reproduce from the *Handbook of the B.A.A. for 1977* columns 6 and 7 of the Satellite Elements and Dimensions table in Appendix IV, his colleague Dr David Clarke for permission to include (with some modification) some material from *Astronomy: Principles and Practice* and *Astronomy: Structure of the Universe* by Archie E Roy and David Clarke (London: Adam Hilger, 1977) and Professor Desmond King-Hele for permission to base figure 10.2 on figure 9 of his paper *A View of Earth and Air*.

The author's indebtedness is also extended to Mrs L Williamson of the Department of Astronomy for her painstaking preparation of the typescript of the book.

Archie E Roy

Department of Astronomy,
The University of Glasgow.

† Roy A E 1965 *The Foundations of Astrodynamics* (New York: Macmillan)

Preface to Third Edition

Celestial mechanics in recent years has continued its rapid progress, particularly in our understanding of the problem of the stability of the Solar System and its subsystems. This third edition has provided an opportunity to review such progress. Again the choice of references and bibliographical items has been dictated not only by the relevance and importance of various studies but also by a given item's possession itself of a reference list useful for the student's further reading. Opportunity has also been taken to correct a few remaining errors to which the author's attention has been drawn. He thanks all those who have helped in this respect.

Thanks are also due to Mrs L Williamson for her help and, as always, her painstaking care in typing the changes and new material for this edition.

January 1988

Archie E Roy

Department of Physics and Astronomy,
The University of Glasgow.

1 The Restless Universe

1.1 Introduction

The myriads of objects making up the universe are never still. From the largest galaxy (containing some 250 000 million times the mass of the Sun) down to the smallest asteroid (dwarfed by many terrestrial cities in size) they move relative to each other. Sometimes the motions are systematic and essentially repeating, as in the orbital movement of a planet about the Sun or the Moon about the Earth; in other cases there is seemingly no repetition, as when a star escapes from a galaxy and wanders for an astronomically long time in the depths of intergalactic space, its trajectory shaped by the spectral gravitational fingers of distant galaxies.

In a surprisingly large number of cases, however, spread over vast ranges of size and mass, we can talk of the movements as essentially orbital. That this is so, from the revolution of a tiny satellite about Mars, up through the orbital motion of one star about another to the colossal paths traced out by members of a cluster of galaxies, is because of the dominating influence of gravitation. Although the force of gravitation is one of the weakest in the atomic and subatomic level—and indeed can be neglected beside electrostatic and nuclear forces—it inherits the universe on the macroscopic scale where orbital motion is concerned, all other forces such as magnetism operating with much smaller effects except in a few special cases.

In this chapter, we will survey briefly the structure of the universe, paying special attention to the types of motion found in its various parts. We shall be concerned with the physical make-up of its members only insofar as it is relevant to the dynamic picture. An attempt will be made to highlight the specific features of the movements of celestial objects that require explanations. In many cases it will be seen that an understanding of the reasons behind the type of motion observed can shed light on the origin and evolution of the bodies concerned, as for example in the case of the planetary orbits in the Solar System. Thus a study of orbital motion is important in many astronomical fields. In recent years, too, with the advent of artificial satellites and interplanetary probes, a mastery of orbital dynamics or astrodynamics is essential in achieving the research goals for which they were created.

1.2 The Solar System

The bodies of the Solar System, with the exception of most comets, are all contained in a region of space of diameter one thirty-thousandth of the distance to the nearest star; the perturbing effects of other stars are therefore negligible, the Solar System being effectively isolated as far as its internal movements are concerned.

The Sun, planets, satellites, asteroids, meteors and the interplanetary medium therefore form a closed system, its members reacting upon each other. As far as their movements within the Solar System are concerned, the movement of the system itself about the galactic centre is irrelevant.

In Appendices 1–4 are provided the basic data concerning the sizes, masses, distances etc. of the members of the Solar System. To augment this information recourse may be had to the books and other references listed at the end of the chapter. Meanwhile the brief description provided below will suffice for present purposes and will be supplemented in later chapters of this book wherever necessary.

The Sun, a typical star, dominates the System in size and mass. The diameter of Jupiter, the largest planet, is but a tenth of the Sun's diameter, its mass but a thousandth. Because of this predominance in mass, the planets Mercury, Venus, Earth, Mars, Jupiter, Saturn, Uranus, Neptune and Pluto move (to a first and high degree of approximation) as if they were attracted solely by the Sun, their orbits being ellipses of various sizes about the Sun with the latter at one focus. In most cases, the eccentricities of these elliptical orbits are less than 0·1; Mercury and Pluto are exceptions, moving in orbits of eccentricities 0·206 and 0·250 respectively.

The planes of the planetary orbits contain the Sun's centre and are in general inclined to the plane of the Earth's orbit by no more than a few degrees. Mercury and Pluto are again the exceptions because, whereas the angle of inclination for the others is less than four degrees, the angles of inclination for Mercury and Pluto are 7° and 17° 9′ respectively. If a diagram is made of the Solar System with the distance to scale, the planes of the planetary orbits being rotated onto the Earth's orbital plane, figure 1.1 is obtained.

Also inserted in the diagram is the asteroid belt. This is a region between the orbits of Jupiter and Mars occupied by the orbits of thousands of minor planets, the largest of which (Ceres) is 1100 km in diameter.

It should also be noticed that the minimum distance of Pluto from the Sun is less than the average distance of Neptune from the Sun. If it were not for the mutual inclination of their orbital planes and the locking mechanism which keeps conjunctions of the two planets to the aphelion side of Pluto's orbit (chapter 8), the danger of a collision between these two planets would be greatly enhanced.

It is therefore seen that the planetary orbits are almost circular and co-planar. Indeed, from 1978 until just beyond the end of the present century, Neptune will be the farthest planet from the Sun, Pluto being in the vicinity of its point of closest approach to the Sun.

Each planet except Mercury and Venus is attended by one satellite or more. Saturn, in addition to its 17 moons, has a ring system composed of millions of tiny satellites of uncertain nature, moving in coplanar, almost circular orbits about the planet. The Earth has one large satellite (the Moon) almost one-eightieth its mass. Mars has two small satellites with diameters less than 16 km. Jupiter has sixteen moons of which the four so-called Galilean satellites are the largest, being about the same size as or larger than our Moon. The other twelve are much smaller, ranging in diameter from about 160 km downwards. In addition Jupiter possesses a ring system. Saturn's 17 moons include Titan, which is about as large as the planet Mercury, and very much smaller bodies. Uranus possesses five large satellites as well as a number of small ones, all revolving in almost circular, coplanar orbits; it also has a complex system of at least nine rings.

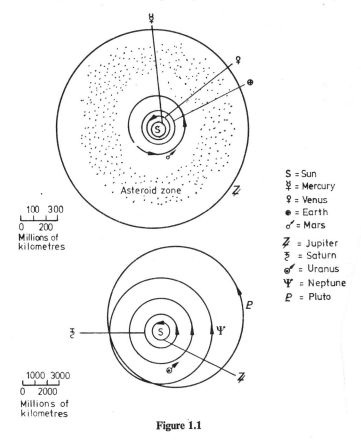

S = Sun
☿ = Mercury
♀ = Venus
● = Earth
♂ = Mars

♃ = Jupiter
♄ = Saturn
♅ = Uranus
♆ = Neptune
♇ = Pluto

Figure 1.1

Neptune has two moons, one of which must be as massive as Titan. The other moon (Nereid) has a highly eccentric orbit and was discovered as late as 1948. The planet pluto has one moon, Charon, discovered in 1978.

Again, most of the satellites revolve about their planets in elliptic orbits of small eccentricity and in almost coplanar systems, though the mean plane in one planet's system of satellites may be very different from that in another's. The directions of rotation of the satellites in their orbits are also (for all but a minority of cases) in the same direction in which the planets revolve about the

3

Sun. The exceptional satellites are thought by some astronomers to have had a different origin from the others.

In general, the planets and satellites also rotate about axes fixed within them, the direction of rotation for most being in the same direction in which the planets revolve about the Sun or the satellites about their primaries. The retrograde satellites are again the exceptions to the rule. The periods of rotation of some satellites are equal to their periods of revolution about their primaries so that they therefore keep the same face turned towards the body about which they revolve.

Conditions at the surface of the bodies in the Solar System vary greatly from body to body. They depend upon the past history of the body, its mass and radius, its distance from the Sun and its period of rotation on its axis.

1.2.1 Kepler's laws

Johannes Kepler (1571–1630), from a study of the mass of observational data on the planets' positions collected by Tycho Brahe (1546–1601), formulated the three laws of planetary motion forever associated with his name. They are:

(i) The orbit of each planet is an ellipse with the Sun at one focus.
(ii) For any planet the rate of description of area by the radius vector joining planet to Sun is constant.
(iii) The cubes of the semimajor axes of the planetary orbits are proportional to the squares of the planets' periods of revolution.

Kepler's first law tells us what the shapes of the planetary orbits are and gives the position of the Sun within them.

Kepler's second law states how the angular velocity of a planet in its orbit varies with its distance from the Sun, being greatest at perihelion and least at aphelion.

Kepler's third law relates the different sizes of the orbits in a system to the periods of revolution of the planets in these orbits.

As far as observational accuracy at the time of their formulation was concerned, Kepler's laws were exact. Even today, they may be taken as very close approximations to the truth. They hold, not only for the system of planets moving about the Sun, but for the various systems of satellites moving about their primaries. Only when the outermost retrograde satellites in the Solar System or close satellites of a nonspherical planet are considered do they fail to describe in their usual highly accurate manner the behaviour of such bodies. Even then, they may be used as a first approximation.

Kepler's laws are in fact a description of a special solution to the gravitational problem of n bodies where (a) all the bodies may be treated as point-masses and (b) all the masses but one are so small that they do not attract each other appreciably, but are attracted solely by the large mass. It so happens that to a high degree of accuracy the system of planets and Sun, and the system of each set of satellites moving about their primary planet, satisfy these conditions. Sir Isaac Newton (1642–1727) was the first to realize this and to treat the problem systematically.

1.2.2 Bode's law

There is an additional interesting feature in the planetary distances from the Sun. This is known as Bode's law, though it has not the same status as Kepler's laws. It is often written as

$$r_n = 0{\cdot}4 + 0{\cdot}3(2^n)$$

where r_n is the mean distance of the planet from the Sun, n taking the values $-\infty, 0, 1, 2, 3, \ldots$. Table 1.1 illustrates the degree to which the law fits the facts.

Table 1.1

Name	n	Distance from the Sun (in units of the Earth's distance)	
		By Bode's law	Actual
Mercury	$-\infty$	0·4	0·39
Venus	0	0·7	0·72
Earth	1	1·0	1·00
Mars	2	1·6	1·52
asteroids	3	2·8	2·80
Jupiter	4	5·2	5·20
Saturn	5	10·0	9·54
Uranus	6	19·6	19·20
Neptune	7	38·8	30·07
Pluto	8	77·2	39·46

When the law was first publicized in 1772, Uranus, Neptune, Pluto and the asteroids were undiscovered. The close fit of Uranus when it was found in 1781 generated confidence in the law and drew attention to the gap that lay between the orbits of Mars and Jupiter. A number of astronomers banded together to make a search for the missing planet. Instead of one large planet being discovered, a number of small bodies (the asteroids) were found whose mean distance turned out to be almost precisely that predicted by Bode's law.

The agreement for Neptune is poor, however, and Pluto does not fit at all, though its position is close to that given by $n=7$. This failure has led people to argue that the law is merely coincidental, having no underlying foundation in physics. Nevertheless some researchers on the origin of the Solar System have arrived at Bode-type laws as a consequence of their theories concerning planetary formation.

Similar laws can be found for the major satellite systems. For example, Miss Blagg generalized Bode's law, and a number of the bodies discovered subsequent to her generalization have been found to fit her version of it.

5

1.2.3 Commensurabilities in mean motion

There exists in the Solar System a remarkable number of approximate commensurabilities in mean motion between pairs of bodies in the planetary and satellite systems. For any planet moving about the Sun, the planet's mean motion may be taken to be its mean angular velocity of revolution. This is obtained by dividing 360° by its mean period of revolution. For example, if n_J, n_S, n_N and n_P are the mean motions in degrees per day of Jupiter, Saturn, Neptune and Pluto respectively, then

$$n_J = 0 \cdot 083\ 091$$

$$n_S = 0 \cdot 033\ 460$$

$$n_N = 0 \cdot 005\ 981$$

$$n_P = 0 \cdot 003\ 979.$$

We then have

$$2n_J - 5n_S = -0 \cdot 001\ 118$$

$$3n_P - 2n_N = -0 \cdot 000\ 025$$

showing how close the ratios of these pairs of mean motions are to simple fractions. A study of the numbers of such commensurabilities was carried out by Roy and Ovenden, who showed that there were many more than could be expected by chance alone.

Triple commensurabilities also exist. If n_1, n_2 and n_3 are the mean motions of Io, Europa and Ganymede (three of the four Galilean satellites of Jupiter) respectively, then in degrees per day,

$$n_1 = 203 \cdot 488\ 992\ 435$$

$$n_2 = 101 \cdot 374\ 761\ 672 \text{ and}$$

$$n_3 = \ \ 50 \cdot 317\ 646\ 290.$$

We then have

$$n_1 - 2n_2 = 0 \cdot 739\ 469\ 091 \text{ and}$$

$$n_2 - 2n_3 = 0 \cdot 739\ 469\ 092$$

giving

$$n_1 - 3n_2 + 2n_3 = 0$$

which is exact to the limit of observational accuracy.

Corresponding to this remarkable commensurability in the mean motions of the satellites, there is an equally exact one in their mean longitudes, viz.

$$l_1 - 3l_2 + 2l_3 = 180°.$$

It will be seen later that there are good grounds for believing that questions of stability underlie the existence of such relationships. At this stage, however, we content ourselves by drawing attention to three other examples of commensurable mean motions.

The first concerns the asteroids. These are a numerous group of bodies revolving about the Sun between the orbits of Mars and Jupiter, though there are a few (usually of highly eccentric orbit) that can approach to within Mercury's orbit or recede as far as Jupiter's. There are also two groups, the Trojans, whose members oscillate about points in Jupiter's orbit. The Trojans are examples of an interesting case, first discovered by Lagrange, of the problem of the gravitational attractions of three bodies. This states that a small body can remain at a corner of an equilateral triangle, the other two corners being occupied by two massive bodies in orbit about each other. The Trojans are distributed between the two possible equilateral points (Jupiter and the Sun being the massive bodies) 60° ahead and 60° behind the heliocentric longitude of Jupiter. The Trojans may be said to be a special case of a commensurability of unity.

In addition, study of the distribution of the orbits of the thousands of other asteroids found to date has shown that certain heliocentric distances are avoided. These distances correspond to mean motions that are commensurable with that of Jupiter (the main disturber of asteroid orbits). Commensurabilities of one-half, one-third, two-fifths and so on are avoided, such gaps in the distribution being referred to as the Kirkwood gaps after their discoverer. On the other hand, there is an accumulation of asteroid orbits near the commensurability of two-thirds, possibly an orbit stable against Jovian disturbances.

The second example also involves Jupiter but in this case that planet is the body fighting to keep its outer satellites from being torn away by the Sun's disturbing gravitational field.

Jupiter is attended by sixteen known satellites. Their mean distances from the planet's centre range from 128 000 km to 24 000 000 km. The four large moons (Io, Europa, Ganymede and Callisto) move in almost circular, coplanar orbits. The others have names but are also numbered in order of discovery. The fifth (Jupiter V) is much smaller and may be only 160 km in diameter. Jupiter VI, VII, X and XIII form a separate group, all having orbits about 11 500 000 km from the centre of Jupiter but with large eccentricities and inclinations. Their orbits, however, are so orientated that the chance of collision with each other is slight. The Sun's gravitational pull disturbs these orbits markedly.

Of the remaining seven, Jupiter VIII, IX, XI and XII move in much larger and retrograde orbits even more strongly perturbed by the Sun. Calculation shows that, if the orbits were direct at such distances, Jupiter could not retain these objects as satellites for more than a short time. They would be pulled away by the Sun to become asteroids pursuing independent orbits about the Sun. The reverse course of events can also take place, with Jupiter capturing asteroids and holding them as satellites for an indefinite time interval. It has been suggested that the four outer Jovian satellites may be captured asteroids that could, under the right conditions, escape from the Jovian system at some time in the future. The remaining satellites, recently discovered, have orbits poorly determined as yet.

The interesting and probably significant fact emerges that these four, as well as the group VI, VII and X, have orbits that are not scattered in size but cluster into three orbital 'spectral lines': VI, VII and X at 11 600 000 km from Jupiter, XII at 20 900 000 km, and VIII, IX and XI at 23 200 000 km. These correspond to mean motions close to seventeen, seven and six times Jupiter's mean motion about the Sun, the major disturber of these moons. Are such commensurable orbits the only relatively stable ones at such distances against solar perturbations?

The last three satellites, all discovered in 1979, are very small. Two, XV and XVI, lie in very similar, almost circular, orbits within that of V (Amalthea) while XIV orbits between the orbits of Amalthea and Io.

The final example takes us to Saturn's rings. These rings lie in the plane of Saturn's equator. The outermost one (known as ring A) has outer and inner radii of 136 000 and 119 800 km respectively. As seen from Earth, it appears separated by a dark space called Cassini's division from ring B (middle one). This ring has outer and inner radii of 117 100 and 90 500 km respectively. Ring C (a hazy, transparent ring sometimes called the crepe ring) is situated just inside ring B. Its inner radius is 74 600 km.

The rings are neither solid nor liquid but consist of numerous small solid particles in orbit about the planet. Their individual orbits are perturbed by the innermost three moons of Saturn: Mimas, Enceladus and Tethys. The major divisions in the rings may be explained by these moons' gravitational effects. Cassini's division (between rings A and B) contains distances where the mean motions of hypothetical particles would be twice that of Mimas and three and four times those of Enceladus and Tethys, while the boundary between rings B and C lies at a distance where the mean motion would be three times that of Mimas. The situation is evidently analogous to that of the Kirkwood gaps in the asteroid region.

In fact the situation is much more complicated than this simple picture would imply. The Voyager spacecraft fly-by of Saturn revealed that the rings known as A, B and C themselves consist of hundreds of ringlets, while the F ring, discovered by Pioneer 11, itself is composed of a number of separate ringlets. Rings D and E also exist. It seems unlikely that this richness of fine-structure phenomena is entirely due to straightforward commensurability mechanisms though undoubtedly the newly-discovered satellites associated with the rings play a major part in producing gravitationally the fine-structure ring phenomena.

1.2.4 Comets and meteors

These are also members of the Solar System, and move in elliptical orbits about the Sun. There is no reliable evidence that comets enter the Solar System from outside; on the contrary, it appears probable that the Sun possesses a roughly spherical shell (of radius up to one-third of the distance to the nearest star) of comets numbering millions. The perturbing action on the distant comets by the nearby stars sends a small number into the region of the planetary orbits where the action of the giant planets, in particular Jupiter, either shrinks their orbits to dimensions shorter than

Pluto's or renders them hyperbolic so that these comets are ejected from the System. For example, Halley's comet revolves about the Sun in an elliptical orbit with a period of 76 years, while a group of comets known as Jupiter's family, comprising some thirty-five members, have periods between three and eight years.

Brook's comet (1889V) is an example of a comet whose orbit was markedly changed by the action of Jupiter. Before its encounter with the planet on July 20th 1886, its period of revolution about the Sun was 29·2 years, its orbit lying outside Jupiter's. After encounter, its period changed to 7·10 years, while its orbit shrank in size to lie completely inside Jupiter's orbit.

Cometary dimensions vary greatly. The bright nucleus of a comet may be several hundreds of kilometres in diameter, while the surrounding head is usually some 130 000 km across. The tail may stretch for many millions of kilometres.

The masses of comets however are small, not exceeding 10^{-6} times the mass of the Earth. They probably consist of aggregations of meteoric stones of various sizes embedded in the ice of ammonia, hydrocarbons, carbon dioxide and water. As the comet approaches the Sun, solar radiation may melt some of the ice and evaporate it so that it and dust particles below a certain size form the comet's tail.

Meteors are closely connected with comets. The bigger ones that enter the Earth's atmosphere at night are visible as 'shooting stars' because of the heat generated due to the conversion of the meteor's kinetic energy.

A *fireball* is an exceptionally bright meteor; if it explodes it is called a *bolide*. If it lands on the Earth's surface it is referred to as a *meteorite*. These are usually predominantly iron in constitution, with some nickel. If they are stony, they resemble terrestrial rock.

The sizes of meteors range from occasional ones of many metres in diameter to microscopic particles about 10^{-4} cm in diameter. Their number increases rapidly with diminishing size. Since they may encounter an artificial satellite or space probe with relative velocities up to 80 km s^{-1}, the kinetic energy associated with a collision with even microscopic meteors is large. For this reason many modern studies have been made, in addition to the classical ones, of the frequency of occurrence of meteors of given size and mass. One of the results of putting artificial satellites into orbit has been an increase in the precision of our figures regarding the probabilities of hits by meteors of given size and mass on space vehicles of various target areas.

Meteors are not distributed uniformly throughout the Solar System but tend to be confined to streams, the orbits of some streams being identical to those of known comets. It is possible that a meteor swarm may be the remains of a totally disrupted comet, or it may be that both comet and swarm originated together. In some swarms the material is distributed throughout the orbit; in others it is still localized in position. When the Earth encounters such a swarm an intense and spectacular meteor shower is observed at night, or is detected by radar from the ionization trails left in the atmosphere.

1.2.5 Conclusions

It is seen that a survey of orbital motion in the Solar System reveals a number of properties and raises many questions to be answered. Meanwhile, we can make such statements as:

(i) Most orbits are approximately elliptic in shape.

(ii) Almost coplanar motion exists in the planetary system and in each satellite system.

(iii) Most orbits and rotations are direct, that is, anticlockwise when viewed from the north side of the ecliptic.

(iv) There exist Kepler's laws.

(v) There possibly exist Bode-type laws of orbital distribution.

(vi) Commensurabilities in mean motion are widespread.

(vii) Groupings of particles in Saturn's rings and bodies in the asteroid region occur, apparently to avoid certain commensurabilities.

(viii) Marked changes can occur in certain cometary and satellite orbits.

Among the questions are:

(*a*) What is the significance of properties (i)–(viii)?

(*b*) How stable are the planetary orbits against their mutual gravitational disturbances?

(*c*) How old are the planets?

(*d*) Can planets collide?

(*e*) Are the retrograde outermost satellites of Jupiter and Saturn captured asteroids?

(*f*) Are most of the other satellite orbits stable over astronomically long intervals of time, even if tidal action is taken into account?

1.3 Stellar Motions

The first indication that stars themselves were not fixed in space relative to each other appeared when Halley announced in 1718 that the present positions of the three brightest stars, Sirius, Aldebaran and Arcturus, differed from those given by the Greek astronomer Hipparchus 19 centuries before. Careful measurements subsequently carried out showed that many more stars had spatial velocities relative to the Sun.

A number of corrections have to be made to the actual observations of angular shift. The observations, made from the Earth's surface, embody effects that have nothing to do with any velocity the star may have relative to the Sun. Corrections for such effects are applied (such as the distorting effect of the Earth's atmosphere, the precessional and nutational movement of the Earth's axis of rotation and the revolution of the Earth about the Sun), giving finally the so-called proper motion and in many cases the star's distance from the Sun (for details see chapter 3). In addition, by using a spectroscope, the star's radial velocity may be measured. Both proper motion and radial velocities are with respect to the Sun's position, the proper motion being the annual angular displacement of the star on the heliocentric celestial sphere.

The first reliable measurement of a star's distance was made by Bessel in 1838. The star 61 Cygni was found to lie at a distance of about 3·33 pc†, about two-thirds of a million times as far from the Sun as the Earth is. In the intervening century and a half, as such information has accumulated about tens of thousands of stars, the sciences of stellar kinematics and stellar dynamics have been developed to account for the observed kinematic behaviour of stars.

If we confine ourselves to the immediate vicinity of the Sun (i.e. to a sphere with a radius of about 10^3 pc, containing some thousands of stars), then it is found that to a first approximation this 'local group' of stars (including the Sun) are in random motion with respect to each other, rather as the members of a flock of birds behave in that within the flock the birds have individual speeds and directions of flight. From the point of view of the Sun, however, a systematic effect is imposed on every star in the local group due to the Sun's intrinsic velocity. Because of this, stars appear to be moving outwards from the direction on the celestial sphere to which the Sun (and the Solar System) is travelling (the solar apex) and closing in towards the antipodal point (the solar antapex). This perspective effect is of the same nature as that experienced by anyone travelling in a car who sees objects ahead separate while those behind close in.

So far we see no indication of any orbital motion where stellar movements are concerned. As far back as the beginning of the nineteenth century, however, the spheroidal shape of the galactic system of stars had been pointed out by Sir William Herschel. His son Sir John Herschel suggested later that such a shape could be due to galactic rotation about an axis at right angles to the galactic equator.

The Galaxy is lens-shaped, with the Sun situated in the equatorial plane about two-thirds of the way out from the centre. The fact that the Milky Way extends in a great circle round the celestial sphere is evidence in support of this. The direction towards the centre lies in the constellation of Sagittarius. Surrounding the disc of the Galaxy and concentric with it is a spherical distribution of globular clusters, each globular cluster being a compact assembly of stars (see section 3.3). Observationally, the vast majority of these clusters appear on one half of the celestial sphere, consistent with the picture of the globular cluster spherical distribution being concentric with the galactic centre and the Sun being far out towards the rim of the galactic disc. In addition to all this, the disc has a central bulge, containing large numbers of stars and dust and gas concentrations. Figure 1.2 shows the shape and dimensions of the Galaxy.

It is not only the stars that have orbital movements about the galactic centre. The dust and gas clouds themselves move, for the most part, in the galactic equatorial plane. The mapping-out of such clouds and the confirmation of the spiral structure of our galaxy has been one of the tasks in recent

† 1 parsec (pc) = $3·083 \times 10^{13}$ km.

11

years of radioastronomy, utilizing the 21 cm radio emission from interstellar hydrogen.

We shall see later that the type of orbital motion pursued by a star or a cloud will depend upon the nature of the gravitational potential dictated by the distribution of material within the Galaxy.

1.3.1 Binary systems

So far we have implied that, apart from the sub-assemblies of stars known as globular clusters, stars pursue their individual orbits through space. For more than half the stars, this is not so.

The discovery of the existence of many pairs of stars, gravitationally bound together, is attributed to Sir William Herschel. In 1782 he published a catalogue of double stars, the criterion for inclusion of a pair of stars in the catalogue being that the stars were almost in the same line of sight. Herschel's intention had been to measure stellar distances by observing the parallactic angular shift of the brighter (and presumably much nearer) member of the pair against the position of the fainter (and presumably much farther) member, such a shift being due to the annual orbital movement of the Earth about the Sun. As the years went by however, he found that the observed proper motions in many cases could only be explained by supposing the stars to be in orbital motion about each other.

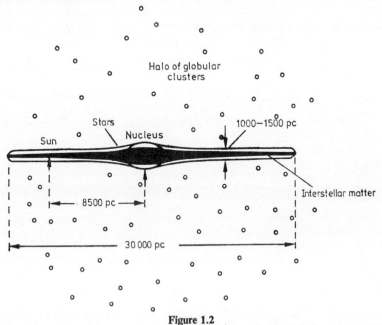

Figure 1.2

A *binary system* is therefore defined as a pair of stars that describe orbits about their common centre of mass, the two components being gravitationally bound together. *Visual binaries* are systems in which both the components can be seen; the members of *spectroscopic binaries* are so close, however, that they can never be resolved in a telescope and are detected by the Doppler effect of their orbital velocities on the spectrum of their light. The third class of binary, the *eclipsing binary*, is again viewed as a single star but, because the members totally or partially eclipse each other, regular diminutions in the star's brightness reveal its double nature. A binary may be both spectroscopic and eclipsing.

In some cases, the binary members are separated by distances thousands of times that separating the Earth from the Sun. In such cases their orbital period may be hundreds of years long. In other cases the two stars are almost in contact, distorting each other's shape by tidal pull, sharing a common atmosphere or transferring material from one component to the other. Their periods can be as short as a few hours.

Widely separated components in binaries have simple elliptical orbits about each other; close binaries have members whose orbits are much more complex. Much of our information about stellar masses, stellar structure and evolution has been derived from a study of binary stars.

With the advent of artificial satellites carrying x-ray telescopes, binaries emitting x-rays have been found, leading to interesting and informative deductions about one or both of the components being neutron stars or black holes and providing valuable tests of relativity and astrophysical theories.

1.3.2 Triple and higher systems of stars

Many investigations have been made to discover the proportion of triple and higher systems of stars among binaries. For example, a visual binary may on closer examination be revealed to be a triple system where one component of the pair is found to be a spectroscopic binary. The number of systems known is sufficiently large for a reliable estimate to be made and it is now accepted that, among multiple systems, the proportion of triple and higher systems lies between one-quarter and one-third. Difficulties arise because of selection effects and the possible inclusion of spurious triples, but widely different research methods still show good agreement.

The same factor (between one-quarter and one-third) seems to hold when the proportion of triples that are quadruple, or quadruples that are quintuple and so on are concerned, though its precision becomes naturally questionable when we appreciate that all the previous difficulties that reduce reliability are enhanced and that small-number statistics are increasingly involved as one advances to larger systems.

When we consider the ratio of periods of revolution within multiple systems it is found that a hierarchy approach, first introduced by Evans, is useful. In figure 1.3 Evans's hierarchy method is applied to (a) a binary system, (b) a triple system, and (c) and (d) two possible quadruple systems.

This family-tree-type procedure is almost self-explanatory. In figure 1.3(b) it represents two distant components, one of which is itself a close binary. Figure 1.3(c) would represent a similar system taken one 'generation' further,

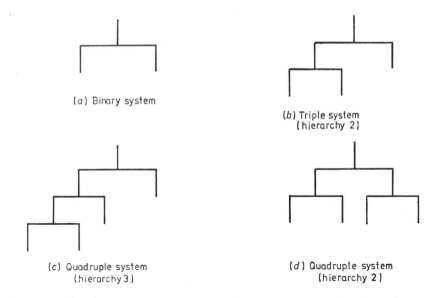

(a) Binary system

(b) Triple system
(hierarchy 2)

(c) Quadruple system
(hierarchy 3)

(d) Quadruple system
(hierarchy 2)

Figure 1.3

where one member of the close binary is itself an even closer binary. Figure 1.3(d), on the other hand, stands for a binary system with widely separated components, each of which is a close binary.

It would appear that the vast majority of triple systems consist of hierarchy-2 arrangements, namely a close binary with a third star at a distance many times (in a number of cases hundreds of times) that of the close binary separation. In quadruple systems, the preference is for two close pairs separated by a distance which is again a large multiple of the close pair components' separations, or a close pair plus two distant companions. Translated into periods of revolution, such configurations mean that in multiple systems the ratios of longer to shorter periods are very large.

The dearth of multiple systems in which all the mutual separations are of the same order is marked, and it will be seen later that recent research in the many-body gravitational problem has shed a great deal of light on the lack of such configurations. Indeed, apart from special cases, such as the Lagrange equilateral triangle configuration, it is found that, in the Solar System and in multiple star systems, the bodies are arranged in hierarchical configurations. This itself implies that such arrangements are inherently more stable gravitationally than any other.

1.3.3 Globular clusters

A globular cluster is a compact star system containing a large number of stars. About 120 globular clusters are known but, from a consideration of the numbers of such systems possessed by nearby galaxies, it is possible that the true number belonging to our galaxy is nearer 1000.

To describe the appearance of a cluster on a time-exposure photographic plate, recourse has been made to the analogies of a swarm of bees or to salt grains poured on to a black sheet. Whatever analogy is used, each cluster seems to consist of anything from 10 000 to 1 000 000 stars, their density (i.e. number per cubic parsec) increasing sharply as one passes from the edge of the cluster into its centre. Numbers are difficult to measure. A short time-exposure photograph loses most of the faint stars in the cluster; on the other hand, a long time-exposure produces a blurred region at the cluster centre where the individual stellar images merge and cannot be counted. Even at the cluster centre however, where the number density may be more than 1000 times the number density of stars in the solar neighbourhood, the chance of a collision between stars is small. Nevertheless, to a human being transported to a planet near a globular cluster centre, the night sky would be awe-inspiring. Instead of a meagre half-dozen first-magnitude stars and a couple of thousand fainter ones, the observer would see as many as 1000 first-magnitude objects with tens of thousands of fainter ones. Indeed it has been estimated that, at the centre of the cluster 47 Tucanae, the starlight would be the equivalent of several thousand full moons.

It has already been mentioned that the system of globular clusters occupies a sphere concentric with the centre of the galactic disc (see figure 1.2). There is some evidence that the number density of clusters increases as the galactic nucleus is approached. Wyatt has remarked that if we plucked out at random all but 150–200 of the stars of a single globular cluster, what would be left would serve as a fair model of the system of globular clusters itself.

The distances of the clusters are reliably measured because the vast majority of them contain variable stars, most of them RR Lyrae stars, the others being Type-II Cepheids. Both kinds of stars may be used as distance indicators. RR Lyrae stars all have much the same absolute brightness; measurement of their mean apparent brightness in a cluster then enables the cluster's distance to be found. For any Cepheid, the period–luminosity relation gives the absolute mean brightness once the period of light fluctuation has been measured; the mean apparent brightness of the Cepheid can then be used to find its distance.

Information about the cluster velocities is derived chiefly from radial velocity measurements utilizing the Doppler formula. The distribution of velocities is compatible with the hypotheses that the Sun is in orbit about the galactic centre and that the globular clusters are themselves orbiting this centre.

Astrophysical theory of stellar structure applied to the Hertzsprung–Russell diagram of a cluster enables a lower limit to be assigned to the cluster's age. It turns out that the ages of globular clusters average 6×10^9 years, with very little dispersion. The system of globular clusters would therefore appear to be stable over an astronomically long time interval.

Problems that have been attacked by many researchers include: (i) the distribution of stars within a globular cluster and the types of orbits pursued; (ii) the possible escape from or capture by the cluster of individual stars; and

(iii) the stability of a stellar system of such size. We shall see later that a number of quite diverse approaches have been developed, complementing each other in some cases and producing insight into this interesting class of dynamical problems.

1.3.4 Galactic or open clusters

Galactic or open clusters consist of systems containing anything between ten and a few thousand stars. For most of them however, the number lies between 50 and 200. They are only roughly spherical in shape, some being quite ragged in outline, and their diameters range between 1·5 and 15 pc. Such clusters are confined close to the galactic disc, unlike the globular clusters. Various estimates of the number of open clusters in our galaxy have been made; they can only be estimates since the dark obscuring clouds in the galactic plane must hide most of them, confined as they are to the vicinity of that plane. At least 800 are known however, and many of the most famous ones such as the Pleiades, the Hyades and the Ursa Major group are near enough for detailed investigation of their stellar members and their proper motions to be carried out.

Unlike the globular clusters, whose ages seem to lie close to 6×10^9 years, the galactic clusters have ages ranging from 2×10^6 years to 6×10^9 years. For example, the ages of the three open clusters h and χ Persei, the Pleiades and the Hyades are 5×10^6, 2×10^7 and 4×10^8 years respectively. Since the age of the Galaxy itself is estimated to be 10^{10} years, it is seen that some open clusters are so young compared with that age that cluster formation must still be taking place. On the other hand, others have ages comparable with that of the Galaxy. These latter clusters must then be dynamically stable against the disruptive gravitational action of the central galactic bulge, nearby dust and gas clouds and of stellar intruders. This may not necessarily be true for all open clusters. It is to be expected that, unlike the highly compact globular clusters with their tens of thousands to millions of members, other open clusters may or may not survive such disturbing influences indefinitely. Questions of the stability of open clusters of different sizes, numbers of members and concentrations of stars have, as for the globular cluster case, attracted many investigators.

1.4 Clusters of Galaxies

The average distance between stars in a galaxy is some millions of times the diameter of an average star. In contrast the average distance between galaxies is some scores of times the equatorial diameter of the average galaxy. In addition, galaxies occur in groups or clusters. Our own galaxy, with its attendants the small and large Magellanic clouds, is part of the *Local Cluster*. This contains about seventeen galaxies, among them the great galaxy in Andromeda with its two satellite galaxies. Other clusters are larger; for example, the Virgo cluster contains several thousand galaxies.

16

Orbital motion of galaxies about each other can therefore exist. Relatively near galaxies can distort each other tidally to the extent (as is seen on many photographic plates) of galactic planes being deformed and bridges of material being created to join the one galaxy to the other. Collisions of galaxies are relatively frequent in the life of a cluster of galaxies, whereas collisions or near encounters of stars within a galaxy are very infrequent.

1.5 Conclusion

We see then that orbital motion, dictated for the most part by gravitational forces, exists up to the largest entities in the observable universe. The problems to be studied may be conveniently if roughly classified in at least two ways:

(i) point-mass problems, in which the finite size of the bodies concerned is irrelevant (e.g. Sun–Jupiter–asteroid),
(ii) extended-mass problems, in which the finite size of at least one of the bodies concerned has to be taken into account (e.g. the orbit of a close artificial satellite about the oblate Earth or the action of two distorted stars in a close binary system upon each other).

An alternative classification is:

(a) the two-body problem, in which two particles attract each other according to Newton's law of gravitation. An exact analytical solution exists for this. An example of this problem is an isolated binary system in which the components are widely separated.
(b) the few-body problem, where at least one more particle is added to the problem but where the total number of bodies remains too few for statistical methods to be applied. No general solution is available. An example is the problem of knowing the planetary orbits in the Solar System for all time.
(c) the many-body problem, in which statistical smoothing methods may be applied to produce solutions applicable not so much to individual members of the problem as to the system itself. This may be called the actuarial approach. An example of this is the globular cluster problem.

Bibliography

The books listed below may be consulted by the reader desirous of more detailed information concerning the Universe.

Audouze J and Israël G (ed) 1985 *The Cambridge Atlas of Astronomy* (London: Cambridge University Press)
Baker R H and Fredrick L W 1971 *Astronomy* (New York: Van Nostrand)
Beaty J K, O'Leary B and Chaikin A (ed) 1981 *The New Solar System* (Cambridge, Mass: Sky Publishing)
Calder N 1969 *The Violent Universe* (London: BBC Publication)
Kaufmann W J III 1978 *Exploration of the Solar System* (New York: Macmillan)

Menzel D H, Whipple F L and de Vaucouleurs G 1970 *Survey of the Universe* (Englewood Cliffs: Prentice-Hall)

Roy A E and Clarke D 1982 *Astronomy: Structure of the Universe* (Bristol: Adam Hilger) 2nd edn

—— 1988 *Astronomy: Principles and Practice* (Bristol: Adam Hilger) 3rd edn

Seeds M A (ed) 1980 *Astronomy—Selected Readings* (Menlo Park: Benjamin/Cummings)

Voigt H H 1974 *Outline of Astronomy* (Holland: Noordhoof Publishers)

Wyatt S P 1964 *Principles of Astronomy* (Boston, Mass.: Allyn and Bacon)

Among the journals devoted to astronomy or regularly containing papers and articles on the subject are: *The Astronomical Journal; The Astrophysical Journal; Journal of Geophysical Research; Monthly Notices, Royal Astronomical Society; Icarus; Nature; Planetary and Space Science; Publications of the Astronomical Society of the Pacific; Science; Sky and Telescope; Celestial Mechanics; Astrophysics and Space Science; Astronomy and Astrophysics.*

2 Coordinate and Time-Keeping Systems

2.1 Introduction

Observing or calculating the position and velocity of any celestial object requires a coordinate system and a system of time measurement. The origins of this search for suitable reference systems go back many thousands of years in astronomy. Originally the Earth was the platform from which all measurements were taken. This situation held until recently, although even before the advent of Martian artificial satellites or the landing of men on the Moon it was often convenient to choose a coordinate system and origin away from Earth. For example, the Sun's centre was chosen where planetary orbital motions were concerned, or a planetary centre in the case of satellite problems, or even the galactic centre in stellar dynamics. In manned spaceflight, the origin can be the ship itself.

The coordinate system likewise depended upon the particular problem involved and could utilize the Earth's equator, or its orbital plane containing the Sun, or a planet's equator or orbital plane, or the galactic equator, and so on. The time system could be based on the movement of the Sun, or on the Earth's rotation, or on what is known as Ephemeris Time, which is related to the movements of the planets round the Sun and of the Moon about the Earth.

In this chapter we consider a number of the concepts concerned with such matters.

2.2 Position on the Earth's Surface

A point on the surface of the Earth is defined by two coordinates, latitude and longitude, based on the equator and a particular meridian passing through the North and South poles and Greenwich, England. The longitude of the point is measured east or west along the equator from the intersection of the Greenwich meridian and the equator to the point where the meridian through the point concerned crosses the equator.

The longitude is usually expressed in time units, related to angular measure by the table

$$360° = 24^h$$

$$1° = 4^m \qquad 1^h = 15°$$

$$1' = 4^s \qquad 1^m = 15'$$

$$1'' = 1/15^s \qquad 1^s = 15''.$$

For example the longitude of Washington DC is $5^h\,08^m\,15\cdot78^s$ west of Greenwich ($77°\,03'\,56\cdot7''$W of Greenwich). Longitude is measured up to 12^h east or west of Greenwich, denoted G in figure 2.1. The latitude of a point is the angular distance north or south of the equator, this angle being measured along the local meridian. For example, Washington DC has a latitude of $38°\,55'\,14\cdot0''$N.

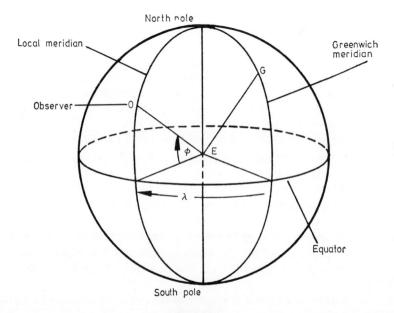

Figure 2.1

Because the Earth is not a sphere the true picture is more complicated than the simple one outlined above, though the latter is accurate enough for calculations of orders of magnitude.

When a plumb-line is suspended by an observer at a point on the Earth's surface its direction makes an angle ϕ with the plane of the Earth's equator. This angle is called the *astronomical latitude*. The point where the plumb-line's direction meets the equatorial plane is not in general the centre of the Earth. The angle between the line joining observer to the Earth's centre and the equatorial plane is the *geocentric latitude* ϕ'.

There is yet a third definition of latitude. Geodetic measurements on the Earth's surface show local irregularities in the direction of gravity, due to variations in density and shape in the Earth's crust. The direction in which a

20

plumb-line hangs is affected by such anomalies and these are referred to as *station error*. The *geodetic* or *geographic latitude* ϕ'' of the observer is the astronomical latitude corrected for station error.

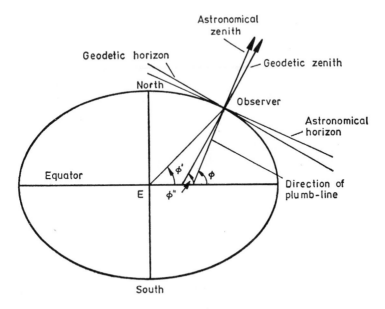

Figure 2.2

The geodetic latitude is therefore referred to a reference spheroid, an oblate spheroid whose surface is defined by the mean ocean level of the Earth. If a and b are the semimajor and semiminor axes of the ellipse of revolution forming the 'geoid', the *flattening* or *ellipticity* γ is given by

$$\gamma = \frac{a-b}{a} = 1 - (1-e^2)^{1/2},$$

where e is the eccentricity of the ellipse.

Various such reference spheroids exist. The dimensions of the Hayford geoid, for example, are:

$$a = 6378 \cdot 388 \text{ km}$$
$$b = 6356 \cdot 912 \text{ km}$$
$$\gamma = 1/297$$

and hence $e = 0 \cdot 081\ 99$.

It may be remarked that the geoid obtained from observations of the changing orbits of Earth satellites departs appreciably from this reference geoid (chapter 10).

The *geocentric longitude* λ is the same as the *geodetic longitude* which is the angular distance east or west measured along the equator from the Greenwich meridian to the meridian of the observer.

2.3 The Horizontal System

This is the most primitive coordinate system and is related to the horizon and to one of the points of intersection of the horizon with the great circle (section 2.9.1) through the north celestial pole and the zenith. The horizontal system of coordinates has the observer at its origin so that it is a strictly local system.

From figure 2.2 it is seen that the *zenith* is obtained by producing upwards the direction in which a plumb-line hangs. The opposite direction leads to the *nadir*. It is a convenient fiction to suppose that a vast sphere of arbitrary radius surrounds the Earth on the inside of which the stars and other heavenly bodies are projected. This sphere is the *celestial sphere*. Since in many astronomical problems the distances of the bodies do not concern us, the radius of the sphere may be chosen as we wish and is often taken to be unity.

The *north and south celestial poles* are the intersections of the Earth's axis of rotation with the celestial sphere. The north celestial pole (above the Earth's North pole) is the point about which, to a northern observer, the heavens appear to revolve once in 24 h. At present the bright star Polaris lies within one degree of this point but, because of precession (section 3.4), it will gradually depart from the north celestial pole, returning to its vicinity in about 26 000 years.

The observer's celestial sphere is shown in figure 2.3, where Z is the zenith, O the observer, P is the north celestial pole and OX the instantaneous direction of a heavenly body. The great circle through Z and P cuts the horizon NESAW in the north (N) and south (S) points. Another great circle WZE at right angles to the great circle NPZS cuts the horizon in the west (W) and east (E) points. The arcs ZN, ZW, ZA etc. are called *verticals*. The points N, E, S and W are the cardinal points.

The two angles that specify the position of X in this system are the *azimuth* $\hat{A}$ and the *altitude â*. Azimuth is defined in a number of ways and care must be taken to find out what definition is followed in any particular use of this system. For example, the azimuth may be defined as the angle between the vertical through the south point and the vertical through the object X, measured westwards along the horizon from 0° to 360°, or the angle between the vertical through the north point and the vertical through the object X, measured eastwards or westwards from 0° to 180° along the horizon. *A third definition commonly used is to measure azimuth from the north point eastwards from 0° to 360°.* This definition will be kept in this book and is in fact similar to the definition of *true bearing*. For an observer in the southern hemisphere, azimuth is measured from the south point eastwards from 0° to 360°.

The *altitude a* of X is the angle measured along the vertical circle through X from the horizon at A to X. It is measured in degrees. An alternative coordinate to altitude is the *zenith distance z*, also measured in degrees, indicated by ZX in figure 2.3. Obviously

$$a = 90° - z.$$

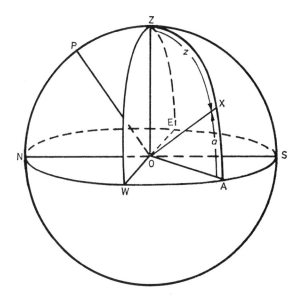

Figure 2.3

The main disadvantage of the horizontal system of coordinates is that it is purely local. Two observers at different points on the Earth's surface will measure different altitudes and azimuths for the same star at the same time. In addition, an observer will find the star's coordinates changing with time as the celestial sphere appears to rotate. Even today, however, many observations are made in the alt–azimuth system, as it is often called. For example, the 250 ft radio telescope at Jodrell Bank, England, moves on an alt–azimuth mounting, a special computer being employed to transform coordinates in this system to equatorial coordinates and vice versa.

2.4 The Equatorial System

If we extend the plane of the Earth's equator it will cut the celestial sphere in a great circle called the *celestial equator*, meeting the observer's horizon in the east and west points. Since the angle between equator and zenith is the observer's latitude it is seen that the altitude of the north celestial pole P is the latitude ϕ of the observer.

Any great semicircle through P and Q, the south celestial pole, is called a meridian. The meridian through the celestial object X is the great semicircle PXBQ cutting the celestial equator in B in figure 2.4.

In particular, the meridian PZTSQ, indicated because of its importance by a heavy line, is the *observer's meridian*.

An observer viewing the sky will note that all natural celestial objects rise in the east, climb in altitude until they transit across the observer's meridian, then decrease in altitude until they set in the west. In contrast most artificial

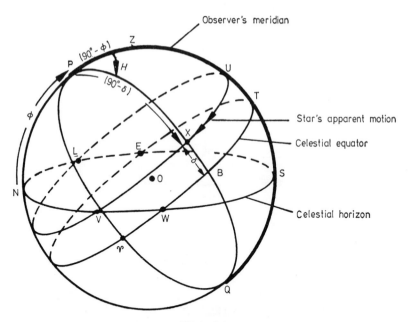

Figure 2.4

satellites at the present time rise in the west and set in the east but these do not concern us at present. A star in fact will follow a small circle (the intersection of a plane not including the centre of the sphere with the sphere) parallel to the celestial equator in the arrow's direction. Such a circle is called a parallel of declination and provides us with one of the two coordinates in the equatorial system. The *declination* δ of the star is the angular distance in degrees from the equator along the meridian through the star. It is measured north and south of the equator from 0° to 90°, being taken as positive when north.

Thus, the star transits at U, sets at V, rises at L and transits again 24 h later. The angle ZPX is called the *hour angle* (HA), H, of the star and is measured from the observer's meridian westwards (for both north- and south-hemisphere observers) to the meridian through the star from 0^h to 24^h or from 0° to 360°. Consequently, the hour angle increases by 24^h each sidereal day for a star (section 2.10.1).

If a point ♈, fixed with respect to the stellar background, is chosen on the equator, its angular distance from the intersection of the meridian through X and the equator will not change in contrast to the changing hour angle of X. In general, all objects may then have their positions on the celestial background specified by their declinations and by the angles between their meridians and the meridian through ♈. The point chosen is the *vernal equinox*, also referred to as the *First Point of Aries*, and the angle between it and the intersection of the meridian through a celestial object and the equator is called the *right ascension* α or RA of the object. Right ascension is measured from 0^h to 24^h or from 0° to 360° along the equator from ♈ eastwards; that

24

is, in the direction opposite to that in which hour angle is measured. This definition again holds for observers in both northern and southern hemispheres. It is advisable in drawing a celestial sphere to (i) mark in the observer's meridian heavily, (ii) mark on the equator a westwards arrow and put HA (hour angle) beside it, and (iii) mark on the equator an eastwards arrow and put RA beside it.

The origin in the equatorial system may be the observer on the surface of the Earth, or the centre of the Earth, or the centre of the Sun, and the celestial spheres based on these origins are referred to as the observer's (or topocentric), the geocentric and the heliocentric celestial spheres respectively. For stellar observations, geocentric equatorial coordinates are used with star catalogues giving right ascensions and declinations referred to the equinox and equator of, say, 1950·0. For planetary orbits heliocentric equatorial coordinates are often used, while the orbits of artificial Earth satellites are customarily referred to a geocentric equatorial celestial sphere, since the major effect of the Earth's gravitational perturbations is due to the equatorial bulge on the Earth.

For distant objects such as stars the size of the Earth is negligible compared to their distances, so that observations of these bodies from any part of the Earth's surface are unaffected by the observer's position. In the case of planets, the Sun, the Moon or a space vehicle, the observer's position on the surface of the Earth is important. The direction in which he sees any of these objects will be different from the direction in which a hypothetical observer stationed at the Earth's centre would see it. Thus in the *'Astronomical Almanac'* and other almanacs, the positions of such natural bodies are tabulated with respect to a geocentric sphere, and observers in given latitudes and longitudes must apply certain corrections to convert from geocentric coordinates to apparent coordinates. A similar procedure is adopted by computing centres for artificial satellites of the Earth. A fuller discussion of such correcting procedures is reserved for chapter 3.

2.5 The Ecliptic System

When the Sun is observed over a long period of time, it is found to possess a second motion in addition to its apparent diurnal movement about the Earth. It moves eastwards (in the direction of increasing right ascension) among the stars at about 1°/day, returning to its original position in one year. Its path is a great circle called the *ecliptic* which lies in the plane of the Earth's orbit about the Sun. This great circle is the fundamental reference plane in the ecliptic system of coordinates. It intersects the celestial equator in the vernal and autumnal equinoxes (First Point of Aries ♈ and Libra ♎) at an angle of 23° 27′, usually denoted by ϵ and referred to as the obliquity of the ecliptic. The pole K of the ecliptic makes the same angle ϵ with the north celestial pole.

In this system the two quantities specifying the position of an object are ecliptic longitude and ecliptic latitude. In figure 2.5 a great circle arc through the pole K of the ecliptic and the celestial object X meets the ecliptic in the point D. Then the *ecliptic longitude* λ is the angle between Υ and D, measured from 0° to 360° or 0ʰ to 24ʰ along the ecliptic in the eastwards direction (i.e. in the direction in which right ascension increases). The *ecliptic latitude* β is measured in degrees from D to X along the great circle arc DX, being measured from 0° to 90° north or south of the ecliptic. It should be noted that K lies in the hemisphere containing the north celestial pole. It should also be noted that ecliptic latitude and longitude are often referred to as *celestial latitude* and *longitude*.

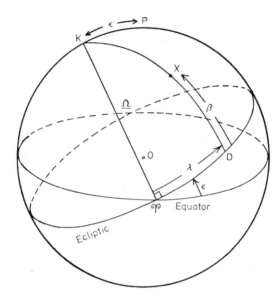

Figure 2.5

The origins most often used with this system of coordinates are the Earth's centre and the Sun's centre, since most of the planets move in planes inclined at only a few degrees to the ecliptic. This system is particularly useful in considering interplanetary missions.

2.6 Elements of the Orbit in Space

In astronomy it is usual to define an orbit and the position of the body describing that orbit by six quantities called the *elements*. Three of 'them define the orientation of the orbit with respect to a set of axes, two of them define the size and shape of the orbit, and the sixth (with the time) defines the position of the body within the orbit at that time. In the case of a planet moving in an elliptic orbit about the Sun, the elements may be defined with respect to a celestial sphere (centred at the Sun), the ecliptic and the First Point of Aries.

In figure 2.6 the plane in which the orbit lies cuts the plane of the ecliptic in a line called the *line of nodes* NN_1. If the direction in which the planet moves in its orbit A_1AP is as indicated by the arrow, N is referred to as the ascending node; N_1 is the descending node. Then the *longitude of the ascending node* Ω is given by ΥN measured along the ecliptic from $0°$ to $360°$.

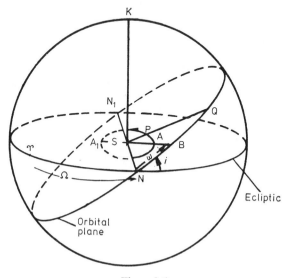

Figure 2.6

The second element is the *inclination i*, which is the angle between the orbital plane and the plane of the ecliptic. These two elements orientate the orbital plane.

The third element orientates the orbit within that plane. Each planetary orbit has a point in it nearest to the Sun called *perihelion*. In the case of elliptic orbits there is a point farthest from the Sun called *aphelion*. The orbits are symmetrical about the line through the Sun's centre and perihelion or in elliptic cases about the *line of apses*, the line joining perihelion A to aphelion A_1. This line passes through the Sun's centre S. The direction of the line of apses therefore fixes the orientation of the orbit. The angular distance from Υ to N, namely Ω, plus the angular distance ω from N to the projection of perihelion A onto the celestial sphere at B, is called the *longitude of perihelion* $\varpi(=\Omega+\omega)$. Note that it is measured from Υ along the ecliptic to N *then along the orbital plane's intersection with the celestial sphere to* B.

The next two elements depend upon the nature of the orbit.

It will be shown later (chapter 4) that the orbit of a particle about another under their mutual gravitational attraction is a conic section (i.e. an ellipse, parabola, or hyperbola) with the second particle at a focus. For the present let the orbit be an ellipse. In this case the two elements defining its size and shape are the semimajor axis and the eccentricity. In the ellipse shown in figure 2.7, the major axis is the distance AA'. The *semimajor axis a* is half of this distance and gives the size of the orbit.

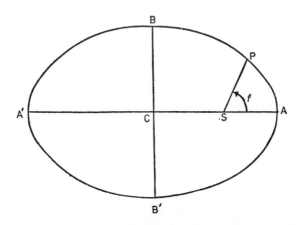

Figure 2.7

The *eccentricity e* is a measure of its departure from a circle. It is related to the distance of a focus S from the centre of the ellipse C by the relation

$$CS = ae.$$

The sixth element is the *time of perihelion passage* τ which is a particular epoch when the body was at perihelion. This epoch, together with any other time, fixes the body's position in the orbit at that time.

These six elements, Ω, ϖ, i and a, e, τ, together with the time, then define the orbit and the position of the body in it.

A further quantity f, the *true anomaly*, is frequently used in orbit work and is defined as the angle at the focus S between the direction of perihelion and the radius vector SP of the body.

If the fundamental reference plane of the coordinate system is changed to the equator, then Ω, ϖ and i take different values while a, e and τ remain unchanged.

If the body is a satellite of the Earth, the fundamental reference plane is the equator and the longitude of the ascending node becomes the right ascension of the ascending node. Taking the place of the longitude of perihelion is a quantity called the argument of perigee (perigee being the point of nearest approach to the Earth's centre in the orbit); this quantity is the angle between the direction of perigee and the ascending node.

If the body is a satellite of a planet, then the reference plane may be the ecliptic, or the planet's equatorial plane, or the plane of the planet's orbit about the Sun, or a plane called the 'proper plane' on which the nodes regress (chapter 5). The point in the body's orbit nearest the planet is often referred

28

to as pericentre òr by prefixing 'peri' to a modification of the planet's name, such as perijove or perisaturnium.

2.7 Rectangular Coordinate Systems

In many astronomical and astronautical problems, positions are computed in rectangular coordinates. A number of such systems are available.

If a reference plane (either the ecliptic or the equator) is chosen, then the x axis can be taken from the centre of the body about which revolution takes place towards the direction of the vernal equinox ♈, the y axis being taken to lie in the reference plane making an angle of 90° with the x axis. The z axis can then be directed towards the pole of the reference plane so that all three axes form a right-handed rectangular coordinate system.

In some problems the origin is taken to lie at the centre of mass of the system of bodies. Such a set is called a barycentric coordinate system.

2.8 Orbital Plane Coordinate Systems

It is often convenient to take a set of rectangular axes in and perpendicular to the orbital plane of the body, with the origin at the centre of the Sun or planet about which the body revolves. We illustrate the various versions of this set with respect to the case of a body moving about the Sun.

The x axis may be taken towards the ascending node N, the y axis being in the orbital plane and 90° from x, while the z axis is taken to be perpen-

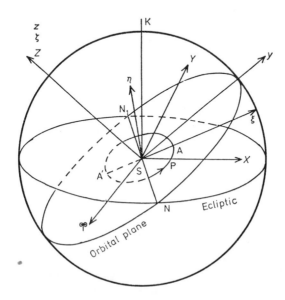

Figure 2.8

dicular to the orbital plane so that the three axes form a right-handed co-ordinate system (see figure 2.8).

Another useful set is to take axis ξ along the line joining Sun to perihelion, axis η at right angles to it and lying in the orbital plane, with axis ζ perpendicular to both.

In a third set, the X axis is taken to pass through the body itself with the Y axis in the orbital plane and at right angles to it, the Z axis being then taken (as usual) perpendicular to the orbit plane. This set constitutes a rotating system since the body is moving in its orbit; it is used in the study of disturbing forces acting on the body.

2.9 Transformation of Systems

It is often necessary to transform from one coordinate system to another. Sometimes the transformation is a translation from one origin to another as well as a rotation of axes; more often the origin remains the same.

Certain transformations can be effected easily by using the fundamental formulae of spherical trigonometry. Other transformations are obtained more easily by the use of vector methods.

2.9.1 The fundamental formulae of spherical trigonometry

The geometry of a sphere is made up of great circles, small circles, and arcs of these figures. All distances along such circles are measured as angles, since for convenience the radius of the sphere is made unity.

A *great circle* is obtained when a plane passing through the centre of the sphere cuts the surface of the sphere.

If the plane does not contain the centre of the sphere, its intersection with the sphere is a *small circle*.

The *poles* of a great circle are those two points of the sphere 90° away from all points on the great circle. Thus in figure 2.9 the poles of the great circle FCD are the points P and Q. Obviously the line joining the poles meets the great circle plane at the centre of the sphere at right angles to it.

Two great circles intersecting at a point include a *spherical angle* defined as the angle between the tangents to the great circles at the point of inter-section. A spherical angle is defined *only* with reference to two intersecting great circles.

The closed figure formed by the arcs of three great circles is called a *spherical triangle* if it possesses the following properties:

(i) Any two sides are together greater than the third side.
(ii) The sum of the three angles is greater than 180°.
(iii) Each spherical angle is less than 180°.

The length of a small circle arc is related simply to the length of an arc of the great circle whose plane is parallel to that of the small circle.

In figure 2.9 the pole P of the great circle FCD is also the pole of the

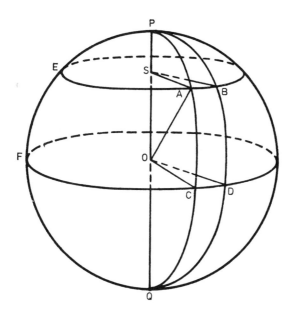

Figure 2.9

small circle EAB whose plane is parallel to that of the great circle FCD. If great circles are drawn through P and the ends A and B of the small circle arc, they will cut the great circle in points C and D. It is then easily shown that

$$AB = CD \cos AC$$

remembering that sides are measured as angles and that the radius of the sphere is unity. An example of the use of this formula is given by considering how far apart two places on the Earth's surface are if they lie on the same parallel of latitude and distance is measured along the parallel. This distance is called the *departure*. In this example we assume the Earth to be spherical. In figure 2.9 the two places are represented by A and B. Angle $A\hat{O}C$ is the latitude ϕ so that

$$AC = BD = \phi.$$

If the longitudes of A and B are $\lambda_A W$ and $\lambda_B W$ respectively, then their difference in longitude is $\lambda_A W - \lambda_B W$ and

$$CD = \text{angle } C\hat{O}D = \lambda_A - \lambda_B.$$

Then

$$AB = CD \cos AC$$

or in other words

departure = difference in longitude × cos (latitude).

Distance on the Earth's surface in such problems is usually measured in nautical miles, a *nautical mile* being the great circle distance subtending an angle of one minute of arc at the Earth's centre. The Earth's surface is not

31

absolutely spherical; consequently the length of the nautical mile varies, but a mean value of 6080 ft is used.

The difference in longitude may now be expressed in minutes of arc, this number being equal to the number of nautical miles. The departure can then be calculated from the formula.

It is to be noted that the difference in longitude is formed algebraically by taking east longitudes to be of opposite sign to west longitudes.

In figure 2.10, ABC is a spherical triangle with sides AB, BC and CA of lengths c, a and b respectively.

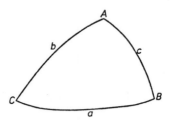

Figure 2.10

There are four formulae, constantly used in astronomy and astrodynamics, which connect sides a, b and c with angles A, B and C. They are:

(i) The cosine formula

$$\cos a = \cos b \cos c + \sin b \sin c \cos A.$$

There are two variations of this, viz.

$$\cos b = \cos c \cos a + \sin c \sin a \cos B$$

$$\cos c = \cos a \cos b + \sin a \sin b \cos C.$$

(ii) The sine formula

$$\frac{\sin A}{\sin a} = \frac{\sin B}{\sin b} = \frac{\sin C}{\sin c}.$$

The latter must be used with care since, in being given, say, a, b and B it is not possible to say whether A or $(180° - A)$ is required unless other information is available.

(iii) The analogue to the cosine formula

$$\sin a \cos B = \cos b \sin c - \sin b \cos c \cos A.$$

There are five variations of this formula.

(iv) The four-parts formula

$$\cos a \cos C = \sin a \cot b - \sin C \cot B$$

32

with five other variations. This formula utilizes four consecutive parts of the spherical triangle.

Proofs of these four important formulae and of a number of less useful ones may be found in the work by Smart and Green (1977) described at the end of this chapter.

2.9.2 Examples in the transformation of systems

Example 1. For a geocentric celestial sphere calculate the hour angle H and declination δ of a body when its azimuth (east of north) and altitude are A and a. Assume the observer has a latitude ϕ.

The required celestial sphere is shown in figure 2.11, where X is the body's position. The other symbols have their usual meanings.

Taking the spherical triangle PZX, the cosine formula gives

$$\sin \delta = \sin \phi \sin a + \cos \phi \cos a \cos A.$$

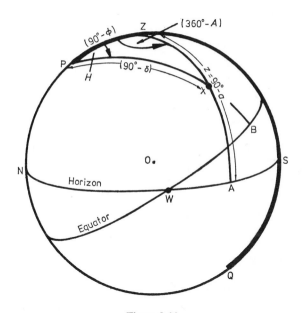

Figure 2.11

This equation enables δ to be calculated.

A second application of the cosine formula gives

$$\sin a = \sin \phi \sin \delta + \cos \phi \cos \delta \cos H$$

or

$$\cos H = \frac{\sin a - \sin \phi \sin \delta}{\cos \phi \cos \delta}$$

giving H since δ is now known.

Alternatively, using the four-parts formula with $(90-a)$, $(360-A)$, $(90-\phi)$ and H, we obtain

$$\sin \phi \cos A = \cos \phi \tan a + \sin A \cot H$$

or

$$\tan H = \frac{\sin A}{\sin \phi \cos A - \cos \phi \tan a}.$$

Example 2. Transfer the ecliptic coordinates (celestial longitude λ and celestial latitude β) of a space vehicle to geocentric equatorial coordinates (right ascension α and declination δ), given that the obliquity of the ecliptic is ϵ.

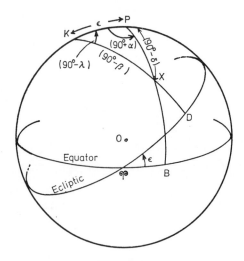

Figure 2.12

In figure 2.12 it is seen that spherical triangle KPX (X being the position of the space vehicle on the celestial sphere) contains the necessary information. Applying in turn the cosine formula, the sine formula and the analogue to the cosine formula, we obtain

$$\sin \delta = \sin \beta \cos \epsilon + \cos \beta \sin \epsilon \sin \lambda \tag{2.1}$$

$$\cos \delta \cos \alpha = \cos \beta \cos \lambda \tag{2.2}$$

$$\cos \delta \sin \alpha = -\sin \beta \sin \epsilon + \cos \beta \cos \epsilon \sin \lambda \tag{2.3}$$

which give α and δ without ambiguity.

The reverse problems in examples 1 and 2 are left as an exercise to the student.

Example 3. Obtain the geocentric distance ρ, right ascension α and declination δ of a space vehicle orbiting the Sun when its heliocentric rectangular coordinates (x, y, z) are known.

This is an important example illustrating a number of principles. Observa-

34

tions of the vehicle from the Earth or communication with it at a given time depend upon a knowledge of the vehicle's geocentric right ascension and declination and upon its distance. On the other hand, for an interplanetary vehicle, its orbit is about the Sun so that the elements of such an orbit are referred to a heliocentric system. These known elements (plus the time) enable its rectangular coordinates with the Sun as origin to be determined. We will see later (chapter 4) how this is done. In this example, we assume that the rectangular coordinates are based on the ecliptic and the direction of the First Point of Aries, and show how they may be transformed to a geocentric distance, right ascension and declination. This particular problem is in fact a standard procedure in astronomy. The reverse problem of determining the elements of the orbit from observations of the body's right ascension and declination is again a standard procedure, but is more difficult and is left until later.

The problem is solved in several stages:

(i) the transformation is made from heliocentric ecliptic rectangular co-ordinates to heliocentric equatorial rectangular coordinates,
(ii) the heliocentric equatorial rectangular coordinates are changed to geocentric equatorial rectangular coordinates,
(iii) the geocentric equatorial rectangular coordinates are changed to geo-centric distance, right ascension and declination.

The methods of these transformations are as follows:

(i) In figure 2.13, V is the position of the vehicle with respect to the Sun S.

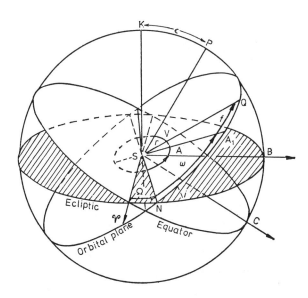

Figure 2.13

35

Its rectangular coordinates referred to axes $S\Upsilon$, SB, SK (forming a right-handed system as shown) are (x, y, z) where

$$SV = r = (x^2 + y^2 + z^2)^{1/2}.$$

SA (where A is perihelion) produced meets the sphere in point A_1 while SV produced meets the sphere in Q.

Then

$$\Upsilon N = \Omega, \quad NA_1 = \omega, \quad A_1 Q = f.$$

By the cosine formula in the spherical triangle $Q\Upsilon N$, where angle $\Upsilon NQ = 180° - i$, we have

$$\cos \Upsilon Q = \cos \Omega \cos (\omega + f) - \sin \Omega \sin (\omega + f) \cos i.$$

But

$$\frac{x}{r} = \cos \Upsilon Q.$$

Hence

$$x = r[\cos \Omega \cos (\omega + f) - \sin \Omega \sin (\omega + f) \cos i]. \qquad (2.4)$$

Similarly, using triangle QNB and the cosine formula and remembering that

$$\frac{y}{r} = \cos BQ$$

we have

$$y = r[\sin \Omega \cos (\omega + f) + \cos \Omega \sin (\omega + f) \cos i]. \qquad (2.5)$$

Finally, using triangle QKN, the cosine formula gives

$$z = r \sin (\omega + f) \sin i. \qquad (2.6)$$

To transform to heliocentric equatorial rectangular coordinates it is noted that in the new set of axes $S\Upsilon$, SC and SP are such that SC lies in the equatorial plane making an angle $90°$ with $S\Upsilon$, while SP is perpendicular to the plane so that the three axes form a right-handed set. Then the new axes SC and SP are obtained from the old axes SB and SK by rotating the latter about S through the angle ϵ. If the heliocentric equatorial rectangular coordinates of the vehicle are (x', y', z'), then

$$x' = x$$
$$y' = y \cos \epsilon - z \sin \epsilon$$
$$z' = y \sin \epsilon + z \cos \epsilon.$$

Using equations (2.4), (2.5) and (2.6) we obtain

$$x' = r[\cos (\omega + f) \cos \Omega - \sin (\omega + f) \sin \Omega \cos i] \qquad (2.7)$$

$$y' = r[\cos (\omega + f) \sin \Omega \cos \epsilon + (\cos i \cos \Omega \cos \epsilon - \sin i \sin \epsilon) \sin (\omega + f)]$$

$$(2.8)$$

36

$$z' = r[\cos{(\omega+f)}\sin{\Omega}\sin{\epsilon} + (\cos{i}\cos{\Omega}\sin{\epsilon} + \sin{i}\cos{\epsilon})\sin{(\omega+f)}].$$

$$(2.9)$$

A set of auxiliary angles may now be defined as follows:

$$\sin{a}\sin{A} = \cos{\Omega}$$

$$\sin{a}\cos{A} = -\cos{i}\sin{\Omega}$$

$$\sin{b}\sin{B} = \sin{\Omega}\cos{\epsilon}$$

$$\sin{b}\cos{B} = \cos{i}\cos{\Omega}\cos{\epsilon} - \sin{i}\sin{\epsilon}$$

$$\sin{c}\sin{C} = \sin{\Omega}\sin{\epsilon}$$

$$\sin{c}\cos{C} = \cos{i}\cos{\Omega}\sin{\epsilon} + \sin{i}\cos{\epsilon}.$$

Then, equations (2.7), (2.8) and (2.9) become

$$x' = r\sin{a}\sin{(A+\omega+f)}$$

$$y' = r\sin{b}\sin{(B+\omega+f)} \qquad (2.10)$$

$$z' = r\sin{c}\sin{(C+\omega+f)}.$$

This form is convenient to use when the rectangular coordinates are required for a number of positions of the vehicle. The auxiliary quantities a, A, b, B, c, C are functions only of the elements Ω, i and of ϵ; they may therefore be calculated once for all positions. The variables r and f must be calculated, however, for each position in a way to be described later (chapter 4). It should however be noted that Ω, i and ω are constant only if the vehicle is in an unperturbed orbit. This situation exists in fact over most of an interplanetary mission conducted in free fall.

(ii) The origin of coordinates is now changed from the Sun's centre to the Earth's centre. Thus, in figure 2.14, if the Earth is taken to be at E, the Sun

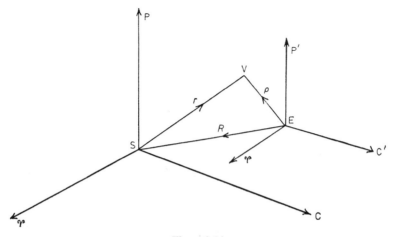

Figure 2.14

at S, and the set of heliocentric equatorial rectangular axes is given by Sϓ, SC and SP, the geocentric equatorial rectangular set of axes is given by Eϓ', EC' and EP', where the plane ϓ'EC' is the plane of the Earth's equator. Let (ξ, η, ζ) be the coordinates of the vehicle V with respect to these axes, where

$$\rho^2 = \xi^2 + \eta^2 + \zeta^2. \tag{2.11}$$

Also let the heliocentric equatorial rectangular coordinates of the Earth be (x_1, y_1, z_1). Then

$$\xi = x' - x_1$$
$$\eta = y' - y_1$$
$$\zeta = z' - z_1.$$

If, then, the Sun's geocentric equatorial rectangular coordinates are (X, Y, Z), we have

$$\left.\begin{aligned} \xi &= x' + X \\ \eta &= y' + Y \\ \zeta &= z' + Z \end{aligned}\right\} \tag{2.12}$$

since

$$X = -x_1, \quad Y = -y_1, \quad Z = -z_1. \tag{2.13}$$

The coordinates of the Sun (X, Y, Z) are tabulated in the 'Astronomical Ephemeris' and other almanacs. Alternatively x_1, y_1, z_1 are obtained from the elements of the Earth's orbit, remembering that since the orbit is in the ecliptic, the inclination is zero. Writing these elements as Ω_1, ϖ_1 ($= \Omega_1 + \omega_1 =$ longitude of perihelion of the Earth's orbit), we obtain from equation (2.10) the three relations

$$x_1 = r_1 \cos(\varpi_1 + f_1)$$
$$y_1 = r_1 \sin(\varpi_1 + f_1) \cos \epsilon$$
$$z_1 = r_1 \sin(\varpi_1 + f_1) \sin \epsilon$$

where the values of the radius vector r_1 and the true anomaly f_1 may be calculated for any time t.

(iii) In figure 2.15 the geocentric celestial sphere is shown with the meridian P'V'H drawn through the projection V' of V (the vehicle's geocentric position) on the celestial sphere.

Then

$$\xi = \rho \cos \text{VE}ϓ' = \rho \cos \text{V'E}ϓ'$$

giving

$$\xi = \rho \cos \text{V'}ϓ'.$$

Similarly

$$\eta = \rho \cos \text{V'C'}$$

and

$$\zeta = \rho \cos \text{V'P'}.$$

38

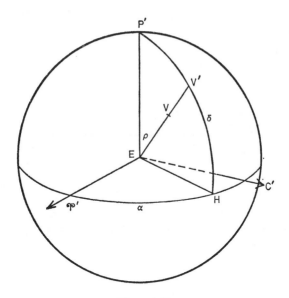

Figure 2.15

Using the spherical triangle ♈'V'H (right-handed at H) and the cosine formula, we obtain the three relations

$$\left.\begin{array}{l} \xi = \rho \cos \alpha \cos \delta \\ \eta = \rho \sin \alpha \cos \delta \\ \zeta = \rho \sin \delta. \end{array}\right\} \tag{2.14}$$

Hence, using equations (2.10), (2.12) and (2.14), we find that

$$\rho \cos \alpha \cos \delta = X + r \sin a \sin (A + \omega + f) \tag{2.15}$$

$$\rho \sin \alpha \cos \delta = Y + r \sin b \sin (B + \omega + f) \tag{2.16}$$

$$\rho \sin \delta = Z + r \sin c \sin (C + \omega + f). \tag{2.17}$$

We have seen that if the elements of the vehicle's orbit are known, the right-hand sides of equations (2.15), (2.16) and (2.17) can be calculated for any time since values of X, Y and Z can be obtained from the 'Astronomical Ephemeris'.

Hence

$$\tan \alpha = \frac{Y + r \sin b \sin (B + \omega + f)}{X + r \sin a \sin (A + \omega + f)}$$

which gives us α.

Also

$$\tan \delta \csc \alpha = \frac{Z + r \sin c \sin (C + \omega + f)}{Y + r \sin b \sin (B + \omega + f)}$$

which gives δ.

Also

ρ = the square root of the sum of the squares of the right-hand sides of equations (2.15), (2.16) and (2.17).

2.10 Galactic Coordinate System

When we consider the distribution and motions of bodies in the Galaxy, it is incongruous in such investigations to use coordinate systems based on the equator or ecliptic. The fact that the Galaxy is lens-shaped, with the Sun in or near to the median plane of this lens, suggests that a convenient reference system would use this plane.

The material (stars, dust and gas) making up the Galaxy is symmetrically distributed on either side of the galactic equator LNA (figure 2.16). The

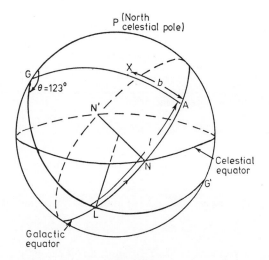

Figure 2.16

galactic equator great circle intersects the celestial equator in the two points N and N′; the former is called the *ascending node*, the latter the *descending node*, since an object travelling along the galactic equator in the direction of increasing right ascension would ascend from southern to northern hemisphere in passing through N. It moves from northern to southern hemisphere in passing through N′. By definition the north and south galactic poles G and G′ lie in the northern and southern hemispheres respectively. Any object X (α, δ) then has a galactic latitude and longitude.

Prior to 1959 the zero from which galactic longitude was measured was the ascending node N (Ohlsson System); since then it has been taken to be L, the point of intersection of the galactic equator by the great semicircle GLG′, where position angle $\theta = PGL = 123°$. By defining L in this way it lies in the direction of the galactic centre as seen from the Sun S. Then the *galactic longitude* of X, namely *l*, is measured along the galactic equator from L to the foot of the meridian from G through X from 0° to 360° in the direction of increasing right ascension. Thus *l*=LNA and the angle PGX is equal to $\theta - l$.

The *galactic latitude* of X (namely *b*) is the object's angular distance north or south of the galactic equator measured from 0° to 90° along the meridian from the north galactic pole G through the object. Thus *b*=arc AX and is north.

To distinguish between the Ohlsson and IAU systems it is usual to label *l* and *b* with superscripts I and II respectively. Thus:

IAU galactic pole ($b^{II}=90$)

$$\alpha = 12^h\ 49\cdot0^m,\ \delta = +27°\ 24\cdot0'\ \text{(Epoch 1950)}$$

$$\alpha = 12^h\ 46\cdot6^m,\ \delta = +27°\ 40\cdot0'\ \text{(Epoch 1900)}.$$

Ohlsson galactic pole ($b^I=90$)

$$\alpha = 12^h\ 40^m,\ \ \ \delta = +28\cdot0°\ \ \ \ \text{(Epoch 1900)}.$$

2.11 Time Measurement

Primitive man based his sense of the passage of time on the growth of hunger or thirst and on impersonal phenomena such as the changing altitude of the Sun during a day, the successive phases of the Moon and the changing seasons. By about 2000 BC more civilized men kept records and systematized the impersonal phenomena into the day, the month and the year. Emphasis was given to the year as a unit of time by their observation that the Sun made one revolution of the stellar background in that period of time.

Since everyday life is geared to daylight the Sun became the body to which the system of timekeeping used by day was bound. The apparent solar day was then the time between successive passages of the Sun over the observer's meridian or the time during which the Sun's hour angle increased by 24^h (360°). In a practical way the Sun was noted to be on the meridian when the shadow cast by a vertical pillar was shortest.

On the other hand, the apparent diurnal rotation of the heavens provided another system of timekeeping called sidereal time, which was based on the rotation of the Earth on its axis. The interval between two successive passages of a star across the observer's meridian was then called a sidereal day. Early on in the history of astronomy it was realized that the difference between the two systems of timekeeping—solar and stellar—was caused by the orbital motion of Sun relative to Earth. Thus, in figure 2.17, if two successive passages of the star over the observer's meridian define a sidereal day (the star being taken to be at an infinite distance effectively from the Earth) the Earth will have rotated the observer O through 360° from O_1 to O_2. In order that one apparent solar day will have elapsed however, the Earth (E) will have to rotate until the observer is at O_3 when the Sun (S) will again be on his meri-

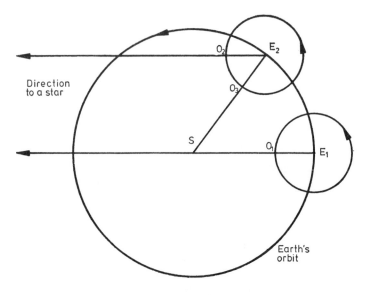

Figure 2.17

dian. Since the Earth's radius vector SE sweeps out about 1°/day and the Earth rotates at an angular velocity of about one degree every 4 min, the sidereal day is consequently about 4 min shorter than the average solar day.

We will now consider these systems in greater detail.

2.11.1 Sidereal time

The First Point of Aries (vernal equinox ♈) is the reference point chosen on the rotating celestial sphere to define the sidereal day (24 sidereal hours). The time between successive passages of the vernal equinox across the observer's meridian is *one sidereal day*. The hour angle of the vernal equinox increases from 0^h to 24^h so that the *local sidereal time* (LST) is defined as the hour angle of the vernal equinox HA (♈). The LST, as its name implies, depends upon the observer's longitude λ on the Earth's surface.

From figure 2.18, it is seen that if X denotes the direction of a celestial object, its right ascension is α and its hour angle is H. Then the local sidereal time is the sum of the hour angle of X and the right ascension of X; that is

$$\text{LST} = \text{HA}(X) + \text{RA}(X). \tag{2.18}$$

This relationship is important because the celestial object may be the Sun, the Moon, a planet, a star, a space vehicle etc.

If the LST is known and the right ascension α and declination δ of the object have been computed for that time, then the hour angle H and declination δ are known at any subsequent time, giving the direction of the object X on the celestial sphere.

In an observatory there are usually one or more clocks keeping the local sidereal time of that longitude. Since the hour angle of a star is zero when it

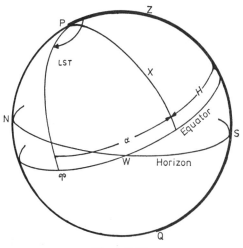

Figure 2.18

transits on the observer's meridian, the star's right ascension α at that instant is the local sidereal time. A careful check on the clock error and rate of change of the error can then be made by observing frequently the sidereal times of transit of well known stars and comparing them with their right ascensions. Such stars are called 'clock stars' and such observations are part of the routine work at any observatory.

In addition, the Greenwich sidereal time is tabulated in the *'Astronomical Almanac'* at frequent epochs of Universal Time (section 2.11.2). Now the time between transits of a celestial object over the Greenwich meridian and the local observer's meridian is equal to the longitude of the local observer as seen in figure 2.19, where the geocentric celestial sphere (north celestial pole P) is shown with the Earth (North Pole p). Greenwich (g) and its zenith (G) is shown, the meridian through G (namely PGB) being the Greenwich observer's meridian. An observer in longitude λW is indicated by O with his zenith and observer's meridian given by Z and PZA. The vernal equinox is shown as γ and a celestial object is indicated transiting at X.

The Greenwich hour angle of X is then $G\hat{P}X$, which is the longitude λW of the observer.

The Greenwich hour angle of the vernal equinox γ, written $H_G(\gamma)$, is $G\hat{P}\gamma$ which is equal to the hour angle $X\hat{P}\gamma$ plus the longitude λW of the observer. In other words,

$$HA_G(\gamma) = HA(\gamma) + \lambda W.$$

But the hour angle of γ is the local sidereal time. We may therefore write

$$GST = LST + \lambda W. \tag{2.19}$$

If γ were any celestial object $*$, we would have

$$HA_G(*) = HA(*) + \lambda W. \tag{2.20}$$

This result is as important as equations (2.18) and (2.19).

43

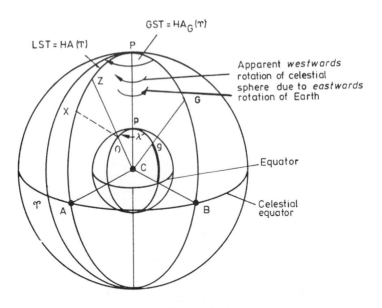

Figure 2.19

It is easily seen that if the longitude is east it is subtracted. This rule is often remembered by the mnemonic

> 'Longitude east, Greenwich least,
> Longitude west, Greenwich "best".'

2.11.2 Mean solar time

If the length of the apparent solar day (the time between two successive passages of the Sun across the observer's meridian) is measured by an accurate sidereal clock it is found to vary throughout the year. There are two main reasons for this:

(i) The Sun's apparent orbit about the Earth is an ellipse in which equal angles are not swept out by the radius vector joining Sun to Earth in equal times.
(ii) The path of the Sun is in the ecliptic which is inclined at an angle of $23\frac{1}{2}°$ approximately to the equator (along which the Sun's hour angle is measured).

Astronomers overcame these irregularities to obtain mean solar time by the following devices.

(i) A fictitious body called the *dynamical mean sun* is introduced which starts off from perigee with the Sun, moves with the mean angular velocity (mean motion) of the Sun and returns to perigee at the same time as the Sun. It also moves in the plane of the ecliptic.
(ii) When this dynamical mean sun, moving in the ecliptic, reaches the vernal equinox ♈, a second fictitious body called the *mean sun* starts off along the

equator with the Sun's mean motion, returning to ♈ with the dynamical mean sun.

Since the mean sun increases its right ascension at a constant rate of about $1°$/day and increases its hour angle by 24^h in one sidereal day, the time between successive passages of the mean sun over the observer's meridian is constant. This interval is called a *mean solar day*.

The relationship between sidereal time and mean solar time is given below.

1 mean solar day $= 24^h\ 03^m\ 56{\cdot}5554^s$ of sidereal time.

1 sidereal day $= 23^h\ 56^m\ 04{\cdot}0905^s$ of mean solar time.

Some astronomical almanacs give tables for the conversion of mean solar time to or from sidereal time. The 'Astronomical Almanac' published in Great Britain and the United States demonstrates how Universal Time (Greenwich Mean Time) may be converted to sidereal time and *vice versa*.

In order to relate the positions of the mean sun and the real Sun, a quantity called the *equation of time* is defined as the difference between the hour angle of the Sun ($\odot$) and the hour angle of the mean sun (MS), or

$$E = \text{HA}(\odot) - \text{HA}(\text{MS}). \tag{2.21}$$

From equation (2.18), namely

$$\text{LST} = \text{HA}(*) + \text{RA}(*),$$

it is seen that

$$E = \text{RA}(\text{MS}) - \text{RA}(\odot). \tag{2.22}$$

The equation of time E is related to the time of ephemeris transit T, tabulated for every day of the year in the 'Astronomical Almanac', by the equation

$$E = 12^h - T.$$

Greenwich Mean Time (GMT) or Universal Time (UT) is based on mean solar time such that

$$\text{GMT} = \text{HA}_G(\text{MS}) \pm 12^h. \tag{2.23}$$

Equation (2.23) implies that a civil day begins when it is mean midnight. GMT (UT) is a convenient time system used in most observatories throughout the world. In civil life, unless the longitude concerned is near the Greenwich meridian, local time systems are used, the surface of the Earth having been divided into *standard time zones* for this purpose.

This convention gives a clock time related approximately to the Sun's position in the sky and also avoids the necessity of a moving observer continually adjusting his watch.

Within each zone the same civil mean time called *Zone Time* (ZT) or *Standard Time* is used and the zones are defined by meridians of longitude, each zone being $15°$ (1^h) wide. The Greenwich Zone (Zone 0) has bounding meridians $0^h\ 30^m$ W and $0^h\ 30^m$ E, and keeps the time of the Greenwich meridian, namely GMT (UT). Zone $+1$ has boundaries $1^h\ 30^m$ W and $0^h\ 30^m$ W, keeping the time of meridian 1^h W. Zone -1 has boundaries $1^h\ 30^m$ E and $0^h\ 30^m$ E, keeping the time of meridian 1^h E. The division of the Earth's surface in this way is continued east and west up to Zones $+12$ and -12.

According to the previous definition both these zones would keep the time of 12^h W which is also 12^h E. The convention is made that the zone from 11^h 30^m W to 12^h W is Zone + 12, while the zone from 11^h 30^m E to 12^h E is Zone − 12. The meridian separating them is called the *International Dateline* where a given day first begins.

It should be added that the actual dateline, for geographical reasons, does not follow faithfully the 12^h meridian but makes local detours to include in one hemisphere parts of countries that would be placed in the other if the Line did not deviate this way. It should also be added that ships crossing the dateline from east to west omit one day, while others crossing from west to east add one day.

In large countries, such as the USA and the USSR, more than one zone is involved. In the United States four time zones are used; the mean times are called Eastern, Central, Mountain and Pacific Times, based on the meridians 5^h, 6^h, 7^h and 8^h west of Greenwich.

The relation between Zone Time and GMT is

$$\text{GMT} = \text{ZT} + \text{longitude,} \tag{2.24}$$

where the longitude of the meridian involved is added when west and subtracted when east (in agreement with the previous rule—see equation (2.20)).

The year used in civil life is based on the *tropical year*, defined as the interval in time between successive passages of the Sun through the vernal equinox. This is 365·2422 mean solar days. For convenience the calendar year contains an integral number of days, either 365 or 366. Every fourth year (called a leap year) has 366 days, excepting those century years (such as 1900 AD) whose number of hundreds (in this case 19) are indivisible by four exactly. These rules give a mean civil year equal in length to 365·2425 mean solar days, a figure very close to the number of mean solar days in a tropical year.

2.11.3 The Julian date

The irregularities in the present calendar, and the change from the Julian to the Gregorian calendar (which took place in different countries at different epochs), makes it difficult to compare lengths of time between observations made many years apart. Again, in the observations of variable stars it is useful to be able to say that the moment of observation occurred so many days and fractions of a day after a definite epoch. The system of *Julian Day Numbers* was therefore introduced to reduce computational labour in such problems and avoid ambiguity. January 1 of the year 4713 BC was chosen, time being measured from that epoch (mean noon on January 1, 4713 BC) by the number of days that have elapsed since then. The *Julian date* is given for every day of the year in the 'Astronomical Almanac.'

Tables also exist for finding the Julian date for any day in any year. For example, the Julian date for June 24, 1962, is 2 437 839·5 when June 24 begins; again the time of an observation made on June 24, 1962, at 18^h GMT is JD 2 437 840·25.

Time may also be measured in *Julian centuries*, each containing exactly 36 525 days.

Orbital data for artificial Earth satellites are often referred to epochs expressed in *Modified Julian Day Numbers* in which the zero point in this system is 17·0 November, 1858. Hence

$$\text{Modified Julian date} = \text{Julian date} - 2\,400\,000\text{·}5 \text{ days.}$$

2.11.4 Ephemeris Time

Both mean solar time and sidereal time are based on the rotation of the Earth on its axis. Until comparatively recently it was thought that, apart from a slow secular increase in the rotation period due to tidal friction, the Earth's period of rotation was constant. A *secular* change is defined to be one that is effectively irreversible, running on from age to age so that its magnitude is proportional to time passed. *Tidal friction* acts as a brake on the Earth's rotation, being due to the Moon's gravitational effect.

The development and use of very accurate clocks revealed that other variations occurred in the period of the Earth's rotation. These small changes in general take place abruptly and are not predictable. Since Universal Time (GMT) is based on observations of the transits of celestial objects made from the irregularly rotating Earth, it must differ from a theoretical time that flows on uniformly. This time is the Newtonian time of celestial mechanics, being the independent variable in the theories of the movements of the Sun, the Moon and the planets. Hence, their positions as published in ephemerides (tables of predicted positions) based on these theories are bound to *Ephemeris Time*.

The value of Ephemeris Time at a given instant is obtained by very accurate observations of abrupt variations in the longitudes of Sun, Moon and planets due to corresponding variations in the Earth's rate of rotation. Clemence estimated that to define Ephemeris Time correctly to one part in 10^{10}, observations of the Moon were required over five years. In practice atomic clocks may be used to give approximate values of Ephemeris Time, their readings being subsequently corrected by long series of astronomical observations. The quantity in fact determined is ΔT, given by

$$\Delta T = \text{Ephemeris Time} - \text{Universal Time.}$$

This quantity is tabulated in the '*Astronomical Almanac*'. At present it is about 55^s.

Various further refinements in time measurement have recently been made, for example International Atomic Time (TAI), related approximately to Ephemeris Time (ET) by the relation $\text{ET} = \text{TAI} + 32\text{·}18^s$, but such refinements are beyond the scope of this text. The interested reader should consult the works by McNally (1974) or Green (1985) described in the reference list.

Problems

In the following problems assume (i) a spherical Earth, (ii) the obliquity of the ecliptic to be 23° 27'.

2.1 Find the departure between two places of the same latitude 60°N, given that their longitudes are (i) 48° 27'W and 27° 11'W, (ii) 32° 19'W and 15° 49'E.

2.2 An aircraft flies at 600 knots ground speed (1 knot=1 nautical mile per hour) between Prestwick (04° 36'W, 55° 31'N) and Gandar (54° 34'W, 49° 00'N) along the great-circle route between these airports. How long does the trip take?

2.3 What is the highest northerly latitude touched by the aircraft in problem 2.2 and when does this occur?

2.4 What are the Sun's approximate right ascensions and declinations on March 21, June 21, September 21 and December 21?

2.5 Draw the celestial sphere for an observer in latitude 60°N, putting in the horizon, equator, zenith, north celestial pole and observer's meridian. If the local sidereal time is 9^h put in the vernal equinox and the ecliptic. The artificial satellite 1960 iota 1 (Echo 1) is observed to have at this instance an altitude of 45° and an azimuth of 315°E of N. Insert the satellite's position in your diagram and estimate (i) Echo's topocentric right ascension and declination, (ii) its topocentric ecliptic longitude and latitude. If the date is March 21, insert the Sun in your diagram.

2.6 Using the data given in problem 2.5, check your estimates of Echo's topocentric right ascension and declination by calculations.

2.7 If a star rises tonight at 10 pm, at what approximate civil time will it rise 30 days hence?

2.8 When the vernal equinox rises in azimuth 90°E of N, find the angle the ecliptic makes with the horizon at that point for an observer in latitude 60°N.

2.9 Show that the point of the horizon at which a star rises is

$$\sin^{-1}(\sec \phi \sin \delta)$$

north of east where ϕ is the observer's latitude and δ is the declination of the star.

2.10 An observation of the Sun was made at approximately 10^h 50^m Zone Time on December 12, the GMT chronometer time being 04^h 49^m 16^s. The zone was −6, the observer's position was 45°N, 92° 30'E and the equation of time (found from the Astronomical Almanac) was +6^m 38^s. Calculate the Sun's hour angle for the observer.

2.11 If the Sun's declination at the time of the observer's observation in problem 2.10 was 23°S, and if the local sidereal time was 16^h 35^m, show on a diagram the position of the ecliptic for the observer at that time.

2.12 A ship steaming eastwards along the parallel of latitude at 15 knots leaves A (44° 30'S, 58° 20'W) at Zone Time 0200 hours on January 3. Find (i) its position B after a voyage of 5 days 6 hours and (ii) the Zone Time, with date, of arrival at B.

2.13 What is the right ascension of the artificial satellite Samos II when it is observed to transit across the observer's meridian at local sidereal time 09^h 23^m 41·6^s?

2.14 The observed times (by a sidereal clock) of consecutive transits of a star whose right ascension is 8^h 21^m 47·4^s are 8^h 22^m 00·8^s and 8^h 21^m 59·7^s. Find the error of the clock at each transit and also its rate.

2.15 In Zone +3 at about 6 pm Zone Time on December 12, a star whose right ascension is 6^h 11^m 12^s was observed. The GMT chronometer time was 21^h 00^m 04^s, the observer's longitude being 46°W. If the Greenwich sidereal time at 0^h GMT on December 13 was 5^h 23^m 07^s, find the hour angle of the star for the observer. (Use the relationship on page 44 between sidereal time and mean solar time or use the Astronomical Almanac if available.)

2.16 Calculate the hour angle of the Sun on June 8, 1962, at San Francisco (longitude 8^h 09^m 43^{s}W) when the Pacific time is 10.30 am. The equation of time is +1^m 14^s.

2.17 Calculate how long the star Altair ($\alpha=19^h$ 48^m 06^s, $\delta=8°$ 43') is above the horizon each day for an observer in latitude 55° 52'N. Is your answer in sidereal time or mean solar time? At what local sidereal time does Altair set in this latitude? At what azimuth does it set?

2.18 Show that the heliocentric equatorial rectangular coordinates of a space vehicle in an interplanetary orbit can be written in the form

$$x = r \sin a \sin (A + \omega + f)$$
$$y = r \sin b \sin (B + \omega + f)$$
$$z = r \sin c \sin (C + \omega + f)$$

and give expressions for the auxiliary angles a, A, b, B, c and C.

2.19 If (λ_1, β_1), (λ_2, β_2) and (λ_3, β_3) are the heliocentric ecliptic longitudes and latitudes of a planet at three points in its orbit, prove that

$$\tan \beta_1 \sin (\lambda_2 - \lambda_3) + \tan \beta_2 \sin (\lambda_3 - \lambda_1) + \tan \beta_3 \sin (\lambda_1 - \lambda_2) = 0.$$

References

Astronomical Almanac (London: HMSO)

Explanatory Supplement to the Astronomical Ephemeris and the American Ephemeris and Nautical Almanac (London: HMSO)

Green R M 1985 *Spherical Astronomy* (London: Cambridge University Press)

McNally D 1974 *Positional Astronomy* (London: Frederick Muller)

Smart W M and Green R M 1977 *Textbook on Spherical Astronomy* (London: Cambridge University Press)

Astronomical Almanac, published yearly, contains predicted positions for the bodies of the Solar System, excepting comets, meteors and all but the four largest asteroids. It also contains data on the brighter stars, sunrise and sunset times, similar times for the Moon, and a number of important tables. The *Astronomical Almanac* is also published yearly by the US Government Printing Office, Washington, DC.

The *Explanatory Supplement to the Astronomical Ephemeris* is a valuable reference book. Not only does it provide the users of *The Astronomical Ephemeris* with a full explanation of the latter's contents and the methods of deriving them; it also gives authoritative treatments of a number of the subjects contained in this and the succeeding chapter.

Before 1981 the *Astronomical Almanac* was called the *Astronomical Ephemeris* and, strictly speaking, the *Explanatory Supplement* refers to the *Astronomical Ephemeris*, which in some respects differs in its contents from those of the *Astronomical Almanac*.

Textbook on Spherical Astronomy, of a mathematical nature, is of moderate difficulty. It discusses the main branches of spherical astronomy from first principles and contains a large number of examples for the student. It is based on the classical text by W M Smart but has also been updated by R M Green. *Spherical Astronomy* by R M Green is a modern work on fundamental astronomy, necessitated by the increase in observational accuracy achieved by modern astrometrical techniques.

Positional Astronomy covers much the same ground as *Textbook on Spherical Astronomy*.

3 The Reduction of Observational Data

3.1 Introduction

A wide armoury of observational techniques is used in noting the direction and distance of any object beyond the Earth's atmosphere. The variety of techniques is dictated by the vast range of object distances, speeds, radiation outputs and sizes. The object (if man-made) may be in close Earth orbit, or at the Moon's distance, or in interplanetary space. It may or may not be transmitting in the radio region and may also be reflecting sunlight. Its observed velocity may range from many degrees per second of time to seconds of arc per hour. If the object is natural and in the Solar System it may be the Sun, the Moon, a planet, a satellite, an asteroid or a comet. It will (if it is not the Sun) reflect sunlight, its brightness depending upon its size, albedo (ability to reflect) and its distance from the Sun and the observer. Its observed velocity with respect to the stellar background can be 13°/day for the Moon, 1°/day for the Sun, or much less for all the others. For stars and other objects in the far reaches of space, their angular speeds are so small that only those nearest to the Solar System can have their transverse motions measured. Much of our knowledge of their movements comes from determination of their radial velocities. In addition their outputs may be predominantly in the visual, radio, X-ray or infrared parts of the spectrum.

Nevertheless, although there is such a bewilderingly large set of ranges of object, distance, speed, radiation output and so on, there are standard reduction techniques to be applied to the observations made of such objects. Such techniques try as far as possible to remove effects due to the observer's position in time and space, thus providing objective observational data that can be compared and utilized by computing centres to provide orbital elements and predictions. In cases of man-made objects (such as artificial satellites), planets, satellites and other objects within the Solar System, such a process using reduced data is called orbit determination and improvement. Reduced data for objects outside the Solar System may be used to compute orbital elements and improve them (in the case of a binary star system) or provide statistical data on the movements of groups of stars leading to an improved knowledge of the structure and dynamics of our Galaxy.

50

3.2 Observational Techniques

Space vehicles are tracked either by optical or electronic means. Typical optical instruments include:

(i) Recording optical tracking instruments which have a small field of view, and which are mounted in the horizontal system, altitude and azimuth being read automatically off graduated circles. These instruments must be calibrated frequently.

(ii) A kinetheodolite, also with a small field of view and set in the alt–azimuth system, being used to track the object and take photographs of it on 35 mm film.

(iii) A ballistic camera of very wide field, taking photographs of the object against the stellar background.

(iv) A Baker–Nunn camera of very wide field, capable of registering objects and stars as faint as magnitude $+17\cdot2$. Hewitt cameras are also used.

(v) Orthodox astronomical telescopes for deep-space objects whose angular velocity is low and whose brightness is less than the limiting magnitude of the Baker–Nunn camera.

In astronomy, the brightness of an object is measured on the magnitude scale. This scale was first introduced in the second century BC in an imprecise way by Hipparchus, who graded the naked-eye stars according to their brightness into six magnitudes: the first consisting of the twenty brightest, the second of the next fifty in order of brightness, until the sixth, which included the faintest stars visible to the naked eye.

Roughly speaking, a star of one magnitude is two and a half times as bright as a star of the next magnitude; the magnitude scale is thus basically logarithmic in character. The system has been rendered precise by the following definition:

If B_1 and B_2 are the brightnesses of two stars and m_1 and m_2 are their magnitudes, then

$$B_1 \propto 10^{-0\cdot4m_1}$$

$$B_2 \propto 10^{-0\cdot4m_2}$$

so that

$$m_1 - m_2 = -\tfrac{5}{2}\log_{10}(B_1/B_2).$$

Hence a difference in magnitude of five gives a brightness ratio of exactly 100. It is to be noted that the greater the magnitude is algebraically, the fainter the object is in brightness. Thus the limiting magnitude (faintest possible object registered) of a Baker–Nunn camera is $+17\cdot2^m$ while the limiting magnitude for the 200 inch Hale telescope at Mount Palomar is $+23\cdot2^m$.

It should also be noted that various magnitude systems exist, depending upon whether the radiation from the object enters the eye, or is allowed to fall on photographic emulsion, or on a photoelectric device.

The concept of an absolute magnitude system is introduced to enable meaningful comparisons of objects' intrinsic luminosities to be made. To get rid of the effect of distance it is customary to state what the magnitude of the object would be at a standard distance. This distance is taken to be 10 pc (see section 1.3). If d is the object's true distance in pc, and M and m are its apparent magnitudes at distances of 10 and d pc respectively, it is easy to see, taking into account that brightness falls off as the square of the distance, that

$$M = m + 5 - 5 \log_{10} d.$$

The quantity M is called the *absolute magnitude* of the object.

Typical electronic instruments include:

(i) Radio telescopes, used either to receive radio signals sent from the space-craft or (if it is near) as radar instruments picking up radar echoes from the craft.

(ii) An interferometer. Two or more antennas in an array of precisely known geometry which in some instrumental designs can be varied. The principle of such a direction-finding system is that a radio signal arriving simultaneously at two points will show a phase difference, depending on the path difference from the signal source to the points. There are well known techniques for finding the direction of the source relative to the receiving points.

(iii) Apparatus capable of detecting Doppler shift. If a source emitting radiation has a velocity v relative to the observer, then the received radiation that normally has a wavelength λ when the velocity relative to the observer is zero will have a measured wavelength λ', where

$$\frac{\lambda' - \lambda}{\lambda} = \frac{v}{c} \tag{3.1}$$

c being the velocity of light. The convention is made that v is negative if the source is approaching and positive if it is receding. Wavelength λ and frequency ν are connected by the well-known relation

$$\nu\lambda = c$$

and so we can rewrite equation (3.1) as

$$\frac{\nu' - \nu}{\nu} = -\frac{v}{c}. \tag{3.2}$$

This change in wavelength and frequency due to relative velocity is called the Doppler effect.

It is seen that electronic apparatus capable of measuring the frequency difference will give the line-of-sight velocity of the object emitting the radio waves. It should be remarked that the above is a gross simplification of a complicated phenomenon.

There are many types of systems based on the Doppler principle. With some, the distance (range) of the object is obtained as well as the line-of-sight velocity (range rate). Accuracies attained with range and range-rate equipment are extremely high.

For natural celestial objects such as planets, stars and galaxies, optical and radio telescopes are used. Most of the work with optical telescopes is now carried out by photography.

Both optical and radio telescopes will obtain the direction coordinates of the object at the time of observation. Unless the radio telescope is used in an interferometric mode with other radio telescopes, the precision with which it pinpoints a celestial object emitting radio waves falls far short of an optical telescope's ability. As part of an interferometer with a long baseline (in some cases thousands of kilometres) however, its accuracy in determining position is as high as the best optical system.

A large radio telescope operating as a radar instrument is capable of measuring accurately the distances of the nearer bodies in the Solar System such as the Moon, Venus, Mars, Mercury, Jupiter and Saturn.

Summarizing all these optical and electronic methods; it is seen that in general the altitude and azimuth of the object (or its position on a photographic plate with respect to a stellar background) is obtained. Its distance from the observer is not usually measured unless Doppler or radar equipment is used. In addition a time is noted at which the observation was made. This time is reduced to Universal Time and then usually to local sidereal time, if not already in that system.

The main corrections to the data to obtain a geocentric equatorial position for the object are now outlined in principle. If the altitude and azimuth of the object are measured, the first corrections applied are known instrumental errors. This entails a frequent calibration of the instrument since such errors are not in general static.

3.3 Refraction

A ray of light entering the Earth's atmosphere is refracted or bent so that the observed altitude of the source of light is increased. Thus in figure 3.1 the ray of light appears to the observer at O to come from the direction C so that the measured zenith distance ζ is $Z\hat{O}C$ while the true zenith distance is $Z\hat{O}B$, where OB is parallel to the original direction in which the ray entered the atmosphere.

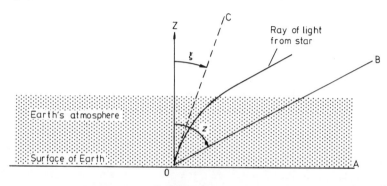

Figure 3.1

Then, assuming the atmosphere to consist of plane parallel layers of different densities, it is easily shown that Snell's law of refraction leads to the relation

$$r = k \tan \zeta = k \cot a \qquad (3.3)$$

where $r = z - \zeta$, and k is about 58·2″. Since the observed altitude a is too large, the angle r is subtracted from it (Smart 1956).

Equation (3.3) is valid for zenith distances less than 45° and is a fairly good approximation up to 70°. Beyond that, a more accurate formula taking into account the curvature of the Earth's surface is required, while for zenith distances near 90° special tables are required.

There are a number of versions of equation (3.3). Among them is Comstock's,

$$r = \frac{983p}{460 + T} \tan \zeta$$

where r is expressed in seconds of arc, p is the barometric pressure in inches of mercury and T is the temperature in degrees Fahrenheit.

For radio measurements refraction depends strongly upon the frequency employed. The lower atmosphere produces refraction effects approximately twice the optical effect, decreasing rapidly with increasing angle of elevation. The ionosphere also refracts radio waves due to induced motion of charged particles in the ionosphere, in amounts dependent on the ion-density gradient. If N is the electron density per cubic centimetre and v is the frequency in kilohertz, then the local effective dielectric constant n (which varies throughout the ionosphere) may be expressed by

$$n = \left(1 - \frac{81N}{v^2}\right)^{1/2}.$$

As height increases above the Earth's surface, the electron density increases then falls off again. N may become so large that n is zero or imaginary. In these cases a radio signal cannot penetrate the ionosphere from the inside or from the outside. In other cases when the frequency is high enough, penetration takes place with bending of the signal. If we assume that the ionosphere consists of concentric shells about the Earth, Snell's law enables the path of the radio signal to be calculated from the relation $n\rho \sin i =$ constant, where ρ is the radius of curvature of the shell of dielectric constant n, and i is the angle of incidence of the signal. Study of ionospheric refraction by comparison of optical and radio tracking of artificial satellites has yielded valuable data.

Having applied the correction for refraction, the topocentric altitude and azimuth may be converted into the topocentric equatorial coordinates hour angle and declination as in section 2.9.2, example 1. The application of the local sidereal time using equation (2.18) enables the topocentric right ascension to be found.

The above procedure is modified if the observations give the position of the object with respect to a stellar background. The directions of the stars whose

54

images appear on the film will be differentially affected by refraction so that suitable corrections must be applied in obtaining the right ascension and declination of the object from the position of its image among the stellar images. Various procedures have been developed in astronomy to correct for this. When such procedures are applied, the equatorial coordinates of the object relative to the observer are obtained. In the section on precession and nutation (section 3.4) the outline of the method is given. An additional allowance for differential refraction must be made when the object is a rocket observed just after take-off. The stellar background will be displaced by refraction due to its light passing through the total thickness of the atmosphere, whereas the rocket's light may have less than 50 km of atmosphere to penetrate.

The observational data can now be said to be expressed in equatorial coordinates with respect to the observer's station on the Earth's surface. It is necessary now to consider more closely the definition of such coordinates.

3.4 Precession and Nutation

Up till now it has been assumed that the planes of the ecliptic and the equator are fixed with respect to the stellar background, in the sense that the right ascensions and declinations of the stars referred to the equator and the vernal equinox (one of the two points where equator and ecliptic intersect) do not change. Due to the gravitational attractions of Sun and Moon on the aspherical Earth, however, the Earth's axis of rotation precesses, so that the north celestial pole P describes a small circle of radius ϵ ($=23\frac{1}{2}°$) about the pole of the ecliptic K in a period of about 26 000 years. The ecliptic remains fixed and the vernal equinox Υ moves backwards along it (that is, in a direction such that the celestial longitudes of stars increase) at a rate of about 50″ per annum. This is called the *luni–solar precession*.

It is seen from figure 3.2 that in general, due to luni–solar precession, the celestial latitude of a star (given by BX) will not change, but that its celestial ·longitude ΥB will change, increasing by about 50″ per annum. Both right ascension and declination, ΥA and AX respectively, will alter in a manner depending upon the star's present RA and DEC. It is easily shown (Smart 1956) that if θ is the luni–solar precession for one year, a star's RA and DEC will change in that time to (α_1, δ_1) where

$$\alpha_1 - \alpha = \theta(\cos \epsilon + \sin \epsilon \sin \alpha \tan \delta) \tag{3.4}$$

$$\delta_1 - \delta = \theta \sin \epsilon \cos \alpha. \tag{3.5}$$

It is to be noted that these formulae are obtained under the assumption that the changes in the coordinates are small.

A further effect due to the Sun and Moon is called nutation, a complicated oscillation of the pole P about the position it would occupy if precession alone acted. Nutation may be broken up into a series of periodic terms depending upon the elements of the orbits of the Sun and Moon about the Earth, their

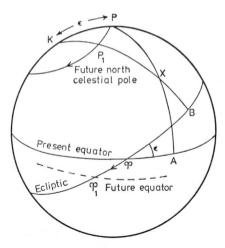

Figure 3.2

periods being small in comparison with that of the luni–solar precession. In addition, due to nutation, the value of the obliquity of the ecliptic oscillates about a mean value.

The planets themselves affect the Earth's orbit, resulting in a slow change in the orientation of the ecliptic. This so-called *planetary precession* decreases the right ascensions of all stars by about 0·13″ per annum.

General precession may now be defined as the combination of luni–solar precession and planetary precession. Due to general precession the ecliptic and equator and the vernal equinox will change. If their positions are taken at, say, the beginning of 1950 (1950·0) they may be regarded as fixed planes of reference. Their changed positions in 1951·0, due to general precession, are called the mean ecliptic, mean equator and mean equinox for 1951·0.

The value χ of the general precession in longitude and the obliquity of the ecliptic ϵ at an epoch t years after 1900 are given by

$$\chi = 50\cdot2564'' + 0\cdot000\ 222''\ t \text{ per annum}$$

and

$$\epsilon = 23°\ 27'\ 08\cdot26'' - 0\cdot4684''\ t.$$

The *mean position* of a star is its RA and DEC referred to the mean equator and equinox of a specified time for a heliocentric celestial sphere (that is, no notice is at present being taken of nutation, aberration, stellar parallax or the star's proper motion, the latter three quantities being defined below).

Equations (3.4) and (3.5) are now generalized to include planetary precession, which decreases right ascension by $l(=0\cdot13'')$ in one year and has no effect on declination.

We obtain for the changes in right ascension and declination in one year due to general precession

$$\alpha_1 - \alpha = (\theta \cos \epsilon - l) + \theta \sin \epsilon \sin \alpha \tan \delta$$

$$\delta_1 - \delta = \theta \sin \epsilon \cos \alpha.$$

Putting
$$m \equiv \theta \cos \epsilon - l, \qquad n \equiv \theta \sin \epsilon$$
we obtain
$$\alpha_1 - \alpha = m + n \sin \alpha \tan \delta \tag{3.6}$$
$$\delta_1 - \delta = n \cos \alpha. \tag{3.7}$$

Both m and n vary slowly with time. Thus
$$m = 3 \cdot 073\ 27^s + 0 \cdot 000\ 018\ 6^s\ (t - 1950)$$
$$n = 20 \cdot 0426'' - 0 \cdot 000\ 085''\ (t - 1950) \tag{3.8}$$
$$= 1 \cdot 336\ 17^s - 0 \cdot 000\ 005\ 7^s\ (t - 1950).$$

For periods longer than 5 years, equations (3.6) and (3.7) are inadequate and a quantity called the annual variation is introduced. If the year is taken as the unit, and $d\alpha/dt$ denotes the rate of change of α due to precession, then from equation (3.6) we have
$$\frac{d\alpha}{dt} = m + n \sin \alpha \tan \delta. \tag{3.9}$$

The rate of change of $d\alpha/dt$ per century is defined as the secular variation s in right ascension. Then, neglecting changes in s itself, we have
$$\alpha - \alpha_0 = t \left[\left(\frac{d\alpha}{dt} \right)_0 + \frac{st}{200} \right] \tag{3.10}$$

where the suffix zero denotes evaluation at the earlier epoch, and t as before is in years.

Also,
$$s = 100 \left(\frac{dm}{dt} + \sin \alpha \tan \delta \frac{dn}{dt} + n \cos \alpha \tan \delta \frac{d\alpha}{dt} + n \sin \alpha \sec^2 \delta \frac{d\delta}{dt} \right). \tag{3.11}$$

Similarly
$$\delta - \delta_0 = t \left[\left(\frac{d\delta}{dt} \right)_0 + \frac{s't}{200} \right] \tag{3.12}$$

where s' is the secular variation in declination given by
$$s' = 100 \left(\cos \alpha \frac{dn}{dt} - n \sin \alpha \frac{d\alpha}{dt} \right). \tag{3.13}$$

In the principal star catalogues are given, together with the secular variations, quantities called the annual variation in right ascension and declination. These latter quantities are the annual precessions $d\alpha/dt$ and $d\delta/dt$ plus the star's proper motion (section 3.6).

The *true position* of a star at any time is its heliocentric right ascension and declination referred to the true equator and equinox of that date. By applying nutation, the mean position computed for that date may be converted to the true position at that date. It has been seen that nutation changes the longitude of a star and also the obliquity of the ecliptic. If $\Delta\psi$ and $\Delta\epsilon$ denote these changes for the date in question, they may be computed. The change $\Delta_1\alpha$ due to $\Delta\psi$ and $\Delta\epsilon$ is then given by

$$\Delta_1\alpha = \Delta\psi(\cos\epsilon + \sin\epsilon\sin\alpha\tan\delta) - \Delta\epsilon\cos\alpha\tan\delta$$

with a similar expression for the change in declination due to nutation at that time.

But the change in RA due to precession from the beginning of that year to the present date (a fraction τ of a year) is $\Delta_2\alpha$ where, using equation (3.6),

$$\Delta_2\alpha = \tau(m + n\sin\alpha\tan\delta).$$

Combining $\Delta_1\alpha$ with $\Delta_2\alpha$ and remembering that

$$m \equiv \theta\cos\epsilon - l$$

$$n \equiv \theta\sin\epsilon$$

we obtain

$$\Delta\alpha = \Delta_1\alpha + \Delta_2\alpha$$

$$= \left(\tau + \frac{\Delta\psi}{\theta}\right)(m + n\sin\alpha\tan\delta) + \frac{l\Delta\psi}{\theta} - \Delta\epsilon\cos\alpha\tan\delta.$$

If we now express m and n in seconds of time, and l, $\Delta\psi$, θ and $\Delta\epsilon$ in seconds of arc, and introduce quantities A, B, E, a and b defined by

$$A \equiv \left(\tau + \frac{\Delta\psi}{\theta}\right), \qquad a \equiv m + n\sin\alpha\tan\delta$$

$$B \equiv -\Delta\epsilon, \qquad b \equiv (1/15)\cos\alpha\tan\delta$$

$$E \equiv \frac{l\Delta\psi}{150}$$

then

$$\Delta\alpha = Aa + Bb + E \tag{3.14}$$

with the right-hand side expressed in seconds of time.

Similarly it is found that

$$\Delta\delta = Aa' + Bb' \tag{3.15}$$

where $a' = n\cos\alpha$, $b' = -\sin\alpha$, and n is in seconds of arc.

The quantities A, B, E are not functions of the star's position, and are tabulated in the almanacs for every day of the year under the heading *Bessel's day numbers* (or *star numbers*). The quantities a, b, a', b' can be computed for the star concerned.

The procedure to obtain the true position of a star at a given epoch (a date in a particular year) from its mean position in a catalogue of epoch 1950·0 is thus as follows:

(i) Calculate the mean coordinates at the beginning of the year in which the date occurs,

(ii) Change these mean coordinates to the true coordinates for the date in question.

There remains one final correction; namely, to change the origin from the Sun's centre to the Earth's centre. This gives the *apparent place* of the star at that instant which is the position on the geocentric celestial sphere with respect to the true equinox and equator at that time. The difference between apparent place and true place is due to aberration and annual stellar parallax (sections 3.5 and 3.7). Anticipating, it is found that except for a very few near stars parallax can be ignored, while the correction due to aberration is of the form

$$\text{RA(apparent)} = \text{RA(true)} + Cc + Dd \qquad (3.16)$$

$$\text{DEC(apparent)} = \text{DEC(true)} + Cc' + Dd' \qquad (3.17)$$

where C and D are tabulated in the almanacs and c, d, c' and d' are functions of the star's position.

The star's geocentric apparent position is now known for the time of observation, in terms of RA and DEC referred to the true equator and equinox at that date.

The reverse procedure is adopted when the positions of the brighter stars are measured. By applying the correction for refraction, the star's geocentric apparent position is found. The application of equations (3.14), (3.15), (3.16) and (3.17) gives the mean coordinates referred to the mean equator and equinox at the beginning of the year in which the observation took place. By applying equations (3.8)–(3.13) the star's mean coordinates can be obtained relative to the equator and equinox of the epoch of the star catalogue in which it appears. Information concerning its proper motion (section 3.6) can then be obtained.

Photography is employed for the measurement of the positions of the fainter stars. On any photographic plate there are usually a number of stars whose coordinates have been determined and catalogued already. They can be used as reference stars with which to obtain the positions of the faint stars.

In practice measurements are made from the negatives on various types of plate-measuring engines, since making a positive inevitably introduces some blurring. The measurements made are of the x and y coordinates of the image with respect to a set of rectangular axes Ox and Oy.

In theory, these axes are chosen such that:

(i) the origin lies on the optical axis of the telescope which corresponds to a given RA and DEC referred to the mean equator and equinox of, say, 1950·0,

(ii) the y axis is the projection of the great circle through the north celestial pole for 1950·0 and the point towards which the telescope is pointing, and

(iii) the x axis is drawn at right angles to the y axis.

In practice, errors enter due to bad orientation, scale error, nonperpendicularity of axes, wrong centre and tilt of the photographic plate's plane to the plane perpendicular to the optical axis. In addition, refraction and aberration produce their effect. Two sets of coordinates are therefore distinguished; the measured coordinates x and y of the star image, and the standard coordinates ξ and η that have to be found, free of the above sources of error. Fortunately, they are connected by the simple equations

$$\xi - x = ax + by + c \tag{3.18}$$

$$\eta - y = dx + ey + f. \tag{3.19}$$

Only in special cases (see Smart 1956) do quadratic terms in x and y have to be introduced. The quantities a, b, c, d, e and f are called the plate constants and have to be calculated.

On the plate will appear a number n of stars whose standard coordinates (ξ_i, η_i) $(i = 1, 2, 3 \dots n)$ are already known since they are already catalogued. If their measured coordinates (x_i, y_i) are obtained from the plate, the plate constants can then be computed by the method of least squares or a similar process from the set of equations

$$\xi_i - x_i = ax_i + by_i + c$$

$$\eta_i - y_i = dx_i + ey_i + f.$$

The standard coordinates (ξ, η) of the star in question can then be calculated from equations (3.18) and (3.19). These can now be transformed into equatorial coordinates α and δ with respect to the observer for the vernal equinox and equator involved, this vernal equinox and equator being the one the reference stars' coordinates are themselves referred to.

The formulae involved in this process are

$$\cot \delta \sin (\alpha - A) = \frac{\xi \sec D}{\eta + \tan D} \tag{3.20}$$

$$\cot \delta \cos (\alpha - A) = \frac{1 - \eta \tan D}{\eta + \tan D} \tag{3.21}$$

$$\tan (\alpha - A) = \frac{\xi \sec D}{1 - \eta \tan D}. \tag{3.22}$$

In these equations, A and D are the right ascension and declination of the theoretical plate centre.

For an object within the Solar System, the star is replaced by the object (planet, satellite, spacecraft), but the principles outlined in this section and in section 3.3 are changed only in detail. It is to be noted that where the instrument used gives the object's altitude and azimuth, the right ascension and declination obtained from these quantities, corrected for refraction, are with respect to the true equator and equinox of the time of observation.

3.5 Aberration

Due to the finite velocity of light, an apparent angular displacement of a star towards the direction of the observer's own motion relative to the star takes place. Thus the annual revolution of the Earth in its orbit produces an annual displacement on the celestial sphere of each star in an ellipse of major axis $\kappa = 20\cdot47''$. It has been seen that this effect is taken care of in the measurement of stellar image positions on a photographic plate.

For stars individually observed, the aberrational displacements in (i) equatorial and (ii) ecliptic coordinates are given as follows, the suffix 1 denoting the star's coordinates affected by aberration:

(i)
$$\alpha_1 - \alpha = Cc + Dd$$
$$\delta_1 - \delta = Cc' + Dd'$$

where

$$C = -\kappa \cos \epsilon \cos \odot, \quad D = -\kappa \sin \odot$$
$$c = \tfrac{1}{15} \cos \alpha \sec \delta, \quad d = \tfrac{1}{15} \sin \alpha \sec \delta$$
$$c' = \tan \epsilon \cos \delta - \sin \alpha \sin \delta$$
$$d' = \cos \alpha \sin \delta.$$

The quantities C and D, functions only of the Sun's longitude $\odot$, are given as $\log C$ and $\log D$ (*Bessel's day numbers* or *Besselian star numbers*) in the almanacs.

(ii)
$$\lambda_1 - \lambda = -\kappa \sec \beta \cos (\odot - \lambda)$$
$$\beta_1 - \beta = -\kappa \sin \beta \sin (\odot - \lambda).$$

where $\odot$ stands for the longitude of the Sun.

In the case of an object in the Solar System, the relative velocity with respect to the observer's position produces an aberrational effect. In general this is different from stellar aberration so that the position given by its image on a photographic plate must be corrected. If the approximate distance and velocity of the object are known, as is usually the case, this correction may easily be made.

It can be shown (Smart and Green 1977) that if v is the relative velocity of the observer and the object, c being the velocity of light, then

$$\Delta\theta = k \sin \theta,$$

where θ is the angle between the direction of the object as viewed by the observer and the direction in which the observer is travelling relative to the object. $\Delta\theta$ is the shift due to aberration in seconds of arc, and $k = 206\,265\,(v/c)$.

Thus, in figure 3.3, the object's true direction OV is displaced by aberration to an apparent direction OV', where at the moment of observation the observer O is travelling with velocity v towards A, relative to V. The velocity v is compounded of the object's velocity relative to the Earth's centre and the observer's rotational velocity on the Earth's surface relative to the same centre.

61

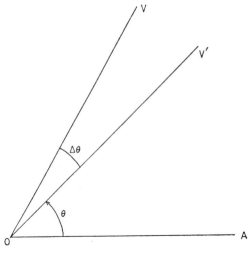

Figure 3.3

The shift $\Delta\theta$ produces shifts $\Delta\alpha$ and $\Delta\delta$ which can then be computed from the geometry of the situation, though it should be noted that k cannot be simply inserted in place of κ in the above equations.

3.6 Proper Motion

The stars have their own intrinsic motions within the Galaxy. Since the Sun is itself a star it also moves in a galactic orbit. These motions reveal themselves by changes in the relative positions of the stars. Even although the relative velocities of the stars in the Sun's neighbourhood are of the order of 20 km s^{-1}, the size of stellar distances is such that the annual changes in direction of even the nearest stars due to their velocities relative to the Sun are usually less than 5″. This annual change in direction of the star is called its proper motion and is known and catalogued for most of the brighter stars.

From photographs taken at intervals of a few years, the shift in right ascension and declination of the star in question may be measured. Due allowance is made for the effect of aberration and parallax (section 3.7), and for precession and nutation according to the procedures sketched in section 3.4.

3.7 Stellar Parallax

The direction of a star as seen from the Earth is not the same as the direction when viewed by a hypothetical observer at the Sun. As the Earth moves in its yearly orbit round the Sun, the geocentric direction (the star's position on a geocentric celestial sphere) changes and traces out what is termed the parallactic ellipse. Thus in figure 3.4, the star X at a distance d is seen from the

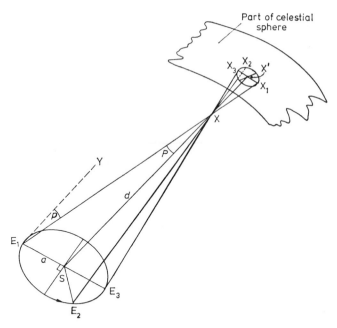

Figure 3.4

Earth at E_1 to lie in the direction E_1X_1 relative to the heliocentric direction SX'. Six months later, the Earth is now at the point E_3 in its orbit and the geocentric direction of the star is E_3X_3.

The parallax of the star is defined to be the angle P subtended at the star by the semimajor axis a of the Earth's orbit taken at right angles to the star's heliocentric direction. Since d is much greater than a, $E_1X=E_3X=SX$. Hence $\sin P=(a/d)$.

The parsec is defined to be the distance at which a celestial object would have a parallax of one second of arc. It is readily seen that $1\ \mathrm{pc}=206\,265$ times the Earth's orbital semimajor axis a. For stars, we may write with sufficient accuracy $P=1/d$, where P is measured in seconds of arc and d in pc. Hence the equation of section 3.2 relating the absolute magnitude M to the apparent magnitude m of a celestial object has the alternative form

$$M=m+5+5\log_{10} P.$$

It may be shown that the observed direction of a star at any instant differs from its heliocentric direction by an angle p, where

$$p=P\sin\theta$$

and θ is the angle between the star's direction and the direction of the Sun. The displacement is towards the Sun. Thus in figure 3.4, $X\hat{E}_2S=\theta$ when the Earth is at E_2.

For the nearest star P is less than 1 second of arc, and indeed only a score of stars are known with parallaxes greater than 0·25 seconds of arc.

3.8 Geocentric Parallax

Theoretically the direction of a celestial object as seen from a station on the Earth's surface (its topocentric direction) is not the same as the geocentric direction of the object. In practice, if the object is a star the directions are indistinguishable; if the object is the Sun, the angle between them can be as great as 8·8″; for the nearest planet the angle can amount to about 32″, while its value for the Moon can be about 1°. For a close artificial satellite, the direction as seen from a station on the Earth's surface can be the best part of 90° different from the satellite's geocentric direction.

The topocentric equatorial coordinates of the object must be transformed now to the centre of the Earth to get rid of this geocentric parallax due to the finite size of the Earth.

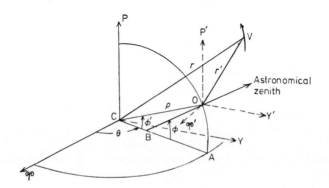

Figure 3.5

In figure 3.5 the observing station at O on the Earth's surface, distance ρ from the Earth's centre C, tracks a satellite V, distance r' from O and r from C. The meridian from the Earth's North pole P through O meets the terrestrial equator ΥA in A where Υ is the direction of the vernal equinox. The direction of Υ as seen from O is OΥ' parallel to CΥ. The geocentric and astronomical latitudes of O are O$\hat{\mathrm{C}}$A (ϕ') and O$\hat{\mathrm{B}}$A (ϕ) respectively.

Now let angle $\Upsilon\hat{\mathrm{C}}$A$=\theta$. As the Earth rotates and carries the observer round with it, angle $\Upsilon\hat{\mathrm{C}}$A increases. But angle $\Upsilon\hat{\mathrm{C}}$A is the LST of the observer; therefore

$$\theta = \text{LST}. \tag{3.23}$$

If a set of non-rotating rectangular axes CΥ, CY and CP are taken as shown, then the coordinates of O ($\bar{x}, \bar{y}, \bar{z}$) are given by

$$\left.\begin{array}{l} \bar{x} = \rho \cos \phi' \cos \theta \\[4pt] \bar{y} = \rho \cos \phi' \sin \theta \\[4pt] \bar{z} = \rho \sin \phi' \end{array}\right\} \tag{3.24}$$

where θ is given by equation (3.23).

64

If the semimajor and semiminor axes of the elliptic cross-section of the Earth (an arc of which is PÔA) are a and b respectively, then it may be shown (Smart and Green 1977) that

$$\tan \phi' = \frac{b^2}{a^2} \tan \phi$$

and that

$$\rho = a \left[\frac{1 - (2e^2 - e^4) \sin^2 \phi}{1 - e^2 \sin^2 \phi} \right]^{1/2}$$

where $e^2 = 1 - b^2/a^2$.

It should be noted that the distance ρ refers to sea level. If the station O is at height h above sea level then ρ should be increased to $(\rho + h)$.

The instantaneous rectangular coordinates $(\bar{x}, \bar{y}, \bar{z})$ of the station can now be computed.

The observed data are the apparent right ascension α' and declination δ' of the vehicle (that is, with respect to a celestial sphere with the observer as origin). The distance r' is not in general known, except approximately, unless range measurements are also being made.

It is desired to obtain the geocentric right ascension α, declination δ and distance r of the vehicle by removing the effects of geocentric parallax. The problem is seen to be analogous to parts (ii) and (iii) of example 3, chapter 2, section 2.9.2.

Take a set of rectangular axes $O\Upsilon'$, OY', OP' through O, parallel to the axes $C\Upsilon$, CY, CP respectively and let the rectangular coordinates of V relative to the set of axes through O be x', y' and z'.

Then

$$\left. \begin{array}{l} x' = r' \cos \delta' \cos \alpha' \\[4pt] y' = r' \cos \delta' \sin \alpha' \\[4pt] z' = r' \sin \delta'. \end{array} \right\} \qquad (3.25)$$

If the geocentric rectangular coordinates of V are x, y and z, then

$$\left. \begin{array}{l} x = r \cos \delta \cos \alpha \\[4pt] y = r \cos \delta \sin \alpha \\[4pt] z = r \sin \delta. \end{array} \right\} \qquad (3.26)$$

Obviously

$$\left. \begin{array}{l} x = \bar{x} + x' \\[4pt] y = \bar{y} + y' \\[4pt] z = \bar{z} + z'. \end{array} \right\} \qquad (3.27)$$

Hence, substituting equations (3.24), (3.25) and (3.26) into equation (3.27), the resulting relations can be solved to give α, δ and r in terms of α', δ' and r'.

Also involved will be the known values of ρ, ϕ' and θ. The three equations are

$$r \cos \delta \cos \alpha = \rho \cos \phi' \cos \theta + r' \cos \delta' \cos \alpha' \tag{3.28}$$

$$r \cos \delta \sin \alpha = \rho \cos \phi' \sin \theta + r' \cos \delta' \sin \alpha' \tag{3.29}$$

$$r \sin \delta = \rho \sin \phi' + r' \sin \delta'. \tag{3.30}$$

In practice it is often more convenient to compute $(\alpha' - \alpha)$ and $(\delta' - \delta)$. Multiplying (3.28) by $\sin \alpha'$ and (3.29) by $\cos \alpha'$ and subtracting gives

$$r \cos \delta \sin (\alpha' - \alpha) = \rho \cos \phi' \sin (\alpha' - \theta). \tag{3.31}$$

Multiplying (3.28) by $\cos \alpha'$ and (3.29) by $\sin \alpha'$ and adding gives

$$r \cos \delta \cos (\alpha' - \alpha) = r' \cos \delta' + \rho \cos \phi' \cos (\alpha' - \theta). \tag{3.32}$$

Dividing (3.31) by (3.32) gives

$$\tan (\alpha' - \alpha) = \frac{(\rho/r') \cos \phi' \sin (\alpha' - \theta)}{\cos \delta' + (\rho/r') \cos \phi' \cos (\alpha' - \theta)}. \tag{3.33}$$

Putting $\cos (\alpha' - \alpha) = 1 - 2 \sin^2 \tfrac{1}{2}(\alpha' - \alpha)$ in equation (3.32) and using equation (3.31) we obtain, after a little reduction

$$r \cos \delta = r' \cos \delta' + \rho \cos \phi' \frac{\cos [\tfrac{1}{2}(\alpha' + \alpha) - \theta]}{\cos \tfrac{1}{2}(\alpha' - \alpha)}.$$

Let the quantities m and γ be defined by

$$m \sin \gamma \equiv \sin \phi'$$

$$m \cos \gamma \equiv \cos \phi' \frac{\cos [\tfrac{1}{2}(\alpha' + \alpha) - \theta]}{\cos \tfrac{1}{2}(\alpha' - \alpha)}.$$

Then

$$r \cos \delta = r' \cos \delta' + \rho m \cos \gamma \tag{3.34}$$

and by equation (3.30)

$$r \sin \delta = r' \sin \delta' + \rho m \sin \gamma. \tag{3.35}$$

Multiplying (3.34) by $\sin \delta'$, (3.35) by $\cos \delta'$, and subtracting gives

$$r \sin (\delta' - \delta) = \rho m \sin (\delta' - \gamma). \tag{3.36}$$

Multiplying (3.34) by $\cos \delta'$, (3.35) by $\sin \delta'$ and adding gives

$$r \cos (\delta' - \delta) = r' + \rho m \cos (\delta' - \gamma). \tag{3.37}$$

Hence, from equations (3.36) and (3.37) we have

$$\tan (\delta' - \delta) = \frac{\rho m \sin (\delta' - \gamma)}{r' + \rho m \cos (\delta' - \gamma)} \tag{3.38}$$

or

$$\tan (\delta' - \delta) = \frac{(\rho/r') \sin \phi' \sin (\delta' - \gamma)}{\sin \gamma + (\rho/r') \sin \phi' \cos (\delta' - \gamma)} \tag{3.39}$$

where

$$\tan \gamma = \frac{\tan \phi' \cos \frac{1}{2}(\alpha' - \alpha)}{\cos \left[\frac{1}{2}(\alpha' + \alpha) - \theta\right]}. \tag{3.40}$$

In a similar fashion using equations (3.34) and (3.35) we may obtain

$$r \cos (\delta' - \delta) = r' \left[1 + \frac{\rho}{r'} \frac{\sin \phi' \cos (\delta' - \gamma)}{\sin \gamma}\right]. \tag{3.41}$$

The four equations (3.33), (3.39), (3.40) and (3.41) are rigorous and give the corrections for geocentric parallax. Several cases may be considered:

(i) Object at distances well beyond the Moon's distance (for example, an interplanetary probe). The corrections $(\alpha' - \alpha)$ and $(\delta' - \delta)$ are much less than $1°$, since ρ/r' is much less than $1/60$.

Then, if $(\alpha' - \alpha)$ is expressed in radians, equation (3.33) may be written, to sufficient accuracy,

$$\alpha' - \alpha = \frac{\rho}{r'} \cos \phi' \sin (\alpha' - \theta) \sec \delta'. \tag{3.42}$$

Similarly, equation (3.39) may be written as

$$\delta' - \delta = \frac{\rho}{r'} \sin \phi' \sin (\delta' - \gamma) \operatorname{cosec} \gamma \tag{3.43}$$

where γ is given by

$$\tan \gamma = \tan \phi' \sec (\alpha' - \theta). \tag{3.44}$$

Also, from equation (3.41)

$$r' - r = -\rho \sin \phi' \cos (\delta' - \gamma) \operatorname{cosec} \gamma. \tag{3.45}$$

To use these equations, the value of r', as well as values of α' and δ' must be known. This is usually satisfied in practice.

(ii) Object at lunar distances (for example, an artificial lunar satellite). Again the quantities involved, namely $(\alpha' - \alpha)$, $(\delta' - \delta)$ and $(r' - r)$ are small corrections. The angles are of order $1°$ or less, while the quantity $(r' - r)$ is of order $1/60$ of the vehicle distance or less. The angles α' and δ' are measured easily and accurately; the range r' is also accurately measured by radar. Hence equations (3.42)–(3.45) may be used as in (i), though the rigorous equations (3.33), (3.39), (3.40) and (3.41) should be used for objects moving between Earth and lunar orbit distance.

(iii) Object at distances similar to the radius of the Earth (for example, an Earth satellite). The rigorous equations must be used.

The quantities $(\alpha' - \alpha)$, $(\delta' - \delta)$ and $(r' - r)$ are no longer small. The range r' may be either measured directly by high-accuracy radar or, if the satellite is in an established orbit, may be known approximately. If neither of these criteria is satisfied then the corrections for geocentric parallax cannot be applied so simply. Observations from at least two places on the Earth's surface are required to obtain a measure of the distance. If two stations O' and O″ observe the satellite simultaneously then they each obtain its apparent position. Let these positions be given by (α', δ') and (α'', δ''). Its geocentric position is (α, δ). If its distances from O′, O″ and the Earth's centre are r', r'' and r, then there are five unknown quantities α, δ, r', r'' and r. Equations (3.28), (3.29) and (3.30), applied first to O′ and then to O″, give six equations in the five unknowns so that they can be determined. In practice, it is unlikely that observations are made simultaneously; the reduction is therefore rather more complicated.

3.9 Review of Procedures

A summary of the procedures in reducing observations may be useful at this stage.

An observation will be made at a station in (i) horizontal coordinates; altitude and azimuth, (ii) equatorial coordinates; hour angle and declination, or (iii) by photographing the object against a stellar background. In all these cases the time at which the observation was made is noted.

The procedure in case (i) can be stated as follows:

(a) Apply known instrument errors.

(b) Apply refraction.

(c) Transform data to hour angle and declination using known station latitude.

(d) Transform time to local sidereal time (if necessary).

(e) Transform hour angle and declination to right ascension and declination using local sidereal time.

(f) Apply aberration correction.

(g) Apply the correction for geocentric parallax using either the measured distance of the object or an estimated distance or observational data from another station or stations.

(h) If desired, transform geocentric RA and DEC for present equator and equinox to a standard equator and equinox.

The procedure in case (ii) is the same as in case (i), omitting step (c). The procedure in case (iii) is as follows:

(a) Measure plate, obtain plate constants and calculate the topocentric RA and DEC of the object for the equinox and equator of the star catalogue used.

(b) Apply the object's aberration correction.

(c) Transform the RA and DEC of the object to the present equator and equinox.

(d) Apply the correction for geocentric parallax as in (g) above.

(e) Transform if desired the geocentric RA and DEC for the present equator and equinox to a standard equator and equinox.

If the object is outside the Solar System, the correction for geocentric parallax is of course irrelevant.

In the case of a body within the Solar System the reduced data, perhaps collected from many observing stations and processed at a central computing station, can then be used to provide elements of the object's orbit or to improve an existing orbit for the object. Predictions from the orbit can then be published or sent to the observing stations for their future operations. A description of the methods involved in determining a body's orbit from a set of observations is reserved for a future chapter.

Problems

3.1 The apparent visual magnitude of Sirius is $-1\cdot58^m$ while that of Procyon is $+0\cdot48^m$. How many times brighter than Procyon is Sirius?

3.2 The distances of Sirius and Procyon are $2\cdot70$ and $3\cdot21$ pc respectively. Calculate their absolute magnitudes.

3.3 The star R Leonis is variable in brightness, its magnitude ranging from $+5\cdot0^m$ to $+10\cdot5^m$. What is the range in brightness?

3.4 At a ground station, transmission from the artificial satellite Ariel 1 was received at $136\cdot4057$ MHz though it was operating on a frequency of $136\cdot4080$ MHz. What was the range rate of the satellite at this instant? (Take the velocity of electromagnetic waves to be 3×10^5 km s^{-1}.)

3.5 A satellite's observed zenith distance was $28°$. Correct this for optical refraction, taking the constant of refraction to be $58\cdot2''$.

3.6 At an observatory in a north latitude, the observed zenith distances at upper and lower transits of a circumpolar star (one that never sets) were $10°\,17'\,24''$ and $56°\,42'\,49''$ respectively, the upper culmination being south of the zenith. Find the latitude of the observatory and the star's declination, taking refraction into account.

3.7 Taking into account only luni–solar precession, find the RA and DEC of a star (i) one-quarter, (ii) one-half of the precessional period hence, if its present RA and DEC are 18^h and $-23°\,27'$ respectively.

3.8 In 1962, the RA and DEC of the star α Aurigae were $5^h\,13^m\,52\cdot7^s$ and $+45°\,57'\,1''$. What were the rates of change of the star's RA and DEC at this time due to precession?

3.9 Show that a star X has no luni–solar precession in RA if $K\hat{X}P$ is a right angle where K and P are the poles of the ecliptic and equator respectively.

3.10 A close Earth-satellite is observed from a ground tracking station Y at local sidereal time $06^h\,00^m$ to have an altitude and azimuth E of N (corrected for refraction and instrument errors) of $45°$ and $265°$ respectively. Its distance found by radar was 225 mi. The astronomical altitude of Y is $55°\,00'$ N. Taking the dimensions given for the Hayford geoid on p20, find the topocentric and the geocentric RA and DEC of the satellite at the moment of observation (neglect aberration).

Reference

Smart W M and Green R M 1977 *Textbook on Spherical Astronomy* (London: Cambridge University Press)

Bibliography

Astronomical Almanac (London: HMSO)
Evans D S 1968 *Observation in Modern Astronomy* (London: English University Press)
Explanatory Supplement to the Astronomical Ephemeris and the American Ephemeris and Nautical Almanac (London: HMSO)
McNally D 1974 *Positional Astronomy* (London: Frederick Muller)
Pausey J L and Bracewell R N 1955 *Radio Astronomy* (London: Oxford University Press)
Smith F G 1974 *Radio Astronomy* (London: Penguin)
Vonbun F O 1962 *NASA Technical Note* D-1178
Woolard E W and Clemence E M 1966 *Spherical Astronomy* (New York: Academic)

4 The Two-Body Problem

4.1 Introduction

The two-body problem, first stated and solved by Newton, asks, 'Given at any time the positions and velocities of two massive particles moving under their mutual gravitational force, the masses also being known, calculate their position and velocities for any other time.'

The importance of the two-body problem lies in two main facts. Firstly, it is the only gravitational problem in dynamics, apart from rather specialized cases in the problem of three bodies, for which we have a complete and general solution. Secondly, a wide variety of practical orbital-motion problems can be treated as approximate two-body problems. The two-body solution may be used to provide approximate orbital parameters and predictions or serve as a starting point for the generation of analytical solutions valid to higher orders of accuracy. Such solutions, called general perturbation theories, will be discussed later.

The orbit of the Moon about the Earth, for example, is (to a first approximation) a two-body problem, as is that of a planet about the Sun. In both cases however, the gravitational actions of other bodies (in the former example the Sun predominantly; in the latter example other planets) disturb the simple two-body picture. Again, the flight of an interplanetary probe from Earth to Mars is a four-body problem—Sun, Earth, Mars and probe. Nevertheless, useful preliminary planning information can be obtained by breaking the flight into three two-body problems:

(i) Earth–probe (near Earth),
(ii) Sun–probe (in interplanetary space),
(iii) Mars–probe (near Mars).

The possession of a complete analytical solution to the two-body problem is therefore valuable; because of this it is treated in detail in this chapter.

4.2 Newton's Laws of Motion

Newton's three laws of motion laid the foundations of the science of dynamics. Though some if not all were implicit in the scientific thought of his time, his explicit formulation of these laws and exploration of their consequences in conjunction with his law of universal gravitation did more to bring into

70

being our modern scientific age than any of his contemporaries' work. They may be stated in the following form:

(i) Every body continues in its state of rest or of uniform motion in a straight line except insofar as it is compelled to change that state by an external impressed force.

(ii) The rate of change of momentum of the body is proportional to the impressed force and takes place in the direction in which the force acts.

(iii) To every action there is an equal and opposite reaction.

Vector notation is a convenient shorthand way of stating dynamical concepts. At any stage, coordinate systems may be introduced as desired by using the relations between the vectors and the components of the concepts relative to the coordinate axes.

Taking a fixed origin O, let $\mathbf{r}$, $\mathbf{v}$ and $\mathbf{a}$ denote the position, velocity and acceleration vectors respectively of a mass m so that

$$\mathbf{v} = \frac{d\mathbf{r}}{dt}$$

and

$$\mathbf{a} = \frac{d\mathbf{v}}{dt} = \frac{d^2\mathbf{r}}{dt^2}.$$

Hence the linear momentum of the mass is $m\mathbf{v}$, and its angular momentum is $m\mathbf{r} \times \mathbf{v} = m\mathbf{r} \times \dot{\mathbf{r}}$.

In vector notation, the relation

$$\frac{d(m\mathbf{v})}{dt} = \mathbf{F} \tag{4.1}$$

summarizes laws (i) and (ii), where $\mathbf{v}$ is the body's velocity, m is its mass and $\mathbf{F}$ is the external force, the unit of force being chosen so that the constant of proportionality is unity. Hence

$$m\frac{d^2\mathbf{r}}{dt^2} = \mathbf{F} \tag{4.2}$$

where it is assumed that the body's mass is constant. It is to be noted that, in the case of a rocket, this assumption is invalid when the rocket motor is in action.

Where more than one force acts, equation (4.2) may be generalized as

$$m\frac{d^2\mathbf{r}}{dt^2} = \sum_{i=1}^{k} \mathbf{F}_i \tag{4.3}$$

where the k forces involved are added vectorially.

4.3 Newton's Law of Gravitation

One of the most far-reaching scientific laws ever formulated, Newton's law of universal gravitation is the basis of celestial mechanics and astrodynamics.

Its consequences were investigated during the two and a half centuries succeeding its formulation by many of the foremost mathematicians and astronomers that have ever lived. Many elegant mathematical methods were evolved to solve the intricate sets of equations that arose from statements of the problems involving mutually attracting systems of masses.

The law itself is stated with deceptive simplicity as follows. Every particle of matter in the universe attracts every other particle of matter with a force directly proportional to the product of the masses and inversely proportional to the square of the distance between them.

Hence, for two particles separated by a distance r, we have the relation

$$F = G \frac{m_1 m_2}{r^2}$$

where F is the force of attraction, m_1 and m_2 are the masses, r is the distance between them and G is the constant of gravitation, often called the constant of universal gravitation.

4.4 The Solution to the Two-Body Problem

In figure 4.1, the force of attraction F_1 on mass m_1 is directed along the vector r towards the mass m_2, while the force F_2 on m_2 is in the opposite direction.

By Newton's third law,

$$F_1 = -F_2. \tag{4.4}$$

Also

$$F_1 = G \frac{m_1 m_2}{r^2} \frac{r}{r}. \tag{4.5}$$

Now let vectors r_1 and r_2 be directed from some fixed reference point O

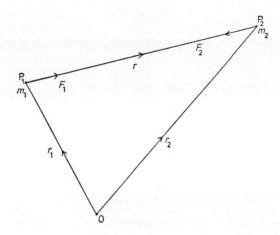

Figure 4.1

to the particles of mass m_1 and mass m_2 respectively. By equations (4.2), (4.4) and (4.5), the equations of motion of the particles under their mutual gravitational attractions are then given by the two equations

$$m_1 \frac{d^2 \mathbf{r}_1}{dt^2} = G \frac{m_1 m_2}{r^2} \frac{\mathbf{r}}{r} \tag{4.6}$$

$$m_2 \frac{d^2 \mathbf{r}_2}{dt^2} = -G \frac{m_1 m_2}{r^2} \frac{\mathbf{r}}{r}. \tag{4.7}$$

Adding equations (4.6) and (4.7) gives

$$m_1 \frac{d^2 \mathbf{r}_1}{dt^2} + m_2 \frac{d^2 \mathbf{r}_2}{dt^2} = 0$$

giving two integrals

$$m_1 \frac{d\mathbf{r}_1}{dt} + m_2 \frac{d\mathbf{r}_2}{dt} = \mathbf{a} \tag{4.8}$$

and

$$m_1 \mathbf{r}_1 + m_2 \mathbf{r}_2 = \mathbf{a}t + \mathbf{b} \tag{4.9}$$

where $\mathbf{a}$ and $\mathbf{b}$ are constant vectors.

But if $\mathbf{R}$ is the position vector of G (the centre of mass of the two masses m_1 and m_2), $\mathbf{R}$ is defined as

$$M\mathbf{R} = m_1 \mathbf{r}_1 + m_2 \mathbf{r}_2$$

where

$$M = m_1 + m_2.$$

Hence by equations (4.8) and (4.9),

$$M \frac{d\mathbf{R}}{dt} = \mathbf{a}, \qquad M\mathbf{R} = \mathbf{a}t + \mathbf{b}.$$

These relations show that the centre of mass of the system moves with constant velocity.

Equations (4.6) and (4.7) may be written as

$$\frac{d^2 \mathbf{r}_1}{dt^2} = G m_2 \frac{\mathbf{r}}{r^3} \tag{4.10}$$

and

$$\frac{d^2 \mathbf{r}_2}{dt^2} = -G m_1 \frac{\mathbf{r}}{r^3}. \tag{4.11}$$

Subtracting equation (4.11) from equation (4.10) gives

$$\frac{d^2}{dt^2} (\mathbf{r}_1 - \mathbf{r}_2) = G(m_1 + m_2) \frac{\mathbf{r}}{r^3}.$$

But

$$\mathbf{r}_1 - \mathbf{r}_2 = -\mathbf{r}$$

and hence

$$\frac{d^2\mathbf{r}}{dt^2} + \frac{\mu\mathbf{r}}{r^3} = 0 \qquad (4.12)$$

where

$$\mu = G(m_1 + m_2).$$

Taking the vector product of $\mathbf{r}$ with equation (4.12) we obtain

$$\mathbf{r} \times \frac{d^2\mathbf{r}}{dt^2} = 0.$$

Integrating, we have

$$\mathbf{r} \times \frac{d\mathbf{r}}{dt} = \mathbf{h} \qquad (4.13)$$

where $\mathbf{h}$ is a constant vector.

This is the angular momentum integral. Since $\mathbf{h}$ is constant, pointing in the same direction for all t, the motion of one body about the other lies in a plane defined by the direction of $\mathbf{h}$.

If polar coordinates r and θ are taken in this plane as in figure 4.2, the velocity components along and perpendicular to the radius vector joining m_1 to m_2 are $\dot{r}$ and $r\dot{\theta}$, where the dot replaces d/dt.

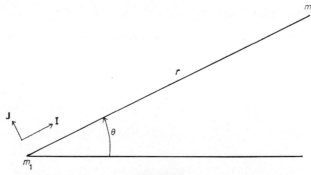

Figure 4.2

Then

$$\dot{\mathbf{r}} = \mathbf{I}\dot{r} + \mathbf{J}r\dot{\theta} \qquad (4.14)$$

where $\mathbf{I}$ and $\mathbf{J}$ are unit vectors along and perpendicular to the radius vector. Hence, by equations (4.13) and (4.14),

$$\mathbf{I}r \times (\mathbf{I}\dot{r} + \mathbf{J}r\dot{\theta}) = r^2\dot{\theta}\mathbf{K} = \mathbf{h}$$

where $\mathbf{K}$ is a unit vector perpendicular to the plane of the orbit. We may then write

$$r^2\dot{\theta} = h \qquad (4.15)$$

where the constant h is seen to be twice the rate of description of area by the radius vector. This is the mathematical form of Kepler's second law.

If the scalar product of $\dot{r}$ with equation (4.12) is now taken, we obtain

$$\dot{\mathbf{r}} \cdot \frac{d^2\mathbf{r}}{dt^2} + \mu \frac{\dot{\mathbf{r}} \cdot \mathbf{r}}{r^3} = 0$$

which may be integrated immediately to give

$$\tfrac{1}{2}\dot{\mathbf{r}} \cdot \dot{\mathbf{r}} - \frac{\mu}{r} = C$$

where C is a constant. That is,

$$\tfrac{1}{2}v^2 - \frac{\mu}{r} = C. \tag{4.16}$$

This is the energy conservation equation of the system. The quantity C is not the total energy; $\tfrac{1}{2}v^2$ is related to the kinetic energy and $-\mu/r$ to the potential energy of the system (see section 4.10).

Referring to figure 4.2 again, and remembering that the components of the acceleration on P_2 along and perpendicular to the radius vector are

$$\ddot{r} - r\dot{\theta}^2 \qquad \text{and} \qquad \frac{1}{r}\frac{d}{dt}(r^2\dot{\theta})$$

respectively, equation (4.12) may be written as

$$\mathbf{I}(\ddot{r} - r\dot{\theta}^2) + \mathbf{J}\left[\frac{1}{r}\frac{d}{dt}(r^2\dot{\theta})\right] + \frac{\mu}{r^3}\mathbf{I}r = 0.$$

Equating coefficients of the vectors, we obtain

$$\ddot{r} - r\dot{\theta}^2 = -\frac{\mu}{r^2} \tag{4.17}$$

$$\frac{1}{r}\frac{d}{dt}(r^2\dot{\theta}) = 0. \tag{4.18}$$

The integration of the second of these equations gives the angular momentum integral

$$r^2\dot{\theta} = h. \tag{4.19}$$

Making the usual substitution of $u = 1/r$ and eliminating the time between equations (4.17) and (4.19) gives us the equation

$$\frac{d^2u}{d\theta^2} + u = \frac{\mu}{h^2}.$$

The general solution of this equation is

$$u = \frac{\mu}{h^2} + A \cos(\theta - \omega) \tag{4.20}$$

where A and ω are the two constants of integration.

Reintroducing r, equation (4.20) becomes

$$r = \frac{h^2/\mu}{1 + (Ah^2/\mu) \cos (\theta - \omega)}.$$

The polar equation of a conic section may be written

$$r = \frac{p}{1 + e \cos (\theta - \omega)} \tag{4.21}$$

so that

$$p = h^2/\mu$$

$$e = Ah^2/\mu.$$

The solution of the two-body problem—a conic section—includes Kepler's first law as a special case. In fact the orbit of one body about the other is classified by the value of the eccentricity e. Thus:

(i) for $0 \leqslant e < 1$ the orbit is an ellipse,
(ii) for $e = 1$ the orbit is a parabola,
(iii) for $e > 1$ the orbit is a hyperbola.

It should be noted that the case $e = 1$ also includes the rectilinear ellipse, parabola and hyperbola (see section 4.8). The case $e = 0$ is the special case of the ellipse of zero eccentricity (i.e. a circle).

These cases will now be examined in detail.

4.5 The Elliptic Orbit

An *ellipse* is the locus of a point which moves so that its distance from a fixed point, the focus, bears a constant ratio (less than 1) to its distance from a fixed line, the directrix.

In figure 4.3 let S be the focus and KL the directrix, with SZ perpendicular to KL.

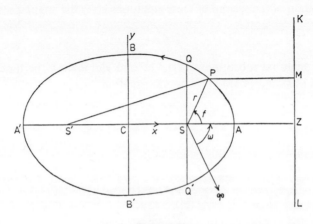

Figure 4.3

Take a point P such that the lengths SP and PM are related by

$$SP/PM = e(<1). \tag{4.22}$$

Then the locus of the point P (i.e. the figure APBA′B′A), as it moves such that equation (4.22) holds with e constant, is an ellipse of eccentricity e, centre C. In this ellipse S′ is the other focus, $AA′ = 2a$ (the major axis), $BB′ = 2b$ (the minor axis) where $b = a (1 - e^2)^{1/2}$, $CS/CA = e$ and $SP + PS′ = 2a$.

In addition, the chord QQ′ through S parallel to the minor axis is called the *latus rectum*; the semilatus rectum SQ ($= SQ′$) having length $p = a(1 - e^2)$.

If cartesian coordinates Cx and Cy are taken as in figure 4.3, the canonic equation of the ellipse is

$$\frac{x^2}{a^2} + \frac{y^2}{b^2} = 1.$$

If polar coordinates r and f are taken such that the length SP is given by r and the angle ASP by f, then the polar equation of the ellipse is

$$r = \frac{p}{1 + e \cos f} \tag{4.23}$$

$$= \frac{a(1 - e^2)}{1 + e \cos f}.$$

Proof of the above statements may be found in any book on conic sections.

In the remainder of this chapter we will apply the two-body solution to orbital motion in the Solar System; but it will be seen later that many of the concepts and results· may be taken over practically unchanged when for example binary stars are treated.

Now let a body P move about the Sun S. The focus S′ is often referred to as the empty focus. For the moment let the orbital plane coincide with the plane of the ecliptic and take the direction of the vernal equinox ♈ as a reference direction as in figure 4.3. Then, if PŜ♈ $= \theta$ and AŜ♈ $= \omega$, the true anomaly $f = \theta - \omega$ and the equation of the body is

$$r = \frac{a(1 - e^2)}{1 + e \cos (\theta - \omega)}. \tag{4.24}$$

It is seen that when $\theta = \omega$ the body is at perihelion with $r = a(1 - e)$. When $\theta = 180° + \omega$, the body is at aphelion with $r = a(1 + e)$.

We had from equation (4.15) the relation

$$r^2 \dot{f} = h \tag{4.25}$$

where h is twice the rate of description of area by the radius vector SP.

Now the area of an ellipse is πab and this must be described in an interval T, the period of the body in its orbit.

Then

$$\frac{2\pi ab}{T} = h$$

or

$$2\pi a^2(1-e^2)^{1/2}=hT.$$

Now by equations (4.21) and (4.24),

$$h^2=\mu a(1-e^2)$$

where

$$\mu=G(m_1+m_2).$$

Eliminating h, we obtain

$$T=2\pi\,(a^3/\mu)^{1/2}. \qquad\qquad\qquad\qquad (4.26)$$

This is an important relationship which shows that the period depends only upon the values of the semimajor axis and the sum of the masses.

If M_S and m_1 are the masses of the Sun and a planet respectively, and T_1 and a_1 are the period and semimajor axis of the planet's orbit about the Sun, then equation (4.26) gives

$$G(M_S+m_1)=4\pi^2\,\frac{a_1{}^3}{T_1{}^2}. \qquad\qquad\qquad (4.27)$$

For another planet of mass m_2 in an orbit of period T_2 and semimajor axis a_2,

$$G(M_S+m_2)=4\pi^2\,\frac{a_2{}^3}{T_2{}^2}. \qquad\qquad\qquad (4.28)$$

Hence by equations (4.27) and (4.28), we have

$$\frac{M_S+m_1}{M_S+m_2}=\left(\frac{a_1}{a_2}\right)^3\left(\frac{T_2}{T_1}\right)^2. \qquad\qquad (4.29)$$

Equation (4.29) is the correct form of Kepler's third law. In fact even for Jupiter, the most massive planet, $m/M_S\sim10^{-3}$, so that the quantity on the left-hand side of equation (4.29) is almost unity.

4.5.1 Measurement of a planet's mass

Any planet that possesses a satellite, natural or artificial, may have its mass measured by a study of the orbit of the satellite.

Let equation (4.27) refer to the Earth's orbit about the Sun, the Earth having mass m_1. Let an artificial satellite of the Earth have period T' and mass m' while its orbital semimajor axis is a'.

Then

$$G(m_1+m')=4\pi^2\,\frac{a'^3}{T'^2}.$$

Hence

$$\frac{m_1+m'}{M_S+m_1}=\left(\frac{a'}{a_1}\right)^3\left(\frac{T_1}{T'}\right)^2. \qquad\qquad (4.30)$$

The mass of the satellite may be neglected compared with the mass of the

Earth, as may the mass of the Earth compared with the Sun's mass. Hence we may write

$$\frac{m_1}{M_S} = \left(\frac{a'}{a_1}\right)^3 \left(\frac{T_1}{T'}\right)^2. \tag{4.31}$$

The quantities on the right-hand side of equation (4.31) may be measured; hence the mass of the Earth in units of the Sun's mass can be found.

Only two planets in the Solar System have no satellites: Mercury and Venus. Their masses have been determined indirectly (and much more inaccurately than the other planetary masses) by their minute effects upon the orbits of these bodies. These *perturbations* change the elements of the planetary orbits very slightly, measurement of such changes yielding values of the masses of the moonless planets. Venus and Mercury have had their masses measured more accurately in recent years by observing the distortions in the orbits of spacecraft such as Mariner 10 caused by their passage through the planets' gravitational fields.

4.5.2 Velocity in an elliptic orbit

Let V be the velocity of the body at the point P in its orbit where $SP=r$. This velocity, acting along the tangent to the ellipse at P, will have components $\dot{r}$ along the radius and $r\dot{f}$ perpendicular to the radius.

Hence

$$V^2 = \dot{r}^2 + r^2 \dot{f}^2. \tag{4.32}$$

By equations (4.23) and (4.25)

$$\dot{r} = \frac{h}{p} e \sin f. \tag{4.33}$$

Also, by equation (4.25)

$$r\dot{f} = \frac{h}{r} = \frac{h}{p}(1 + e \cos f). \tag{4.34}$$

Hence, squaring and adding (4.33) and (4.34), we obtain

$$V^2 = \left(\frac{h}{p}\right)^2 (1 + 2e \cos f + e^2)$$

or

$$V^2 = \left(\frac{h}{p}\right)^2 [2 + 2e \cos f - (1 - e^2)]. \tag{4.35}$$

Using (4.23), equation (4.35) becomes

$$V^2 = \frac{2h^2}{rp} - \left(\frac{h}{p}\right)^2 (1 - e^2).$$

But $h^2/\mu = p = a(1 - e^2)$; hence

$$V^2 = \mu\left(\frac{2}{r} - \frac{1}{a}\right). \tag{4.36}$$

79

It is seen that at perihelion, V is greatest, since, putting $r=a(1-e)$,

$$V_P{}^2=\frac{\mu}{a}\left(\frac{1+e}{1-e}\right).$$

At aphelion V is least, where on putting $r=a(1+e)$,

$$V_A{}^2=\frac{\mu}{a}\left(\frac{1-e}{1+e}\right).$$

Hence $V_A V_P=\mu/a=$ constant.

It should also be noted from equation (4.36) that V is a function only of the radius r. By rearranging (4.36) we obtain

$$a=\frac{\mu}{(2\mu/r)-V^2}. \tag{4.37}$$

But

$$T=2\pi\left(\frac{a^3}{\mu}\right)^{1/2}. \tag{4.38}$$

Hence

$$T=2\pi\mu\left(\frac{2\mu}{r}-V^2\right)^{-3/2}. \tag{4.39}$$

These relations highlight some interesting properties of elliptical motion. It is seen that the semimajor axis is a function of the radius vector and the square of the velocity. If therefore a body of mass m_1 is projected at a given distance r from another body of mass m_2 with velocity V, the semimajor axis of the orbit is independent of the direction of projection and depends only on the magnitude of the velocity. In figure 4.4 all the orbits have the same initial radius vector SP and the same initial velocity magnitude V though the directions of projection are different. All orbits have the same semimajor axis a given by equation (4.37).

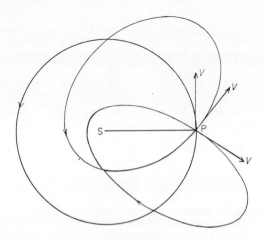

Figure 4.4

It is also seen from equations (4.38) or (4.39) that the periods in these orbits must also be the same. If particles were projected from P simultaneously into these covelocity orbits, they would all pass through P together on return though the orbits they pursued were quite different in shape.

The velocity in an elliptic orbit may be usefully resolved into two components, both constant in magnitude. One component is perpendicular to the radius vector and so varies in direction; the other is perpendicular to the major axis and so is constant both in magnitude and direction.

In figure 4.5, the velocity V may be resolved into components: (i) $\dot{r}$, along $SP=PF$, and (ii) rf, perpendicular to $SP=PD$. The required components are then HE and PH.

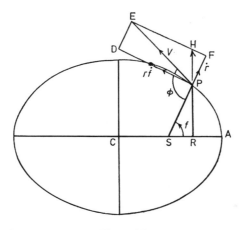

Figure 4.5

Now

$$HE = PD - FH$$

$$= rf - \dot{r} \cot f.$$

Hence, using equations (4.33) and (4.34),

$$HE = h/p. \qquad (4.40)$$

Now $PH = PF \operatorname{cosec} f = \dot{r} \operatorname{cosec} f$. Hence, using (4.33),

$$PH = eh/p. \qquad (4.41)$$

It is to be noted that if the orbit is a circle, e is zero and the component that remains is the circular velocity V_c given by

$$V_c = h/p = (\mu/a)^{1/2} \qquad (4.42)$$

where a is the radius of the circular orbit.

4.5.3 The angle between velocity and radius vectors

In figure 4.5 let ϕ be the angle between the velocity vector PE and the radius vector SP. Then

$$S\hat{P}E = 90° + D\hat{P}E.$$

Now

Also

$$\left.\begin{array}{l} \cos D\hat{P}E = DP/PE = r\dot{f}/V. \\ \sin D\hat{P}E = \dot{r}/V. \end{array}\right\} \tag{4.43}$$

From the first of (4.43) and using (4.25) and (4.36) we obtain

$$\cos D\hat{P}E = \frac{h}{r\mu^{1/2}\left(\dfrac{2}{r} - \dfrac{1}{a}\right)^{1/2}}.$$

Applying the relation

$$h^2 = \mu a(1 - e^2)$$

it is found, after a little reduction, that

$$\cos D\hat{P}E = \left[\frac{a^2(1-e^2)}{r(2a-r)}\right]^{1/2}.$$

Hence,

$$\phi = 90° + \cos^{-1}\left[\frac{a^2(1-e^2)}{r(2a-r)}\right]^{1/2}$$

or

$$\sin\phi = \left[\frac{a^2(1-e^2)}{r(2a-r)}\right]^{1/2}. \tag{4.44}$$

Rearranging (4.44) we have

$$e = \left[1 - \frac{r}{a^2}(2a-r)\sin^2\phi\right]^{1/2}. \tag{4.45}$$

Using equations (4.24), (4.25), (4.43) and (4.53), the following useful relations between ϕ, f and E may be easily established:

$$\left.\begin{array}{l} \sin\phi = \left(\dfrac{1-e^2}{1-e^2\cos^2 E}\right)^{1/2} = \dfrac{1+e\cos f}{(1+e^2+2e\cos f)^{1/2}} \\[4mm] \cos\phi = -\dfrac{e\sin E}{(1-e^2\cos^2 E)^{1/2}} = -\dfrac{e\sin f}{(1+e^2+2e\cos f)^{1/2}}. \end{array}\right\} \tag{4.46}$$

The quantity E is the so-called eccentric anomaly and is defined in the following section.

4.5.4 The mean, eccentric and true anomalies

We now consider three quantities and the relations among them that are of importance in the elliptical orbit case. They are the mean, eccentric and true anomalies.

Since the radius vector turns through 2π radians in the orbital period T, the mean angular velocity (mean motion) n is given by

$$n = 2\pi/T. \tag{4.47}$$

The relation

$$\frac{2\pi a^2 (1-e^2)^{1/2}}{T} = h$$

may therefore be written as

$$na^2(1-e^2)^{1/2} = h. \tag{4.48}$$

If τ is the time of perihelion passage, the angle swept by a radius vector rotating about S with mean angular velocity n in the interval $(t-\tau)$ will be M, where

$$M = n(t-\tau). \tag{4.49}$$

M, defined in this way, is called the *mean anomaly*.

If a circle is described on AA' as diameter, as shown in figure 4.6, and the line through P on the ellipse perpendicular to the major axis AA' is produced

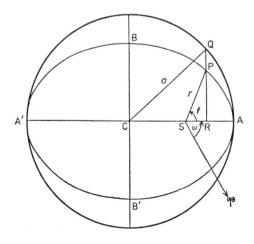

Figure 4.6

to meet the circle in Q, the angle QĈA, usually denoted by E and called the *eccentric anomaly*, is related to the true anomaly f.

Now

$$SR = CR - CS = a \cos E - ae.$$

But

$$SR = r \cos f$$

and hence

$$r \cos f = a (\cos E - e). \tag{4.50}$$

83

Also, by a property of ellipses and eccentric circles,

$$PR/QR = b/a. \tag{4.51}$$

Hence

$$r \sin f = b \sin E.$$

Or

$$r \sin f = a(1 - e^2)^{1/2} \sin E. \tag{4.52}$$

Squaring and adding (4.50) and (4.52), we obtain after a little reduction

$$r = a(1 - e \cos E). \tag{4.53}$$

Now

$$r \cos f = r \left(1 - 2 \sin^2 \frac{f}{2}\right).$$

Hence

$$2r \sin^2 (f/2) = r(1 - \cos f). \tag{4.54}$$

Using equations (4.50) and (4.53) we obtain

$$2r \sin^2 (f/2) = a(1 + e)(1 - \cos E). \tag{4.55}$$

Similarly

$$2r \cos^2 (f/2) = a(1 - e)(1 + \cos E). \tag{4.56}$$

Dividing (4.55) by (4.56) we finally obtain

$$\tan\left(\frac{f}{2}\right) = \left(\frac{1+e}{1-e}\right)^{1/2} \tan\left(\frac{E}{2}\right). \tag{4.57}$$

The eccentric anomaly E and the mean anomaly M are related by an important equation called Kepler's equation, which we now derive.

By Kepler's second law,

$$\frac{\text{area SPA}}{\text{area of ellipse}} = \frac{t - \tau}{T}$$

or

$$\text{area SPA} = \frac{\pi ab(t - \tau)}{T}$$

or, using equation (4.47),

$$\text{area SPA} = \tfrac{1}{2}abM. \tag{4.58}$$

Now

$$\text{area SPA} = \text{area SPR} + \text{area RPA}.$$

$$\text{Area RPA} = \frac{b}{a} (\text{area QRA})$$

which is obtained by dividing these areas into thin strips parallel to the minor axis and using the property described in equation (4.51). Then

$$\text{area SPA} = \text{area SPR} + \frac{b}{a} (\text{area QRA})$$

$$= \text{area SPR} + \frac{b}{a} (\text{area QCA} - \text{area QCR})$$

$$= \tfrac{1}{2}r^2 \sin f \cos f + \frac{b}{a} (\tfrac{1}{2}a^2 E - \tfrac{1}{2}a^2 \sin E \cos E)$$

$$= \tfrac{1}{2}ab(E - e \sin E) \tag{4.59}$$

using (4.50) and (4.52).

Comparing (4.58) and (4.59) it is seen that

$$E - e \sin E = M = n(t - \tau). \tag{4.60}$$

This is Kepler's equation. It should be noted that both E and M are in circular measure.

4.5.5 The solution of Kepler's equation

In some astronomical and astrodynamical applications of Kepler's equation, the mean anomaly M is required when a value of the eccentric anomaly E is given, the eccentricity being known. M is found without trouble from equation (4.60). More often, however, M is given, e being known, and the corresponding value of E is required. It is obtained by using one of the dozens of methods of successive approximations that have been devised for the solution of Kepler's equation by mathematicians and astronomers from Kepler himself onwards.

The usual method of procedure is to obtain an approximate value of E that nearly satisfies equation (4.60) by inspection or by special tables or by a graphical method (Moulton 1914, Astrand 1890, Bauschinger 1901). Where the eccentricity is smaller than 0·1, a suitable starting value of E (say E_0) is obtained by simply taking $E_0 = M$; otherwise, tables or graphs are required. Let the starting value in either case be E_0 so that the true value E is given by

$$E = E_0 + \Delta E_0.$$

where ΔE_0 is a small fraction of E_0.

Then, substituting in (4.60), we obtain

$$(E_0 + \Delta E_0) - e \sin (E_0 + \Delta E_0) = M. \tag{4.61}$$

Expanding and neglecting all but zero and first-order terms, equation (4.61) becomes

$$E_0 - e \sin E_0 + \Delta E_0 (1 - e \cos E_0) = M$$

or

$$\Delta E_0 = \frac{M - (E_0 - e \sin E_0)}{1 - e \cos E_0}$$

from which E_0 can be calculated. Then E_1 (where $E_1 = E_0 + \Delta E_0$) is a more accurate value of E and the process can be repeated as often as is necessary.

An alternative method uses the following scheme: writing Kepler's equation in the form

$$E = M + e \sin E$$

and obtaining a first approximation E_0 for E as usual, proceed further as indicated below.

$$E_0$$

$$E_1 = M + e \sin E_0$$

$$E_2 = M + e \sin E_1 \qquad (4.62)$$

$$E_3 = M + e \sin E_2$$

etc.

Example: Calculate to the nearest 10″ the value of the eccentric anomaly E of Jupiter five years after its perihelion passage, given that Jupiter's period T and eccentricity e are 11·8622 years and 0·04844 respectively.

Since we want the mean anomaly M in degrees and since $(t - \tau)$ is given in years we require the mean motion in degrees per year.

Now M increases by 360° in T years. Therefore the mean motion n is 360°/T, so that

$$M = 5n = \frac{5 \times 360°}{11·8622} = 151°44'33\cdot0''.$$

Now E is of the order of M in size (i.e. $\sim 540\,000''$). To find E in degrees correct to the nearest 10″ therefore requires five significant figures. Six-figure logarithm tables should be used if an electronic calculator is not available.

(i) *First approximation*: Since e is small we may take

$$E_1 = M = 151°44'30''.$$

(ii) *Second approximation*: Before using Kepler's Equation in the form

$$E = M + e \sin E$$

we express circular measure in degrees

$$e \text{ radians} \equiv \frac{180}{\pi} e \text{ degrees}.$$

Hence

$$E° = M° + \left(\frac{180\, e}{\pi} \right) \sin E°.$$

It is found that $(180\, e/\pi) = 2\cdot77541$ in this example, so that

$$E_2° = M° + 2\cdot77541° \sin M°$$

$$= 151°44'30'' + 1°18'50''$$

or

$$E_2° = 153°03'20''.$$

(iii) *Third approximation*:

$$E_3 = M + e \sin E_2$$

gives

$$E_3 = 151°44'30'' + 2 \cdot 77541° \sin 153°03'20''$$

$$= 151°44'30'' + 1°15'27 \cdot 4''$$

that is

$$E_3 = 152°59'57 \cdot 4''$$

(iv) *Fourth approximation*: It is found that

$$E_4 = 153°00'06''$$

which is the required answer.

4.5.6 The equation of the centre

It is possible to express the true anomaly f as a series in terms of the eccentricity e and the mean anomaly M.

It is easily seen that if the scheme given by equation (4.62) is followed analytically rather than numerically, E_0 being taken to be M and the angles being expanded to the appropriate powers of e, there results the following series:

$$E = M + \left(e - \frac{e^3}{8}\right) \sin M + \frac{1}{2} e^2 \sin 2M + \frac{3}{8} e^3 \sin 3M + O\,(e^4) \quad (4.63)$$

where $O\,(e^4)$ denotes terms of the order of e^4 and higher.

Again, using equation (4.57) it can be shown (Smart 1956) that a series for f in terms of e and E may be found. This series is

$$f = E + \left(e + \frac{1}{4} e^3\right) \sin E + \frac{1}{4} e^2 \sin 2E + \frac{1}{12} e^3 \sin 3E + O\,(e^4). \quad (4.64)$$

Equations (4.63) and (4.64) may then be combined to give the *equation of the centre*, namely

$$f - M = \left(2e - \frac{1}{4} e^3\right) \sin M + \frac{5}{4} e^2 \sin 2M + \frac{13}{12} e^3 \sin 3M + O\,(e^4). \quad (4.65)$$

Thus when e and M are given, the true anomaly may be found directly from equation (4.63). The use of such a series, however, is limited to orbits of small eccentricity.

Various sets of tables have been published giving the true anomaly f or $(r/a) \cos f$ and $(r/a) \sin f$ for various eccentricities (Schlesinger and Udick 1912, Stracke 1928). In particular Cayley's tables (Cayley 1861) give developments of various often used functions in elliptic motion.

4.5.7 Position of a body in an elliptic orbit

There are two problems in orbital work that are encountered frequently both in astronomy and in astrodynamics. One problem is to obtain the position and velocity of the body, given the elements and the time; the other is to obtain the elements of the orbit, given the position and velocity and the time.

An example of the latter problem is the case where a probe is injected into a solar orbit from the Earth with a given position and velocity relative to the Sun at a given time and it is desired to obtain the elements a, e and τ of the elliptic orbit. As an example of the former problem, it may be desired to find the body's position some time after it has been injected into the solar orbit, the orbital elements now being known.

The formulae established in previous sections that are of use in these problems are collected below.

$$r = \frac{a(1-e^2)}{1+e\cos f} \tag{4.66}$$

$$r = a(1 - e \cos E) \tag{4.67}$$

$$\tan\left(\frac{f}{2}\right) = \left(\frac{1+e}{1-e}\right)^{1/2} \tan\left(\frac{E}{2}\right) \tag{4.68}$$

$$M = E - e \sin E \tag{4.69}$$

$$M = n(t - \tau) \tag{4.70}$$

$$h^2 = \mu a(1 - e^2) \tag{4.71}$$

$$V^2 = \mu\left(\frac{2}{r} - \frac{1}{a}\right) \tag{4.72}$$

$$T = 2\pi/n \tag{4.73}$$

$$n = \mu^{1/2} a^{-3/2} \tag{4.74}$$

$$\sin \phi = \left[\frac{a^2(1-e^2)}{r(2a-r)}\right]^{1/2}. \tag{4.75}$$

It is to be noted that in these formulae a, e and τ are three of the elements of the elliptic orbit. It is assumed that μ is known. Given these elements and the time in question, the position and velocity of the body in its orbit may then be found as follows:

(i) Calculate n by equation (4.74).
(ii) Use n in equation (4.70) to find M.
(iii) Solve Kepler's equation (4.69) to obtain E.
(iv) Obtain r from (4.67).
(v) Check r by recalculating it, using equations (4.68) and (4.66).
(vi) Calculate V from equation (4.72).
(vii) Calculate ϕ from equation (4.75).

In the other problem, where it is assumed that V, r, ϕ, t and μ are given, the procedure is as follows:

(i) From equation (4.72) calculate a.

88

(ii) From equation (4.75) calculate e.
(iii) Obtain E from equation (4.67).
(iv) Use equation (4.66) to calculate f.
(v) Check f by recalculating it from equation (4.68).
(vi) From equation (4.69) find M.
(vii) Use n and M in equation (4.70) to obtain τ.

4.6 The Parabolic Orbit

In *this* type of two-body motion (where $e=1$) the orbit is open, the second body approaching the first from infinity until, at its nearest approach when the relative velocity is a maximum, it begins to recede to infinity as in figure 4.7.

The equation of the parabolic orbit is obtained by putting $e=1$ in equation (4.21), whence

$$r = \frac{p}{1+\cos f} \qquad (4.76)$$

where, as before, p and f are the semilatus rectum and true anomaly respectively.

The integral of areas is

$$r^2\dot{f}=h \qquad (4.77)$$

where $p=h^2/\mu$. It is seen that when $f=0$,

$$r=p/2.$$

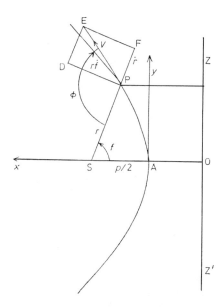

Figure 4.7

89

The canonic equation of the parabola, referred to cartesian coordinate axes Ax and Ay as shown in figure 4.7, is

$$y^2 = 2px.$$

The velocity V of the body in a parabolic orbit is given by considering as before

$$V^2 = \dot{r}^2 + r^2 \dot{f}^2.$$

Differentiating (4.76) and using it with (4.77) it is seen that V is given by the simple relation

$$V^2 = 2\mu/r. \qquad (4.78)$$

An interesting relationship between circular and parabolic velocity exists here. Referring to section 4.5.2 it was seen that the velocity V_c in a circular orbit of radius a was given by

$$V_c^2 = \mu/a. \qquad (4.79)$$

If the body is now given an impulse so that its velocity becomes V given by

$$V^2 = 2\mu/a \qquad (4.80)$$

it will enter a parabolic orbit that will take it to infinity. It will reach infinity with zero velocity (put $r = \infty$ in equation (4.78)) so that parabolic velocity is an alternative name for escape velocity. It is seen from equations (4.79) and (4.80) that

$$\text{escape velocity} = \text{circular velocity} \times \sqrt{2}. \qquad (4.81)$$

This is a useful relationship to remember.
Now equation (4.76) may be written as

$$r = \frac{p}{2} \sec^2 \frac{f}{2} = \frac{p}{2} \left(1 + \tan^2 \frac{f}{2} \right).$$

Hence (4.77) gives

$$\left(\frac{p}{2} \right)^2 \dot{f} \sec^4 \frac{f}{2} = (p\mu)^{1/2}$$

or

$$4 \left(\frac{\mu}{p^3} \right)^{1/2} dt = \left(\sec^2 \frac{f}{2} + \sec^2 \frac{f}{2} \tan^2 \frac{f}{2} \right) df.$$

Integrating, we obtain

$$2 \left(\frac{\mu}{p^3} \right)^{1/2} (t - \tau) = \tan \frac{f}{2} + \frac{1}{3} \tan^3 \frac{f}{2} \qquad (4.82)$$

where τ is the time of perihelion passage.
If we define $\bar{n}$ by the equation

$$\bar{n}^2 p^3 = \mu$$

which should be compared with (4.74), and let

$$D=\tan \frac{f}{2}$$

we may write equation (4.82) as

$$D+\frac{D^3}{3}=2\bar{n}(t-\tau). \qquad (4.83)$$

Equations (4.82) and (4.83) are versions of Barker's equation, which has been extensively used in studies of the orbits of comets and is now being used in astrodynamics. Tables have been constructed enabling f to be found by interpolation when $t-\tau$ is given, or vice versa (Watson 1892).

To solve Barker's equation, which is a cubic in $\tan (f/2)$, let

$$\tan \frac{f}{2}=2 \cot 2\theta=\cot \theta-\tan \theta$$

so that

$$\tan^3 \frac{f}{2}=-3 \tan \frac{f}{2}+\cot^3 \theta-\tan^3 \theta.$$

Then (4.82) becomes

$$\cot^3 \theta-\tan^3 \theta=6\bar{n}(t-\tau).$$

Now define s by

$$\cot \theta=\left(\cot \frac{s}{2}\right)^{1/3}$$

whence

$$\cot s=3\bar{n}(t-\tau).$$

The procedure is therefore to apply the equations (4.84) below in the order in which they appear.

$$\left.\begin{array}{c} \cot s=3\bar{n}(t-\tau) \\[2mm] \cot \theta=\left(\cot \dfrac{s}{2}\right)^{1/3} \\[2mm] \tan \dfrac{f}{2}=2 \cot 2\theta. \end{array}\right\} \qquad (4.84)$$

Having obtained $\tan (f/2)$, r is obtained from the relation

$$r=\frac{p}{2}\left(1+\tan^2 \frac{f}{2}\right). \qquad (4.85)$$

The velocity is found from equation (4.78).

For the angle ϕ between velocity vector and radius vector, it is seen that

$$\phi=90°+\cos^{-1}\left(\frac{rf^\cdot}{V}\right)$$

which on using equations (4.77) and (4.78) reduces to

$$\phi=90^\circ+\cos^{-1}\left(\frac{p}{2r}\right)^{1/2}. \tag{4.86}$$

This may be written as

$$p=2r\sin^2\phi. \tag{4.87}$$

Hence, given the elements p and τ of a parabolic orbit with μ and a time t, it is a straightforward matter to calculate r, V and ϕ for that time.

Conversely, given μ, r, V, ϕ and t for a parabolic orbit, the elements p and τ may be found by applying (4.87), (4.85) and (4.82) in turn.

4.7 The Hyperbolic Orbit

In astronomy the use of hyperbolic orbits has been confined chiefly to comet and meteor work; in astrodynamics such orbits are frequently of interest. For example, to put a probe into an interplanetary orbit requires energy such that its orbit with respect to the Earth is a hyperbola until it recedes to about one million kilometres.

In figure 4.8, if P moves so that SP/PM$=e$, where e is constant and greater

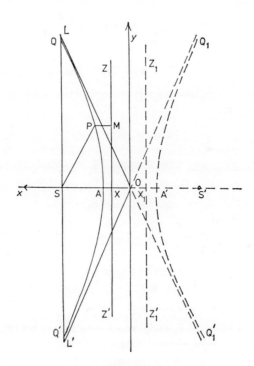

Figure 4.8

than unity and the straight line Z'XZ is perpendicular to SX, P sweeps out the hyperbola Q'AQ whose polar equation is

$$r = \frac{p}{1 + e \cos f} = \frac{a(e^2 - 1)}{1 + e \cos f}. \tag{4.88}$$

When $f = 0$, $r = a(e - 1)$.

The canonic equation of the hyperbola with respect to a cartesian set of axes Ox and Oy is

$$\frac{x^2}{a^2} - \frac{y^2}{b^2} = 1$$

where

$$b^2 = a^2(e^2 - 1).$$

Obviously a hyperbola $Q_1'A'Q_1$ may be swept out by a similarly moving point P_1 about S', but this hyperbola does not concern us since a given particle traverses only one branch.

When r becomes infinite,

$$1 + e \cos f = 0. \tag{4.89}$$

Letting the value of f be f_0 when (4.89) holds, we have

$$f_0 = \cos^{-1}\left(-\frac{1}{e}\right)$$

which means that the true anomaly can only vary from

$$-\pi + \cos^{-1}\left(\frac{1}{e}\right) \qquad \text{to} \qquad +\pi - \cos^{-1}\left(\frac{1}{e}\right).$$

At these limits, the straight lines OL' and OL touch the hyperbola tangentially, being asymptotes to it. The angles SOL and SOL' are therefore of magnitude $\pi - \cos^{-1}(1/e)$. The asymptotes are also defined by the relation

$$\tan \psi = \pm b/a$$

where ψ is angle LÔX in figure 4.8. The semilatus rectum is given by

$$SQ = p = a(e^2 - 1).$$

Again, as in the elliptic case, these statements are proved in any book on conic sections.

4.7.1 Velocity in a hyperbolic orbit

The integral of area in the hyperbolic case is

$$r^2 \dot{f} = h = \mu^{1/2} p^{1/2}. \tag{4.90}$$

Proceeding as in section 4.5.2 and using equations (4.88) and (4.90), it is found that the velocity V is given by

$$V^2 = \mu\left(\frac{2}{r} + \frac{1}{a}\right). \tag{4.91}$$

93

At perihelion

$$V_P{}^2 = \frac{\mu}{a}\left(\frac{e+1}{e-1}\right). \tag{4.92}$$

It should be noted that when $r = \infty$,

$$V^2 = \mu/a. \tag{4.93}$$

In other words, the body reaches infinity with a nonzero velocity.

The velocity in hyperbolic orbits may be split into two components h/p and eh/p perpendicular to the radius vector and to the axis SA respectively. This follows immediately from the fact that the equation for both elliptic and hyperbolic orbits is

$$r = \frac{p}{1 + e \cos f}$$

so that the appropriate analysis of section 4.6.3 holds. Indeed the same resolution holds for parabolic orbits where e is put equal to unity, and it is then seen that both components are equal in magnitude.

The angle between velocity vector and radius vector is obtained as in the elliptic case. Thus

$$\phi = 90° + \cos^{-1}\left(\frac{r\dot{f}}{V}\right)$$

which reduces, using equations (4.90) and (4.91), to

$$\phi = 90° + \cos^{-1}\left[\frac{a^2(e^2-1)}{r(2a+r)}\right]^{1/2}. \tag{4.94}$$

Rearranging equation (4.94), we obtain

$$e = \left[1 + \frac{r}{a^2}(2a+r)\sin^2\phi\right]^{1/2}. \tag{4.95}$$

As in the elliptic case, the following relations between ϕ, f and F are easily obtained:

$$\left.\begin{array}{l} \sin\phi = \left(\dfrac{e^2-1}{e^2\cosh^2 F-1}\right)^{1/2} = \dfrac{1+e\cos f}{(1+e^2+2e\cos f)^{1/2}} \\[4mm] \cos\phi = -\dfrac{e\sinh F}{(e^2\cosh^2 F-1)^{1/2}} = -\dfrac{e\sin f}{(1+e^2+2e\cos f)^{1/2}} \end{array}\right\} \tag{4.96}$$

where F, a quantity analogous to the eccentric anomaly, is introduced in the following section.

4.7.2 Position in the hyperbolic orbit

In the hyperbolic orbit there are equations analogous to all the elliptic equations from (4.66) to (4.75), except (4.73).

Let v be a quantity defined by

$$v^2 a^3 = \mu.$$
(4.97)

Also from equation (4.88),

$$\cos f = \frac{a(e^2 - 1)}{er} - \frac{1}{e}$$

so that

$$\frac{df}{dr} = \frac{a(e^2 - 1)^{1/2}}{r[(a+r)^2 - a^2 e^2]^{1/2}}.$$
(4.98)

From equations (4.90) and (4.97) we obtain

$$v\frac{dt}{dr} = \frac{r}{a[(a+r)^2 - a^2 e^2]^{1/2}}.$$
(4.99)

We now define the variable F, analogous to the elliptic eccentric anomaly E, by the relation

$$r = a(e \cosh F - 1).$$
(4.100)

Then

$$\cosh F = \frac{a+r}{ae}$$

and equation (4.99) becomes

$$v\frac{dt}{dF} = e \cosh F - 1.$$

Integrating, we obtain

$$e \sinh F - F = M = v(t - \tau)$$
(4.101)

which is analogous to Kepler's equation.

The solution of equation (4.101) provides problems similar to that of solving Kepler's equation in elliptic motion. The first problem is finding an approximate value of F. One suitable method consists in plotting

$$y_1 = \frac{M+F}{e}$$

against

$$y_2 = \sinh F.$$

Their intersection at $y_1 = y_2$ provides the sought-for approximation to F for the given M. Alternatively, we may proceed as follows; we note that for $F < 2 \cdot 5$, we may write

$$(e \sinh F) - F \simeq \frac{eF^3}{6} + (e-1)F = M.$$

This cubic in F, i.e.

$$F^3 + \frac{6(e-1)}{e} F - \frac{6M}{e} = 0$$
(4.102)

95

has the solution

$$F = Y \sinh w$$

where

$$\sinh 3w = 3M/Y(e-1) \quad \text{and} \quad Y = [8(e-1)/e]^{1/2}$$

It should be noted that if $e \sim 1$, a solution should be obtained from

$$F^3 = 6M/e.$$

When $F > 2 \cdot 5$, we may use

$$F \simeq \ln(2M/e) \tag{4.103}$$

where ln denotes the natural logarithm.

The choice of equation (4.102) or (4.103) is dictated by the value of M corresponding to $F = 2 \cdot 5$; this value is given approximately by

$$m = 5e - 5/2.$$

For

$$F < 2 \cdot 5, \quad M < (5e - 5/2)$$

$$F > 2 \cdot 5, \quad M > (5e - 5/2).$$

Having found F, obtain the Gudermannian function q of F, where

$$\tan\left(45° + \frac{q}{2}\right) = \exp F.$$

Then f is obtained from

$$\sin f = \frac{(e^2 - 1)^{1/2} \sin q}{e - \cos q}$$

$$\cos f = \frac{e \cos q - 1}{e - \cos q}.$$

This method avoids the use of hyperbolic functions. Alternatively, by equations (4.88) and (4.100) we have

$$\frac{(e^2 - 1)}{1 + e \cos f} = e \cosh F - 1. \tag{4.104}$$

Applying the formulae

$$\cos f = \frac{1 - \tan^2(f/2)}{1 + \tan^2(f/2)}, \quad \cosh F = \frac{1 + \tanh^2(F/2)}{1 - \tanh^2(F/2)}$$

it is found that (4.104) gives

$$\tan\frac{f}{2} = \left(\frac{e+1}{e-1}\right)^{1/2} \tanh\frac{F}{2}. \tag{4.105}$$

The equations found in this and sections 4.7 and 4.7.1 may then be used to find the position of the body in its orbit at any time given the elements a, e and τ, or to find the elements from a given position and velocity in a manner analogous to that exhibited in section 4.5.7.

4.8 The Rectilinear Orbit

Let us suppose that in an elliptic orbit we keep the major axis constant in length and let the eccentricity tend to unity. Then the ellipse becomes more and more elongated, with the perihelion distance $a(1-e)$ tending to zero. In the limit, the ellipse becomes a line segment connecting both foci. This is called a rectilinear ellipse. Similar limiting processes obtain the rectilinear parabola and hyperbola, where each is a line from the focus along the axis of symmetry to infinity. In the rectilinear ellipse, the line is traversed so that maximum velocity occurs at one focus and zero velocity at the other; in the rectilinear parabola maximum velocity occurs at the focus and zero velocity at infinity; in the rectilinear hyperbola maximum velocity occurs at the focus with some velocity remaining at infinity.

Such orbits may seem unrealistic and of no practical value but this is by no means the case. For example, in many elliptic and hyperbolic cometary orbits, the value of e is so close to unity that the comet's orbital behaviour closely approximates to the behaviour of a body in a rectilinear ellipse or hyperbola. In astrodynamics bodies in many problems behave very much as if they followed rectilinear hyperbolas.

The following equations relating time, position and velocity in the two-body problem are valid when the motion is rectilinear (i.e. when $V=dr/dt$):

(i) The rectilinear ellipse

$$\left.\begin{array}{l} M=n(t-\tau) \\[4pt] M=E-\sin E \\[4pt] r=a(1-\cos E) \\[4pt] r\dot{r}=a^{1/2}\mu^{1/2}\sin E. \end{array}\right\} \tag{4.106}$$

(ii) The rectilinear parabola

$$\left.\begin{array}{l} M=\mu^{1/2}(t-\tau) \\[4pt] M=B^3/6 \\[4pt] r=B^2/2 \\[4pt] r\dot{r}=\mu^{1/2}B. \end{array}\right\} \tag{4.107}$$

(iii) The rectilinear hyperbola

$$\left.\begin{array}{l} M=v(t-\tau) \\[4pt] M=(\sinh F)-F \\[4pt] r=a[(\cosh F)-1] \\[4pt] r\dot{r}=a^{1/2}\mu^{1/2}\sinh F. \end{array}\right\} \tag{4.108}$$

Equations (4.107) may be derived easily from equation (4.78) since we now have $V = dr/dt$.

Tables based on slightly different versions of equations (4.104), (4.107) and (4.108) have been constructed by Herrick (1953). Herrick's tables are useful in rectilinear motion enabling position and velocity to be determined from the time by direct interpolation, without the aid of series expansions or successive approximations. They may also be used for near-rectilinear motion, a case often found in astrodynamics. The method of procedure for such a use may be sketched out by considering the elliptic case when e is nearly unity in Kepler's equation:

$$M = E - e \sin E$$

$$= (E - \sin E) + (1 - e) \sin E \qquad (4.109)$$

$$= (E - \sin E) + \epsilon \sin E.$$

With $e \sim 1$, ϵ is small and the departure of (4.109) from the rectilinear equation ($M = E - \sin E$) is of the same nature as the departure of Kepler's equation from the 'circular' equation ($M = E$) when e is nearly zero. Thus equation (4.109) may be solved by a method of successive approximations such as those given in section 4.6.5. Looking up Herrick's tables, the value of E for $E - \sin E = M$ is obtained. If this value is E_0 and the true value required is E, where

$$E = E_0 + \Delta E_0$$

then equation (4.109) becomes

$$(E_0 + \Delta E_0) - \sin (E_0 + \Delta E_0) + \epsilon \sin E_0 = M.$$

Expanding and collecting terms we obtain, on neglect of higher orders,

$$\Delta E_0 = - \frac{\epsilon \sin E_0}{1 - \cos E_0}.$$

The process can obviously be continued to provide a more accurate value of E if necessary. Similar procedures may be adopted in the near-rectilinear parabolic and hyperbolic cases. For details of time saving and efficient techniques the reader is referred to Herrick's tables (1953).

4.9 Barycentric Orbits

In figure 4.9, P_1 and P_2 are as before (section 4.4) the positions of the two particles of mass m_1 and m_2, O is a fixed reference point, and G is the centre of mass of m_1 and m_2 defined by

$$M\mathbf{R} = m_1 \mathbf{r}_1 + m_2 \mathbf{r}_2$$

where M is the sum of m_1 and m_2.

Let the vectors from G to P_1 and P_2 be $\mathbf{R}_1$ and $\mathbf{R}_2$ respectively. Then

$$m_1 R_1 = m_2 R_2, \qquad R_1 + R_2 = r$$

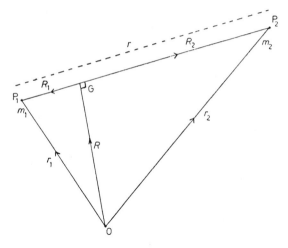

Figure 4.9

so that we have

$$R_1 = \left(\frac{m_2}{M}\right) r$$

$$R_2 = \left(\frac{m_1}{M}\right) r.$$

It was seen in section 4.4 that the centre of mass travels with constant velocity through space and, by Kepler's second law, the radius vector r sweeps out equal areas in equal times. For the relative orbit (one body about the other), we therefore have

$$r^2 \dot{f} = h.$$

But P_1GP_2 must always be a straight line; the radius vectors of the orbits of m_1 and m_2 about G (the barycentric orbits) must therefore also obey Kepler's second law, such that

$$R_1^2 \dot{f} = h_1$$

$$R_2^2 \dot{f} = h_2.$$

But

$$R_1^2 = \left(\frac{m_2}{M}\right)^2 r^2$$

so that

$$h_1 = \left(\frac{m_2}{M}\right)^2 r^2 \dot{f} = \left(\frac{m_2}{M}\right)^2 h.$$

Similarly

$$h_2 = \left(\frac{m_1}{M}\right)^2 h.$$

The barycentric orbits of m_1 and m_2 are therefore geometrically similar to each other and to their relative orbit. Hence in elliptic motion for example,

99

if a is the semimajor axis in the relative orbit, a_1 and a_2 being the semimajor axes in the barycentric orbits where $a_1+a_2=a$, we have

$$a_1=\left(\frac{m_2}{M}\right)a$$

$$a_2=\left(\frac{m_1}{M}\right)a.$$

Because of their geometrical similarity the orbits have equal eccentricities and equal periods.

It is seen that, where the mass of one particle is very small compared with the mass of the other, the relative orbit of the smaller about the larger is almost the size of the former's barycentric orbit, while the latter's barycentric orbit becomes very small.

4.10 Classification of Orbits with Respect to the Energy Constant

In the motion of one particle about the other, we derived the energy conservation equation (4.16)

$$\tfrac{1}{2}V^2-\frac{\mu}{r}=C \qquad\qquad (4.110)$$

where $\mu=G(m_1+m_2)$.

If V_1 and V_2 are the velocities of the masses m_1 and m_2 with respect to the centre of mass (taken to be at rest), the total energy E of the system is given by

$$E=\frac{1}{2}m_1V_1{}^2+\frac{1}{2}m_2V_2{}^2-\frac{Gm_1m_2}{r}$$

where the sum of the first two terms is the kinetic energy and $-Gm_1m_2/r$ is the potential energy of the system.

Now by the results of the previous section

$$V_1{}^2=\dot{R}_1{}^2+R_1{}^2\dot{f}_1{}^2=\left(\frac{m_2}{M}\right)^2(\dot{r}^2+r^2\dot{f}^2)=\left(\frac{m_2}{M}\right)^2V^2.$$

Similarly

$$V_2{}^2=\left(\frac{m_1}{M}\right)^2V^2$$

so it is easily seen that

$$E=\frac{m_1m_2}{m_1+m_2}\left(\frac{1}{2}V^2-\frac{\mu}{r}\right)=\left(\frac{m_1m_2}{m_1+m_2}\right)C$$

by using equation (4.110).

In astrodynamics, if m_1 is the mass of a vehicle and m_2 the mass of a planet, we can write

$$E=m_1\left(\frac{1}{2}V^2-\frac{\mu}{r}\right)=m_1C$$

where $\mu=Gm_2$, since m_1 is very much smaller than m_2.

Hence C becomes the total energy of the vehicle, $\frac{1}{2}V^2$ the kinetic energy and $-\mu/r$ the potential energy of the vehicle, all per unit mass.

We can classify the resultant orbit into ellipse, parabola or hyperbola according to the value of the energy C of the vehicle. This is useful in astrodynamics where it is often necessary to know the energy required to break out of a circular orbit about a planet and achieve escape velocity; that is to turn the planetocentric orbit into a parabola or hyperbola.

It is seen that the velocity V for a given distance is the deciding factor. Thus we had:

(i) for an ellipse, $V^2=\mu[(2/r)-(1/a)]$; hence $C=-\mu/2a$.
(ii) for a parabola, $V^2=2\mu/r$; hence $C=0$.
(iii) for a hyperbola, $V^2=\mu[(2/r)+(1/a)]$; hence $C=\mu/2a$.

$$(4.111)$$

Hence for a closed orbit, the total energy (kinetic plus potential) must be negative; for escape to just take place, the velocity must be increased until the total energy is zero; for an energy greater than zero an escape along a hyperbola takes place. In particular, for break-out from a circular orbit where $V^2=\mu/r$, the velocity must be increased to $\sqrt{2}\times$ (circular velocity).

4.11 The Orbit in Space

So far we have not considered in this chapter the orientation of the orbit in space. The three quantities necessary to take care of the orientation have already been introduced (section 2.6), namely the elements known as the longitude of the ascending node Ω, the longitude of perihelion ϖ (if the orbit is about the Sun) and the inclination i.

Since a great deal of computation in astrodynamics is done in rectangular coordinates x, y, z it is necessary to consider their relationship to the elements and the initial conditions of position and velocity in the orbit.

Let a spacecraft V be in orbit about the Sun S, its radius vector SV and true anomaly VŜA having values r and f at time t. If a set of axes Sξ and Sη are taken in the plane of the orbit with Sξ along the major axis towards perihelion and Sη perpendicular to the major axis, then the coordinates of V relative to this set of axes are ξ and η, given by

$$\xi=r\cos f, \qquad \eta=r\sin f.$$

If rectangular axes Sx, Sy and Sz are taken with Sx in the direction of the vernal equinox ♈, Sy in the plane of the ecliptic 90° from Sx and Sz in the direction of the north pole of the ecliptic, then by equations (2.4), (2.5) and

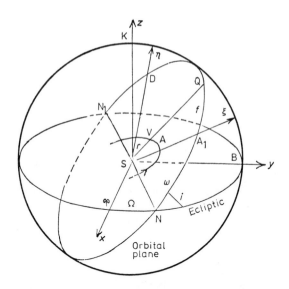

Figure 4.10

(2.6) the coordinates of V are (x, y, z) given by

$$x = r[\cos \Omega \cos (\omega + f) - \sin \Omega \sin (\omega + f) \cos i]$$

$$y = r[\sin \Omega \cos (\omega + f) + \cos \Omega \sin (\omega + f) \cos i]$$

$$z = r \sin (\omega + f) \sin i.$$

The radius vector may be obtained for a given time by using

$$r = \frac{p}{1 + e \cos f}$$

where p and e have the values associated with the given orbit and the true anomaly is computed according to the procedures outlined in previous sections of this chapter.

Now alternatively, if (l_1, m_1, n_1) and (l_2, m_2, n_2) are the direction cosines of $S\xi$ and $S\eta$ with respect to axes Sx, Sy and Sz, then

$$\left. \begin{array}{l} x = l_1 \xi + l_2 \eta \\ y = m_1 \xi + m_2 \eta \\ z = n_1 \xi + n_2 \eta. \end{array} \right\} \tag{4.112}$$

Also

$$\left. \begin{array}{l} \dot{x} = l_1 \dot{\xi} + l_2 \dot{\eta} \\ \dot{y} = m_1 \dot{\xi} + m_2 \dot{\eta} \\ \dot{z} = n_1 \dot{\xi} + n_2 \dot{\eta}. \end{array} \right\} \tag{4.113}$$

102

From triangles $A_1 \Upsilon N$, $A_1 BN$ and $A_1 KN$ we have

$$\left. \begin{aligned} l_1 &= \cos A_1 \Upsilon = \cos \Omega \cos \omega - \sin \Omega \sin \omega \cos i \\ m_1 &= \cos A_1 B = \sin \Omega \cos \omega + \cos \Omega \sin \omega \cos i \\ n_1 &= \cos A_1 K = \sin \omega \sin i. \end{aligned} \right\} \tag{4.114}$$

From triangles $D \Upsilon N$, DBN and DKN we have

$$\left. \begin{aligned} l_2 &= \cos D \Upsilon = -\cos \Omega \sin \omega - \sin \Omega \cos \omega \cos i \\ m_2 &= \cos DB = -\sin \Omega \sin \omega + \cos \Omega \cos \omega \cos i \\ n_2 &= \cos DK = \cos \omega \sin i. \end{aligned} \right\} \tag{4.115}$$

Hence, for a given set of elements and the time, the coordinates (x, y, z) and the velocity components $(\dot{x}, \dot{y}, \dot{z})$ can be computed.

For example, in the case of elliptic motion we have

$$\xi = r \cos f = a(\cos E - e)$$

$$\eta = r \sin f = a(1 - e^2)^{1/2} \sin E$$

giving

$$x = a l_1 \cos E + b l_2 \sin E - a e l_1$$

$$y = a m_1 \cos E + b m_2 \sin E - a e m_1$$

$$z = a n_1 \cos E + b n_2 \sin E - a e n_1.$$

Also,

$$\dot{x} = \frac{na}{r} (b l_2 \cos E - a l_1 \sin E)$$

$$\dot{y} = \frac{na}{r} (b m_2 \cos E - a m_1 \sin E)$$

$$\dot{z} = \frac{na}{r} (b n_2 \cos E - a n_1 \sin E).$$

We now consider the reverse problem, namely the derivation of a set of elements from a given position and velocity at a given time.

Let the position have coordinates (x, y, z) and the velocity components be $(\dot{x}, \dot{y}, \dot{z})$ at the time.

Then

$$r^2 = x^2 + y^2 + z^2 \tag{4.116}$$

$$V^2 = \dot{x}^2 + \dot{y}^2 + \dot{z}^2. \tag{4.117}$$

If $\mathbf{i}$, $\mathbf{j}$ and $\mathbf{k}$ are unit vectors along $S\Upsilon$, SB and SK, then

$$\mathbf{r} = \mathbf{i}x + \mathbf{j}y + \mathbf{k}z$$

$$\mathbf{V} = \mathbf{i}\dot{x} + \mathbf{j}\dot{y} + \mathbf{k}\dot{z}.$$

Hence

$$\mathbf{r} \times \mathbf{V} = \mathbf{h}$$

where the components of **h** are given by

$$x\dot{y}-y\dot{x}=h_z$$

$$y\dot{z}-z\dot{y}=h_x$$

$$z\dot{x}-x\dot{z}=h_y.$$

h_x, h_y and h_z being the constants of angular momentum in the yz, zx and xy planes respectively.

Then

$$h^2=\mu p=h_x{}^2+h_y{}^2+h_z{}^2. \tag{4.118}$$

From (4.118) we obtain p, since μ is known.

The type of conic section the orbit follows is determined from the energy equation (4.110), namely

$$\frac{1}{2}V^2-\frac{\mu}{r}=C$$

by computing C and using equations (4.111). When the type of conic section has been found, the appropriate set of relations can be used.

Thus if the orbit is an ellipse, we have

$$V^2=\mu\left(\frac{2}{r}-\frac{1}{a}\right).$$

Hence a is obtained. Also

$$p=a(1-e^2)$$

and hence e is obtained. In addition, projecting h on to the three planes xy, yz and zx, we obtain

$$h\cos i=h_z \tag{4.119}$$

$$h\sin i\sin \Omega=\pm h_x \tag{4.120}$$

$$h\sin i\cos \Omega=\mp h_y \tag{4.121}$$

giving

$$\tan \Omega=(-h_x/h_y)$$

and

$$\cos i=\frac{h_z}{(h_x{}^2+h_y{}^2+h_z{}^2)^{1/2}}=\frac{h_z}{h}.$$

Hence (4.119), (4.120) and (4.121) give i and Ω, the upper or lower sign being taken in equations (4.120) and (4.121) according to whether i is less than or greater than 90° (i.e. h_z is positive or negative).

By equations (2.4), (2.5) and (2.6),

$$\sin (\omega+f)=\frac{z}{r}\operatorname{cosec} i$$

and

$$\cos (\omega+f)=\frac{1}{r}(x\cos \Omega+y\sin \Omega)$$

giving $(\omega+f)$ unambiguously.

If $i=0$, the equations used are

$$\sin(\omega+f)=\frac{1}{r}(y \cos \Omega - x \sin \Omega)$$

and

$$\cos(\omega+f)=\frac{1}{r}(x \cos \Omega + y \sin \Omega)$$

again giving $(\omega+f)$.

But from

$$r=\frac{h^2/\mu}{1+e \cos f}$$

we can compute f and hence ω is obtained.

There remains to be found the time of perihelion passage τ.
In the elliptic case the eccentric anomaly E is obtained from

$$\tan\frac{E}{2}=\left(\frac{1-e}{1+e}\right)^{1/2}\tan\frac{f}{2} \quad \text{or} \quad r=a(1-e \cos E).$$

But

$$M=n(t-\tau)=E-e \sin E$$

giving τ, since t, n, E and e are known.

In the hyperbolic case the procedure is similar, equations (4.100) and (4.101) or (4.105) being used. In the parabolic case equation (4.82) is used.

4.12 The f and g Series

The equation

$$\frac{d^2\mathbf{r}}{dt^2}+\mu\frac{\mathbf{r}}{r^3}=0 \qquad (4.122)$$

where $\mu=G(m_1+m_2)$ may be solved in a time series, the coefficients of the various powers of time being functions of the constants μ, $\mathbf{r}_0$ and $(d\mathbf{r}/dt)_0$, the last two being the values of $\mathbf{r}$ and $d\mathbf{r}/dt$ at $t=0$.

We first introduce τ as an independent variable, where

$$\tau=\mu^{1/2}t.$$

Then equation (4.122) becomes

$$\frac{d^2\mathbf{r}}{d\tau^2}+\frac{\mathbf{r}}{r^3}=0. \qquad (4.123)$$

To obtain a series we differentiate equation (4.123) to obtain the higher derivatives, and use (4.123) to eliminate all derivatives of $\mathbf{r}$ higher than $d\mathbf{r}/d\tau$

from the right-hand sides. The values of $\mathbf{r}$ and $d\mathbf{r}/d\tau$ at $\tau=0$ are then inserted. We thus obtain

$$\mathbf{r}_0 = \mathbf{r}_0 \qquad (\dot{\mathbf{r}}_0 = \dot{\mathbf{r}}_0)$$

$$\ddot{\mathbf{r}}_0 = -\frac{\mathbf{r}_0}{r_0{}^3}$$

$$\dddot{\mathbf{r}}_0 = -\left(\frac{\dot{\mathbf{r}}_0}{r_0{}^3} - \frac{3\mathbf{r}_0}{r_0{}^4}\frac{\dot{\mathbf{r}}_0 \cdot \mathbf{r}_0}{r_0}\right)$$

$$\ddddot{\mathbf{r}}_0 = -\left\{\left[\frac{2\mu}{r_0{}^6} - \frac{3\dot{\mathbf{r}}_0 \cdot \dot{\mathbf{r}}_0}{r_0{}^5} + \frac{15(\dot{\mathbf{r}}_0 \cdot \mathbf{r}_0)^2}{r_0{}^7}\right]\mathbf{r}_0 - 6\frac{\dot{\mathbf{r}}_0 \cdot \mathbf{r}_0}{r_0{}^5}\dot{\mathbf{r}}_0\right\}$$

and so on, where $\dot{\mathbf{r}}_0 \equiv (d\mathbf{r}/d\tau)_0$ etc.

Now define constants s, u and w by

$$s = \frac{\dot{\mathbf{r}}_0 \cdot \mathbf{r}_0}{r_0{}^2}; \qquad u = \frac{1}{r_0{}^3}; \qquad w = \frac{\dot{\mathbf{r}}_0 \cdot \dot{\mathbf{r}}_0}{r_0{}^2}.$$

It is then seen that the Taylor series

$$\mathbf{r} = \mathbf{r}_0 + \dot{\mathbf{r}}_0\tau + \frac{1}{2!}\ddot{\mathbf{r}}_0\tau^2 + \frac{1}{3!}\dddot{\mathbf{r}}_0\tau^3 + \frac{1}{4!}\ddddot{\mathbf{r}}_0\tau^4 + \dots$$

where the coefficients of the powers of τ are given by

$$\mathbf{r}_0 = \mathbf{r}_0$$

$$\dot{\mathbf{r}}_0 = \dot{\mathbf{r}}_0$$

$$\tfrac{1}{2}\ddot{\mathbf{r}}_0 = -\tfrac{1}{2}u\mathbf{r}_0$$

$$\tfrac{1}{6}\dddot{\mathbf{r}}_0 = \tfrac{1}{2}us\mathbf{r}_0 - \tfrac{1}{6}u\dot{\mathbf{r}}_0$$

$$\tfrac{1}{24}\ddddot{\mathbf{r}}_0 = \tfrac{1}{24}u(3w - 2u - 15s^2)\mathbf{r}_0 + \tfrac{1}{4}us\dot{\mathbf{r}}_0$$

and so forth is the solution of the equation.

In fact, we may write

$$\mathbf{r} = f\mathbf{r}_0 + g\dot{\mathbf{r}}_0,$$

where

$$f = 1 - \frac{1}{2}u\tau^2 + \frac{1}{2}us\tau^3 + \frac{u}{24}(3w - 2u - 15s^2)\tau^4 - \frac{1}{8}us(3w - 2u - 7s^2)\tau^5 + \dots$$

and

$$g = \tau - \frac{1}{6}u\tau^3 + \frac{1}{4}us\tau^4 + \frac{1}{120}u(9w - 8u - 45s^2)\tau^5 + \dots$$

correct to order τ^5.

If τ is small, the f and g series converge rapidly and can be very useful, for example in the determination of orbits (chapter 13). Since equation (4.12) is nonlinear, however, the higher coefficients of τ become cumbersome. The use of the series is therefore restricted to values of τ so small that the higher terms may be neglected. It may be remarked however that Sconzo $et\ al$ (1965)

106

have given explicit expressions for the f and g coefficients up to τ^{27} by using a formal symbol manipulation on a computer.

In using the series it must be remembered that τ is in a time scale such that $\mu = 1$.

In sections 5.5 and 5.6 it was seen that Kepler's equation could be solved by an iterative numerical procedure or by an analytical procedure, producing the so-called equation of the centre. There similarly exist numerical procedures enabling the values of the higher coefficients of τ to be found without explicit knowledge of their analytical form. When a computer is available it is advantageous to use one of these methods, which are known as recurrence relation procedures.

4.13 The Use of Recurrence Relations

Steffenson (1956, 1957) suggested and applied a procedure which allowed the recursive calculation of the derivatives needed to use a Taylor's series. Various versions of this procedure have been used by several authors. The original equation of motion is modified by the introduction of auxiliary variables, so chosen that the equation and the differential equations of the auxiliary variables are quadratic on their right-hand sides.

Thus, if we introduce the set of variables (only one of several possible sets) $u = r^{-3}$, $w = r^{-2}$, $\sigma = ws$ and $s = \mathbf{r} \cdot \dot{\mathbf{r}}$ it is readily seen that equation (4.123) may be reduced to the following set

$$
\left.
\begin{aligned}
\ddot{\mathbf{r}} &= -r u \\
\dot{u} &= -3 u \sigma \\
\dot{w} &= -2 u \sigma \\
s &= \mathbf{r} \cdot \dot{\mathbf{r}} \\
\sigma &= w s.
\end{aligned}
\right\} \tag{4.124}
$$

The right-hand sides of these equations are all of quadratic form. Substituting the infinite series

$$
\left.
\begin{aligned}
\mathbf{r} &= \sum_{i=0}^{\infty} \mathbf{r}_i \tau^i \\
u &= \sum_{i=0}^{\infty} u_i \tau^i \\
w &= \sum_{i=0}^{\infty} w_i \tau^i \\
s &= \sum_{i=0}^{\infty} s_i \tau^i \\
\sigma &= \sum_{i=0}^{\infty} \sigma_i \tau^i
\end{aligned}
\right\} \tag{4.125}
$$

into equation (4.124) and equating the constant coefficients of powers of τ, we obtain the set of recurrence relations

$$
\left.
\begin{aligned}
r_{j+2} &= -\frac{1}{(j+1)(j+2)}\sum_{i=0}^{j} r_i u_{j-i} \\
u_j &= -\frac{3}{j}\sum_{i=0}^{j-1} u_i \sigma_{j-i-1} \\
w_j &= -\frac{2}{j}\sum_{i=0}^{j-1} w_i \sigma_{j-i-1} \\
s_j &= \sum_{i=0}^{j} (i+1) r_{i+1} \cdot r_{j-1} \\
\sigma_j &= \sum_{i=0}^{j} w_i s_{j-i}.
\end{aligned}
\right\}
\qquad (4.126)
$$

From the initial conditions of position and velocity, starting values of u, w, s and σ are obtained. From the set of equations (4.126), the higher derivatives of u, w, s, σ and r may be computed step by step. Although the procedure may seem cumbersome and time-consuming, it is far more efficient in practice on a computer than getting it to evaluate the increasingly complicated explicit expressions for the higher-order terms in the f and g series.

A notable improvement in the set of recurrence relations (4.124) is obtained as follows.

Let

$$ ur^n = 1 $$

where n is a positive integer, $n > 0$.

Then differentiating, we obtain after a little reduction,

$$ \dot{u}r^2 = -n\, us $$

where $s = r_0 \cdot \dot{r}$ as before.

A further differentiation provides

$$ \ddot{u}r^2 = -(n+2)\dot{u}s - n\, u\dot{s} \qquad\qquad \dot{s} = r \cdot \ddot{r} + \dot{r} \cdot \dot{r}. $$

The process may be continued in the same way as earlier, using the relevant infinite series. The advantages are (i) the reduction of the number of auxiliary variables from four to two (ii) the generalization of the integral power of r from 3 to n which is useful when a potential such as the Earth's is expanded in a series which involves a number of powers of r. The appearance of r in each of the successive derivatives of u is a minor disadvantage easily overcome.

For more information the student should consult the work by Herrick (1971, 1972) or the series of papers (Roy *et al* 1972, Moran 1973, Roy and Moran 1973, Moran *et al* 1973, Emslie and Walker 1979).

4.14 Universal Variables

It has been seen in this chapter that special sets of formulae exist for elliptic, parabolic and hyperbolic motion as well as for the three corresponding cases of rectilinear motion. Even in elliptic motion itself a number of the formulae break down when the eccentricity approaches zero (i.e. when the orbit tends to a circle). In section 4.12 for example, in deriving the orbital elements from a given position and velocity, the equation

$$r = \frac{h^2/\mu}{(1+e\cos f)}$$

cannot be used in the circular case to obtain f from a knowledge of r, h and μ. In the circular case $e=0$ and there is no perihelion or time of perihelion passage. Even if e is slightly greater than zero, the use of orthodox elliptic formulae would lead to very inaccurate determination of e, w and τ.

In a different context, the same problem exists when the inclination tends to zero; in that case the longitude of the ascending node Ω becomes indeterminate and other formulae must be used to overcome this problem (see section 4.11, also section 7.5).

Various attempts have been made to provide sets of universal or unified formulae that can be used with all kinds of two-body conic-section orbital motion, the distinction between the universal and unified sets being that the former can be applied even if e tends to zero whereas the latter cannot. It is not within the scope of this work to describe these attempts. The student should refer to Herrick (1971, 1972) for a full discussion of universal and unified variables and parameters.

Problems

Take the necessary data from the appendices.

4.1 From equations (4.17) and (4.19) derive the equation

$$\frac{d^2u}{d\theta^2} + u = \frac{\mu}{h^2}.$$

4.2 Halley's comet moves in an elliptical orbit of eccentricity 0·9673. Compare its velocities, both linear and angular, at perihelion and aphelion.

4.3 Obtain the equation of the centre from the series (4.63) and (4.64), correct to O (e^3).

4.4 Find the perihelion distance of that comet which, moving in a parabolic orbit in the plane of the ecliptic, remains the longest time within the Earth's orbit (assumed circular).

4.5 Prove that the mean anomaly M and the true anomaly f in elliptic motion are related by the equation

$$M = (1-e^2)^{3/2} \int_0^f \frac{df}{(1+e\cos f)^2}$$

Hence deduce that, to O (e^2),

$$M = f - 2e\sin f + \tfrac{3}{4} e^2 \sin 2f$$

$$f = M + 2e\sin M + \tfrac{5}{4} e^2 \sin 2M.$$

4.6 A space vehicle is moving in an elliptical orbit of period T under the attraction of the Sun, mass M. The motors are fired momentarily so that its orbital speed V is suddenly increased by the increment ΔV. Show that the resulting change ΔT in period is given by

$$\Delta T = 3(2\pi GM)^{-2/3}T^{5/3}V\Delta V.$$

4.7 A minor planet is moving in an orbit of eccentricity 0·21654 and period 4·3856 years. Calculate the eccentric anomaly 1·2841 years after perihelion passage, correct to 1′ of arc.

4.8 A rocket leaves the Earth's atmosphere just before burn-out (thrust terminated), which occurs at a height of 640 km. At this instant its geocentric velocity is 10·4 km s^{-1}. In what direction must it be travelling to achieve maximum distance from the Earth's centre? Calculate this distance. If the direction of travel of the rocket at burn-out has made an angle of 88° with the geocentric radius vector of the rocket, calculate the period of the rocket's orbit.

4.9 When first injected into orbit, artificial Earth satellite Sputnik 16 had a semimajor axis of 1·0478 Earth radii and a period of 90·54 minutes. Calculate the mass of the Earth in units of the Sun's mass.

4.10 On January 10·0 1963, the heliocentric ecliptic rectangular coordinates of position and velocity of an interplanetary probe were $x=0·68$, $y=0·52$, $z=0·18$ and $\dot{x}=-2·2$, $\dot{y}=28·1$, $\dot{z}=2·6$ respectively; the distance being measured in units of the Earth's semimajor axis, the velocity in km s^{-1}. Find the elements of the Earth's orbit.

References

Astrand J J 1890 *Huelftafeln zur Leichten und Genauen Aufloesung des Keplerischen Problems* (*Auxiliary Tables for Simple and Accurate Solution of Kepler's Problems*) (Leipzig: Engelmann)†

Bauschinger J 1901 *Tafeln zur Theoretischen Astronomie* (*Tables on Theoretical Astronomy*) (Leipzig: Engelmann)†

Cayley A 1861 *Mem. R. Astron. Soc.* **29** 191†

Emslie A G and Walker I W 1979 *Cel. Mech.* **19** 147

Herrick S 1953 *Tables for Rocket and Comet Orbits* AMS **20** (Washington : National Bureau of Standards)

—— 1971, 1972 *Astrodynamics* vols 1 and 2 (London: Van Nostrand)‡

Moran P E 1973 *Cel. Mech.* **7** 122

Moran P E, Roy A E and Black W 1973 *Cel. Mech.* **8** 405

Moulton F R 1914 *An Introduction to Celestial Mechanics* (New York: Macmillan)

Roy A E and Moran P E 1973 *Cel. Mech.* **7** 236

Roy A E, Moran P E and Black W 1972 *Cel. Mech.* **6** 468

Schlesinger F and Udick S 1912 *Tables for the True Anomaly in Elliptic Orbits* **2** No. 17 (Publications of the Allegheny Observatory)†

Sconzo P, Le Shak A R and Tobey R 1965 *Astron. J.* **70** 269

Smart W M and Green R M 1977 *Textbook on Spherical Astronomy* (London: Cambridge University Press)

Steffensen J F 1956 *K. Danske Vidensk. Selsk. Mat.-Fys. Meddr* **30** number 18

—— 1957 *K. Danske Vidensk. Selsk. Mat.-Fys. Meddr* **31** number 3

Stracke G 1928 *Tafeln der Elliptischen Koordinaten C=(r/a) cos v und S=(r/a) sin v fuer Exzentrizitaetswinkel von 0° bis 25°* (*Tables of the Elliptical Coordinates C=(r/a) cos v and S=(r/a) sin v for Eccentricity angles from 0° to 25°*) (Berlin: Veroeffentlichen des Astronomisches Recheninstituts)†

Watson J C 1892 *Theoretical Astronomy* (Philadelphia: Lippincott)

† These references contain tables for the solution of two-body motion.
‡ This reference contains, among other tables, those for the solution of Barker's equation.

5 The Many-Body Problem

5.1 Introduction

The many-body problem was first formulated precisely by Newton. In its form where the objects involved are point masses it may be stated as follows: Given at any time the positions and velocities of three or more massive particles moving under their mutual gravitational forces, the masses also being known, calculate their positions and velocities for any other time.

The problem is more complicated when the bodies' shapes and internal constitutions have to be taken into account as in the Earth–Moon–Sun problem. The point-mass many-body problem has inspired (and frustrated!) many eminent astronomers and mathematicians in the last three centuries. It is perhaps not obvious that even the three-body problem is of a much higher degree of complexity than the two-body problem. If we consider, however, that each body is subject to a complicated variable gravitational field due to its attraction by the other two such that close encounters with either may be brought about, the result of each near-collision being an entirely new type of orbit, we see that it would require a general formula of unimaginable complexity to describe all the consequences of all such encounters.

In point of fact, several general and useful statements may be made concerning the many-body problem, such statements being embodied in the ten known integrals of the motion. These integrals were known to Euler; since then no further integrals have been discovered or are likely to be. In addition, particular solutions of the three-body problem were found by Lagrange which are of interest in astrodynamics as well as in astronomy. These solutions exist when certain relationships hold among the initial conditions.

Further progress has been mainly in studying special problems where approximations of various kinds may be utilized. For example, in the circular-restricted three-body problem, two massive particles move in undisturbed circular orbits about their common centre of mass while they attract a particle of mass so small that it cannot appreciably affect their circular orbits. It is possible to draw certain conclusions about the resulting orbit of the particle of infinitesimal mass and to establish the existence of families of periodic orbits of this test particle. Many of Poincaré's epoch-making researches were devoted to this problem; one of immediate interest when we consider that the Earth, the Moon and a space vehicle in Earth–Moon space constitute an approximate example of this three-body case.

111

It has also been seen that the planets move in almost perfectly elliptical orbits about the Sun, since the mutual attraction between the planets is so much smaller than the Sun's attraction upon them. This two-body approximation has been the starting point in many attempts to obtain theories of the planets' motions. In the two-body solution (termed the reference orbit) the elements are constant; if they are now supposed to vary because of the mutual gravitational attractions of the planets, their differential equations may be set up and solved. The resulting expressions for the elements (in general long sums of sines, cosines and secular terms) can be used to obtain a more accurate approximation still. In practice this method is rapidly convergent though laborious, it being only rarely necessary to go beyond the third approximation. Such analytical expressions, valid for a given period of time, are called *general perturbations*. They enable some deductions to be made regarding the past and future states of the planetary system though it must be emphasized that no results valid for an arbitrarily long time may be obtained in this way. The method of general perturbations has also been applied to satellite systems, to asteroids disturbed by Jupiter, and to the orbits of artificial satellites. It is in fact a powerful tool in astrodynamics since the analytical expressions clearly exhibit the various forces at work (for example, the oblateness effect of the Earth on a satellite).

A different approach to the many-body problem is that of using *special perturbations*, a tool which most workers in celestial mechanics before the days of high-speed computers shrank away from, since it involved the step-by-step numerical integration of the differential equations of motion from the initial epoch to the epoch at which the bodies' positions were desired. Its great advantage, however, is that it is applicable to any orbit involving any number of bodies, and nowadays special perturbations are applied to all sorts of astrodynamical problems, especially since many of these problems fall into regions in which special perturbation theories are absent. One such case is that of a lunar circumnavigation, where the orbit of the vehicle in the Earth–Moon field can be adequately treated only by special perturbations. The main disadvantage of this method is that it rarely leads to any general formulae; in addition, though they may be of no interest to the worker, the body's positions at all intermediate steps must be computed in order to arrive at the final configuration.

Perturbations may also be divided into two further classes; *periodic* and *secular*. Any disturbance of the reference orbit that is repeated with a given period of revolution is termed a periodic perturbation and is usually the result of recurrent similar configurations of the bodies involved. Since these are unlikely to occur exactly, such a periodic perturbation (a short-period one) is often bound up with cyclic behaviour of a much longer period so that one speaks of a long-period perturbation.

A secular perturbation causes a change proportional to the time; for example, the advance of perihelion or the retrogression of the ascending node of a planetary orbit. In many cases it is difficult to distinguish between very long-period perturbations and secular perturbations since the time over which observations have been made is short compared with the suspected long period.

Finally, we should note that a distinction should be made in the n-body problem between the few-body and the many-body problem. In the Solar System we are concerned with the few-body problem where orbits have to be calculated precisely and too few bodies are involved to enable statistical or hydrodynamical approaches to be tried. In a stellar system we have a many-body problem, allowing us to utilize such methods. A description of them is however retained till a later chapter.

5.2 The Equations of Motion in the Many-Body Problem

We now set up the equations of motion of n massive particles of masses m_i $(i=1, 2, \ldots n)$ whose radius vectors from an unaccelerated point O are $\mathbf{R}_i$, while their mutual radius vectors are given by $\mathbf{r}_{ij}$ where

$$\mathbf{r}_{ij} = \mathbf{R}_j - \mathbf{R}_i. \tag{5.1}$$

From Newton's laws of motion and the law of gravitation, we therefore have

$$m_i \ddot{\mathbf{R}}_i = G \sum_{j=1}^{n} \frac{m_i m_j}{r_{ij}^3} \mathbf{r}_{ij}. \qquad (j \neq i, \quad i=1, 2, \ldots n) \tag{5.2}$$

It is to be noted that $\mathbf{r}_{ij}$ implies that the vector between m_i and m_j is directed from m_i to m_j.

Thus

$$\mathbf{r}_{ij} = -\mathbf{r}_{ji}. \tag{5.3}$$

The set of equations (5.2) are the required equations of motion, G being the constant of gravitation.

5.3 The Ten Known Integrals and Their Meanings

Summing the equations (5.2) and using (5.3) we obtain

$$\sum_{i=1}^{n} m_i \ddot{\mathbf{R}}_i = 0.$$

Integrating twice gives

$$\sum_{i=1}^{n} m_i \dot{\mathbf{R}}_i = \mathbf{a} \tag{5.4}$$

and

$$\sum_{i=1}^{n} m_i \mathbf{R}_i = \mathbf{a}t + \mathbf{b}. \tag{5.5}$$

Now by definition the centre of mass of the system has a radius vector $\mathbf{R}$ where

$$M\mathbf{R} = \sum_{i=1}^{n} m_i \mathbf{R}_i$$

and

$$M = \sum_{i=1}^{n} m_i.$$

Hence by equations (5.4) and (5.5),

$$\mathbf{R} = (\mathbf{a}t + \mathbf{b})/M \qquad (5.6)$$

and

$$\dot{\mathbf{R}} = \mathbf{a}/M. \qquad (5.7)$$

Relations (5.6) and (5.7) state that the centre of mass of the system moves through space with constant velocity.

If (5.6) and (5.7) are resolved with respect to a set of three unaccelerated rectangular axes through O, we obtain six constants of integration a_x, a_y, a_z, b_x, b_y and b_z.

Taking the vector product of $\mathbf{R}_i$ and $\ddot{\mathbf{R}}_i$ for each of the set (5.2) and summing, we obtain

$$\sum_{i=1}^{n} m_i \mathbf{R}_i \times \ddot{\mathbf{R}}_i = G \sum_{i=1}^{n} \sum_{j=1}^{n} \frac{m_i m_j}{r_{ij}^3} \mathbf{R}_i \times \mathbf{r}_{ij} \qquad j \neq i. \qquad (5.8)$$

Now

$$\mathbf{R}_i \times \mathbf{r}_{ij} = \mathbf{R}_i \times (\mathbf{R}_j - \mathbf{R}_i) = \mathbf{R}_i \times \mathbf{R}_j.$$

Also

$$\mathbf{R}_j \times \mathbf{r}_{ji} = \mathbf{R}_j \times \mathbf{R}_i = -\mathbf{R}_i \times \mathbf{R}_j.$$

Hence the right-hand side of (5.8) reduces in pairs to zero, giving

$$\sum_{i=1}^{n} m_i \mathbf{R}_i \times \ddot{\mathbf{R}}_i = 0.$$

Integrating we obtain

$$\sum_{i=1}^{n} m_i \mathbf{R}_i \times \dot{\mathbf{R}}_i = \mathbf{C}. \qquad (5.9)$$

Equation (5.9) states that the sum of the moments of momenta or angular momenta of the masses in the system is a constant. The constant vector $\mathbf{C}$ defines a plane called the *invariable plane* of Laplace. It has been suggested that this fixed plane should be used in the planetary system as a fundamental reference plane instead of the plane of the ecliptic but, although the accuracy of our knowledge of its position is high, it is not such as to justify this change. At present it is inclined at about one and a half degrees to the plane of the ecliptic and lies between the orbital planes of Jupiter and Saturn, the two most massive bodies among the planets.

If relation (5.9) is resolved with respect to the set of unaccelerated rectangular axes through O, the following three 'integrals of area' are obtained:

$$\sum_{i=1}^{n} m_i(x_i\dot{y}_i - y_i\dot{x}_i) = C_1$$

$$\sum_{i=1}^{n} m_i(y_i\dot{z}_i - z_i\dot{y}_i) = C_2$$

$$\sum_{i=1}^{n} m_i(z_i\dot{x}_i - x_i\dot{z}_i) = C_3$$

where

$$C^2 = C_1^2 + C_2^2 + C_3^2$$

giving three more constants of integration C_1, C_2, C_3 to add to the six already obtained. Thus the sums of the angular momenta of the n masses about each of the axes of reference are constants.

The tenth constant is obtained by taking the scalar product of $\dot{\mathbf{R}}_i$ with equation (5.2) in i and summing over all i. Then

$$\sum_{i=1}^{n} m_i \dot{\mathbf{R}}_i \cdot \ddot{\mathbf{R}}_i = G \sum_{i=1}^{n} \sum_{j=1}^{n} \frac{m_i m_j}{r_{ij}^3} \dot{\mathbf{R}}_i \cdot \mathbf{r}_{ij}. \quad (j \neq i) \qquad (5.10)$$

Now

$$\dot{\mathbf{R}}_i \cdot \mathbf{r}_{ij} = \dot{\mathbf{R}}_i \cdot (\mathbf{R}_j - \mathbf{R}_i) \qquad (5.11)$$

while

$$\dot{\mathbf{R}}_j \cdot \mathbf{r}_{ji} = \dot{\mathbf{R}}_j \cdot (\mathbf{R}_i - \mathbf{R}_j). \qquad (5.12)$$

Adding (5.11) and (5.12) we have

$$\dot{\mathbf{R}}_i \cdot \mathbf{r}_{ij} + \dot{\mathbf{R}}_j \cdot \mathbf{r}_{ji} = -(\dot{\mathbf{R}}_j - \dot{\mathbf{R}}_i) \cdot (\mathbf{R}_j - \mathbf{R}_i).$$

Hence, using equation (5.1), equation (5.10) integrates to give

$$\frac{1}{2} \sum_{i=1}^{n} m_i \dot{\mathbf{R}}_i \cdot \dot{\mathbf{R}}_i - \frac{1}{2} G \sum_{i=1}^{n} \sum_{j=1}^{n} \frac{m_i m_j}{r_{ij}} = E. \quad (j \neq i) \qquad (5.13)$$

Now the velocity of the ith mass is V_i, where

$$V_i^2 = \dot{\mathbf{R}}_i \cdot \dot{\mathbf{R}}_i$$

Also, by putting

$$U = \frac{1}{2} G \sum_{i=1}^{n} \sum_{j=1}^{n} \frac{m_i m_j}{r_{ij}}$$

equation (5.13) becomes

$$T - U = E \qquad (5.13)$$

where

$$T = \frac{1}{2} \sum_{i=1}^{n} m_i V_i^2.$$

The first term in equation (5.13) (namely T) is the kinetic energy of the system while $-U$ is its potential energy. Hence (5.13) states that the total energy of the system of n particles is a constant E, which is the tenth constant

115

of integration. Thus while neither the total kinetic energy nor the total potential energy of the system is constant and there is a continual 'trade-off' among the bodies of kinetic energy and potential energy, the total energy remains invariant with time. Systems of constant total energy, to which the present system belongs, are called *conservative systems*.

No further integrals have ever been discovered. Indeed Bruns and Poincaré proved that apart from the energy integral, the integrals of area and the centre-of-mass integrals, no other integrals of the many-body problem exist that give equations involving only algebraic or integral functions of the co-ordinates and velocities of the bodies valid for all masses, and which satisfy the equations of motion.

5.4 The Force Function

We consider more closely in this section the function U defined by

$$U = \frac{1}{2} G \sum_{i=1}^{n} \sum_{j=1}^{n} \frac{m_i m_j}{r_{ij}}. \qquad (i \neq j)$$

A symmetrical function of all the masses and their mutual distance apart, neither time nor the particles' radius vectors from the origin enter U explicitly. It is indeed these properties of U that enable the ten integrals to be obtained. The first nine integrals result from the property that U is invariant with respect to rotations of the axes or translations of the origin. The energy integral arises because U does not contain the time explicitly (though it is of course a function of time through the r_{ij}).

If we introduce the unit vectors **i**, **j** and **k** along the axes Ox, Oy and Oz, then the gradient of U is given by

$$\nabla U \equiv \operatorname{grad} U = \mathbf{i}\frac{\partial U}{\partial x} + \mathbf{j}\frac{\partial U}{\partial y} + \mathbf{k}\frac{\partial U}{\partial z}.$$

The symbol ∇ (pronounced 'nabla' or 'del') denotes the grad operator where

$$\nabla \equiv \operatorname{grad} = \mathbf{i}\frac{\partial}{\partial x} + \mathbf{j}\frac{\partial}{\partial y} + \mathbf{k}\frac{\partial}{\partial z}.$$

And since

$$\ddot{\mathbf{R}}_i = \mathbf{i}\ddot{x}_i + \mathbf{j}\ddot{y}_i + \mathbf{k}\ddot{z}_i$$

it is seen that for the particle of mass m_i

$$m_i \ddot{\mathbf{R}}_i = \operatorname{grad}_i U \qquad (5.14)$$

where

$$\operatorname{grad}_i U = \mathbf{i}\frac{\partial U}{\partial x_i} + \mathbf{j}\frac{\partial U}{\partial y_i} + \mathbf{k}\frac{\partial U}{\partial z_i}.$$

Hence, equating coefficients of the unit vectors,

$$
\left.
\begin{aligned}
m_i \ddot{x}_i &= \frac{\partial U}{\partial x_i} \\[6pt]
m_i \ddot{y}_i &= \frac{\partial U}{\partial y_i} \\[6pt]
m_i \ddot{z}_i &= \frac{\partial U}{\partial z_i}.
\end{aligned}
\right\}
\tag{5.15}
$$

The set of equations (5.15) are the equations of motion of the particle of mass m_i in rectangular coordinates; U is consequently called the force function because the partial derivatives of U with respect to the coordinates give the components of the forces acting on the particles.

We now show that the potential energy of the system is indeed $-U$.

Let the particles be so situated that there is an infinite distance between any two of them. Suppose the mass m_1 is fixed with radius vector $\mathbf{R}_1$. Let mass m_2 be moved from infinity to position $\mathbf{R}_2$ along a path s. Then if at any point on the path the force required to move the particle along a small element of the curve ds is $\mathbf{F}$, the work done is

$$\mathbf{F} \cdot ds.$$

The total work done is the line integral

$$\int_s \mathbf{F} \cdot ds.$$

But if $\mathbf{F}$ is the gravitational attraction of m_1 on m_2, then

$$\mathbf{F} = \nabla U_2$$

where

$$U_2 = G \frac{m_1 m_2}{r_{12}}.$$

Hence

$$\int \mathbf{F} \cdot ds = \int_s \nabla U_2 \cdot ds$$

where

$$ds = \mathbf{i}\,dx + \mathbf{j}\,dy + \mathbf{k}\,dz.$$

Thus

$$
\begin{aligned}
\int_s \mathbf{F} \cdot ds &= \int_{R_\infty}^{R_1} \left(\frac{\partial U_2}{\partial x}\,dx + \frac{\partial U_2}{\partial y}\,dy + \frac{\partial U_2}{\partial z}\,dz \right) \\[6pt]
&= U_2(\mathbf{R}_2) - U_2(\mathbf{R}_\infty) \\[6pt]
&= U_2(\mathbf{R}_2) \\[6pt]
&= G\frac{m_1 m_2}{r_{12}}
\end{aligned}
$$

since

$$U_2(\mathbf{R}_\infty)=G\frac{m_1m_2}{\infty}=0.$$

Consider now the particle m_3 brought to a position $\mathbf{R}_3$ by the forces of attraction of m_1 and m_2, supposed fixed at positions $\mathbf{R}_1$ and $\mathbf{R}_2$.

The total work done is W, given by

$$W=\int_S \mathbf{F}\cdot d\mathbf{s}=\int_S \nabla U_3\cdot d\mathbf{s}$$

where

$$U_3=G\left(\frac{m_1m_3}{r_{13}}+\frac{m_2m_3}{r_{23}}\right).$$

Thus by the previous argument the work done in bringing the particle m_3 to a position $\mathbf{R}_3$ is

$$W=U_3(\mathbf{R}_3)=G\left(\frac{m_1m_3}{r_{13}}+\frac{m_2m_3}{r_{23}}\right).$$

The total work done in assembling the three particles is therefore

$$U=\frac{1}{2}G\sum_{i=1}^{3}\sum_{j=1}^{3}\frac{m_im_j}{r_{ij}}.\qquad (j\neq i)$$

It is then obvious that for a system of n particles, the work done in assembling it so that the particles are brought to finite distances from each other is

$$U=\frac{1}{2}G\sum_{i=1}^{n}\sum_{j=1}^{n}\frac{m_im_j}{r_{ij}}.\qquad (j\neq i)$$

But the potential energy of the system is the work done in moving the system to a state of complete dispersion so that $-U$ is the potential energy.

5.5 The Virial Theorem

Let I be the moment of inertia of the system, defined by

$$I=\sum_{i=1}^{n}m_i\mathbf{R}_i^2.$$

If we differentiate twice with respect to time we obtain

$$\ddot{I}=2\sum_{i=1}^{n}m_i\dot{\mathbf{R}}_i^2+2\sum_{i=1}^{n}m_i\mathbf{R}_i\cdot\ddot{\mathbf{R}}_i$$

or

$$\ddot{I}=4T+2\sum_{i=1}^{n}\mathbf{R}_i\cdot\nabla_i U. \qquad (5.16)$$

Now

$$\sum_{i=1}^{n}\mathbf{R}_i\cdot\nabla_i U=\sum_{i=1}^{n}\left(x_i\frac{\partial U}{\partial x_i}+y_i\frac{\partial U}{\partial y_i}+z_i\frac{\partial U}{\partial z_i}\right).$$

118

Also U is a homogeneous function of all the coordinates of order -1. Hence by Euler's theorem,

$$\sum_{i=1}^{n} \mathbf{R}_i \cdot \nabla_i U = -U.$$

Hence equation (5.16) becomes

$$\ddot{I} = 4T - 2U.$$

But

$$T - U = C$$

so that

$$\ddot{I} = 2U + 4C = 2T + 2C.$$

Now

$$\ddot{I} = \frac{d^2}{dt^2} \sum_{i=1}^{n} m_i \mathbf{R}_i{}^2.$$

Both U and T are positive so that if C is positive, $\ddot{I}$ is positive and I increases indefinitely. If this is so, at least one of the particles will escape from the system. If no escape is to take place, C must be negative and such that $\ddot{I}$ is negative; but this by no means is sufficient to render the system stable.

5.6 The Mirror Theorem

The form of the equations of motion enables two more statements to be made. The first is the mirror theorem, stated as follows: if n point masses are acted upon by their mutual gravitational forces only, and at a certain epoch each radius vector from the centre of mass of the system is perpendicular to every velocity vector, then the orbit of each mass after that epoch is a mirror image of its orbit prior to that epoch. Such a configuration of radius and velocity vectors is called a *mirror configuration*.

The second statement is a corollary of the first: if n point masses are moving under their mutual gravitational forces only, their orbits are periodic if at two separate epochs a mirror configuration occurs. We may remark that the orbital motions of a system of bodies are periodic if, at periodic intervals of time, the same relative configuration of radius and velocity vectors occurs with no change of scale.

A rigorous proof of the mirror theorem (Roy and Ovenden 1955) is easy to supply if we note that in the equations of motion velocities do not appear. Thus if time were to be reversed the bodies would return along their previous paths. If a mirror configuration occurs at an epoch, each particle's orbit beyond the epoch is not only continuous with the orbit before the epoch, but the forces on it at any subsequent time 'reverse' the effect of the forces upon it at the corresponding times before the epoch.

There are only two possible mirror configurations:

(i) when all the point masses lie in a plane, all the velocity vectors being at right angles to the plane and therefore parallel to each other,

(ii) when all the point masses lie on a straight line, all the velocity vectors being at right angles to that line but not necessarily parallel to each other.

The proof of the periodicity statement is trivial; if mirror configurations A and B occur at $t = -t_0$ and $t = 0$, then A occurs again at $t = +t_0$, B at $t = +2t_0$ and so on. Hence the orbits are periodic, with period $2t_0$.

In fact the theorem, its periodicity corollary and the two distinct mirror configurations would appear, according to Marchal, to have been first formulated by Poincaré in the last part of the 19th century and subsequently rediscovered by Roy and Ovenden.

5.7 Reassessment of the Many-Body Problem

The set of differential equations of the many-body problem ($n = 3$ or more) is one of $3n$ second-order equations, so that $6n$ constants of integration are required to specify completely the behaviour of the particles. Of these $6n$ only ten have been found.

It is possible to reduce the order of the problem by using the ten integrals obtained; the origin may be transferred to the centre of mass of the system, and with the aid of the area integrals and the energy integral a set of equations of order $(6n - 10)$ results. If the time is eliminated by taking one of the other variables as the independent variable and use is made of the so-called elimination of the nodes (due to Jacobi), the problem may be reduced to order $(6n - 12)$. In spite of this, it is seen that even for the three-body problem there remains a set of equations to be solved of order 6. Though a general solution of the three-body problem was finally obtained in 1912 by Sundman, it is so complicated and the series obtained so slowly converging that it is useless for practical purposes.

It should be noted that the integrals of area and energy can be used to check numerical investigation of conservative systems. If a long numerical investigation is carried out, such as the calculation some years ago of the coordinates of the five outer planets for a period of one million years, the computation at intervals of the energy of the system (putting the calculated coordinates and momenta into the energy integral) will afford a means of sampling the accumulation of rounding-off error.

But to obtain further progress, recourse has had to be made to special or general perturbation methods. The possibility of developing satisfactory general perturbation theories hinges on a very important theorem by Cauchy, which states in essence that if at any time a set of point masses are at finite distances from each other, their differential equations possess a solution in the sense that the particles' coordinates and velocities may be represented by convergent series expansions for a finite time interval beyond that epoch.

Before describing such methods however, we consider the particular solutions in the problem of three bodies given by Lagrange. The treatment in the next section is based on a treatment of the problem by Danby (1962).

5.8 Lagrange's Solutions of the Three-Body Problem

There exist cases where the geometrical form of the three-body configuration does not change although the scale can change and the figure can rotate. In one case the three particles are at the vertices of an equilateral triangle; in the other case they are collinear. In 1772, Lagrange showed that three particles of arbitrary mass could exist in such solutions if the following conditions held:

(i) the resultant force on each mass passed through the centre of mass of the system,

(ii) this resultant force was directly proportional to the distance of each mass from the centre of mass, and

(iii) the initial velocity vectors were proportional in magnitude to the respective distances of the particles from the centre of mass and made equal angles with the radius vectors to the particles from the centre of mass.

The equations of motion of the three bodies are, from equations (5.1)–(5.3),

$$m_i\ddot{\mathbf{R}}_i = G \sum_{j=1}^{3} \frac{m_i m_j}{r_{ij}{}^3} \mathbf{r}_{ij} \qquad (j \neq i, \quad i = 1, 2, 3) \tag{5.17}$$

where

$$\mathbf{r}_{ij} = \mathbf{R}_j - \mathbf{R}_i, \quad \mathbf{r}_{ij} = -\mathbf{r}_{ji}$$

and the three masses are m_1, m_2 and m_3.

Using the six centre-of-mass integrals we may transfer the origin from which the radius vectors $\mathbf{R}_i$ are drawn to the centre of mass.

Then

$$\sum_{i=1}^{3} m_i \mathbf{R}_i = 0. \tag{5.18}$$

By equation (5.18), we obtain

$$(m_1 + m_2 + m_3)\mathbf{R}_1 + m_2(\mathbf{R}_2 - \mathbf{R}_1) + m_3(\mathbf{R}_3 - \mathbf{R}_1) = 0$$

or

$$M\mathbf{R}_1 = -m_2\mathbf{r}_{12} - m_3\mathbf{r}_{13} \tag{5.19}$$

where

$$M = m_1 + m_2 + m_3.$$

Squaring equation (5.19), we obtain

$$M^2 R_1{}^2 = m_2{}^2 r_{12}{}^2 + m_3{}^2 r_{13}{}^2 + 2m_2 m_3 \mathbf{r}_{12} \cdot \mathbf{r}_{13}. \tag{5.20}$$

If the shape of the configuration does not alter, the relative distances r_{12}, r_{23} and r_{31} are given by

$$\frac{r_{12}}{(r_{12})_0} = \frac{r_{23}}{(r_{23})_0} = \frac{r_{31}}{(r_{31})_0} = f(t) \tag{5.21}$$

where $(r_{ij})_0$ denotes the value of r_{ij} at $t=0$, the epoch when the particles are placed in the required configuration.

Also, if the angle between r_{12} and r_{13} in (5.20) is to be constant we must have

$$\theta_1 = \theta_2 = \theta_3 = \theta(t) \tag{5.22}$$

where θ_i is the angular velocity of the particle of mass m_i about the centre of mass.

However, by the angular momentum integral (5.9), the total angular momentum of the system about the origin is a constant vector **C**.

Then

$$\mathbf{C} = \sum_{i=1}^{3} m_i \mathbf{R}_i \times \dot{\mathbf{R}}_i. \tag{5.23}$$

Using equations (5.20), (5.21) and (5.22), we obtain

$$MR_1{}^2 = [f(t)]^2[m_2{}^2(r_{12})_0{}^2 + m_3{}^2(r_{13})_0{}^2 + 2m_2m_3(r_{12})_0(r_{13})_0 \cos \alpha_1] \tag{5.24}$$

where α_1, the angle between r_{12} and r_{13}, is constant.

Hence

$$R_1 = (R_1)_0 f(t)$$

or, in general,

$$R_i = (R_i)_0 f(t). \tag{5.25}$$

From (5.22) and (5.25) we find that

$$C = \sum_{i=1}^{3} [m_i(R_i)_0{}^2] f^2 \theta. \tag{5.26}$$

Relation (5.26), indicating that the angular momentum of each particle about the centre of mass is constant, shows that the force acting on each mass passes through the centre of mass.

If F_i is the force per unit mass acting on the mass m_i, its equation of motion is

$$m_i F_i = m_i(\ddot{R}_i - R_i \theta_i{}^2).$$

Then by (5.22) and (5.25),

$$m_i F_i = m_i[\ddot{f}(R_i)_0 - R_i \theta^2]$$

or

$$m_i F_i = R_i m_i[\ddot{f}/f - \theta^2].$$

Hence

$$F_1 : F_2 : F_3 = R_1 : R_2 : R_3. \tag{5.27}$$

We now consider the two cases that satisfy the above conditions. We have

$$\mathbf{R}_i \times \mathbf{F}_i = 0$$

or

$$\mathbf{R}_i \times \ddot{\mathbf{R}}_i = 0.$$

If we take the vector product of $\mathbf{R}_1$ with the left- and right-hand sides of (5.17), we obtain (when $i = 1$)

$$\mathbf{R}_1 \times \left(m_2 \frac{\mathbf{R}_2}{r_{12}{}^3} + m_3 \frac{\mathbf{R}_3}{r_{13}{}^3} \right) = 0. \tag{5.28}$$

122

Applying equation (5.18), equation (5.28) becomes

$$m_2 \mathbf{R}_1 \times \mathbf{R}_2 \left(\frac{1}{r_{12}^3} - \frac{1}{r_{13}^3} \right) = 0.$$

There are of course two similar equations for the other particles. This set exhibits immediately the two conditions that must hold if the set is to be satisfied: these are either

$$r_{12} = r_{23} = r_{31} = r$$

which gives the equilateral triangle solution, or

$$\mathbf{R}_1 \times \mathbf{R}_2 = \mathbf{R}_2 \times \mathbf{R}_3 = \mathbf{R}_3 \times \mathbf{R}_1 = 0$$

which puts the particles on a straight line. These two cases are the only ones possible.

In the former case, the first equation of (5.17) becomes

$$m_1 \ddot{\mathbf{R}}_1 = \frac{Gm_1}{r^3} (m_2 \mathbf{r}_{12} + m_3 \mathbf{r}_{13}).$$

Using equation (5.19), we obtain

$$\ddot{\mathbf{R}}_1 + \frac{GM}{r^3} \mathbf{R}_1 = 0. \qquad (5.29)$$

Now in (5.20), the angle between $\mathbf{r}_{12}$ and $\mathbf{r}_{13}$ is $60°$ in this case, so that (5.20) becomes

$$M^2 R_1^2 = (m_2^2 + m_3^2 + m_2 m_3) r^2.$$

Substituting in (5.29) for r, there results

$$\ddot{\mathbf{R}}_1 + GM_1 \frac{\mathbf{R}_1}{R_1^3} = 0 \qquad (5.30)$$

where

$$M_1 = \frac{(m_2^2 + m_3^2 + m_2 m_3)^{3/2}}{(m_1 + m_2 + m_3)^2}. \qquad (5.31)$$

Hence by (5.30), which is the two-body equation of motion (see equation (4.12)), the particle of mass m_1 moves about the centre of mass in an orbit (ellipse, parabola or hyperbola, depending upon the initial velocities) as if it was of unit mass and a mass M_1 were placed there. A corresponding result is obtained for each of the other particles. As long as the initial conditions already stated are satisfied, the figure remains an equilateral triangle though its size may oscillate or grow indefinitely.

In the latter case (i.e. the collinear solution), if we take the line to be the x axis, the force acting on m_1 is

$$F_1 = m_2 \frac{(x_2 - x_1)}{x_{12}^3} + m_3 \frac{(x_3 - x_1)}{x_{13}^3}.$$

But by equation (5.25),
$$x_i = (x_i)_0 f(t)$$
so that
$$F_1 = \frac{1}{f^2}\left[m_2\frac{(x_2-x_1)}{x_{12}{}^3}+m_3\frac{(x_3-x_1)}{x_{13}{}^3}\right]_0 = \frac{\text{constant}}{f^2}.$$

Since f is proportional to the distance, m_1 is acted upon by an inverse-square-law central force. Its orbit is therefore a conic section, as are the orbits of the other two particles.

The condition
$$F_1:F_2:F_3 = x_1:x_2:x_3$$

is now imposed. The x axis is supposed to rotate with angular velocity $\dot\theta$ and we want solutions that satisfy

$$Ax_1 = m_2\frac{x_2-x_1}{x_{12}{}^3}+m_3\frac{x_3-x_1}{x_{13}{}^3} \qquad (5.32)$$

$$Ax_2 = m_3\frac{x_3-x_2}{x_{23}{}^3}+m_1\frac{x_1-x_2}{x_{12}{}^3} \qquad (5.33)$$

$$Ax_3 = m_1\frac{x_1-x_3}{x_{13}{}^3}+m_2\frac{x_2-x_3}{x_{23}{}^3} \qquad (5.34)$$

where A is a constant that depends upon the initial conditions.

The three particles can be arranged in the orders 321, 231 and 213. If we take the first case (as in figure 5.1), we are looking for a positive value of X such that
$$X = \frac{x_2-x_3}{x_1-x_2}.$$
Then
$$\frac{x_1-x_3}{x_1-x_2} = 1 + X.$$

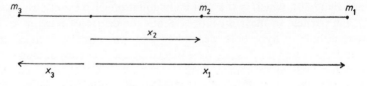

Figure 5.1

Subtract (5.33) from (5.32) to give

$$Ax_{12} = -\frac{(m_1+m_2)}{x_{12}{}^2}+m_3\left(\frac{1}{x_{23}{}^2}-\frac{1}{x_{13}{}^2}\right). \qquad (5.35)$$

124

Subtract (5.34) from (5.33) to give

$$Ax_{23} = -\frac{(m_2 + m_3)}{x_{23}^2} + m_1 \left(\frac{1}{x_{12}^2} - \frac{1}{x_{13}^2} \right).$$ (5.36)

Substituting for X in (5.35) and (5.36), eliminating Ax_{12}^3 between the resulting equations and arranging in powers of X, there results Lagrange's quintic equation

$$(m_1 + m_2)X^5 + (3m_1 + 2m_2)X^4 + (3m_1 + m_2)X^3$$
$$- (m_2 + 3m_3)X^2 - (2m_2 + 3m_3)X - (m_2 + m_3) = 0.$$ (5.37)

By Descartes' rule of signs there is only one positive root, since the co-efficients of the powers of X change sign only once.

Hence this positive value of X obtained from (5.37) defines uniquely the distribution of the three particles in the order chosen. It is obvious that by taking the other two orders (namely 231 and 213) two more distinct straight-line solutions for the particles could be obtained.

5.9 General Remarks on the Lagrange Solutions

If there is no change of scale the solutions are called *stationary* and the relative distances do not alter; the system also rotates in a plane about the centre of mass with constant angular velocity.

If two particles at A and B of masses m_1 and m_2 are taken as points of reference, then we see that there are five points at which the third may be placed. The points L_1, L_2, L_3, L_4 and L_5 are called the Lagrange points and are shown in figure 5.2.

Both the equilateral triangle and the straight line solutions were considered to be interesting but purely academic solutions to the three-body problem for a long time after they were found. It seemed highly unlikely that in nature such unusual formations could exist. In fact both solutions are realized in the Solar System.

About the points L_4 and L_5 with respect to the Sun and Jupiter there are some 12 asteroids (the Trojans) in oscillation, each one with the Sun and Jupiter providing an example of the equilateral triangle solution (see section 1.2.3). A Trojan can wander some 20° or more from the points L_4 and L_5 (the angle being measured from the Sun) but still remain in general for a long time in orbit about L_4 or L_5 (its point of libration). Again, in the Earth–Moon system it has been suggested by Kordelewski that the points L_4 and L_5 are occupied by meteoric particles, visible under the best seeing conditions as faint nebulosities. The Voyager missions to Saturn led to the discovery of other cases in nature of the equilateral triangle solution (see section 8.5).

With respect to the straight line solution it appears that the Gegenschein, a faintly visible light observed after sunset in the plane of the ecliptic in a

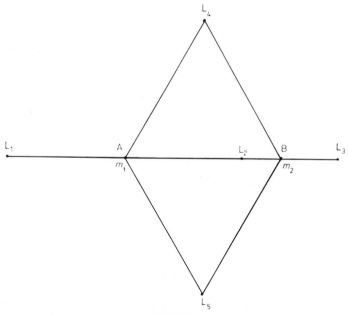

Figure 5.2

direction opposite to that of the Sun, may be due to the Sun's illumination of a further accumulation of meteoric particles in the Lagrange point L_3. In this case the masses m_1 and m_2 refer to Sun and Earth respectively.

In a later section the question of the stability of such libration points will be investigated, it being of practical interest to determine whether a small 'nudge' given to a particle at a Lagrange point will cause it to depart to greater and greater distances from it or merely cause it to oscillate about the point.

Finally it may be remarked that in the general n-body case $(n > 3)$ there also exist special solutions consisting of regular polyhedra formed by the mass points that are the counterparts of the Lagrange solutions.

5.10 The Circular Restricted Three-Body Problem

In an effort to obtain insight into the possible types of motion in the three-body problem a great deal of study has been made by Poincaré, Hill and others of the so-called circular restricted three-body problem, where two massive particles move in circles about their centre of mass and attract (but are *not* attracted by) a third particle of infinitesimal mass. The orbits and masses of the two massive particles being known, the problem is to determine the possible movements of the third particle given the coordinates and velocities of the system at some epoch.

The general three-body problem is thus reduced from nine second-order differential equations to three second-order ones; that is, a reduction from 18 to six. If the problem is restricted further, the test particle being constrained to move in the orbital plane of the two massive bodies, there are only two second-order equations so that the problem is of order 4. This particular variation is called the coplanar circular restricted three-body problem. It is therefore understandable that, although in setting up this problem the ten available integrals have had perforce to be jettisoned, a great deal of analytical and numerical work should have been expended on both the three-dimensional and the coplanar–circular-restricted three-body problem.

An integral of the motion (first obtained by Jacobi) can be found which is valuable in gaining information about the behaviour of the tiny particle.

5.10.1 Jacobi's integral

Let the unit of mass be such that the sum of the masses of the two particles is unity, their masses being $1-\mu$ and μ where $\mu \leqslant \frac{1}{2}$. We also choose the unit of distance to be their constant separation; the unit of time is so chosen that the gravitational constant G is also unity.

Now the mean angular velocity (or mean motion) of the two bodies is n where

$$n^2 a^3 = G(m_1 + m_2).$$

by relation (4.74). It is then seen that because of the units chosen the angular velocity of the two particles of finite mass is also unity.

If the coordinates of the masses $(1-\mu)$ and μ are (ξ_1, η_1, ζ_1) and (ξ_2, η_2, ζ_2) respectively, referred to non-rotating axes ξ, η, ζ with the centre of mass of the two finite bodies as origin, and the coordinates of the test particle are (ξ, η, ζ), the equations of motion of this particle are

$$
\left.
\begin{aligned}
\ddot{\xi} &= (1-\mu)\frac{\xi_1-\xi}{r_1^3} + \mu\frac{\xi_2-\xi}{r_2^3} \\[4pt]
\ddot{\eta} &= (1-\mu)\frac{\eta_1-\eta}{r_1^3} + \mu\frac{\eta_2-\eta}{r_2^3} \\[4pt]
\ddot{\zeta} &= (1-\mu)\frac{\zeta_1-\zeta}{r_1^3} + \mu\frac{\zeta_2-\zeta}{r_2^3}
\end{aligned}
\right\}
\tag{5.38}
$$

where

$$r_1^2 = (\xi_1-\xi)^2 + (\eta_1-\eta)^2 + (\zeta_1-\zeta)^2$$

and

$$r_2^2 = (\xi_2-\xi)^2 + (\eta_2-\eta)^2 + (\zeta_2-\zeta)^2.$$

If the ζ axis is perpendicular to the plane of rotation of the two massive particles, $\zeta_1 = \zeta_2 = 0$.

We now take a set of axes x, y and z having the same origin as before, but with the x and y axes rotating (with angular velocity unity about the z axis which coincides with the ζ axis) perpendicular to the plane of the paper in figure 5.3.

The direction of the x axis can be chosen such that the two massive particles P_1 and P_2 always lie on it, having coordinates $(-x_1, 0, 0)$ and $(x_2, 0, 0)$ respectively, such that

$$x_2 - x_1 = 1.$$

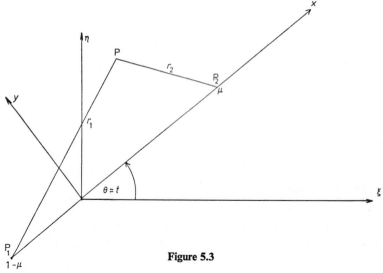

Figure 5.3

In addition, in the units chosen,

$$x_1 = -\mu$$
$$x_2 = 1 - \mu.$$

Hence

$$r_1^2 = (x_1 - x)^2 + y^2 + z^2$$

and

$$r_2^2 = (x_2 - x)^2 + y^2 + z^2$$

where (x, y, z) are the coordinates of the infinitesimal particle with respect to the rotating axes. They are connected to the old coordinates by the relations

$$\left. \begin{array}{l} \xi = x \cos t - y \sin t \\ \eta = x \sin t + y \cos t \\ \zeta = z \end{array} \right\} \tag{5.39}$$

with similar equations for the coordinates of the two bodies of finite mass.

Differentiating (5.39) twice and substituting the resulting expression into (5.38), we obtain

$$\left. \begin{array}{l} (\ddot{x} - 2\dot{y} - x) \cos t - (\ddot{y} + 2\dot{x} - y) \sin t \\ \qquad = -\left[(1-\mu)\dfrac{x_1-x}{r_1^3} + \mu\dfrac{x_2-x}{r_2^3} \right] \cos t + \left[\dfrac{1-\mu}{r_1^3} + \dfrac{\mu}{r_2^3} \right] y \sin t \\[4mm] (\ddot{x} - 2\dot{y} - x) \sin t + (\ddot{y} + 2\dot{x} - y) \cos t \\ \qquad = -\left[(1-\mu)\dfrac{x_1-x}{r_1^3} + \mu\dfrac{x_2-x}{r_2^3} \right] \sin t - \left[\dfrac{1-\mu}{r_1^3} + \dfrac{\mu}{r_2^3} \right] y \cos t \\[4mm] \qquad\qquad \ddot{z} = -\left[\dfrac{(1-\mu)}{r_1^3} + \dfrac{\mu}{r_2^3} \right] z. \end{array} \right\} \tag{5.40}$$

128

If we multiply the first of equations (5.40) by cos t, the second by sin t and add, then multiply the first by $-\sin t$, the second by cos t and add, we obtain two equations which with the third of equations (5.40) form the set (5.41) below:

$$
\left.
\begin{aligned}
\ddot{x}-2\dot{y}-x &= -(1-\mu)\frac{x-x_1}{r_1^3}-\mu\frac{x-x_2}{r_2^3} \\
\ddot{y}+2\dot{x}-y &= -\left(\frac{1-\mu}{r_1^3}+\frac{\mu}{r_2^3}\right)y \\
\ddot{z} &= -\left(\frac{1-\mu}{r_1^3}+\frac{\mu}{r_2^3}\right)z.
\end{aligned}
\right\}
\qquad (5.41)
$$

These equations, which do not involve the independent variable t explicitly, are the equations of motion of the infinitesimal body with respect to the set of rotating coordinates.

Let a function U be defined by

$$
U=\frac{1}{2}(x^2+y^2)+\frac{1-\mu}{r_1}+\frac{\mu}{r_2}.
$$

It is then readily seen that the set (5.41) may be written as

$$
\ddot{x}-2\dot{y}=\frac{\partial U}{\partial x} \qquad (5.42)
$$

$$
\ddot{y}+2\dot{x}=\frac{\partial U}{\partial y} \qquad (5.43)
$$

$$
\ddot{z}=\frac{\partial U}{\partial z}. \qquad (5.44)
$$

If we multiply (5.42) by $\dot{x}$, (5.43) by $\dot{y}$ and (5.44) by $\dot{z}$ and add, we obtain

$$
\dot{x}\ddot{x}+\dot{y}\ddot{y}+\dot{z}\ddot{z}=\frac{\partial U}{\partial x}\dot{x}+\frac{\partial U}{\partial y}\dot{y}+\frac{\partial U}{\partial z}\dot{z}
$$

which is a perfect differential since U is a function of x, y and z alone.

Integrating, we therefore obtain

$$
\dot{x}^2+\dot{y}^2+\dot{z}^2=2U-C \qquad (5.45)
$$

where C is a constant of integration.

The left-hand side is the square of the velocity of the particle of infinitesimal mass in the rotating frame. If we denote it by V^2, then

$$
V^2=2U-C \qquad (5.46)
$$

or

$$
\dot{x}^2+\dot{y}^2+\dot{z}^2=x^2+y^2+\frac{2(1-\mu)}{r_1}+\frac{2\mu}{r_2}-C. \qquad (5.47)
$$

This is Jacobi's integral, sometimes called the integral of relative energy.

It is the only one that can be obtained in the circular restricted three-body problem.

The integral may, of course, be expressed in terms of the coordinates and velocity components in the nonrotating coordinate system. If this is done, we obtain

$$\dot{\xi}^2 + \dot{\eta}^2 + \dot{\zeta}^2 - 2(\xi\dot{\eta} - \eta\dot{\xi}) = 2\left(\frac{1-\mu}{r_1} + \frac{\mu}{r_2}\right) - C. \tag{5.48}$$

5.10.2 Tisserand's criterion

It happens on occasion that a comet will make a close approach to Jupiter or one of the other planets. The consequence of such an encounter can be a drastic change in the elements of its orbit. Unless such a comet had been tracked visually or had had its orbit computed numerically throughout the period in question, it might not be possible to identify it after the encounter as the same comet observed before the encounter, unless some property of its heliocentric orbit remained unaffected by the planetary disturbance.

Such a property was discovered by Tisserand by assuming that in the Sun–planet–comet case there was an approximate example of the circular restricted three-body problem, the comet playing the part of the infinitesimal particle. The planet most often involved in such problems is Jupiter on account of its great mass and distance from the Sun. While its orbit is not strictly circular, its eccentricity is small enough to regard its neglect as justified.

Jacobi's integral then shows that something does remain the same throughout the encounter, namely the constant C. If this quantity (computed by using the elements of the two comets in question) is found to be approximately the same, the two comets are probably two appearances of the same one; it is then worthwhile to conduct a step-by-step integration to verify this.

It is in fact more convenient to replace the coordinates and velocity components in equation (5.48) by the elements themselves.

In the case of Jupiter and the Sun we find that $\mu \sim 10^{-3}$, so the centre of the Sun may be taken to be the origin without sensible error. If r and h are the heliocentric radius vector of the comet and the constant of area in the two-body Sun–comet problem, while a, e and i are respectively its semimajor axis, eccentricity, and inclination of its orbital plane to that of Jupiter's orbit about the Sun, then

$$\dot{\xi}^2 + \dot{\eta}^2 + \dot{\zeta}^2 = \frac{2}{r} - \frac{1}{a}$$

$$\xi\dot{\eta} - \eta\dot{\xi} = h \cos i$$

and

$$h^2 = a(1 - e^2)$$

using the results of sections 4.6 and 4.6.2, and remembering that in the units adopted

$$G(m_{Sun} + m_{comet}) = 1.$$

130

Hence equation (5.48) becomes

$$\frac{2}{r}-\frac{1}{a}-2a^{1/2}(1-e^2)^{1/2}\cos i=\frac{2}{r_1}-\frac{2\mu}{r_2}-C. \qquad (5.49)$$

Now r is very nearly equal to r_1; also the heliocentric elements are determined when the comet is far from Jupiter so that we can neglect the second term on the right-hand side of equation (5.49). We then obtain

$$\frac{1}{2a}+a^{1/2}(1-e^2)^{1/2}\cos i=C' \qquad (5.50)$$

where C' is a constant.

If then the relevant elements of the two comets are a_0, e_0, i_0 and a_1, e_1, i_1, they are related by the equation

$$\frac{1}{2a_0}+a_0^{1/2}(1-e_0^2)^{1/2}\cos i_0=\frac{1}{2a_1}+a_1^{1/2}(1-e_1^2)^{1/2}\cos i_1 \qquad (5.51)$$

This is Tisserand's criterion. It must be remembered that the unit of length is the Sun–Jupiter distance while the unit of mass is the Sun's mass; also that the time scale is such that Jupiter revolves about the Sun with angular velocity unity. It must also be noted that Tisserand's criterion is only approximately valid. Nevertheless, if substitution of the two sets of elements in (5.51) give a marked inequality, it is safe to say that they do not belong to a single comet.

5.10.3 Surfaces of zero velocity

Jacobi's integral was

$$V^2=2U-C$$

or

$$(\dot{x}^2+\dot{y}^2+\dot{z}^2)=(x^2+y^2)+\frac{2(1-\mu)}{r_1}+\frac{2\mu}{r_2}-C \qquad (5.52)$$

where

$$r_1^2=(x-x_1)^2+y^2+z^2$$

and

$$r_2^2=(x-x_2)^2+y^2+z^2.$$

This is a relation between the square of the velocity and the coordinates of the infinitesimal particle with respect to the set of rotating axes. If the particle's velocity becomes zero, we have

$$2U=C$$

or

$$(x^2+y^2)+\frac{2(1-\mu)}{r_1}+\frac{2\mu}{r_2}=C \qquad (5.53)$$

where C is a constant determined from the initial conditions.

Equation (5.53) is important in this problem in that it defines for a given value of C the boundaries of regions in which the particle must be found. These regions are those for which $2U > C$, since otherwise V^2 would be negative, giving imaginary values for the velocity.

Equation (5.53), called Hill's limiting surface, does not tell us anything about the orbits of the particle within the volumes of space available to it; to obtain such information the other integrals of the problem would have to be found. We can, however, study the behaviour of Hill's limiting surface for various values of C.

If both C and $(x^2 + y^2)$ are large, then by equation (5.53) we have

$$x^2 + y^2 \sim C_1$$

which is the equation of a circle. If however C is large $(= C_1)$ and either r_1 or r_2 is very small, the surfaces become separate ovals enclosing $(1 - \mu)$ and μ. This case is sketched in figure 5.4(a), the z axis being taken to be perpendicular to the plane of the paper. The volume of space in which the particle's velocity would be imaginary (and therefore inaccessible to the particle) is shaded. But if the particle starts off originally within one of the ovals or outside the almost circular contour surrounding both (the intersection with the xy plane of a cylinder parallel to the z axis, it must be noted), the particle must remain there since the three possible regions are separated by the 'forbidden' region.

If C now decreases, the inner ovals expand while the outer surface (of almost circular cross section) shrinks. For a certain value of C (say C_2) the inner ovals meet at the double point L_2 where they have common tangents. This is illustrated in figure 5.4(b). A slight decrease in C now results in the ovals coalescing to form a dumbbell-shaped surface with a narrow neck through which it is possible for the particle to escape from the vicinity of one finite mass to the other, though it is still not possible for the particle to reach the outer region (figure 5.4(c)). For a further decrease, the inner region meets the outer at a double point L_3 (figure 5.4(d)) and then, as C is decreased still further, a new double point L_1 is obtained while the widening of the neck about L_3 enables the particle to wander out of the region about the two finite masses into the outer space (figure 5.4(e)). As the process continues, the regions inaccessible to the particle in the xy plane shrink until they vanish at two points L_4 and L_5 (figure 5.4(f)).

Now by the rules of analytical geometry, double points are places where the partial derivatives of a function vanish. In this case the function is f, given by

$$f = x^2 + y^2 + \frac{2(1-\mu)}{r_1} + \frac{2\mu}{r_2} - C = 2U - C.$$

Hence

$$\frac{\partial U}{\partial x} = \frac{\partial U}{\partial y} = \frac{\partial U}{\partial z} = 0. \tag{5.54}$$

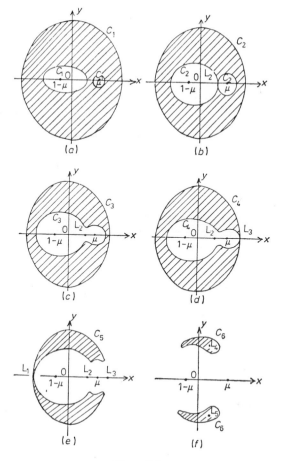

Figure 5.4

But we had as the equations of motion of the particle the relations

$$
\left.
\begin{aligned}
\ddot{x} - 2\dot{y} &= \frac{\partial U}{\partial x} \\[2mm]
\ddot{y} + 2\dot{x} &= \frac{\partial U}{\partial y} \\[2mm]
\ddot{z} &= \frac{\partial U}{\partial z}.
\end{aligned}
\right\}
\tag{5.55}
$$

Since the surfaces are places where the particle has zero velocity (i.e. $\dot{x} = \dot{y} = \dot{z} = 0$), by equations (5.54) and (5.55) we have $\ddot{x} = \ddot{y} = \ddot{z} = 0$.

This statement may then be interpreted as saying that at the five double points L_1, L_2, L_3, L_4 and L_5 no resultant force acts on the particle. Placed at any one of these points it would remain there. Such points are consequently the Lagrange points previously obtained.

The behaviour of the surfaces of zero velocity with changing C in the xz and yz planes are sketched in figures 5.5 and 5.6 where the values of C are the same as those used in figure 5.4.

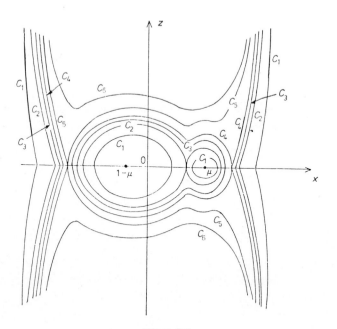

Figure 5.5

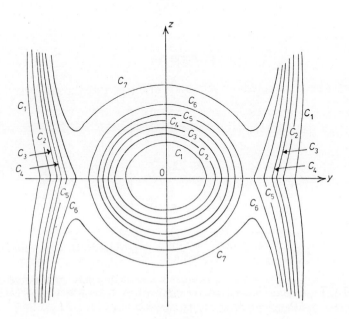

Figure 5.6

Several remarks should be made here. It should be noted that we can in this circular restricted three-body problem use the surface of zero velocity to state categorically in what regions the particle can move. If the constant C confines the particle to the oval about the mass μ for example, we do not know whether or not it will collide with μ but we can at least say that it will *never* cross the surface of zero velocity.

If the two finite bodies move in ellipses about their common centre of mass (the elliptical restricted three-body problem), there is no Jacobi integral but it is tempting to suppose (as has been done by many) that if the eccentricity of the elliptical orbit of one finite mass about the other is small, then the results of the circular problem may apply for a long time to the elliptical problem. This is pure supposition and can be shown to be so (Ovenden and Roy 1961). The most one can say is that predictions from the Jacobi integral can be applied for a time interval of the order of a few times the period of the two finite bodies.

5.10.4 The stability of the libration points

We consider now what happens to the infinitesimal particle if it is displaced a little from one of the Lagrangian points. This would occur if some mass other than the two finite ones on occasion perturbed the particle. We can suppose too that as well as the displacement the particle is given a small velocity. If the resultant motion of the particle is a rapid departure from the vicinity of the point we can call such a position of equilibrium an *unstable* one; if however the particle merely oscillates about the point it is said to be a *stable* position. This method of investigating the stability of a solution by small displacements has been applied frequently in celestial mechanics.

In rotating coordinates, let the position of a Lagrangian point be (x_0, y_0) and let the particle be displaced to the point $(x_0+\xi, y_0+\eta, \zeta)$, receiving a velocity with components $(\dot{\xi}, \dot{\eta}, \dot{\zeta})$. Then substituting these quantities into the equations of motion of the particle (5.55) and expanding in a Taylor's series we obtain

$$\ddot{\xi}-2\dot{\eta}=\xi\left(\frac{\partial^2 U}{\partial x^2}\right)_0+\eta\left(\frac{\partial^2 U}{\partial x\,\partial y}\right)_0+\zeta\left(\frac{\partial^2 U}{\partial x\,\partial z}\right)_0+\ldots$$

$$\ddot{\eta}+2\dot{\xi}=\xi\left(\frac{\partial^2 U}{\partial y\,\partial x}\right)_0+\eta\left(\frac{\partial^2 U}{\partial y^2}\right)_0+\zeta\left(\frac{\partial^2 U}{\partial y\,\partial z}\right)_0+\ldots$$

$$\ddot{\zeta}=\xi\left(\frac{\partial^2 U}{\partial z\,\partial x}\right)_0+\eta\left(\frac{\partial^2 U}{\partial z\,\partial y}\right)_0+\zeta\left(\frac{\partial^2 U}{\partial z^2}\right)_0+\ldots$$

where the suffix zero means that after the partial differentiation of U is accomplished, x, y and z in it are set equal to x_0, y_0 and z_0 respectively.

Now if the displacements ξ, η and ζ are small, we may neglect terms involving squares, products and higher-degree terms in ξ, η and ζ, and so

135

the equations become

$$\left.\begin{array}{c} \ddot{\xi}-2\dot{\eta}=\xi U_{xx}+\eta U_{xy}+\zeta U_{xz} \\ \ddot{\eta}+2\dot{\xi}=\xi U_{yx}+\eta U_{yy}+\zeta U_{yz} \\ \ddot{\zeta}=\xi U_{zx}+\eta U_{zy}+\zeta U_{zz} \end{array}\right\} \qquad (5.56)$$

where

$$U_{xx}\equiv\left(\frac{\partial^2 U}{\partial x^2}\right)_0 \text{ etc.}$$

and the U are constant since they are evaluated at the Lagrange point.

Consider for the moment the two-dimensional case in the xy plane. Then

$$\left.\begin{array}{c} \ddot{\xi}-2\dot{\eta}=\xi U_{xx}+\eta U_{xy} \\ \ddot{\eta}+2\dot{\xi}=\xi U_{yx}+\eta U_{yy}. \end{array}\right\} \qquad (5.57)$$

These are linear differential equations with constant coefficients, the general solution of which may be written as

$$\xi=\sum_{i=1}^{4}\alpha_i\exp\left(\lambda_i t\right), \qquad \eta=\sum_{i=1}^{4}\beta_i\exp\left(\lambda_i t\right)$$

where the α_i are constants of integration, the β_i being constants dependent upon them and the constants appearing in the differential equations. The λ_i are the roots of the characteristic determinant set equal to zero obtained from equation (5.57) rewritten as

$$\left.\begin{array}{c} (\mathrm{D}^2-U_{xx})\xi-(2\mathrm{D}+U_{xy})\eta=0 \\ (2\mathrm{D}-U_{xy})\xi+(\mathrm{D}^2-U_{yy})\eta=0 \end{array}\right\} \qquad (5.58)$$

where

$$\mathrm{D}\equiv\frac{\mathrm{d}^2}{\mathrm{d}t^2}.$$

The determinant, obtained by substituting

$$\xi=\alpha e^{\lambda t}, \qquad \eta=\beta e^{\lambda t}$$

into equation (5.58) is

$$\begin{vmatrix} \lambda^2-U_{xx} & -2\lambda-U_{xy} \\ 2\lambda-U_{xy} & \lambda^2-U_{yy} \end{vmatrix}=0$$

or

$$(\lambda^2-U_{xx})(\lambda^2-U_{yy})+4\lambda^2-(U_{xy})^2=0.$$

Hence

$$\lambda^4+(4-U_{xx}-U_{yy})\lambda^2+U_{xx}U_{yy}-U_{xy}^2=0. \qquad (5.59)$$

If all the λ_i obtained from equation (5.59) are pure imaginary numbers, then ξ and η are periodic and thus give stable periodic solutions in the vicinity of x_0, y_0. If, however, any of the λ_i are real or complex numbers, then ξ and η increase with time so that the solution is unstable.

136

It can happen, however, that the solution contains constant terms in the place of exponentials. The solution is then stable if the remaining exponentials are purely imaginary.

We can now consider the Lagrange points in detail.

Now

$$U = \frac{1}{2}(x^2 + y^2) + \frac{1-\mu}{r_1} + \frac{\mu}{r_2}$$

where

$$r_i^2 = (x - x_i)^2 + y^2 + z^2 \qquad (i = 1, 2).$$

By then defining the quantities A, B and C as

$$A = \frac{1-\mu}{r_1^3} + \frac{\mu}{r_2^3}$$

$$B = 3\left(\frac{1-\mu}{r_1^5} + \frac{\mu}{r_2^5}\right)$$

and

$$C = 3\left[\frac{1-\mu}{r_1^5}(x_0 - x_1) + \frac{\mu}{r_2^5}(x_0 - x_2)\right]$$

we find that

$$U_{xx} = 1 - A + 3(1 - \mu)\frac{(x_0 - x_1)^2}{r_1^5} + 3\mu\frac{(x_0 - x_2)^2}{r_2^5}$$

$$U_{yy} = 1 - A + By_0^2$$

$$U_{zz} = -A + Bz_0^2$$

while

$$U_{xy} = Cy_0, \qquad U_{xz} = Cz_0, \qquad U_{yz} = By_0z_0.$$

In the straight line solution, $y_0 = z_0 = 0$, so that

$$r_i^2 = (x_0 - x_i)^2. \qquad (i = 1, 2)$$

Hence

$$U_{xy} = U_{xz} = U_{yz} = 0$$

and the equations of motion for a small displacement become

$$\ddot{\xi} - 2\dot{\eta} = \xi U_{xx} = \xi(1 + 2A)$$

$$\ddot{\eta} + 2\dot{\xi} = \eta U_{yy} = \eta(1 - A)$$

$$\ddot{\zeta} = -A\zeta.$$

The ζ equation is independent of the first two, being the equation for simple harmonic motion since A is positive. Hence its solution is

$$\zeta = C_1 \cos A^{1/2}t + C_2 \sin A^{1/2}t,$$

showing that the oscillation in the z direction is finite and small with period $2\pi A^{-1/2}$.

137

Applying the values for U_{xx}, U_{xy} and U_{yy} in equation (5.59) we obtain

$$\lambda^4 + [4 - (1+2A) - (1-A)]\lambda^2 + (1+2A)(1-A) = 0$$

giving

$$\lambda^4 + (2-A)\lambda^2 + (1+A-2A^2) = 0 \qquad (5.60)$$

where

$$A = \frac{1-\mu}{r_1{}^3} + \frac{\mu}{r_2{}^3}.$$

Now there are three values of A corresponding to the three Lagrangian points L_1, L_2, L_3 (see figure 5.4), obtained from the three quintic equations of which (5.37) is one. It can be shown that for all three values

$$1 + A - 2A^2 < 0$$

for values of μ up to its limit of $\frac{1}{2}$. Hence the four roots of equation (5.60) consist of two real roots, numerically equal but opposite in sign, and two conjugate pure imaginary roots. Hence the solution for the straight-line case is unstable. At the same time, by carefully selecting the initial values of ξ, η and ζ the motions can be rendered periodic, the particle moving about the Lagrangian point in an elliptical path. In general however, the collinear case must be considered to be unstable; Abhyanker found by numerical integration that a particle does not complete more than two revolutions about L_2 or L_3 before wandering off (Abhyanker 1959).

We now consider the equilateral triangle solutions giving the Lagrange points L_4 and L_5. Here $r_1 = r_2 = r_3 = r = 1$, so that

$$x = \tfrac{1}{2} - \mu, \qquad y = \pm\frac{\sqrt{3}}{2} \qquad \text{and} \qquad z = 0.$$

Then taking the point L_4 we have

$$U_{xx} = \frac{3}{4}, \qquad U_{zz} = -1$$

$$U_{yy} = \frac{9}{4}, \qquad U_{xy} = \frac{3\sqrt{3}}{4}(1-2\mu)$$

while

$$U_{xz} = U_{yz} = 0.$$

The equations of motion for a small displacement therefore become

$$\ddot{\xi} - 2\dot{\eta} = \xi U_{xx} + \eta U_{xy} = \frac{3}{4}\xi + \frac{3\sqrt{3}}{4}(1-2\mu)\eta$$

$$\ddot{\eta} + 2\dot{\xi} = \xi U_{xy} + \eta U_{yy} = +\frac{3\sqrt{3}}{4}(1-2\mu)\xi + \frac{9}{4}\eta$$

$$\ddot{\zeta} = \zeta U_{zz} = -\zeta.$$

Again the oscillation in the z direction is stable, being given by

$$\zeta = C_3 \cos t + C_4 \sin t$$

138

where C_3 and C_4 are constants of integration and the period is the same as that of the revolution of the finite bodies, namely 2π.

Applying (5.59) as before we obtain

$$\lambda^4 + \lambda^2 + \frac{27}{4}\mu(1-\mu) = 0.$$

The condition that the four roots of this biquadratic are pure imaginary roots in conjugate pairs is that

$$1 - 27\mu(1-\mu) \geqslant 0.$$

Rewriting this inequality as

$$1 - 27\mu(1-\mu) = \epsilon$$

we have

$$\mu = \frac{1}{2} \pm \sqrt{\frac{23+4\epsilon}{108}}.$$

Now since $\mu \leqslant \frac{1}{2}$, the negative sign must be taken. When $\epsilon = 0$, $\mu = 0.0385$, so that for stability,

$$\mu < 0.0385.$$

This condition being satisfied, there then exists in the immediate vicinity of the libration point L_4 (and L_5) periodic orbits for a particle placed there. With respect to Jupiter and the Sun $\mu \sim 0.001$, so that the condition is satisfied and we find the Trojan asteroids oscillating about the Lagrange points. For the Earth–Moon system $\mu \sim 0.01$, again satisfying the condition, though the problem is further complicated by the effect of the Sun. We shall return to this system later.

5.10.5 Periodic orbits

The non-existence of uniform integrals apart from the Jacobi integral makes it impossible to obtain the totality of solutions of the restricted problem, and attention was directed very early towards the study of periodic orbits in the problem. According to Poincaré's conjecture, such orbits are dense in the set of all possible solutions of the problem that are bounded in phase space. It was hoped that their discovery and study would be sufficient for a qualitative description of all possible solutions, while their periodicity made their determination and the study of their properties easier.

By phase space, we mean the $6n$-dimensional space defined by the $6n$ coordinates and velocities of the n bodies. In the general n-body problem there are 10 integral relations among the $6n$ quantities and so the phase space can be reduced to $(6n-10)$ dimensions. In the three-(spatial) dimensional restricted three-body problem, where the particles' coordinates and velocity components are related by the Jacobi integral, the phase space can be reduced to five dimensions. Restrict the trajectory of the particle to the orbital plane of the two massive bodies and the phase space is reduced to three dimensions.

A point in phase space defines the state of the system at a given time t. As time passes, the point traces out a trajectory in phase space which must not be confused with the physical trajectories of any of the particles in real space. The phase-space trajectory is defined by the equations of motion and the starting conditions: in the case of the circular restricted coplanar three-body problem these are x_0, y_0, $\dot{x}_0$, $\dot{y}_0$ at time t_0 though the Jacobi integral gives a relationship among them, viz

$$f(x, y, \dot{x}, \dot{y}) = C.$$

In this relation the masses and separation of the two finite bodies will also appear as parameters. If the initial conditions are changed to $x_0 + \delta x_0$, $y_0 + \delta y_0$, $\dot{x}_0 + \delta \dot{x}_0$, $\dot{y}_0 + \delta \dot{y}_0$ a new trajectory is defined.

In the restricted problem we speak of orbits as being periodic when the motion of the infinitesimal particle is periodic with respect to the rotating coordinate system. Poincaré, in his classical work on the restricted problem, considered the study of periodic orbits as a matter of the greatest importance and a starting point for attacking the problem of classifying the solutions. His famous conjecture emphasizes the importance he attached to periodic orbits. It states that if a particular solution of the restricted problem is given, we can always find a periodic solution (possibly with a very long period) with the property that at all times its difference from the original solution is as small as we please. In terms of the phase space this is equivalent to saying that given a point in this space there is always another point, as close to the first as we want, which represents a periodic orbit. He did limit the application of his conjecture to the set of all possible solutions bounded in phase space; that is to say he excluded escape or collision orbits of the particle.

The task is then to give a complete 'global' picture of the properties of the circular restricted three-body problem for any value of the mass parameter μ (the ratio of the mass of the smaller of the two massive bodies to the total mass of the system). For a given value of μ, families of periodic orbits are searched for. Theoretically, it is possible to work with a solution for $\mu = 0$ and, by analytic continuation for positive values of μ, to prove the existence of periodic orbits in the restricted problem. This approach goes back to Poincaré (1895) but has been used by many other workers. Poincaré, in his analytic continuation approach, classified the periodic orbits of the restricted problem into three kinds. The first kind (*première sorte*) are those that are generated from two-body circular orbits ($e = 0$, $i = 0$) while the second kind (*deuxième sorte*) are generated by two-body elliptical orbits ($e \neq 0$, $i = 0$). Periodic orbits of the third kind (*troisième sorte*) are again generated from two-body orbits but with a nonzero inclination of the infinitesimal particle with respect to the plane of motion of the primaries ($e = 0$, $i \neq 0$). In other words the first two classes belong to the coplanar circular restricted problem, the third class belonging to the three-dimensional circular restricted problem.

Other approaches are analytic–numerical, or numerical, utilizing suitable numerical integration procedures to search for the families of periodic orbits. Apart from the pioneering work of G H Darwin and E Strömgren, the most

140

complete studies of periodic orbits in the restricted problem are by Hénon (1965a, b), Broucke (1968) and Hénon (1969), who dealt with the cases $\mu=0.5$, 0.012 and 0 respectively. The particular study for $\mu=0$ does not imply the two-body problem but refers to Hill's form of the restricted three-body problem, obtained by a special limiting process taken to zero.

Other workers such as Rabe (1961, 1962), Deprit and Henrard (1965, 1967) have carried out studies with values of the mass parameter $\mu=0.00095$ (the Sun–Jupiter system) and $\mu=0.012$ (the Earth–Moon system). Additional studies for the Sun–Jupiter system have been carried out by Carpenter and Stumpff (1968), Colombo *et al* (1970), Sinclair (1970), Schanzle (1967), Message (1959a, b), Frangakis (1973), Markellos (1974a, b) and Markellos *et al* (1974, 1975a, b).

The motivation for studying periodic orbits can therefore be said to stem from the following facts:

(i) they appear to be significant in nature,
(ii) they can be used as reference orbits (as implied by Poincaré's conjecture),
(iii) they are possible to obtain and classify (as in Poincaré's analytic continuation and classification into three kinds),
(iv) they are possible to find accurately and in a short time because integration is required for a finite time, the period.

5.10.6 The search for symmetric periodic orbits

A solution

$$s=s(t; \mu; C), \qquad s=(x, y, z, \dot{x}, \dot{y}, \dot{z}) \tag{5.61}$$

of the equations of motion (5.55) will be periodic if an equation

$$s(t_0; \mu)=s(t_0+T; \mu) \tag{5.62}$$

holds true for any value of t_0 and a fixed value of T. This value of T, the period, corresponds to the first instance in time after t_0 for which (5.62) is true. It follows that

$$s(t_0+nT)=s(t_0) \tag{5.63}$$

so that the solution can be considered periodic of period $nT=T^*$, where n is any integer.

Periodicity can be discussed in terms of mirror configurations (section 5.6). Applying the periodicity theorem to the three-dimensional restricted problem, it is seen that there are two types of mirror configurations:

(a) the third body is in the (x, z) plane and its velocity vector is perpendicular to that plane, or
(b) the third body is on the x axis and its velocity vector is perpendicular to that axis.

The two cases are shown in figure (5.7). Periodicity of an orbit is established by the periodicity theorem above if this orbit reaches a mirror configuration twice.

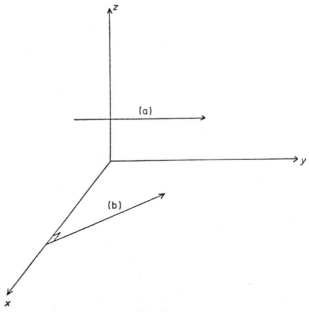

Figure 5.7

Goudas (1961) has used combinations of cases (a) and (b) to find periodic orbits in three dimensions. These orbits are simply or doubly symmetric depending on which combination of (a) and (b) has been used.

In the planar restricted problem a search is made for symmetric periodic orbits by seeking to establish a mirror configuration of the type (b) twice. The velocity vector of the third body will in both cases be perpendicular to the x axis and will always lie in the (x, y) plane. Such orbits will be symmetric with respect to the x axis.

We start with a set of initial conditions satisfying a mirror configuration; by varying these conditions in such a manner that the mirror configuration is preserved, we seek to reach a second mirror configuration. In any admissible set $(x_0, 0, 0, \dot{y}_0)$ of initial conditions only two variables are free to vary while the other two are kept fixed and equal to zero for the preservation of the mirror configuration. The usual procedure is a differential corrections method. Let

$$
\left.
\begin{aligned}
y &= f(x_0, 0, 0, \dot{y}_0) \\
\dot{x} &= g(x_0, 0, 0, \dot{y}_0)
\end{aligned}
\right\}
\tag{5.64}
$$

be the values of y, $\dot{x}$ at an epoch corresponding to the period of the periodic orbit sought, and let

$$
\left.
\begin{aligned}
y &= f(x_0 + \Delta x_0, 0, 0, \dot{y}_0 + \Delta \dot{y}_0) \\
\dot{x} &= g(x_0 + \Delta x_0, 0, 0, \dot{y}_0 + \Delta \dot{y}_0)
\end{aligned}
\right\}
\tag{5.65}
$$

be the corresponding values for a 'corrected' set of initial conditions, where

142

Δx_0, $\Delta \dot{y}_0$ are the corrections. We can linearize the system described in (5.65) by means of a Taylor series expansion around $(x_0, 0, 0, \dot{y}_0)$ and obtain the corrections by solving the system that results when we impose the conditions for periodicity, $y = \dot{x} = 0$. The procedure can be repeated to produce more accurate results in each iteration until the required tolerance is met.

Omitting the zeros inside the parentheses, we can write

$$\left.\begin{array}{c} f(x_0, \dot{y}_0) + \dfrac{\partial f}{\partial x_0} \Delta x_0 + \dfrac{\partial f}{\partial \dot{y}_0} \Delta \dot{y}_0 = 0 \\[4mm] g(x_0, \dot{y}_0) + \dfrac{\partial g}{\partial x_0} \Delta x_0 + \dfrac{\partial g}{\partial \dot{y}_0} \Delta \dot{y}_0 = 0 \end{array}\right\} \tag{5.66}$$

or, using the values of the functions f and g obtained in (5.64),

$$\left.\begin{array}{c} a\Delta x_0 + b\Delta \dot{y}_0 = -\dot{x} \\[2mm] c\Delta x_0 + d\Delta \dot{y}_0 = -y \end{array}\right\} \tag{5.67}$$

where

$$a = \frac{\partial f}{\partial x_0}, \quad b = \frac{\partial f}{\partial \dot{y}_0}, \quad c = \frac{\partial g}{\partial x_0}, \quad d = \frac{\partial g}{\partial \dot{y}_0}. \tag{5.68}$$

Solving the equations (5.66) will give the required corrections to the initial conditions x_0 and $\dot{y}_0$.

We may remark that the search can be simplified considerably by reducing it down to a one-dimensional search. In the set of equations (5.64) the first equation will be

$$y = 0 = f(x_0, 0, 0, \dot{y}_0)$$

provided we define the functions f and g to be the values of y and $\dot{x}$ at the pth crossing of the orbit with the x axis.

We are then left with only one condition to satisfy; this is

$$\dot{x} = g(x_0, 0, 0, \dot{y}_0) = 0. \tag{5.69}$$

One of the two free variables x_0 and $\dot{y}_0$ can be kept fixed (say $\dot{y}_0$) and we have reduced the search to the problem of finding a zero of a single-variable function

$$\dot{x} = g(x_0). \tag{5.70}$$

The Jacobi integral can be usefully employed here. Indeed, solving (5.47) for $\dot{y}$, we obtain

$$\dot{y}^2 = 2U - C - \dot{x}^2$$

or

$$\dot{y} = \pm [2U - C - \dot{x}^2]^{1/2}. \tag{5.71}$$

The search is now one-dimensional along the x axis, since any admissible set of initial conditions can be written as

$$s_0 = \{x_0, 0, 0, \dot{y}_0(x_0, C)\}. \tag{5.72}$$

Because of (5.71), in which the minus sign is invariably chosen to ensure that the correspondence of a $\dot{y}$ to an x is unique, $\dot{x}$ is kept equal to zero and the Jacobi constant is fixed. For a given value of the Jacobi constant C the equations of motion are integrated numerically and the sign of the function

$$\dot{x} = \dot{x}(t, x_0, c) \qquad (5.73)$$

is recorded at the pth crossing of the x axis. This function is continuous with respect to the initial conditions and a change of its sign for two values of the variable x_0 indicates the existence of a zero of the function in the interval defined by these two values.

When a zero has been found a second mirror configuration has been established (since $y=0$ at any crossing of the x-axis) and with it a periodic orbit. This orbit will close at the $2p$th crossing of the x axis provided that a mirror configuration was not reached at any instance before the pth crossing in the course of the orbit.

5.10.7 Examples of some families of periodic orbits

The total number of periodic orbits discovered and studied to date is enormous and in this section only a few examples can be given. An exhaustive study between 1913 and 1939 was made by E Strömgren and the Copenhagen school of the $\mu=1/2$ coplanar restricted case where both massive particles have unit mass and unit separation. Hence their special problem is commonly called the Copenhagen problem. The configuration of periodic orbits is therefore symmetric with respect to the y axis (rotating coordinates, origin at centre of mass of two unit masses). Although this study was invaluable in exploring the evolution of periodic orbits within families, it was restricted to the special case $\mu=1/2$. Now we have seen that stable periodic orbits about the Lagrange equilateral triangle points exist for $\mu \leqslant 0.0385$ (known as Routh's value), and so a study of the Copenhagen problem cannot be sufficient in itself. Properties of solutions of the restricted problem depend upon the value of the mass parameter μ. There are cert. values (for example Routh's value) on one side of which special orbits exist, or where a group of orbits changes character. A complete 'global' picture of the properties of the problem for many values of μ is therefore required.

In the Copenhagen problem there are many families of periodic orbits. Only one is considered here in any detail but it gives an understanding of what is meant by a family, and what is meant by the evolution of orbits within a family, and increases our insight into the method of search. In figures 5.8 (a–c) the characteristics and development of the class (f) family of the Copenhagen problem are shown.

This is a set of retrograde periodic orbits round one of the two equal masses, say P_1. Since both masses are equal, there is no distinction between considering the orbits as planetary or satellite in nature.

The orbits are generated from tiny circular orbits round P_1. As the orbits increase in size by starting them off from greater and greater distances on the

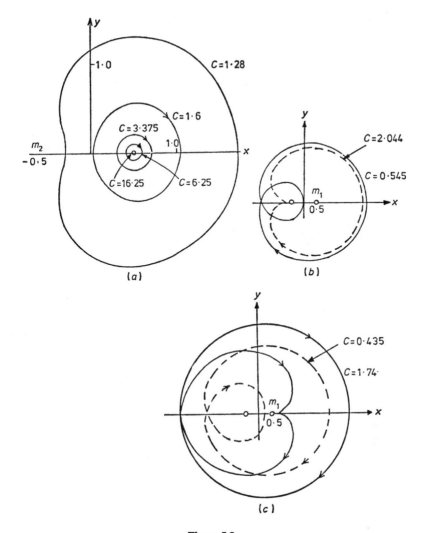

Figure 5.8

positive x axis from P_1 and P_2, they evolve from oval to kidney-shaped, becoming more and more distorted from circles until a collision orbit is reached and the particle collides with P_2. This orbit is of course also an ejection orbit and ends the first phase of the development. Figure 5.8 (b) shows the second phase. From the previous collision orbit, orbits develop showing a loop about P_2 instead of a collision and ejection cusp at P_2. This loop grows and distorts from orbit to orbit until the second phase ends with a collision at P_2. A new oval appears, grows and a new collision occurs. This process is repeated indefinitely.

The calculation of the Jacobi constant C from orbit to orbit of the family shows that, as expected, it falls in value rapidly at first from its infinite size for the first infinitesimal orbit about P_1, reaching a value 2·044 at the first collision with P_2, and a value 1·74 when collision occurs with P_1.

In figure 5.9 (*a, b*) the first phase of periodic orbits from class (g) of the

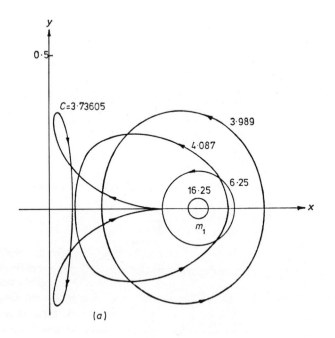

(*a*)

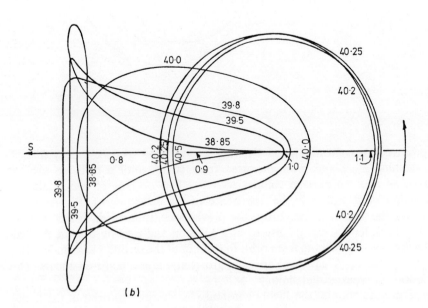

(*b*)

Figure 5.9

Copenhagen problem is compared with Darwin's family A of satellites. Class (g) consists of direct periodic orbits around P_2.

Darwin's computations were carried out for a value of $\mu = 1/11$. The bodies S and J (masses 10 and 1 respectively) were of unit distance apart. The perturbations of S on orbits about J are therefore much stronger than in the Copenhagen problem. Nevertheless, the resemblances between orbits of figures 5.9 (a, b) are close.

The American and Russian lunar space research programmes inspired large computational and analytic searches for periodic orbits in the Earth–Moon system ($\mu \simeq 1/82 \cdot 3$). Many families of orbits were found, many of extremely complicated shape. Of deep interest were those that gave close approaches to both Earth and Moon.

5.10.8 Stability of periodic orbits

As a rule of thumb, the feeling that the more extravagantly shaped an orbit is the less stable it will be is probably a reasonable one. But in fact the meaning of the stability concept has to be looked at more closely before we can make valid judgements.

In the early part of this chapter, we examined the stability of the five Lagrange points in the restricted three-body case. If the coordinates and velocity components of the particle at a Lagrange point were given small increments, would the particle merely oscillate about the point or depart rapidly from it? The former and latter cases were termed stable and unstable respectively. In the treatment we linearized the equations of variation and solved them, examining the roots of the characteristic determinant to discover whether or not the Lagrange solution was stable.

Stability can also be defined rigorously for orbits that are exactly periodic. This is done traditionally by means of Poincaré's characteristic exponents. The integration of an extra set of equations, the variational equations, is required for a rigorous definition and determination of the stability of periodic orbits. Firstly, we consider the concept known as *the surface of section*.

5.10.9 The surface of section

Consider the differential equations of motion of an autonomous system in the form

$$\dot{x}_i = X_i(x_1, x_2 \ldots x_m) \qquad (i = 1, 2 \ldots m) \qquad (5.74)$$

where $m = 2n$ and n is the number of degrees of freedom of the system. Let $\mathbf{x}_0 = (x_{01}, x_{02} \ldots x_{0m})$ be the vector representing the values of the variables at the epoch, which we can take to be $t = 0$. The solution of equation (5.74) may then be written in the form

$$x_i = x_i(x_{01}, x_{02} \ldots x_{0m}, t). \qquad (5.75)$$

We speak of the curves (5.75) in the phase space ($x_1 \ldots x_m$) as the 'characteristics', while we speak of the projections of these curves in the position space $(x_1, x_2 \ldots x_n)$ as the trajectories of the moving particle.

The characteristics (5.75) define a transformation $\mathbf{T}(t)$ from $\mathbf{x}_0$ to $\mathbf{x} = (x_1 \ldots x_m)$, which we may represent as

$$\mathbf{x} = \mathbf{T}(t)\mathbf{x}_0. \tag{5.76}$$

The operator $\mathbf{T}(t)$ transforms the point $\mathbf{x}_0$ in phase space occupied by the particle at time $t_0 = 0$ into the point $\mathbf{x}$ occupied by the particle at time, t. For the restricted problem the Jacobian matrix of the transformation $\mathbf{T}(t)$ has unit determinant

$$\det \mathbf{J} = \det \frac{\partial(x_i)}{\partial(x_{0j})} = 1. \tag{5.77}$$

This is due to the fact that for autonomous systems the property

$$\frac{d\mathbf{J}}{dt} = \Delta \mathbf{J} \tag{5.78}$$

is satisfied, where Δ is the divergence of the vector field $x = (x_1 \ldots x_m)$, while for Hamiltonian systems we have

$$\Delta = 0.$$

On the other hand, for $t = 0$ we have $\mathbf{J} = \mathbf{I}$ (the unit 4×4 matrix). Liouville's famous theorem follows from the above: 'the volume of the phase space is invariant under the transformation defined by the equation'. In this respect it is usually said that 'the fluid is incompressible'. If we pick a region in phase space and measure its m-dimensional volume, and then follow what happens to this region as it moves along with the state trajectories in phase space, we find (i) that the state trajectories (called streamlines) do not intersect: through each and every point in phase space only one streamline passes, and (ii) no matter how deformed the region becomes, its m-dimensional volume remains the same. This theorem is important in hydrodynamics and stellar dynamics.

Coming back to the restricted problem, we assume that a two-dimensional surface can be constructed in phase space such that the characteristics mentioned above cut it at least once in a fixed interval of time. Poincaré and Birkhoff studied these intersections with this 'surface of section' and saw that, as time varied and each characteristic intersected the surface of section at various points, the surface transformed into itself. In the planar-restricted problem $n = 2$ and $m = 2n = 4$, and the use of the Jacobi integral allows the reduction of the dimensions of the phase space to three. In other words, a three-dimensional subspace of the phase space corresponding to a fixed value of the Jacobi constant can be studied. This three-dimensional subspace contains two-dimensional surfaces that can be considered as *surfaces of section*. For instance, in the phase space

$$(x_1, x_2, x_3 = \dot{x}_1, x_4 = \dot{x}_2)$$

we can replace x_4 by a value of the Jacobi constant C and examine the three-dimensional space

$$(x_1, x_2, x_3, C)$$

for C fixed.

If we further set $x_2=0$ we arrive at the surface of section (x_1, x_3). We cān define a mapping M^p

$$(x_{01}, x_{03}) \xrightarrow{\text{M}^p} (x_1, x_3) \tag{5.79}$$

which takes a point in the (x_1, x_3) plane to another point in the same plane. Once the dynamical system is associated with this transformation of the surface of section into itself, its properties become the properties of this transformation. The periodicity of certain solutions of the dynamical system becomes the property of invariance of certain points of this surface under the transformation M^p. For example, the fixed points of M^p for which $x_3=x_{03}=0$ are the symmetric periodic orbits that can be found by the search method described in section 5.10.5. They are called 'symmetric periodic orbits of order p'.

The use of the surface-of-section approach was first carried out by Hénon and Heiles (1964) in relation to the existence of a third integral of motion in a galaxy (suggested by Contopoulos). It was also used by Hénon (1966a, b) in relation to the possible existence of such an integral under certain conditions in the restricted problem and the study of global stability properties of this problem.

5.10.10 The stability matrix

We can now return to the concept of stability. Of central importance in this respect is the Jacobian matrix of the transformation $\mathbf{T}(t)$ in the relation (5.76),

$$\mathbf{x} = \mathbf{T}(t)\mathbf{x}_0.$$

Let $\mathbf{x}_0$ be the initial state that corresponds to a periodic orbit of period T and let $\Delta\mathbf{x}_0$ be a small increment in this initial state. If we define the vector $\mathbf{y}$ by

$$\mathbf{y}(\mathbf{x}_0+\Delta\mathbf{x}_0; t) = \mathbf{x}(\mathbf{x}_0+\Delta\mathbf{x}_0; t) - \mathbf{x}(\mathbf{x}_0, t)$$

in the phase space, then it may be shown that

$$\mathbf{y}(\mathbf{x}_0+\Delta\mathbf{x}_0; t+T) = \Delta(\mathbf{x}_0; t+T)\, \mathbf{y}(\mathbf{x}_0+\Delta\mathbf{x}_0; 0) \tag{5.80}$$

where

$$\Delta(\mathbf{x}_0; t) = \frac{\partial(x_i)}{\partial(x_{0j})}. \tag{5.81}$$

In addition we have

$$\Delta(\mathbf{x}_0; t+T) = \Delta(\mathbf{x}_0, t)\, \Delta(\mathbf{x}_0, T) \tag{5.82}$$

and (5.80) becomes

$$\mathbf{y}(\mathbf{x}_0+\Delta\mathbf{x}_0; t+T) = \Delta(\mathbf{x}_0, t)\, \Delta(\mathbf{x}_0, T)\mathbf{y}(\mathbf{x}_0+\Delta\mathbf{x}_0; 0). \tag{5.83}$$

149

In general it can be shown that

$$\mathbf{y}(\mathbf{x}_0 + \Delta\mathbf{x}_0; pT) = \Delta^p(\mathbf{x}_0, T)\mathbf{y}(\mathbf{x}_0 + \Delta\mathbf{x}_0; 0) \tag{5.84}$$

and this relation implies that the 'distance' $\mathbf{y}(\mathbf{x}_0 + \Delta\mathbf{x}_0; t)$ between the periodic orbit and the nearby aperiodic one depends to first order only on the matrix $\Delta(\mathbf{x}_0; T)$ and its higher powers.

It is through considerations of this nature that a rigorous mathematical definition of the stability of a periodic orbit can be arrived at. Many properties of the matrix $\Delta(\mathbf{x}_0; T)$ can be demonstrated. In stability studies of the restricted problem the eigenvalues of this matrix are sought; traditionally the quantity concentrated on is the trace of the matrix. It may be shown that, for the restricted problem, two of the eigenvalues are equal to unity and the other two have the product unity (Pars 1965). Here we content ourselves by simply stating the relations between the trace of $\Delta(T)$, its eigenvalues, and Poincaré's characteristic exponents α, $-\alpha$.

$$\text{Tr } \Delta(T) = 1 + 1 + \lambda + \frac{1}{\lambda}$$

$$\lambda + \frac{1}{\lambda} = \exp(\alpha T) + \exp(-\alpha T) = 2\cosh(\alpha T).$$

By these and other methods, the stability of periodic orbit solutions in the restricted problem have been studied. Many orbits are highly unstable but regions of stability are shown to exist. In such a region the disturbance of the particle form any point of its periodic orbit, accompanied by a slight change in its velocity, simply produces a new trajectory that departs by only a small amount from the old one during an arbitrarily long time.

5.11 The General Three-Body Problem

It might be thought that, apart from the known integrals and the virial theorem, no general statements can be made on the three-body problem, especially since the totality of solutions in even the restricted problem is not yet explored. In fact, when the restriction that the two finite masses in the restricted problem move in circular orbits about their common centre of mass is relaxed to the extent that they may move in Keplerian ellipses, we also lose the Jacobi integral. Nonetheless, work in recent years—mainly in extended numerical integrations of the general three-body problem utilizing wide spectra of starting conditions and masses—has enabled certain statements to be made about three-body systems in general. In a sense, we now have the actuarist's ability to make precise statements about a population of human beings as time passes—what percentage will die within the next year, and so on. We have his limitations too in his inability to single out the individual human beings who will make up that percentage. We will see also, that by a suitable combination of the angular momentum and energy integrals, a time-invariant statement analogous to the Jacobi integral in the restricted three-body problem may in fact be made in the case of certain general three-body problems.

Szebehely (1967) introduced a useful system of clarification of the dynamic behaviour of the general three-body problem. Before using it, we set up the equations of motion and define certain quantities.

Let $i = 1, 2, 3$ denote the three bodies. Let I be the moment of inertia of the system, T the total kinetic energy, U the force function, C the total energy of the system. Take $\mathbf{r}_i$ as the position vector of the ith body, of mass m_i, and take $\mathbf{r}_{ij} = \mathbf{r}_j - \mathbf{r}_i$ as the position vector of the jth body with respect to the ith. The equations of motion are then

$$m_i \ddot{\mathbf{r}}_i = \nabla_i U \tag{5.85}$$

where the force function U is defined in the usual way by

$$U = \tfrac{1}{2} G \sum_{i=1}^{3} \sum_{j=1}^{3} \frac{m_i m_j}{r_{ij}}$$

G being the constant of gravitation and ∇_i the grad operator of the ith body.

From these equations we have the 10 integrals, including the energy relation

$$T - U = C.$$

The moment of inertia I is given by

$$I = \sum_{i=1}^{3} m_i r_i^2 .$$

Now we know by the virial theorem that for positive energy $(C > 0)$ the system must split up, since in this case

$$\frac{\mathrm{d}^2}{\mathrm{d}t^2} \left(\sum_{i=1}^{3} m_i r_i^2 \right) = 2U + 4C$$

or

$$\frac{\mathrm{d}^2 I}{\mathrm{d}t^2} > 0.$$

Then either one mass recedes to an infinite distance (the other two forming a binary), or all three depart on hyperbolic orbits. Szebehely terms the former occurrence *escape* (sometimes called *hyperbolic–elliptic*); the latter he calls *explosion*.

5.11.1 The case $C < 0$

The case of total negative energy $(C < 0)$ is more complicated and is best split into a number of classes, though it may be remarked that in any system one class of dynamic behaviour does not necessarily preclude another. In interplay the masses follow complicated trajectories, including close approaches to each other so that on many occasions $|\mathbf{r}_{ij}| < r$, a small distance. This may be followed by *ejection*, when two bodies form a binary while the third departs with elliptic velocity relative to the centre of mass of the binary. If the third body achieves escape velocity it will recede indefinitely, so that this event may also be classed as *escape* (or *hyperbolic–elliptic*).

If the semimajor axis of the third body's perturbed elliptic motion about the binary's centre of mass is large compared with the binary component's separation, the configuration is relatively stable; we may recall that such a configuration is common in triple stellar systems. Szebehely classifies this as *revolution*.

The Lagrange special solutions are of course *equilibrium configurations* but all are unstable (none of the masses is infinitesimal apart from the unlikelihood of the other two having a μ value below Routh's value). Hence if a triple system was set up for any of the Lagrange solutions it would immediately pass into the *interplay* mode.

Periodic orbits are known in the general three-body problem for $C < 0$ but are unstable.

5.11.2 The case for $C = 0$

The case $C = 0$ is a special one. Separating the ranges of total positive from total negative energies, it is unlikely to occur in nature. It can give hyperbolic–parabolic (i.e. *explosion*), or hyperbolic–elliptic (i.e. *escape*) cases.

Summing up, we give a table modelled on one drawn up by Szebehely of the possible modes of behaviour. If there is no escape or explosion the moment of inertia I remains bounded; otherwise $I \to \infty$.

Table 5.1

Sign of C	Behaviour of I	Class of behaviour
Positive	Tends to ∞	Hyperbolic, explosion Hyperbolic–parabolic, explosion Hyperbolic–elliptic, escape
Zero	Tends to ∞	Hyperbolic–elliptic, escape Parabolic, explosion
Negative	Less than a maximum I_0	Interplay Ejection Revolution $\Big\}$ all bounded Equilibrium Periodic orbit Oscillating
	Tends to ∞	Hyperbolic–elliptic, escape Parabolic–elliptic, escape

What the table does not state is the established fact that the vast majority of initial triple configurations end up in escape (after a sufficiently long time) in the hyperbolic–elliptic class. This result is immediately relevant to the understanding of the ratios of the numbers of single, binary and triple stellar systems found in the Galaxy. It is also found that when a triple system breaks up it is the particle with the smallest mass that is usually ejected.

5.11.3 Jacobian coordinates

We introduce a form of the equations of motion of the general three-body problem that is found to be extremely useful in both a lunar problem (for example Earth–Moon–Sun) and the typical triple stellar system problem.

If we let C be the centre of mass of the particles P_1 and P_2 (figure 5.10), then the vector CP_3 (ρ) is taken with the vector P_1P_2 (r) as the position vectors. This set of variables was first introduced by Jacobi and Lagrange.

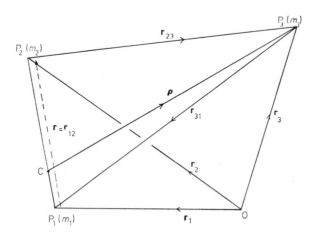

Figure 5.10

Now the relative equations of motion of the three particles may be obtained from equations (5.85) by dividing each by m_i ($i=1, 2, 3$) using the grad operator and using the fact that $\mathbf{r}_{ij}=\mathbf{r}_j-\mathbf{r}_i$. We obtain

$$\left.\begin{array}{l} \ddot{\mathbf{r}}_{12}+GM\dfrac{\mathbf{r}_{12}}{r_{12}{}^3}=Gm_3\,\mathbf{F} \\[2.5ex] \ddot{\mathbf{r}}_{23}+GM\dfrac{\mathbf{r}_{23}}{r_{23}{}^3}=Gm_1\,\mathbf{F} \\[2.5ex] \ddot{\mathbf{r}}_{31}+GM\dfrac{\mathbf{r}_{31}}{r_{31}{}^3}=Gm_2\,\mathbf{F} \end{array}\right\} \qquad (5.86)$$

where

$$M=m_1+m_2+m_3$$

and

$$\mathbf{F}=\frac{\mathbf{r}_{12}}{r_{12}{}^3}+\frac{\mathbf{r}_{23}}{r_{23}{}^3}+\frac{\mathbf{r}_{31}}{r_{31}{}^3}. \qquad (5.87)$$

Now $\mathbf{r}=\mathbf{r}_{12}$, and also $\boldsymbol{\rho}=(m_1/\mu)\mathbf{r}+\mathbf{r}_{23}=(-m_2/\mu)\mathbf{r}-\mathbf{r}_{31}$ (where $\mu=m_1+m_2$), since the vector sum of the sides of a triangle is zero. Then from the

153

first of equations (5.86), we have

$$\ddot{\mathbf{r}} = -G\mu\frac{\mathbf{r}}{r^3} + Gm_3\left(\frac{\mathbf{r}_{23}}{r_{23}{}^3} + \frac{\mathbf{r}_{31}}{r_{31}{}^3}\right)$$

or

$$\ddot{\mathbf{r}} = -G\mu\frac{\mathbf{r}}{r^3} + Gm_3\left(\frac{\boldsymbol{\rho} - \dfrac{m_1}{\mu}\mathbf{r}}{\left|\boldsymbol{\rho} - \dfrac{m_1}{\mu}\mathbf{r}\right|^3} - \frac{\boldsymbol{\rho} + \dfrac{m_2}{\mu}\mathbf{r}}{\left|\boldsymbol{\rho} + \dfrac{m_2}{\mu}\mathbf{r}\right|^3}\right). \tag{5.88}$$

Also

$$\ddot{\boldsymbol{\rho}} = \frac{m_1}{\mu}\ddot{\mathbf{r}} + \ddot{\mathbf{r}}_{23}.$$

Hence, using the second of equations (5.86) and equation (5.88), we have after a little reduction

$$\ddot{\boldsymbol{\rho}} = -\frac{GM}{\mu}\left[m_1\frac{\left(\boldsymbol{\rho} + \dfrac{m_2}{\mu}\mathbf{r}\right)}{\left|\boldsymbol{\rho} + \dfrac{m_2}{\mu}\mathbf{r}\right|^3} + m_2\frac{\left(\boldsymbol{\rho} - \dfrac{m_1}{\mu}\mathbf{r}\right)}{\left|\boldsymbol{\rho} - \dfrac{m_1}{\mu}\mathbf{r}\right|^3}\right]. \tag{5.89}$$

Following Szebehely we define the vector $f(\mathbf{x})$ by $f(\mathbf{x}) = G\mathbf{x}|\mathbf{x}|^{-3}$ and write $\nu = m_1/\mu$, $\nu^* = m_2/\mu$. Then equations (5.88) and (5.89) may be written as

$$\ddot{\mathbf{r}} + \mu f(\mathbf{r}) = (M - \mu)[f(\boldsymbol{\rho} - \nu\mathbf{r}) - f(\boldsymbol{\rho} + \nu^*\mathbf{r})] \tag{5.90}$$

and

$$\ddot{\boldsymbol{\rho}} = -M[\nu^* f(\boldsymbol{\rho} - \nu\mathbf{r}) + \nu f(\boldsymbol{\rho} + \nu^*\mathbf{r})]. \tag{5.91}$$

Equations (5.90) and (5.91) in the Jacobi coordinates form a 12th-order system, the reduction from 18th order to 12th having been essentially effected by the use of the six centre-of-mass integrals. There therefore remain the energy and angular momentum integrals. Their formulation using relations (5.90) and (5.91) is left as an exercise for the reader.

Equations (5.90) and (5.91) may be put in a neater form which will be of immediate use later when we consider the lunar problem (chapter 9) and the three-body stellar problem (chapter 14). Define

$$U = G\left(\frac{m_1 m_2}{r} + \frac{m_2 m_3}{r_{23}} + \frac{m_3 m_1}{r_{31}}\right) = U(\mathbf{r}, \boldsymbol{\rho}). \tag{5.92}$$

It is then readily seen that equations (5.90) and (5.91) take the form

$$\ddot{\mathbf{r}} = \frac{1}{g_1}\frac{\partial U}{\partial \mathbf{r}}, \qquad \ddot{\boldsymbol{\rho}} = \frac{1}{g_2}\frac{\partial U}{\partial \boldsymbol{\rho}} \tag{5.93}$$

where

$$g_1 = \frac{m_1 m_2}{\mu}, \qquad g_2 = \frac{m_3 \mu}{M}. \tag{5.94}$$

5.12 Jacobian Coordinates for the Many-body Problem

The Solar System planetary and satellite systems demonstrate a hierarchical arrangement of the orbit sizes, with few exceptions. In addition, multiple stars (triples, quadruples and so on) likewise favour hierarchical arrangements. Such hierarchical arrangements may be termed *simple* or *general* with the simple case as a special form of the general. The classical Jacobi coordinate system, first generated for simple hierarchical dynamical systems (HDS) can be easily generalized for application to the general HDS. The fact that n-body systems in nature are invariably found in such HDS must say something about their inherent stability and it will be seen that the Jacobi coordinate system exhibits readily why this is so, the bodies being shown to perform disturbed Keplerian orbits.

5.12.1 The equations of motion of the simple n-body HDS

Let n point masses P_i, of masses m_i, have radius vectors $\mathbf{R}_i$, $(i = 1, 2, \ldots, n)$ with respect to an origin O in an inertial system (figure 5.11). Then the mutual radius vector joining P_i to P_j is $\mathbf{r}_{ij}$, where $\mathbf{r}_{ij} = \mathbf{R}_j - \mathbf{R}_i$.

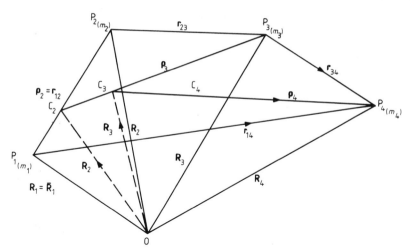

Figure 5.11

Let the vectors $\boldsymbol{\rho}_i$ be defined such that

$$\boldsymbol{\rho}_2 = \mathbf{r}_{12}$$

$$\boldsymbol{\rho}_3 = \text{vector } C_2 P_3$$

where C_2 is the centre of mass of P_1 and P_2,

$$\boldsymbol{\rho}_4 = \text{vector } C_3 P_4$$

where C_3 is the centre of mass of P_1, P_2 and P_3, and so on to vector $\boldsymbol{\rho}_n$, where

$$\boldsymbol{\rho}_n = \text{vector } C_{n-1} P_n$$

C_{n-1} being the centre of mass of all the masses except P_n.

155

The system is now termed a simple hierarchical dynamical system if we further take

$$|\boldsymbol{\rho}_i| < |\boldsymbol{\rho}_{i+1}|.$$

Let the radius vector OC_i be $\bar{\mathbf{R}}_i$.

Then, obviously,

$$\boldsymbol{\rho}_i = \mathbf{R}_i - \bar{\mathbf{R}}_{i-1}. \tag{5.95}$$

The equations of motion of the bodies in the inertial system under Newton's law of gravitation and his three laws of motion are thus:

$$m_i \ddot{\mathbf{R}}_i = \boldsymbol{\nabla}_i' U \qquad (i = 1, \ldots, n) \tag{5.96}$$

where

$$U = \tfrac{1}{2} G \sum_{i=1}^{n} \sum_{j=1}^{n} \frac{m_i m_j}{r_{ij}} \qquad (j \neq i)$$

$$\boldsymbol{\nabla}_i' = \mathbf{i} \frac{\partial}{\partial X_i} + \mathbf{j} \frac{\partial}{\partial Y_i} + \mathbf{k} \frac{\partial}{\partial Z_i}, \qquad \mathbf{i, j, k} \text{ being unit vectors,}$$

and $\mathbf{R}_i = (X_i, Y_i, Z_i)$. Defining

$$M_i = \sum_{j=0}^{i} m_j \qquad (m_0 = 0) \tag{5.97}$$

and using (5.95) and (5.96), we obtain, after some algebra, the equations of motion in a Jacobi coordinate system, namely

$$\frac{m_i M_{i-1}}{M_i} \ddot{\boldsymbol{\rho}}_i = \boldsymbol{\nabla}_i U \qquad (i = 1, 2, \ldots, n) \tag{5.98}$$

where

$$\boldsymbol{\nabla}_i \equiv \frac{\partial}{\partial \boldsymbol{\rho}_i} = \mathbf{i} \frac{\partial}{\partial x_i} + \mathbf{j} \frac{\partial}{\partial y_i} + \mathbf{k} \frac{\partial}{\partial z_i}$$

and $\boldsymbol{\rho}_i = (x_i, y_i, z_i)$.

From equations (5.98) the usual integrals of energy and angular momentum may be formed. In essence, we have already used the system's centre-of-mass integrals in forming the equations of motion in a Jacobi coordinate system.

We now express U as a function of the ρ's. It may be easily shown that

$$\mathbf{r}_{kl} = \boldsymbol{\rho}_l - \boldsymbol{\rho}_k + \sum_{j=k}^{l-1} \frac{m_j}{M_j} \boldsymbol{\rho}_j. \tag{5.99}$$

The relationship may be used to obtain U as a function of the $\boldsymbol{\rho}_i$. An expansion of U in terms of the ratios ρ_i / ρ_j ($i = 2, 3, \ldots, n-1, j = 3, \ldots, n$; $j > 1$), where $\rho_i / \rho_j < 1$, may then be applied, yielding, correct to the second

156

order in ρ_i / ρ_j,

$$\ddot{\boldsymbol{\rho}}_i = GM_i \nabla_i \left[\frac{1}{\rho_i} \left(1 + \sum_{k=1}^{i-1} \epsilon^{ki} P_2(C_{ki}) + \sum_{l=1+i}^{n} \epsilon_{li} P_2(C_{il}) \right) \right] \qquad (5.100)$$

where ϵ^{ki} and ϵ_{li} are small quantities given by

$$\epsilon^{ki} = \frac{m_k M_{k-1}}{M_k M_{i-1}} \alpha_{ki}^2, \qquad \epsilon_{li} = \frac{m_l}{M_i} \alpha_{il}^3 \qquad (i = 2, \ldots, n).$$

In these expressions

$$\alpha_{ij} = \rho_i / \rho_j < 1, \qquad C_{ij} = \frac{\boldsymbol{\rho}_i \cdot \boldsymbol{\rho}_j}{\rho_i \rho_j}, \qquad \rho_i = |\boldsymbol{\rho}_i|$$

while $P_2(x)$ is the Legendre polynomial of order 2 in x.

On examination it is seen that the first term of the right-hand side of each of equations (5.100) represents the undisturbed elliptic motion of the ith mass about the mass centre of the subsystem of masses $m_1, m_2, \ldots, m_{i-1}$. The other terms, and of course the higher-order terms neglected, provide the perturbations of the Keplerian orbit.

5.12.2 The equations of motion of the general n-body HDS

Following Walker (1983), let n point masses be arranged in the system shown in figure 5.12, with $n = 2^q$, q being an integer.

Define parameters M_{ij}, a and b by the relations

$$M_{ij} = \sum_{k=a}^{b} m_k \qquad (5.101)$$

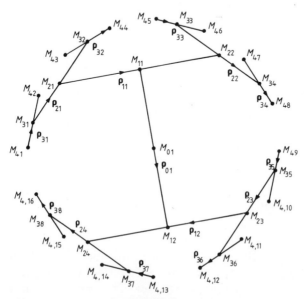

Figure 5.12

where

$$a = (j-1)2^{q-i}+1$$

and

$$b = j2^{2-i}.$$

The parameter M_{ij} denotes the jth *subsystem* in *level i* of the whole n-body system.

Consider, for example, the case $q = 3$. Then we have an eight-body system with the following values of M_{ij}:

$$M_{01} = \sum_{k=1}^{8} m_k$$

$$M_{11} = \sum_{k=1}^{4} m_k, \qquad M_{12} = \sum_{k=5}^{8} m_k$$

$$M_{21} = m_1 + m_2, \qquad M_{22} = m_3 + m_4, \qquad M_{23} = m_5 + m_6$$

$$M_{24} = m_7 + m_8$$

$$M_{3k} = m_k \qquad (k = 1, 2, \ldots, 8).$$

Thus M_{01} is the sum of all the masses in the system. It also represents the zeroth level and that subsystem which is the system itself. It is convenient to take M_{01} in figure 5.12 to represent also the position of the mass centre of the system.

The first level contains $2^1(=2)$ subsystems, the numbering of the masses in M_{11} and M_{12} showing that we are dealing with two separate quadruple systems. Again it is convenient to allow M_{11} and M_{12} to denote the positions of the mass centres of these two quadruple systems.

Progressing in this way to the second level in which there are $2^2(=4)$ subsystems, each M_{2i} ($i = 1, 2, 3, 4$) can denote the mass centres of these subsystems which are binary systems in the case $q = 3$.

The third level contains $2^3(=8)$ subsystems but now the M_{3i} are the eight masses themselves.

In general, then, a system of $n = 2^q$ bodies may be described in this way as consisting of $(q+1)$ levels, with the kth level containing 2^k subsystems, each subsystem in level k being made up of 2^{q-k} bodies.

It should be noted, however, that a general HDS may also be *filled* or *unfilled*. If it is filled, then in the highest level, namely the qth, all the M_{qi} are individual masses. An unfilled system will have one or more of the M_{ki}, $k < q$, representing individual masses.

For example, in figure 5.13, we have a nine-body HDS, the masses being represented by $M_{41}, M_{42}, M_{32}, M_{22}, M_{35}, M_{36}, M_{37}, M_{4,15}, M_{4,16}$.

We now obtain the equations of motion of the general HDS in a generalized Jacobi coordinate system.

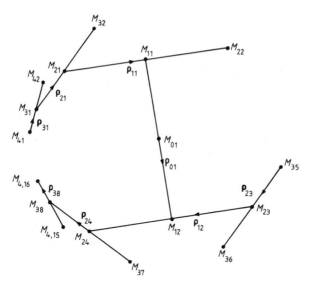

Figure 5.13

Let $\bar{\mathbf{R}}_{ij}$ be the position vectors of the M_{ij}, measured from O in an inertial system. Thus

$$\bar{\mathbf{R}}_{ij} = OM_{ij}.$$

Then,

$$\bar{\mathbf{R}}_{ij} = \frac{1}{M_{ij}} \sum_{h=a}^{b} m_h \mathbf{R}_h \tag{5.102}$$

where

$$\mathbf{R}_h = OP_h$$

P_h being the position of the body of mass m_h and a and b being defined as in (5.101).

Defining the vector $\boldsymbol{\rho}_{ij}$ by

$$\boldsymbol{\rho}_{ij} = M_{i+1,2j-1}M_{i+1,2j} = OM_{i+1,2j} - OM_{i+1,2j-1}$$

we have

$$\boldsymbol{\rho}_{ij} = \bar{\mathbf{R}}_{i+1,2j} - \bar{\mathbf{R}}_{i+1,2j-1}. \tag{5.103}$$

Using equations (5.96) and (5.102) and differentiating equation (5.103) twice, we find that

$$\ddot{\boldsymbol{\rho}}_{ij} = \frac{1}{M_{i+1,2j}} \sum_{g=\alpha}^{\beta} \boldsymbol{\nabla}'_g U - \frac{1}{M_{i+1,2j-1}} \sum_{h=\gamma}^{\delta} \boldsymbol{\nabla}'_h U \tag{5.104}$$

159

where

$$\alpha = (2j-1)2^{q-i-1}+1$$

$$\beta = 2j2^{q-i-1}$$

$$\gamma = (2j-2)2^{q-i-1}+1$$

$$\delta = (2j-1)2^{q-i-1}$$

and $\mathbf{\nabla}'_g$ is the gradient operator associated with $\mathbf{R}_g$.

After a little reduction, equations (5.104) may be transformed to give the required equations of motion in generalized Jacobi coordinates, namely

$$\frac{M_{i+1,2j}M_{i+1,2j-1}}{M_{ij}}\ddot{\mathbf{\rho}}_{ij} = \mathbf{\nabla}_{ij}U \qquad (5.105)$$

where $i=0,\dots,q-1$, $j=1,\dots,2^i$, and $\mathbf{\nabla}_{ij}$ is the gradient operator associated with $\mathbf{\rho}_{ij}$.

The force function U is now expanded in a manner analogous to the way in which it was expanded in Section 5.12.1, the expansion being now carried out in terms of the ratios α^{ij}_{kl}, defined by

$$\alpha^{ij}_{kl} = \rho_{ij}/\rho_{kl}$$

where $i=0,1,\dots,q-1$; $j=1,2,\dots,2^i$; $k=0,1,\dots,q-1$; $l=1,2,\dots,2^k$; $k<i$ and all α^{ij}_{kl} are less than unity.

The expansion involves expressing $\mathbf{r}_{kl}$ as a function of the $\mathbf{\rho}_{ij}$. After some algebra, details of which may be found in the paper by Walker (1982), the resulting expression is found to be

$$\mathbf{r}_{kl} = \sum_{h=0}^{q-1}\left((-1)^{g_h(l)}\frac{M_{q-h,f(g_h(l))}}{M_{q-h-1,g_{h+1}(l)}}\mathbf{\rho}_{q-h-1,g_{h+1}(l)}\right.$$

$$\left. -(-1)^{g_h(k)}\frac{M_{q-h,f(g_h(k))}}{M_{q-h-1,g_{h+1}(k)}}\mathbf{\rho}_{q-h-1,g_{h+1}(k)}\right), \qquad (5.106)$$

where

$$f(j)=j-(-1)^j, \qquad g(j)=\text{int}\left(\frac{j+1}{2}\right)$$

int (x) denoting the integer part of x.

Applying expression (5.106) to the expression for U, namely

$$U = \tfrac{1}{2}G\sum_{i=1}^{n}\sum_{j=1}^{n}\frac{m_i m_j}{r_{ij}} \qquad (j\neq i)$$

and expanding, it is found that, correct to the second order in the ratios of

160

the smaller to the larger radius vectors, namely the α^{ij}_{kl},

$$U = \tfrac{1}{2}G \sum_{l=1}^{n} \sum_{k=1}^{n} \frac{M_{qk}M_q}{M_{q-a-1,g_{a+1}(l)}}$$

$$\times \left\{ 1 + \sum_{h=0}^{a-1} \left[F^2_{qhl}(\alpha^{q-h-1,g_{h+1}(l)}_{q-a-1,g_{a+1}(l)})^2 P_2(C^{q-h-1,g_{h+1}(l)}_{q-a-1,g_{a+1}(l)}) \right. \right.$$

$$\left. \left. + F^2_{qhk}(\alpha^{q-h-1,g_{h+1}(k)}_{q-a-1,g_{a+1}(k)})^2 P_2(C^{q-h-1,g_{h+1}(k)}_{q-a-1,g_{a+1}(k)}) \right] \right\} \tag{5.107}$$

where

$$F_{qhl} = (-1)^{g_h(l)} \frac{M_{q-h,f(g_h(l))}}{M_{q-h-1,g_{h+1}(l)}}$$

$$\rho_{i,j} = |\boldsymbol{\rho}_{i,j}|$$

$$\alpha^{i,j}_{k,l} = \rho_{i,j}/\rho_{k,l}$$

$$C^{i,j}_{k,l} = \frac{\boldsymbol{\rho}_{i,j} \cdot \boldsymbol{\rho}_{k,l}}{\rho_{i,j}\rho_{k,l}}$$

and $P_2(x)$ is the Legendre polynomial of order 2 in x.

Inspection of (5.105) and (5.107) shows that the first term on the right-hand side of (5.107) provides the unperturbed Keplerian motion of the $\boldsymbol{\rho}_{ij}$ radius vectors. The other terms in (5.107) and, of course, the terms neglected in the expansion, provide the perturbations in the Keplerian orbits.

5.12.3 An unambiguous nomenclature for a general HDS

Consider the nine-body system in figure 5.13. A short-hand description of this unfilled five-level general HDS is obviously desirable. It is provided unambiguously by the formula 9(5(3,2),4), arrived at by progressively breaking down the nine-body system until it is composed of a number of simple HDS. Thus the nine-body system is composed of a five-body system ($M_{4,16}$, $M_{4,15}$, M_{37}, M_{36}, M_{35}) and a four-body system (M_{41}, M_{42}, M_{32}, M_{22}). The latter is already a simple HDS but the former can be further broken down into a three-body ($M_{4,16}$, $M_{4,15}$, M_{37}) and a two-body (M_{36}, M_{35}) system.

The filled sixteen-body general HDS of figure 5.12 is a 16(8(4(2,2), 4(2,2)), 8(4,(2,2), 4(2,2))) system while the multiple star Castor is a 6(4(2,2),2) system, consisting as it does of three close binaries (say A, B and C) where the centres of mass of A and B orbit each other while C orbits the centre of mass of A and B at a distance far greater from it than A is separated from B.

5.13 The Hierarchical Three-body Stability Criterion

Let the three-body system consist of three finite point-masses P_1, P_2 and P_3 of masses m_1, m_2 and m_3, respectively. Suppose P_2 is in orbit about P_1, with P_3

in orbit about the centre of mass C_{12} of P_1 and P_2. Then equations (5.90) and (5.91) give the behaviour of P_2 with respect to P_1 and of P_3 with respect to C_{12}. Let $|\rho| > |\mathbf{r}|$.

Such a system may be termed a hierarchical dynamical system, consisting as it does of a binary (P_1-P_2) about which a third body orbits.

In recent years a number of authors (see, for example, Marchal and Saari 1975, Zare 1976, 1977, Szebehely and Zare 1977) have shown that it is possible to establish a condition enabling a decision to be made about the permanency or otherwise of the binary. This is analogous to the use of surfaces of zero velocity in the restricted three-body problem to investigate whether or not the massless particle must remain in orbit about one of the massive particles.

For example let the energy and angular momentum integrals be formed from equations (5.90) and (5.91). Let the total energy be E and the total angular momentum vector be $\mathbf{C}$. Then it may be shown that the stability or otherwise of the binary is controlled by the value of the parameter $S = |\mathbf{C}^2|E$. The value of S is of course known from the initial values of the masses and the position and velocity components appearing in the energy and angular momentum relations. If S is smaller than or equal to a critical quantity, S_{cr}, which can be computed from the values of the three masses applied to the Lagrange collinear solution of the three-body problem (section 5.8) then the binary part of the configuration cannot be broken up by the third mass. If, however, $S > S_{cr}$, then break-up may occur. The criterion, $S \leqslant S_{cr}$, may therefore be usefully applied to any general three-body problem of the hierarchical type (binary plus third body) found in nature. Examples of these are the triple stellar systems, planet–moon–Sun, Sun–Jupiter–Saturn, but in each case, although the general three-body problem model is a close approximation to the system found in nature, the presence of other perturbing bodies cannot be totally disregarded. We will return to this topic in chapter 8.

Problems

5.1 Show that, if an exact solution of the n-body problem were available, an infinite number of other solutions could be generated from it by multiplying all the linear dimensions by a constant factor D and all the time intervals by $D^{3/2}$.

5.2 In the two-body problem, what form do the surfaces of zero velocity take for the orbit of one body with respect to another? What type of orbit must the body have if it is to touch the surface of zero velocity?

5.3 A system of n particles of masses m_i $(i=1, 2, \ldots n)$ moves under the action of a law of gravitation such that the force of attraction between each pair of particles is directly proportional to the product of their masses and directly proportional to the distance between them. Show that under such a law the orbit of any particle about any other particle is an ellipse with the other particle at the centre of the ellipse, and that the orbit of any particle with respect to the centre of mass of the system is also an ellipse.

5.4 In the system of problem 5.3, what is the period of such orbits? How does the centre of mass of the system behave?

References

Abhyanker K D 1959 *Astron. J* **64** 163
Broucke R 1968 *JPL Tech. Rep.* 32-1168
Carpenter L and Stumpff K 1968 *Astron. Notes* **291** 25
Colombo G, Franklin F A and Munford C M 1970 *Astron. J.* **73** 111
Danby J M A 1962 *Fundamentals of Celestial Mechanics* (New York: Macmillan) ch 8
Deprit A and Henrard J 1965 *Astron. J.* **70** 271
—— 1967 *Astron. J.* **72** 158
Frangakis C 1973 *Astrophys. and Space Sci.* **23** 17
Goudas C L 1961 *Bull. de la Soc. Math. de Grece; Nouv. Ser.* **2** 1
Hénon M 1965a *Ann. Astrophys.* **28** 499, 992
—— 1965b *Ann. Astrophys.* **1** part 1 p 49, part 2 p 57
—— 1966a *Bull. Astron.* **1** part 1 p 49
—— 1966b *Bull. Astron.* **1** part 2 p 57
—— 1969 *Astron. Astrophys.* **1** 223
—— 1970 *Astron. Astrophys.* **9** 24
—— 1973 *Cel. Mech.* **8** 269
Hénon M and Heiles C 1964 *Astrophys. J.* **69** 73
Marchal C and Saari D 1975 *Cel. Mech.* **12** 115
Markellos V V 1974a *Cel. Mech.* **9** 365
—— 1974b *Cel. Mech.* **10** 87
Markellos V V, Black W and Moran P E 1974 *Cel. Mech.* **9** 507
Markellos V V, Moran P E and Black W 1975a *Astrophys. and Space Sci.* **33** 129
—— 1975b *Astrophys. and Space Sci.* **33** 385
Message P J 1959 *Astron. J.* **64** 226
Ovenden M W and Roy A E 1961 *Mon. Not. R. Astron. Soc.* **123** 1
Pars L A 1965 *Treatise on Analytical Dynamics* (London: Heinemann)
Poincaré H 1895 *Les Méthodes Nouvelles de la Mécanique Celeste* (Paris: Gauthier–Villars)
Rabe E 1961 *Astron. J.* **66** 500
—— 1962 *Astron. J.* **67** 382
Roy A E and Ovenden M W 1955 *Mon. Not. R. Astron. Soc.* **115** 297
Schanzle A F 1967 *Astron. J.* **72** 149
Sinclair A T 1970 *Mon. Not. R. Astron. Soc.* **148** 289
Szebehely V 1967 *Theory of Orbits* (New York: Academic)
Szebehely V and Zare K 1977 *Astron. Astrophys.* **58** 145
Walker I W 1983 *Cel. Mech.* **29** 149
Zare K 1976 *Cel. Mech.* **14** 73
—— 1977 *Cel. Mech.* **16** 35

Bibliography

Brouwer D and Clemence G M 1961 *Methods of Celestial Mechanics* (New York and London: Academic)
Giacaglia G E D (ed) 1969 *Periodic Orbits, Stability and Resonances* (Dordrecht: Reidel)
Plummer H C 1918 *An Introductory Treatise on Dynamical Astronomy* (London: Cambridge University Press)
Rutherford D E 1948 *Vector Methods* (London and Edinburgh: Oliver & Boyd, New York: Interscience)
Smart W M 1953 *Celestial Mechanics* (London and New York: Longmans)
Sterne T E 1960 *An Introduction to Celestial Mechanics* (New York: Interscience)
Tapley B D and Szebehely V (ed) 1973 *Recent Advances in Dynamical Astronomy* (Dordrecht: Reidel)
Tisserand F 1889 *Traité de la Mécanique Céleste* (Paris: Gauthier-Villars)
Whittaker E T 1959 *A Treatise on the Analytical Dynamics of Particles and Rigid Bodies* (Cambridge: Cambridge University Press)

6 General Perturbations

6.1 The Nature of the Problem

It has been seen that whereas the two-body problem can be solved completely, the many-body problem, apart from special cases and a few general results, is insoluble in the sense that analytical expressions describing the behaviour of the bodies for all time cannot be obtained. Even the two-body case, where one of the bodies is of arbitrary shape and mass distribution, cannot in general be solved in closed form.

Later in this chapter, however, it will be seen that in the case of a planet in the Solar System, and in the case of the motion of a close satellite about a nonspherical planet, a potential function U can be formed such that

$$U = U_0 + R$$

where U_0 is the potential function due to the point-mass two-body problem and R is a potential function due either to any other attracting masses in the system, or to the oblateness of the planet about which the body revolves. The effect of R (the so-called disturbing function) is usually at least an order of magnitude smaller than that due to U_0. If it is, then either general or special perturbation methods may be used to obtain the future behaviour of the body to any desired degree of accuracy; if it is not, as happens in close approaches of a comet to Jupiter or at certain stages in an Earth–Moon voyage, then the methods of special perturbations described in the next chapter must be used.

Many general perturbation theories make use of the fact that the two-body orbit of the body due to U_0 only changes slowly due to R, and they attempt to obtain analytical expressions for the changes in the orbital elements due to R valid within a certain time interval. If the elements of the orbit (let it be an ellipse) are a_0, e_0, i_0, Ω_0, ω_0 and τ_0 at time t_0, the ellipse with these elements is called the *osculating ellipse* while the elements are referred to as the *osculating elements* at time t_0. The velocity of the disturbed planet at this time in its osculating ellipse is equal to its velocity in the actual orbit.

Because of the presence of R the elements at a future time t_1 will be a_1, e_1, i_1, Ω_1, ω_1 and τ_1 and the quantities $(a_1 - a_0)$ etc are the *perturbations* of the elements in the interval $(t_1 - t_0)$. It is obvious that corresponding to these perturbations in the elements there are perturbations in the coordinates and velocity components. If the two-body formulae of chapter 4 were used to obtain the position (x, y, z) and velocity $(\dot{x}, \dot{y}, \dot{z})$ at time t_1 from the osculating elements at time t_0, these quantities would differ from the corresponding quantities (x', y', z') and $(\dot{x}', \dot{y}', \dot{z}')$ computed for time t_1 from the osculating elements at that time. The differences $(x-x')$ etc are the perturbations in the coordinates, etc.

The power of using the two-body conic-section solution as an intermediate orbit lies in its close approximation, at least for a considerable time, to the actual orbit of the body. Attempts have been made to use even closer approximations to the actual orbit as intermediate orbits, a notable example being that used by Hill in his lunar theory. In the case of an artificial satellite it is possible, as we shall see later, to choose as a first approximation an orbit that is a far more accurate description of the motion than a simple Keplerian ellipse.

General perturbations are useful not only in giving future positions of the body, but also because they enable the source of certain observed perturbations to be discovered. This is because the various parts of the disturbing function enter the analytical expressions explicitly. For example, the discovery of the Earth's pear shape by O'Keefe, Eckels, and Squires was made from a study of long-period perturbations of the orbit of Earth satellite 1958 ($\beta 2$) due to the third harmonic in the Earth's gravitational potential.

In the sections that follow we consider the method of the variation of parameters since it exhibits the basic ideas and results of general perturbation theory. We also note several useful methods of splitting up the disturbing force.

6.2 The Equations of Relative Motion

For our discussion in later sections of special and general perturbation methods, it is useful to have the differential equations of relative motion of n bodies $(n > 2)$ where the origin is taken to be the centre of one of the bodies.

We had (from section 5.2) the relation

$$m_i \ddot{\mathbf{R}}_i = G \sum_{j=1}^{n} \frac{m_i m_j}{r_{ij}^3} \mathbf{r}_{ij} \quad (j \neq i, \quad i = 1, 2 \ldots n)$$

where

$$\mathbf{r}_{ij} = \mathbf{R}_j - \mathbf{R}_i = -\mathbf{r}_{ji}.$$

Let the reference body be that of mass m_1. Its equation of motion is then

$$\ddot{\mathbf{R}}_1 = G \sum_{j=2}^{n} \frac{m_j}{r_{1j}^3} \mathbf{r}_{1j} \tag{6.1}$$

while the equation of motion of another particle of mass m_i is

$$\ddot{\mathbf{R}}_i = G \sum_{j=1}^{n} \frac{m_j}{r_{ij}^3} \mathbf{r}_{ij}. \quad (i \neq 1, \quad j \neq i) \tag{6.2}$$

Subtracting (6.1) from (6.2) we obtain

$$\ddot{\mathbf{r}}_{1i}=\ddot{\mathbf{R}}_i-\ddot{\mathbf{R}}_1=G\left[-(m_i+m_1)\frac{\mathbf{r}_{1i}}{r_{1i}{}^3}+\sum_{j=2}^{n}m_j\frac{\mathbf{r}_{ij}}{r_{ij}{}^3}-\sum_{j=2}^{n}m_j\frac{\mathbf{r}_{1j}}{r_{1j}{}^3}\right]$$

where again the case $i=j$ is not included in the summation. Now

$$\mathbf{r}_{ij}=\mathbf{r}_{1j}-\mathbf{r}_{1i}$$

so that

$$r_{ij}{}^3=[(\mathbf{r}_{1j}-\mathbf{r}_{1i})\bullet(\mathbf{r}_{1j}-\mathbf{r}_{1i})]^{3/2}.$$

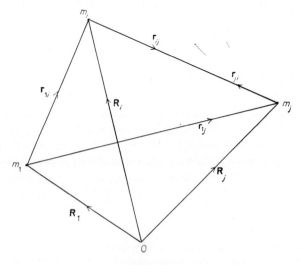

Figure 6.1

Hence

$$\ddot{\mathbf{r}}_{1i}+G(m_1+m_i)\frac{\mathbf{r}_{1i}}{r_{1i}{}^3}=G\sum_{j=2}^{n}m_j\left(\frac{\mathbf{r}_{1j}-\mathbf{r}_{1i}}{r_{ij}{}^3}-\frac{\mathbf{r}_{1j}}{r_{1j}{}^3}\right). \qquad (j\neq i)$$

Dropping the suffix 1, we have

$$\ddot{\mathbf{r}}_{i}+G(m+m_i)\frac{\mathbf{r}_{i}}{r_{i}{}^3}=G\sum_{j=2}^{n}m_j\left(\frac{\mathbf{r}_{j}-\mathbf{r}_{i}}{r_{ij}{}^3}-\frac{\mathbf{r}_{j}}{r_{j}{}^3}\right). \qquad (j\neq i) \qquad (6.3)$$

This is the equation of motion of mass m_i relative to the mass m. The set of such equations $i=2, 3\ldots n$ is the set of required equations of relative motion of the system. It is seen that:

(i) if the other masses m_j ($j\neq i$) do not exist or are vanishingly small, the right-hand side of the equation may be made zero, giving the two-body equation of motion of a mass m_i about a mass m,

(ii) the terms on the right-hand side indicated by the first term in the bracket are the accelerations on mass m_i due to the masses m_j ($j\neq i$),

166

(iii) the other terms on the right-hand side are the negative of the accelerations on the mass m due to the masses m_j $(j \neq i)$.

The right-hand side therefore consists of the perturbations by the masses m_j $(j \neq i)$ on the orbit of m_i about m. In the planetary system, m is the Sun's mass, with m_j/m no more than 10^{-3} even for Jupiter, so for that reason alone the right-hand sides are of small effect.

If we consider the three-body system Sun, Earth and Moon with the Earth as origin, the Moon as mass m_i and the Sun having mass m_j, then

$$m_j \sim 330\,000\,(m+m_i)$$

and it is found that the Sun's force on the Moon is much greater than the Earth's force on the Moon; yet the Moon still revolves about the Earth. Some other factor must therefore be involved to explain this at first computation seemingly paradoxical behaviour of the Earth's satellite. On examining equation (6.3) it is seen that it is the difference of the attractive force of the Sun on the Earth and on the Moon that operates on the right-hand side of the equation. Because both Moon and Earth are at almost the same distance from the Sun, this difference is small compared with the term due to the Earth itself and can be treated as a perturbation of the two-body orbit of the Moon about the Earth.

The two cases (the planets moving about the Sun and disturbing each other's heliocentric orbit, and the Moon in its geocentric orbit disturbed by the Sun) illustrate two entirely different types of problem solved by different applications of general perturbation theory. In the former, the procedure is to use the ratio of the mass of a disturbing planet to that of the Sun as a small quantity, expanding in successive powers of this, while in the latter the ratio of the satellite–planet distance to the Sun–planet distance is essentially the small quantity that is used in the expansion. As mentioned above, even for the case of Jupiter as the disturbing planet $m_j/m \sim 10^{-3}$, while in the Earth–Moon–Sun system $r_{\mathrm{Moon}}/r_{\mathrm{Sun}} \sim 1/400$. In addition to these expansions, auxiliary expansions in powers and products of the eccentricities and inclinations are involved.

In the artificial satellite case the main perturbing effects are due to the nonspherical components of the Earth's gravitational field and to drag by the Earth's atmosphere.

6.3 The Disturbing Function

Let a scalar function R_i be defined by

$$R_i = G \sum_{j=2}^{n} m_j \left(\frac{1}{r_{ij}} - \frac{\mathbf{r}_i \cdot \mathbf{r}_j}{r_j^3} \right) \quad (j \neq i) \tag{6.4}$$

where

$$r_{ij} = [(\mathbf{r}_j - \mathbf{r}_i) \cdot (\mathbf{r}_j - \mathbf{r}_i)]^{1/2}$$

$$= [(x_j - x_i)^2 + (y_j - y_i)^2 + (z_j - z_i)^2]^{1/2}$$

and

$$r_j = (\mathbf{r}_j \cdot \mathbf{r}_j)^{1/2} = (x_j{}^2 + y_j{}^2 + z_j{}^2)^{1/2}.$$

Then

$$\nabla_i R_i = \left(\mathbf{i}\frac{\partial}{\partial x_i} + \mathbf{j}\frac{\partial}{\partial y_i} + \mathbf{k}\frac{\partial}{\partial z_i}\right) \left[G \sum_{j=2}^{n} m_j \left(\frac{1}{r_{ij}} - \frac{\mathbf{r}_i \cdot \mathbf{r}_j}{r_j{}^3}\right)\right]$$

$$= G \sum_{j=2}^{n} m_j \left(\frac{\mathbf{r}_j - \mathbf{r}_i}{r_{ij}{}^3} - \frac{\mathbf{r}_j}{r_j{}^3}\right) \qquad (j \neq i)$$

since $\mathbf{r}_j$ is not a function of x_i, y_i and z_i. Also

$$\nabla_i\left(\frac{1}{r_i}\right) = -\left(\frac{\mathbf{r}_i}{r_i{}^3}\right)$$

and hence equation (6.3) may be written as

$$\ddot{\mathbf{r}}_i = \nabla_i(U_i + R_i) \tag{6.5}$$

where

$$U_i = \frac{G(m + m_i)}{r_i}.$$

The function R_i is called the disturbing function and the treatment of it is the major problem in general perturbations. For each body of mass m_i there is of course a different disturbing function R_i defined by equation (6.4) above.

6.4 The Sphere of Influence

In the case of the near approach of a comet or a space vehicle to a planet, the *sphere of influence* (or sphere of activity) is an almost spherical surface centred on the planet, within which it is more convenient to take the comet or vehicle's planetocentric orbit and consider it as disturbed by the Sun. In the case of the Earth–Moon system a lunar probe will enter a similar sphere of influence about the Moon.

The size of a given sphere may be arrived at from the following considerations. Let the planet P, Sun S and vehicle V have masses m, M and m' where $m \ll M$ and m' is vanishingly small with respect to either. Then by equation (6.3) we have the equation of motion of the vehicle relative to the Sun given by

$$\ddot{\mathbf{r}}_V + G(M + m')\frac{\mathbf{r}_V}{r_V{}^3} = Gm\left(\frac{\mathbf{r}_P - \mathbf{r}_V}{r_{VP}{}^3} - \frac{\mathbf{r}_P}{r_P{}^3}\right). \tag{6.6}$$

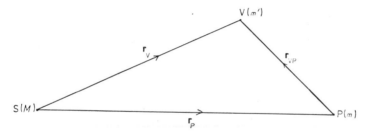

Figure 6.2

The equation of motion of the vehicle relative to the planet is also given by (6.3) and is

$$\ddot{\mathbf{r}}_{VP} + G(m+m')\frac{\mathbf{r}_{VP}}{r_{VP}^3} = GM\left(\frac{-\mathbf{r}_P - \mathbf{r}_{VP}}{r_V^3} + \frac{\mathbf{r}_P}{r_P^3}\right).$$ (6.7)

Neglecting the mass m' and noting that

$$\left.\begin{array}{l}\mathbf{r}_{VP} = \mathbf{r}_V - \mathbf{r}_P \\ \mathbf{r}_V = \mathbf{r}_P + \mathbf{r}_{VP}\end{array}\right\}.$$

we may write equations (6.6) and (6.7) as

$$\ddot{\mathbf{r}}_V + GM\frac{\mathbf{r}_V}{r_V^3} = -Gm\left(\frac{\mathbf{r}_{VP}}{r_{VP}^3} + \frac{\mathbf{r}_P}{r_P^3}\right)$$ (6.8)

and

$$\ddot{\mathbf{r}}_{VP} + Gm\frac{\mathbf{r}_{VP}}{r_{VP}^3} = -GM\left(\frac{\mathbf{r}_V}{r_V^3} - \frac{\mathbf{r}_P}{r_P^3}\right).$$ (6.9)

Introducing $\mathbf{A}_S$, $\mathbf{P}_P$, $\mathbf{A}_P$ and $\mathbf{P}_S$ by the relations

$$\mathbf{A}_S = GM\frac{\mathbf{r}_V}{r_V^3}$$

$$\mathbf{A}_P = Gm\frac{\mathbf{r}_{VP}}{r_{VP}^3}$$

$$\mathbf{P}_P = -Gm\left(\frac{\mathbf{r}_{VP}}{r_{VP}^3} + \frac{\mathbf{r}_P}{r_P^3}\right)$$

$$\mathbf{P}_S = -GM\left(\frac{\mathbf{r}_V}{r_V^3} - \frac{\mathbf{r}_P}{r_P^3}\right)$$

we have

$$\ddot{\mathbf{r}}_V + \mathbf{A}_S = \mathbf{P}_P$$

and

$$\ddot{\mathbf{r}}_{VP} + \mathbf{A}_P = \mathbf{P}_S.$$

The ratios $|\mathbf{P}_P|/|\mathbf{A}_S|$ and $|\mathbf{P}_S|/|\mathbf{A}_P|$ give respectively the order of magnitude of the perturbation of the planet on the two-body heliocentric orbit and that of the Sun on the two-body planetocentric orbit. The sphere of influence

169

is taken to be the surface about the planet where these ratios are equal. Outside the surface $|\mathbf{P_P}|/|\mathbf{A_S}|$ is smaller than $|\mathbf{P_S}|/|\mathbf{A_P}|$ so that it is more convenient to consider the vehicle's heliocentric orbit disturbed by the planet; within the surface, the ratio $|\mathbf{P_P}|/|\mathbf{A_S}|$ is larger than $|\mathbf{P_S}|/|\mathbf{A_P}|$, showing that it is better in this region to consider the planetocentric orbit disturbed by the Sun.

In practice r_{VP} is always much less than r_V and r_P in magnitude, and Tisserand showed that the surface was therefore almost spherical, its radius r_A being given by

$$r_A = \left(\frac{m}{M}\right)^{2/5} r_P. \tag{6.10}$$

In the case of the Earth–Moon system, the radius of the Moon's sphere of activity is given by

$$r'_A = \left(\frac{m'}{M'}\right)^{2/5} r_M \tag{6.11}$$

where r_M is the Moon's geocentric distance while m' and M' are the masses of the Moon and the Earth respectively. The sizes of the spheres of influence of the planets are listed in table 12.1 in chapter 12.

A more refined criterion leads to two spheres of influence and is of use in feasibility studies in astrodynamics. If we may neglect the perturbation of the planet on the vehicle when it is less than a certain small fraction ϵ_P of the two-body heliocentric acceleration, then

$$|\mathbf{P_P}| = \epsilon_P |\mathbf{A_S}|$$

defines an outer sphere of influence beyond which the ordinary two-body equations may be used; again the relation

$$|\mathbf{P_S}| = \epsilon_S |\mathbf{A_P}|$$

gives a second *inner* sphere of influence within which the perturbation due to the Sun is less than ϵ_S times the planetary two-body acceleration, so the ordinary two-body equations for planet and vehicle may be used. Within the shell bounded by the two spheres, some form of general or special perturbation method would be utilized to complete the vehicle's path across the thickness of the shell unless, as happens in some feasibility studies, the particular problem has conditions that show that the probe does not spend long enough in the shell to depart appreciably from a conic-section orbit.

To derive simple and useful formulae for ϵ_P and ϵ_S from equations (6.8) and (6.9), we note that if the vehicle V lies on the planet–Sun line between the two massive bodies we have the relation

$$|\epsilon_P| = \frac{m}{M}\left(\frac{r^3}{x}\right)\left(\frac{x_P - x}{\rho^3} - \frac{x_P}{r_P^3}\right)$$

where the heliocentric x axis is assumed to lie along the Sun–planet radius vector, and x_P and x are the heliocentric x coordinates of planet and vehicle respectively. Note also the relations

$$r_P = |\mathbf{r}_P|$$

$$\rho = |\mathbf{r}_{VP}|$$

$$r = |\mathbf{r}_V|.$$

Also,

$$|\epsilon_S| = \frac{M}{m}\left(\frac{\rho^3}{X}\right)\left(\frac{X_S - X}{r^3} - \frac{X_S}{r_P^3}\right)$$

where the planetocentric x axis is assumed to lie along the planet–Sun radius vector, while X_S and X are the planetocentric x coordinates of Sun and vehicle respectively.

Now letting

$$x = r$$

$$X = \rho$$

$$x_P = X_S = r_P$$

$$\rho/r = d$$

$$r_P = \rho + r$$

and putting $m/M = m'$, we obtain

$$|\epsilon_P| = \frac{m'}{d^2}\left[1 - \left(\frac{d}{1+d}\right)^2\right]$$

$$|\epsilon_S| = \frac{d^2}{m'}[1 - (1+d)^{-2}]$$

which give values of $|\epsilon_P|$ and $|\epsilon_S|$ for values of $d = \rho/r$.

In the Sun–Earth system for example, the radii of outer and inner spheres about the Earth are 0·0178 and 0·0027 astronomical units (AU) respectively if $\epsilon = 0·01$, as against 0·0062 AU computed from the single relation (6.10). Table 12.2 in Chapter 12 should also be consulted.

6.5 The Potential of a Body of Arbitrary Shape

If (as shown in figure 6.3) we have two particles P_1 and P of masses M_1 and m, then as before

$$M_1\ddot{\mathbf{R}}_1 = GM_1 m\frac{\mathbf{r}_1}{r_1^3}, \quad m\ddot{\mathbf{R}} = -GM_1 m\frac{\mathbf{r}_1}{r_1^3}$$

where

$$\mathbf{r}_1 = \mathbf{R} - \mathbf{R}_1.$$

171

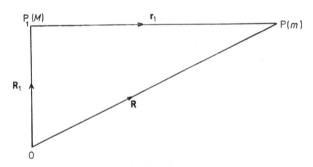

Figure 6.3

The *potential* at P per unit mass due to the presence of P_1 of mass M_1 is then defined to be U, where

$$U = \frac{GM_1}{r_1}.$$

Thus

$$m\ddot{\mathbf{R}} = m\nabla U$$

where

$$\nabla \equiv \mathbf{i}\frac{\partial}{\partial x} + \mathbf{j}\frac{\partial}{\partial y} + \mathbf{k}\frac{\partial}{\partial z}.$$

If now there are a number of masses M_i ($i = 1, 2 \ldots n$) distributed throughout a finite volume, we may take the potential at P to be given by

$$U = G \sum_{i=1}^{n} \frac{M_i}{r_i}.$$

Then

$$m\ddot{\mathbf{R}} = m\nabla U.$$

U is often referred to as the Newtonian potential. So far we have considered only point-masses in the many-body problem; we now consider the case where one or more of the masses are solid bodies of finite size. For simplicity we consider the potential at a point due to one solid body of arbitrary shape and mass distribution, the point being taken to be outside the body. Let the point in figure 6.4 be P, distance r from the centre of mass O of the body. The potential at P due to an element of mass ΔM at a point Q in the body distant ρ from O is then given by

$$\Delta U = G \frac{\Delta M}{PQ}$$

and thus the potential at P due to the whole body is

$$U = G \int \frac{dM}{PQ}$$

the integral being taken over the whole body.

172

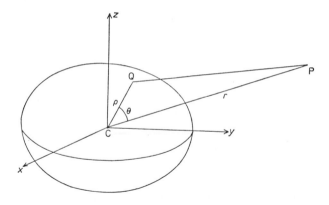

Figure 6.4

Let the coordinates of P and Q be (x, y, z) and (ξ, η, ζ) respectively with respect to a set of rectangular axes (x, y, z) with origin O and fixed in the body. Then

$$PQ^2 = (x-\xi)^2 + (y-\eta)^2 + (z-\zeta)^2 \tag{6.12}$$

$$r^2 = x^2 + y^2 + z^2 \tag{6.13}$$

and

$$\rho^2 = \xi^2 + \eta^2 + \zeta^2. \tag{6.14}$$

From (6.12), (6.13) and (6.14) we have

$$PQ^2 = r^2\left[1 - 2\left(\frac{x\xi+y\eta+z\zeta}{r\rho}\right)\frac{\rho}{r} + \left(\frac{\rho}{r}\right)^2\right].$$

Introducing α, q and θ by the relations

$$\alpha = \rho/r, \quad q = \cos\theta = \frac{x\xi+y\eta+z\zeta}{r\rho}$$

where it is seen that θ is angle $P\hat{O}Q$, we may write

$$PQ^2 = r^2(1 - 2q\alpha + \alpha^2).$$

Then

$$U = G\int\frac{dM}{r(1 - 2q\alpha + \alpha^2)^{1/2}}.$$

Now by definition $\alpha < 1$ and $q \leqslant 1$, so the square root may be expanded in a series to give

$$U = \frac{G}{r}\left(\int P_0\,dM + \int P_1\alpha\,dM + \int P_2\alpha^2\,dM + \ldots + \int P_n\alpha^n\,dM \ldots\right) \tag{6.15}$$

173

where r may be taken outside the integral sign and where

$$P_0 = 1$$
$$P_1 = q$$
$$P_2 = \tfrac{1}{2}(3q^2 - 1)$$
$$P_3 = \tfrac{1}{2}(5q^3 - 3q)$$

etc.

The P_i are Legendre polynomials, functions that occur frequently in mathematical physics (see Appendix II).

We may now write equation (6.15) as

$$U = U_0 + U_1 + U_2 + U_3 + \ldots$$

where

$$U_0 = \frac{G}{r} \int P_0 \, dM$$

$$U_1 = \frac{G}{r} \int P_1 \alpha \, dM$$

$$U_2 = \frac{G}{r} \int P_2 \alpha^2 \, dM$$

etc.

The task is now to evaluate these integrals, as follows:

(i)
$$U_0 = \frac{G}{r} \int dM = \frac{GM}{r}.$$

(ii)
$$U_1 = \frac{G}{r} \int q\alpha \, dM = \frac{G}{r^2} \int \left(\frac{x\xi + y\eta + z\zeta}{r} \right) dM$$

$$= \frac{G}{r^3} \left(x \int \xi \, dM + y \int \eta \, dM + z \int \zeta \, dM \right).$$

Now O is the centre of mass of the body, defined such that

$$\int \xi \, dM = \int \eta \, dM = \int \zeta \, dM = 0.$$

Hence

$$U_1 = 0.$$

(iii)
$$U_2 = \frac{1}{2} \frac{G}{r^3} \int (3q^2 - 1)\rho^2 \, dM$$

$$= \frac{G}{2r^3} \left(3 \int X^2 \, dM - \int \rho^2 \, dM \right)$$

174

where X is the projection of ρ on OP. If ρ makes projections Y and Z on two other axes that together with OP form a rectangular set, then

$$U_2 = \frac{G}{2r^3} \int \left[2\rho^2 - 3(Y^2 + Z^2) \right] dM. \tag{6.16}$$

Now the moments of inertia of the body about the axes Ox, Oy, Oz and OP are respectively

$$A = \int (y^2 + z^2)\, dM$$

$$B = \int (z^2 + x^2)\, dM$$

$$C = \int (x^2 + y^2)\, dM$$

and

$$I = \int (Y^2 + Z^2)\, dM.$$

Hence (6.16) becomes

$$U_2 = \frac{G}{2r^3}(A + B + C - 3I).$$

Most celestial bodies are very nearly spherical, so the U_2 part of the potential is small compared with U_0.

The expression given by U_0, U_1 and U_2, namely

$$U = \frac{GM}{r} + \frac{G}{2r^3}(A + B + C - 3I)$$

is called MacCullagh's formula and is sufficiently accurate for most astronomical purposes.

If the body were a sphere, then

$$A = B = C = I$$

so that

$$U = \frac{GM}{r}.$$

This is the potential of a point mass M, indicating that a sphere of mass M with a radially symmetrical density distribution behaves as far as its potential is concerned as if its mass were concentrated at its centre. This result was first obtained by Newton.

(iv) $$U_3 = \frac{G}{2r^4} \int (5q^3 - 3q)\rho^3\, dM.$$

This leads to a complicated expression containing integrals of the form

$$\int \xi^a \eta^b \zeta^c\, dM$$

where a, b and c are positive integers and

$$a+b+c=3.$$

If the body is symmetrical about all three coordinate planes (for example a homogeneous ellipsoid with three unequal axes) all the integrals vanish so that U_3 is zero. Indeed, all odd U vanish in this case so that

$$U_3=U_5=U_7=\ldots=0.$$

Artificial satellite studies have established that the Earth departs slightly from this condition, being slightly pear-shaped so that U_3 is almost but not quite zero.

Proceeding in this way, it may be shown that the potential at any point of a finite body can be expressed as the sum of various potential functions of the point's position and the body's shape and mass distribution. Since the potential functions other than the one of zero order ($U_0=GM/r$) are factored by various inverse powers of the distance of the point from the body's centre of mass, it is now seen that since in addition the Sun, planets and satellites are substantially spherical, their treatment as point masses is valid to a very high degree of approximation. Indeed the term U_2 enters only when we are dealing with the motions of satellites of oblate planets or with precession and nutation; terms U_3 and higher are used only with close artificial satellites.

It is useful to introduce polar coordinates r, λ, ϕ, where r is already defined and λ and ϕ are the point's (or satellite's) longitude and latitude. Then

$$x=r \cos \lambda \cos \phi$$

$$y=r \sin \lambda \cos \phi$$

$$z=r \sin \phi.$$

The expression for U_2 becomes, after a little reduction,

$$U_2=\frac{G}{r^3}\left[\left(C-\frac{A+B}{2}\right)\left(\frac{1}{2}-\frac{3}{2}\sin^2\phi\right)-\frac{3}{4}(A-B)\cos^2\phi\cos 2\lambda\right].$$

If the body is rotationally symmetrical about the z axis, but not necessarily symmetrical with respect to the equator (that is, it may be pear-shaped), A is equal to B and U_2 becomes

$$U_2=\frac{G}{r^3}(C-A)\left(\frac{1}{2}-\frac{3}{2}\sin^2\phi\right).$$

The Earth's potential at a distance r from its centre of mass may in fact be approximated by the expression

$$U=\frac{GM}{r}\left[1+J\left(\frac{R}{r}\right)^2\left(\frac{1}{3}-\sin^2\phi\right)+H\left(\frac{R}{r}\right)^3\left(\frac{3}{5}-\sin^2\phi\right)\sin\phi\right.$$

$$\left.+K\left(\frac{R}{r}\right)^4\left(\frac{1}{10}-\sin^2\phi+\frac{7}{6}\sin^4\phi\right)+\ldots\right] \quad (6.17)$$

where the constants J, H and K are called the coefficients of the second, third and fourth harmonics of the Earth's gravitational potential; R is the Earth's equatorial radius, and M is the Earth's mass.

If it is assumed that the Earth is a spheroid, then its potential may be written as a series of spherical harmonics of the form

$$U=\frac{GM}{r}\left[1-\sum_{n=2}^{\infty} J_n\left(\frac{R}{r}\right)^n P_n(\sin\phi)\right]$$

where $P_n(\sin\phi)$ is the Legendre polynomial. The first three of these polynomials are

$$P_0(\sin\phi)=1, \quad P_1(\sin\phi)=\sin\phi, \quad P_2(\sin\phi)=\tfrac{1}{2}(3\sin^2\phi-1).$$

The origin here is the centre of mass.

This general result was first obtained by Laplace. It is seen that it corresponds to equation (6.17) with

$$J=\frac{3}{2}J_2, \quad H=\frac{5}{2}J_3, \quad K=-\frac{15}{4}J_4.$$

It will be seen in chapter 10 how a study of the orbits of artificial satellites enables the value of these and the higher-order constants to be found (see also Appendix II).

6.6 Potential at a Point Within a Sphere

We shall find in chapter 15 that we require the expression for the gravitational potential of a massive sphere at a point within it.

Consider first the attraction at O of a spherical shell of density ρ, defined by two concentric spheres of different radii (figure 6.5). Let a cone with

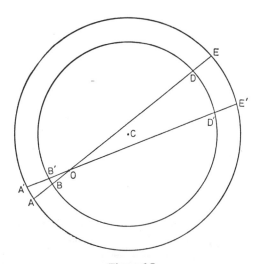

Figure 6.5

vertex O cut the shell as shown, defining two frusta A'B'BA and D'E'ED. If the cone has a small solid angle $d\omega$, then if OB$=r$, the mass of the frusta ABB'A' is $\rho r^2 d\omega$BA and its force of attraction per unit mass at O is $G\rho d\omega$BA. Similarly the attraction per unit mass at O is $G\rho d\omega$DE. But AB$=$DE, since any chord (for example ABODE) makes equal intercepts on the concentric spheres. The attractions are thus equal and opposite. By taking cones in every direction about O, the resultant attraction at O is seen to be zero. Since O is any point inside the shell, the attraction of the shell throughout its interior must be zero. It follows that the potential must be constant at every point and so must equal the potential at the centre C of the shell. By definition, if m is the mass of the shell and a is its radius, the potential is Gm/a.

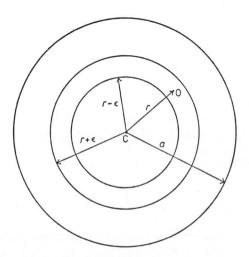

Figure 6.6

A solid sphere can be considered to be made up of concentric shells. Take a point O distant r from its centre. Let a thin shell of matter of thickness 2ϵ and mid-radius r be removed so that O lies within the cavity formed (figure 6.6). Those shells external to the cavity thus exert no force on O, while the shells internal to the cavity exert an attraction as if their mass were gathered at the centre C. Hence, if ρ is the density of the material, the attraction at O is given by the expression

$$\lim_{\epsilon \to 0} \frac{4}{3} G\pi\rho(r-\epsilon)^3/r^2 = \frac{4}{3} G\pi\rho r.$$

We see then that the attraction of a uniform solid sphere at a point inside it is directly proportional to its distance from the centre.

To obtain the gravitational potential at O we recall that the potential due to a sphere of radius $(r-\epsilon)$ and density ρ at a point outside it at a distance

178

$(r-\epsilon)$ from the centre is

$$\frac{4}{3}G\rho\pi(r-\epsilon)^3/(r-\epsilon)=\frac{4}{3}G\rho\pi(r-\epsilon)^2. \qquad (6.18)$$

Let s be the radius of a shell of thickness ds, where $s>r+\epsilon$. Then its mass is $4\pi s^2\rho\,ds$ and the potential it produces at O is, by the previous result for the potential of a spherical shell at a point within it, equal to $4\pi G\rho s\,ds$. For the potential due to all such shells we integrate, giving

$$\int_{r+\epsilon}^{a}4\pi G\rho s\,ds=2\pi G\rho[a^2-(r+\epsilon)^2]. \qquad (6.19)$$

Combining equations (6.18) and (6.19) and taking the limit as ϵ tends to zero, we obtain for the potential of a uniform sphere of radius a at a point within it and distant r from its centre the expression $(2/3)G\pi\rho(3a^2-r^2)$.

In passing, it may be noted that the attraction of a uniform shell bounded by two similar ellipsoids at a point inside the shell is also zero.

6.7 The Method of the Variation of Parameters

Let us consider the case of a planet P of mass m, moving about the Sun of mass M and being disturbed in its heliocentric orbit by a second planet P_1 of mass m_1. Then by equation (6.3) the equation of motion of the planet P is

$$\ddot{\mathbf{r}}+G(M+m)\frac{\mathbf{r}}{r^3}=Gm_1\left(\frac{\mathbf{r}_1-\mathbf{r}}{\rho^3}-\frac{\mathbf{r}_1}{r_1^3}\right) \qquad (6.20)$$

where $\mathbf{r}$, $\mathbf{r}_1$ are the heliocentric radius vectors of planets P and P_1 and

$$\rho=[(\mathbf{r}_1-\mathbf{r})\cdot(\mathbf{r}_1-\mathbf{r})]^{1/2}.$$

We may write equation (6.20), following equation 6.3, as

$$\ddot{\mathbf{r}}=\nabla U=\nabla(U_0+R) \qquad (6.21)$$

where

$$U_0=\frac{G(M+m)}{r}$$

$$R=Gm_1\left(\frac{1}{\rho}-\frac{\mathbf{r}\cdot\mathbf{r}_1}{r_1^3}\right)$$

and

$$\nabla\equiv\mathbf{i}\frac{\partial}{\partial x}+\mathbf{j}\frac{\partial}{\partial y}+\mathbf{k}\frac{\partial}{\partial z}.$$

The corresponding equation of motion of planet P_1 is

$$\ddot{\mathbf{r}}_1+G(M+m_1)\frac{\mathbf{r}_1}{r_1^3}=Gm\left(\frac{\mathbf{r}-\mathbf{r}_1}{\rho^3}-\frac{\mathbf{r}}{r^3}\right) \qquad (6.22)$$

179

or

$$\ddot{\mathbf{r}}_1 = \nabla_1 U_1 = \nabla_1 (U_0' + R_1) \qquad (6.23)$$

where

$$U_0' = \frac{G(M + m_1)}{r_1}$$

and

$$R_1 = Gm \left(\frac{1}{\rho} - \frac{\mathbf{r}_1 \cdot \mathbf{r}}{r^3} \right).$$

If the left-hand sides of equations (6.20) and (6.22) are set equal to zero as a first approximation, the resulting two-body problems may be solved as in chapter 4 giving the Keplerian elliptic undisturbed orbits of the planets about the Sun, each orbit being defined by the six elements.

The coordinates of the planet P may then be written as

$$x = x(a, e, i; \Omega, \omega, \tau; t)$$

$$y = y(a, e, i; \Omega, \omega, \tau; t)$$

$$z = z(a, e, i; \Omega, \omega, \tau; t)$$

where the right-hand sides are functions of the six elements and the time (see section 4.12). Different functions express the velocity components as functions of the elements and the time:

$$\dot{x} = \dot{x}(a, e, i; \Omega, \omega, \tau; t)$$

$$\dot{y} = \dot{y}(a, e, i; \Omega, \omega, \tau; t)$$

$$\dot{z} = \dot{z}(a, e, i; \Omega, \omega, \tau; t).$$

There are corresponding functions for the other planet.

The method of the variation of parameters (in this case the parameters are the elements) supposes that the expressions for the coordinates are now differentiated, the elements now being considered to be variables, and inserted back into equations (6.20) and (6.22) since the variations in the elements are considered to be caused by the so-far neglected right-hand sides of these equations. This process is to be carried out to obtain the differential equations of the elements. Thus we have three equations of the form

$$\frac{dx}{dt} = \sum_{i=1}^{6} \frac{\partial x}{\partial \alpha_i} \frac{d\alpha_i}{dt} + \frac{\partial x}{\partial t} \qquad (6.24)$$

where α_i is any one of the six elements.

Now the equations so far solved are

$$\frac{\partial^2 \mathbf{r}}{\partial t^2} = \nabla U_0, \qquad \frac{\partial^2 \mathbf{r}_1}{\partial t^2} = \nabla_1 U_0' \qquad (6.25)$$

where the partial differential sign $\partial^2/\partial t^2$ signifies that the elements are constants in the solutions of these equations. These give the osculating orbits of the two planets.

180

At any instant t we may suppose that

$$\frac{d\mathbf{r}}{dt} = \frac{\partial \mathbf{r}}{\partial t}, \quad \frac{d\mathbf{r}_1}{dt} = \frac{\partial \mathbf{r}_1}{\partial t} \tag{6.26}$$

which means that the actual velocity vectors at time t are given by differentiating the elliptic formulae, keeping the instantaneous values of the elements constant as implied by the partial differential signs.

By equation (6.24), we therefore obtain

$$\sum_{i=1}^{6} \frac{\partial x}{\partial \alpha_i} \frac{d\alpha_i}{dt} = 0 \tag{6.27}$$

with two similar equations in y and z. Equations (6.27) provide us with three functional relationships for each planet.

Now differentiate the x component equation of the set (6.26) and obtain

$$\frac{d^2 x}{dt^2} = \frac{\partial}{\partial t}\left(\frac{dx}{dt}\right) + \sum_{j=1}^{6} \frac{\partial}{\partial \alpha_j}\left(\frac{dx}{dt}\right)\frac{d\alpha_j}{dt}.$$

But by equation (6.26)

$$\frac{dx}{dt} = \frac{\partial x}{\partial t}$$

and therefore

$$\frac{d^2 x}{dt^2} = \frac{\partial^2 x}{\partial t^2} + \sum_{j=1}^{6} \frac{\partial \dot{x}}{\partial \alpha_j}\frac{d\alpha_j}{dt}.$$

From equations (6.21) and (6.25) we have

$$\frac{d^2 x}{dt^2} = \frac{\partial U_0}{\partial x} + \frac{\partial R}{\partial x}$$

and

$$\frac{\partial^2 x}{\partial t^2} = \frac{\partial U_0}{\partial x}.$$

Hence we may write

$$\sum_{j=1}^{6} \frac{\partial \dot{x}}{\partial \alpha_j}\frac{d\alpha_j}{dt} = \frac{\partial R}{\partial x} \tag{6.28}$$

with two similar equations in y and z.

The six equations (6.27) and (6.28) are then transformed to obtain the six first-order differential equations giving the rates of change of the elements, a transformation first carried out by Lagrange. The resulting equations are

181

$$\frac{da}{dt} = \frac{2}{na}\frac{\partial R}{\partial \chi}$$

$$\frac{de}{dt} = \frac{1}{na^2e}\left\{(1-e^2)\frac{\partial R}{\partial \chi} - \sqrt{(1-e^2)}\frac{\partial R}{\partial \omega}\right\}$$

$$\frac{d\chi}{dt} = -\frac{(1-e^2)}{na^2e}\frac{\partial R}{\partial e} - \frac{2}{na}\left(\frac{\partial R}{\partial a}\right)$$

$$\frac{d\Omega}{dt} = \frac{1}{na^2\sqrt{(1-e^2)}\sin i}\frac{\partial R}{\partial i}$$

$$\frac{d\omega}{dt} = \frac{\sqrt{(1-e^2)}}{na^2e}\frac{\partial R}{\partial e} - \frac{\cot i}{na^2\sqrt{(1-e^2)}}\frac{\partial R}{\partial i}$$

$$\frac{di}{dt} = \frac{1}{na^2\sqrt{(1-e^2)}}\left\{\cot i\frac{\partial R}{\partial \omega} - \mathrm{cosec}\, i\frac{\partial R}{\partial \Omega}\right\}$$

(6.29)

where $n^2a^3 = G(M+m)$ and $\chi = -n\tau$.

These equations are one form of *Lagrange's planetary equations*. There is obviously a corresponding set for the planet of mass m_1.

It should be noted that these equations are rigorous. Although they were originally derived for a perturbation given by another planet, they hold when R is due to many other causes, such as the shape and distribution of mass within a planet acting upon a close satellite. The analytical form of R will depend of course upon the force at work.

A further transformation is to replace the elements ω and χ by ϖ and ϵ where, in the case of a planet, ϖ is the longitude of perihelion (see section 2.6) and the quantity ϵ is called the *mean longitude at the epoch*. This latter quantity is defined in the following way.

The *true longitude L* of the planet, measured from Υ to the ascending node N and then along the great circle that is the intersection of the orbital plane with the celestial sphere, is given by

$$L = \Omega + \omega + f = \varpi + f$$

where f is the true anomaly.

The *mean longitude l* of the planet is given by

$$l = \varpi + M = \varpi + n(t - \tau)$$

where M is the mean anomaly and n is the mean motion as before. Then ϵ is defined by

$$l = nt + \epsilon = \varpi + n(t - \tau).$$

Hence ϵ is the planet's mean longitude at the instant from which time is measured.

The disturbing function R, which was originally expressed in terms of the elements a, e, i, Ω, ω, χ of both planets, now becomes a function R_1 of the elements a, e, i, Ω, ϖ, ϵ where $\varpi = \Omega + \omega$ and $\epsilon = \Omega + \omega + \chi$, since $\chi = -n\tau$.

Then

$$\frac{\partial R}{\partial \Omega} = \frac{\partial R_1}{\partial \Omega} + \frac{\partial R_1}{\partial \varpi} + \frac{\partial R_1}{\partial \epsilon}$$

$$\frac{\partial R}{\partial \omega} = \frac{\partial R_1}{\partial \varpi} + \frac{\partial R_1}{\partial \epsilon}$$

$$\frac{\partial R}{\partial \chi} = \frac{\partial R_1}{\partial \epsilon}.$$

Substituting in the set (6.29), we obtain

$$\frac{da}{dt} = \frac{2}{na}\frac{\partial R}{\partial \epsilon}$$

$$\frac{de}{dt} = -\frac{\sqrt{1-e^2}}{na^2 e}(1-\sqrt{1-e^2})\frac{\partial R}{\partial \epsilon} - \frac{\sqrt{1-e^2}}{na^2 e}\frac{\partial R}{\partial \varpi}$$

$$\frac{d\epsilon}{dt} = -\frac{2}{na}\frac{\partial R}{\partial a} + \frac{\sqrt{1-e^2}(1-\sqrt{1-e^2})}{na^2 e}\frac{\partial R}{\partial e} + \frac{\tan i/2}{na^2\sqrt{1-e^2}}\frac{\partial R}{\partial i}$$

$$\frac{d\Omega}{dt} = \frac{1}{na^2\sqrt{1-e^2}\sin i}\frac{\partial R}{\partial i}$$

$$\frac{d\varpi}{dt} = \frac{\sqrt{1-e^2}}{na^2 e}\frac{\partial R}{\partial e} + \frac{\tan i/2}{na^2\sqrt{1-e^2}}\frac{\partial R}{\partial i}$$

$$\frac{di}{dt} = -\frac{\tan i/2}{na^2\sqrt{1-e^2}}\left(\frac{\partial R}{\partial \epsilon} + \frac{\partial R}{\partial \varpi}\right) - \frac{1}{na^2\sqrt{1-e^2}\sin i}\frac{\partial R}{\partial \Omega}$$

$$(6.30)$$

where the suffix 1 is now omitted.

It may be remarked here that these equations become inconvenient to use if e or i is very small, since e and $\sin i$ appear in some of the denominators on the right-hand sides.

If however the quantities h, k, p and q are defined by the relations

$$h = e \sin \varpi, \qquad k = e \cos \varpi$$

$$p = \tan i \sin \Omega, \qquad q = \tan i \cos \Omega$$

they may be used to form equations for h, k, $\dot p$ and $\dot q$ replacing the equations (6.30) for $\dot e$, $\dot \varpi$, $\dot i$ and $\dot \Omega$.

6.7.1 Modification of the mean longitude at the epoch

A more serious inconvenience in the use of the Lagrange planetary equations in the form (6.30) arises in the following manner.

If the planetary disturbing function R is expanded to give a series of periodic terms, it is found that it takes the form

$$R = Gm_1 \sum P \cos Q + Gm_1 \sum P' \cos Q' \tag{6.31}$$

where the elements a, e and i for both planets appear in the coefficients P and P' while the elements Ω, ϖ and ϵ for both planets appear in the arguments, such that

$$Q = j\Omega + j_1\Omega_1 + k\varpi + k_1\varpi_1 \qquad (6.32)$$

while

$$Q' = h(nt + \epsilon) + h_1(n_1t + \epsilon_1) + j\Omega + j_1\Omega_1 + k\varpi + k_1\varpi_1$$

$$= hl + h_1l_1 + j\Omega + j_1\Omega_1 + k\varpi + k_1\varpi_1$$

where h, h_1, j, j_1, k, k_1 are integers.

In particular, since n and n_1 are functions of a and a_1 respectively through the relation $n^2a^3 = \mu$, it follows that a and a_1 are present explicitly in the coefficients and implicitly in the arguments.

Now in the Lagrange planetary equation for $d\epsilon/dt$, the partial derivative $\partial R/\partial a$ appears. The term in which it occurs is

$$-\frac{2}{na}\frac{\partial R}{\partial a} = -\frac{2}{na}\left(\frac{\partial R}{\partial a}\right) - \frac{2}{na}\frac{\partial R}{\partial l}\frac{\partial l}{\partial n}\frac{dn}{da}$$

where the brackets denote the part of R that arises from the explicit appearance of a in the P and P' coefficients. Then

$$-\frac{2}{na}\frac{\partial R}{\partial a} = -\frac{2}{na}\left(\frac{\partial R}{\partial a}\right) - \frac{2t}{na}\frac{dn}{da}\frac{\partial R}{\partial l}. \qquad (6.33)$$

The variations in the elements are generally small for a considerable interval of time about the osculating epoch, and the method used in solving the set of equations (6.30) is one of successive approximations. Having obtained the partial derivatives of R, the first approximation to the solution is obtained by integrating the resulting equations, the elements being kept constant on the right-hand sides. Hence by equations (6.31) and (6.32) it is seen that the expression

$$-\frac{2}{na}\frac{\partial R}{\partial a} = -\frac{2}{na}\left(\frac{\partial R}{\partial a}\right) - \frac{2t}{na}\frac{dn}{da}\frac{\partial R}{\partial l}$$

will give rise, in the first-order solution to the differential equation for ϵ, to a series where the time appears as a factor in the coefficients of the periodic terms comprising it. These unwelcome mixed terms, as they are called, are avoided as follows.

From equations (6.31) and (6.32) we have

$$\frac{\partial R}{\partial l} = \frac{\partial R}{\partial \epsilon}.$$

Also

$$\frac{dn}{dt} = \frac{dn}{da}\frac{da}{dt} = \frac{2}{na}\frac{dn}{da}\frac{\partial R}{\partial \epsilon}$$

which is obtained by using the first equation of (6.30). Hence equation (6.33) becomes

$$-\frac{2}{na}\frac{\partial R}{\partial a}=-\frac{2}{na}\left(\frac{\partial R}{\partial a}\right)-\frac{t\,dn}{dt}.$$

The equation for ϵ may therefore be written

$$\frac{d\epsilon}{dt}=-\frac{2}{na}\left(\frac{\partial R}{\partial a}\right)-t\frac{dn}{dt}+\frac{(1-e^2)^{1/2}(1-\sqrt{1-e^2})}{na^2e}\frac{\partial R}{\partial e}+\frac{\tan i/2}{na^2\sqrt{1-e^2}}\frac{\partial R}{\partial i}.$$

Let ϵ' be defined by the relation

$$\frac{d\epsilon'}{dt}=\frac{d\epsilon}{dt}+t\frac{dn}{dt}.$$

Then

$$\frac{d\epsilon'}{dt}=-\frac{2}{na}\left(\frac{\partial R}{\partial a}\right)+\frac{(1-e^2)^{1/2}(1-\sqrt{1-e^2})}{na^2e}\frac{\partial R}{\partial e}+\frac{\tan i/2}{na^2\sqrt{1-e^2}}\frac{\partial R}{\partial i}$$

which on integration is found to be without the troublesome mixed terms.
    Now

$$\frac{d\epsilon'}{dt}=\frac{d\epsilon}{dt}+\frac{d}{dt}(nt)-n$$

so that

$$l=nt+\epsilon=\int n\,dt+\epsilon'.$$

Introducing ρ, defined by

$$\rho=\int n\,dt \tag{6.34}$$

we have $l=\rho+\epsilon'$. By equation (6.34) we have $(d\rho/dt)=n$ and also

$$\frac{d^2\rho}{dt^2}=\frac{dn}{dt}=-\frac{3n}{2a}\frac{da}{dt}$$

which gives

$$\frac{d^2\rho}{dt^2}=-\frac{3}{a^2}\frac{\partial R}{\partial \epsilon}. \tag{6.35}$$

We may then use equation (6.30) without change if we note that:

(i) ϵ now means ϵ' so that

$$l=\rho+\epsilon \tag{6.36}$$

(ii) in the term $\partial R/\partial a$ the mean motion n is not to be considered as a function of a when the partial differentiation is carried out,

(iii) that equations (6.35) and (6.36) are added to the set (6.30).

These seeming complications are more than offset by the advantage of eliminating the possibility of having mixed terms. This device is also introduced in artificial satellite theory.

6.7.2 The solution of Lagrange's planetary equations

It has been noted that since the perturbing acceleration is small compared with that due to the two-body potential function, the changes in the orbital elements are small over considerable periods of time. To a first approximation therefore, we may consider the right-hand sides of the equations (6.30) to be functions only of t.

Now

$$R = Gm_1\left(\sum P \cos Q + \sum P' \cos Q'\right)$$

where P, P' are functions of a, a_1, e, e_1, i and i_1,

$$Q = j\Omega + j_1\Omega_1 + k\varpi + k_1\varpi_1$$

and

$$Q' = h(\rho + \epsilon) + h_1(\rho_1 + \epsilon_1) + j\Omega + j_1\Omega_1 + k\varpi + k_1\varpi_1.$$

In the first approximation $\rho = n_0 t$ and $\rho_1 = n_{10}t$, where n_0 and n_{10} are the osculating values of the mean motions at the epoch.

It is then seen that if α is any element, the equations (6.30) may be straightforwardly integrated to give $\alpha = \alpha_0 + \lambda t + $ periodic terms, where λ is a non-zero constant for all elements (except a where it is zero). It is found that the series of periodic terms in the expressions for a, e and i are cosines; those in the expressions for Ω, ϖ and ϵ are sines. For example, the equation for Ω

$$\frac{d\Omega}{dt} = \frac{1}{na^2\sqrt{1-e^2}\sin i}\frac{\partial R}{\partial i}$$

is now written

$$\frac{d\Omega}{dt} = \frac{Gm_1}{n_0 a_0^2 \sqrt{1-e_0^2}\sin i_0}\left(\sum \frac{\partial P}{\partial i}\cos Q + \sum \frac{\partial P'}{\partial i}\cos Q'\right)$$

which on integration gives

$$\Omega = \Omega_0 + \frac{Gm_1}{n_0 a_0^2 \sqrt{1-e_0^2}\sin i_0}\left[\left(\sum \frac{\partial P}{\partial i}\cos Q\right)t + \sum \frac{1}{hn_0 + h_1 n_{10}}\frac{\partial P'}{\partial i}\sin Q'\right]$$

where h and h_1 are integers.

In the case of the equation for the semimajor axis

$$\frac{da}{dt} = -\frac{2Gm_1}{n_0 a_0}\left(\sum hP' \sin Q'\right)$$

where h is an integer, giving $a = a_0 + $ periodic terms.

In this method there are so far no mathematical subtleties, though it is evident that the algebra can be tedious. The real drudgery begins when one proceeds to a second approximation. In the first approximation to the solution of the Lagrange planetary equations the perturbations due to each disturbing planet are independent of those due to the other disturbing planets. If however we proceed to the second approximation, then the perturbations on a planet of mass m of the second order due to a planet of mass m_1 include

186

terms with factors m_1^2 and mm_1; if there is more than one disturbing planet the problem is even more involved. If a third planet of mass m_2 is present then terms with factors m_1m_2 will also appear in the second-order perturbations of the orbit of the planet of mass m.

We do no more now than sketch out the method of obtaining the second-order perturbations in the planetary case where there are only two planets. Let an element α be given by

$$\alpha = \alpha_0 + \Delta_1\alpha + \Delta_2\alpha + \Delta_3\alpha + \ldots$$

where $\Delta_1\alpha$, $\Delta_2\alpha$ etc denote the perturbations in α of the first, second etc orders respectively. The procedure is to expand the right-hand sides of the planetary equations in a Taylor series and collect the terms of the various orders of small quantities. For example, taking the equation for Ω we had

$$\frac{d\Omega}{dt} = \frac{1}{na^2\sqrt{1-e^2}\sin i}\frac{\partial R}{\partial i}$$

or

$$\frac{d\Omega}{dt} = \frac{Gm_1}{na^2\sqrt{1-e^2}\sin i}\left(\sum\frac{\partial P}{\partial i}\cos Q + \sum\frac{\partial P'}{\partial i}\cos Q'\right).$$

Let a function f of the six elements of planet P and the six of planet P_1 be defined by

$$f = \frac{Gm_1}{na^2\sqrt{1-e^2}\sin i}\left(\sum\frac{\partial P}{\partial i}\cos Q + \sum\frac{\partial P'}{\partial i}\cos Q'\right).$$

Expanding f by Taylor's theorem and taking α_i to denote any element, we then have

$$\frac{d}{dt}(\Omega_0 + \Delta_1\Omega + \Delta_2\Omega + \ldots) = f_0 + \sum_i\left(\frac{\partial f}{\partial \alpha_i}\right)_0\Delta_1\alpha_i$$

$$+ \frac{1}{2}\sum_i\sum_j\left(\frac{\partial^2 f}{\partial \alpha_i\partial \alpha_j}\right)_0\Delta_1\alpha_i\Delta_1\alpha_j + \ldots.$$

In the above equation the following points must be made:

(i) f_0 means that in the function f only the osculating values of the elements are used,

(ii) the second term indicates that after forming the quantity $\partial f/\partial \alpha_i$ and evaluating it for the osculating elements (as indicated by the bracket and suffix zero) it is multiplied by the appropriate series already obtained in the first order for $\Delta_1\alpha_i$; the summation sign then indicates that all such products are included,

(iii) the third term indicates that in higher orders, cross-products of the first-order series enter as well as the second partial differentials of function f.

Equating the various orders and remembering that the zero order corresponds to the two-body problem with constant elements, we obtain

$$\frac{d\Omega_0}{dt}=0 \tag{6.37}$$

$$\frac{d(\Delta_1\Omega)}{dt}=f_0 \tag{6.38}$$

$$\frac{d(\Delta_2\Omega)}{dt}=\sum_i \left(\frac{\partial f}{\partial \alpha_i}\right)_0 \Delta_1\alpha_i \tag{6.39}$$

$$\frac{d(\Delta_3\Omega)}{dt}=\frac{1}{2}\sum_i \sum_j \left(\frac{\partial^2 f}{\partial \alpha_i \partial \alpha_j}\right)_0 \Delta_1\alpha_i \Delta_1\alpha_j \tag{6.40}$$

etc.

A similar series of equations results for every other element (and for ρ, obtained from equation (6.35)). The first-order solutions, obtained from all the equations of the type (6.38) and a knowledge of the values of the osculating elements, now enable the solutions of the second-order equations of the type (6.39) to be obtained, giving the second-order perturbations. It is obvious that the process may be continued to higher orders than the second—it is also obvious that the amount of labour increases manifold with each succeeding order. Fortunately, with the exception of the mutual perturbations of the giant planets Jupiter and Saturn it is not necessary to include terms of the third order.

Including perturbations of the second order in the masses, it is found that the solutions giving the elements are of the form

$$\alpha=\alpha_0+\lambda_1 t+\lambda_2 t^2+\text{periodic terms}+[t\times(\text{periodic terms})]$$

where α is any element and α_0, λ_1, λ_2 are constants. (If α is the semimajor axis, however, $\lambda_1=\lambda_2=0$.) It is seen that for all elements except a, not only secular terms but also secular accelerations appear; while for all elements including a, mixed terms are present.

The convergence of such series and their application to the question of the stability of the Solar System have been the subjects of many studies.

It might appear that, although to the second order there are no secular terms in the semimajor axes of the planetary orbits, the presence of secular terms in the eccentricities indicates that the System is basically unstable. This is not so. Even though secular or mixed terms appear, we can say nothing at all about either convergence or stability. In this connection, Sterne has pointed out that the function $\sin(1+a)t$ may be written as a series, viz.

$$\sin t+at\cos t-\frac{1}{2}a^2 t^2 \sin t-\frac{1}{6}a^3 t^3 \cos t+\ldots$$

which is convergent for all t in spite of its mixed terms, while

$$\sin bt = bt - \frac{1}{6}b^3 t^3 + \frac{1}{120}b^5 t^5 - \dots$$

is also convergent for all t though 'secular' terms appear on its right-hand side.

Moreover, it should be remembered that in the application of the method of the variation of parameters sketched above, the use of a Taylor's expansion was justified by assuming that the perturbations of the first order $\Delta_1\Omega$, $\Delta_1 e$ etc were so small that squares, products and higher powers could be neglected. The presence of secular terms in these expressions means, however, that the series obtained can be accurate only over a certain range of time; on that account alone, no statement can be made about the stability of the Solar System from such methods. We shall return to this question in a later chapter. As a method of obtaining accurately the changes in a planetary or an artificial satellite orbit over a considerable time interval, the general perturbation method of the variation of parameters is nonetheless a very useful one.

6.7.3 Short- and long-period inequalities

It has been seen that in the planetary case the disturbing function when expanded is of the form given by equations (6.31) and (6.32). The resulting integration to obtain the first-order perturbations gave periodic terms of the form

$$\frac{A(h, h_1)}{hn + h_1 n_1} \sin[(hn + h_1 n_1)t + B]$$

where

$$B = h\epsilon + h_1\epsilon_1 + j\Omega + j_1\Omega_1 + k\varpi + k_1\varpi_1.$$

$A(h, h_1)$ is a constant, its magnitude being given by the magnitude of the eccentricities and inclinations of which it is a function. As h and h_1 increase, the order of size of $A(h, h_1)$ diminishes rapidly.

Now h is positive while h_1 may be positive or negative. The period T of such a term is given by

$$T = \frac{2\pi}{hn + h_1 n_1}$$

while its amplitude is $A(h, h_1)/(hn + h_1 n_1)$. The mean motions n and n_1 are known quantities derived from observations and are given to so many significant figures. It is always possible to find two integers h and h_1 such that

$$hn + h_1 n_1 < \nu$$

where ν is arbitrarily small.

Normally, most values of h and h_1 are such that $(hn + h_1 n_1)$ is not a particularly small quantity in comparison with hn or $h_1 n_1$ and the periods of such terms are of the same order as the orbital periods of the two planets concerned. Such terms are referred to as *short-period inequalities*.

189

Of more interest are those terms in which a pair of values of h and h_1 make $(hn + h_1 n_1)$ small.

The function $A(h, h_1)$ of the eccentricities and inclinations, it has been seen, is very small if h or h_1 is large so that the amplitude $A(h, h_1)/(hn + h_1 n_1)$ of the libration will not in general become large if $(hn + h_1 n_1)$ becomes small. If, however, a small $(hn + h_1 n_1)$ is obtained for small integral values of h and h_1, the amplitude will be large.

Two orbits where the ratio of the mean motions of the bodies are approximately given by a vulgar fraction in this way are said to be *commensurable*. This phenomenon can give rise to a *long-period inequality* of large amplitude. A striking case of such an inequality exists in the mutual perturbations of Jupiter and Saturn. For these planets $n = 0 \cdot 083091°$ and $n_1 = 0 \cdot 033460°$ per mean solar day respectively.

Putting $h = -2$ and $h_1 = 5$, we have

$$hn + h_1 n_1 = 0 \cdot 001118°.$$

The period of the resulting perturbation is about 900 years. Its effects are most evident in the mean longitude of the planets.

We had $l = \rho + \epsilon$, so that the first-order perturbation in l, written $\Delta_1 l$, is given by

$$\Delta_1 l = \Delta_1 \rho + \Delta_1 \epsilon.$$

Now $\Delta_1 \epsilon$ in its periodic terms gives rise to the short- and long-period inequalities discussed above. The more interesting effect arises from $\Delta_1 \rho$.

By definition

$$\rho = \int n \, dt$$

and hence

$$\rho_0 + \Delta_1 \rho + \Delta_2 \rho + \ldots = \int (n_0 + \Delta_1 n + \Delta_2 n + \ldots) \, dt$$

so that

$$\Delta_1 \rho = \int \Delta_1 n \, dt.$$

Now

$$n^2 a^3 = \mu$$

and thus

$$n_0 + \Delta_1 n + \Delta_2 n + \ldots = \mu^{1/2} (a_0 + \Delta_1 a + \Delta_2 a + \ldots)^{-3/2}$$

giving, on expansion by the binomial theorem,

$$n_0 + \Delta_1 n + \Delta_2 n + \ldots$$

$$= \mu^{1/2} a_0^{-3/2} \left\{ 1 - \frac{3}{2} \frac{\Delta_1 a}{a_0} + \left[\frac{15}{8} \left(\frac{\Delta_1 a}{a_0} \right)^2 - \frac{3}{2} \frac{\Delta_2 a}{a_0} \right] + \ldots \right\}.$$

Hence

$$\Delta_1 \rho = \int \left(-\frac{3}{2} \frac{\mu^{1/2} \Delta_1 a}{a_0^{5/2}} \right) dt$$

$$= -\frac{3n_0}{2a_0} \int \Delta_1 a \, dt.$$

Now $\Delta_1 a$ is of the form

$$\sum \frac{A(h, h_1)}{hn + h_1 n_1} \cos [(hn + h_1 n_1)t + B]$$

where

$$B = h\epsilon + h_1 \epsilon_1 + j\Omega + j_1 \Omega_1 + k\varpi + k_1 \varpi_1.$$

Hence $\Delta_1 \rho$ is of the form

$$\sum \frac{A(h, h_1)}{(hn + h_1 n_1)^2} \sin [(hn + h_1 n_1)t + B].$$

The amplitude of a long-period inequality in the mean longitude is therefore much enhanced by the presence of the square of the small quantity in the denominator. In the case of Jupiter and Saturn, the mean longitudes of these bodies can vary by 21′ and 49′ respectively because of such perturbations.

6.7.4 The resolution of the disturbing force

In the forms of the planetary equations so far discussed the right-hand sides contain the partial derivatives of the disturbing function R with respect to the elements. It has been mentioned without proof that the disturbing function can be expanded by suitable procedures into a series of the form

$$R = Gm_1 \sum P \cos Q + Gm_1 \sum P' \cos Q'$$

where P, Q, P', Q' have the meanings already attached to them. Once the partial derivatives are obtained, integration gives the long and complicated series for each element. To compute numerical values from such series is time-consuming, especially where the eccentricities are large and require that the development be carried out to high powers of e.

A method due to Gauss enables this work to be short-circuited, obtaining the differential equations for the elements in terms of three mutually perpendicular components of the disturbing acceleration. It should be noted that in celestial mechanics and astrodynamics the right-hand side of the equation of relative motion

$$\ddot{\mathbf{r}} + G(M + m)\frac{\mathbf{r}}{r^3} = Gm_1 \left(\frac{\mathbf{r}_1 - \mathbf{r}}{\rho^3} - \frac{\mathbf{r}_1}{r_1^3} \right)$$

is strictly speaking the disturbing acceleration, though it is often referred to as the disturbing force. The three components are S, T and W, where:

S is the radial component directed outwards along the planet's heliocentric radius vector from the planet,

T is the transverse component in the orbital plane, at right angles to S such that it makes an angle less than 90° with the velocity vector,
W is the component perpendicular to the orbital plane and positive towards the north side of the plane.

To introduce S, T and W into the right-hand sides of equations (6.30) we require expressions for $\partial R/\partial \sigma$ in terms of S, T and W, where σ is any element.
It is found that

$$\frac{\partial R}{\partial a}=\frac{r}{a}S$$

$$\frac{\partial R}{\partial e}=-aS\cos f+r\sin f\left(\frac{1}{1-e^2}+\frac{a}{r}\right)T$$

$$\frac{\partial R}{\partial i}=rW\sin u$$

$$\frac{\partial R}{\partial \Omega}=-2rT\sin^2(i/2)-rW\cos u\sin i$$

$$\frac{\partial R}{\partial \epsilon}=\frac{aeS\sin f}{\sqrt{1-e^2}}+\frac{Ta^2\sqrt{1-e^2}}{r}$$

$$\frac{\partial R}{\partial \varpi}=-\frac{\partial R}{\partial \epsilon}+rT$$

where $u=\varpi-\Omega+f$, and f is the true anomaly.
Substituting these expressions into (6.30) we obtain

$$\left.\begin{aligned}
\frac{da}{dt}&=\frac{2}{n\sqrt{1-e^2}}\left(Se\sin f+\frac{pT}{r}\right)\\[2mm]
\frac{de}{dt}&=\frac{\sqrt{1-e^2}}{na}[S\sin f+T(\cos E+\cos f)]\\[2mm]
\frac{di}{dt}&=\frac{Wr\cos u}{na^2\sqrt{1-e^2}}\\[2mm]
\frac{d\Omega}{dt}&=\frac{Wr\sin u}{na^2\sqrt{1-e^2}\sin i}\\[2mm]
\frac{d\varpi}{dt}&=\frac{\sqrt{1-e^2}}{nae}\left[-S\cos f+T\left(1+\frac{r}{p}\right)\sin f\right]+2\frac{d\Omega}{dt}\sin^2\frac{i}{2}\\[2mm]
\frac{d\epsilon}{dt}&=\frac{e^2}{1+\sqrt{1-e^2}}\frac{d\varpi}{dt}+2\frac{d\Omega}{dt}(1-e^2)^{1/2}\left(\sin^2\frac{i}{2}\right)-\frac{2rS}{na^2}
\end{aligned}\right\}\quad(6.41)$$

where E is the eccentric anomaly and $p=a(1-e^2)$. In addition we have

$$\frac{d^2\rho}{dt^2}=-\frac{3}{a\sqrt{1-e^2}}\left(Se\sin f+\frac{pT}{r}\right).$$

It should be noted that the element ϵ is the one defined in section 6.7.1, and is such that the mean longitude l is given by

$$l = p + \epsilon = \int n \, dt + \epsilon.$$

It should also be noted that equations (6.41) as given above would have the same form even if the components of the forces could not be expressed as the differentials of a single function. They therefore hold if, for example, the disturbance is due to drag.

Equations (6.41) are often used in special perturbations for the components S, T and W can be computed at any instant as follows; for one disturbing planet P_1,

$$R = Gm_1 \left(\frac{1}{\Delta} - \frac{xx_1 + yy_1 + zz_1}{r_1^3} \right)$$

where (x, y, z) and (x_1, y_1, z_1) are the heliocentric rectangular coordinates of planets P and P_1 respectively and

$$\Delta^2 = (x - x_1^2) + (y - y_1)^2 + (z - z_1)^2.$$

Then

$$\frac{\partial R}{\partial x} = -Gm_1 \left(\frac{x - x_1}{\Delta^3} + \frac{x_1}{r_1^3} \right)$$

with two similar equations in y and z.

In figure 6.7 it is seen that the components S, T and W form a right-handed rectangular set of axes. Let the direction cosines of these axes be (l_S, m_S, n_S), (l_T, m_T, n_T) and (l_W, m_W, n_W) with respect to OX, OY and OZ.

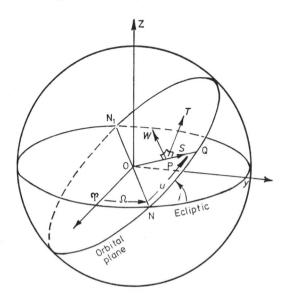

Figure 6.7

Then

$$\frac{\partial R}{\partial x}=l_S S+l_T T+l_W W$$

$$\frac{\partial R}{\partial y}=m_S S+m_T T+m_W W$$

$$\frac{\partial R}{\partial z}=n_S S+n_T T+n_W W.$$

From these we deduce that

$$S=l_S\frac{\partial R}{\partial x}+m_S\frac{\partial R}{\partial y}+n_S\frac{\partial R}{\partial z}$$

$$T=l_T\frac{\partial R}{\partial x}+m_T\frac{\partial R}{\partial y}+n_T\frac{\partial R}{\partial z}$$

$$W=l_W\frac{\partial R}{\partial x}+m_W\frac{\partial R}{\partial y}+n_W\frac{\partial R}{\partial z}.$$

But it is readily seen from figure 6.7 that the direction cosines are expressed by means of the cosine formula in terms of the quantities Ω, ω, i and f. For example

$$l_S=\cos\Omega\cos(\varpi-\Omega+f)-\sin\Omega\sin(\varpi-\Omega+f)\cos i.$$

Hence values of S, T and W may be computed at any time when the elements of P's orbit and the positions of P and P_1 are known.

It is sometimes useful to resolve the perturbing acceleration in a different way by introducing components of the perturbing acceleration T' tangential to the orbit in the direction of motion, and N perpendicular to the tangent (taken to be positive when directed to the interior of the orbit). The tangential component T' and the normal component N replace the components S and T used above. The orthogonal component W is retained as the third component.

It may be easily shown by using equations (4.46) that

$$\left.\begin{aligned}T&=\frac{T'(1+e\cos f)}{\sqrt{1+e^2+2e\cos f}}+\frac{Ne\sin f}{\sqrt{1+e^2+2e\cos f}}\\[2mm]S&=\frac{T'e\sin f}{\sqrt{1+e^2+2e\cos f}}-\frac{N(1+e\cos f)}{\sqrt{1+e^2+2e\cos f}}\end{aligned}\right\}\qquad(6.42)$$

where f is the true anomaly.

This particular resolution is useful in the discussion of drag upon an artificial satellite. If drag is considered to be a negative tangential component and is taken to be the only perturbing force, inspection of equations (6.41) and (6.42) shows that neither Ω nor i changes while the semimajor axis a continually decreases. The other elements suffer changes that will be considered in more detail in chapter 10.

6.8 Lagrange's Equations of Motion

A particular form of the equations of motion due to Lagrange is often used, which involves the concept of *generalized coordinates*. Suppose we have a system of n particles whose coordinates are (x_i, y_i, z_i), where $i = 1, 2 \ldots n$. Let these coordinates be expressible as functions of $3n$ generalized coordinates q_r $(r = 1, 2 \ldots 3n)$ and possibly of the time t.

Thus

$$
\left.
\begin{aligned}
x_i &= x_i(q_r, t) \\
y_i &= y_i(q_r, t) \\
z_i &= z_i(q_r, t)
\end{aligned}
\right\}
\quad
\begin{aligned}
&(i = 1, 2 \ldots n) \\
&(r = 1, 2 \ldots 3n).
\end{aligned}
\tag{6.43}
$$

Then

$$
\dot{x}_i = \frac{\partial x_i}{\partial t} + \sum_{r=1}^{3n} \frac{\partial x_i}{\partial q_r} \dot{q}_r
\tag{6.44}
$$

with similar equations in the y_i and the z_i.

We have then for a particular q (say q_k)

$$
\frac{\partial \dot{x}_i}{\partial \dot{q}_k} = \frac{\partial x_i}{\partial q_k}.
\tag{6.45}
$$

In addition the equations of motion of the n particles are

$$
\left.
\begin{aligned}
m_i \ddot{x}_i &= \frac{\partial U}{\partial x_i} \\[1em]
m_i \ddot{y}_i &= \frac{\partial U}{\partial y_i} \\[1em]
m_i \ddot{z}_i &= \frac{\partial U}{\partial z_i}
\end{aligned}
\right\}
\quad (i = 1, 2 \ldots n).
$$

where U is the force function, or the negative of the potential energy (see section 5.4).

If T is the kinetic energy of the whole system,

$$
T = \frac{1}{2} \sum_{i=1}^{n} m_i(\dot{x}_i{}^2 + \dot{y}_i{}^2 + \dot{z}_i{}^2).
\tag{6.46}
$$

The substitution of the set of equations of the form (6.44) into (6.46) then transforms T to a function $T(q_r, \dot{q}_r, t)$, where $r = 1, 2 \ldots 3n$.

The application of the transformations (6.43) to U, which is given by a function $U(x_i, y_i, z_i)$ $(i = 1, 2 \ldots n)$, changes U to a function $U(q_r, t)$ $(r = 1, 2 \ldots 3n)$. Hence

$$
\frac{\partial T}{\partial \dot{q}_k} = \sum_{i=1}^{n} m_i \left(\dot{x}_i \frac{\partial \dot{x}_i}{\partial \dot{q}_k} + \dot{y}_i \frac{\partial \dot{y}_i}{\partial \dot{q}_k} + \dot{z}_i \frac{\partial \dot{z}_i}{\partial \dot{q}_k} \right)
$$

or

$$\frac{\partial T}{\partial \dot{q}_k} = \sum_{i=1}^{n} m_i \left(\dot{x}_i \frac{\partial x_i}{\partial q_k} + \dot{y}_i \frac{\partial y_i}{\partial q_k} + \dot{z}_i \frac{\partial z_i}{\partial q_k} \right) \qquad (6.47)$$

using (6.45).

Differentiating (6.47) and using (6.45) again, we obtain

$$\frac{\mathrm{d}}{\mathrm{d}t} \left(\frac{\partial T}{\partial \dot{q}_k} \right) = \sum_{i=1}^{n} m_i \left[\left(\ddot{x}_i \frac{\partial x_i}{\partial q_k} + \ddot{y}_i \frac{\partial y_i}{\partial q_k} + \ddot{z}_i \frac{\partial z_i}{\partial q_k} \right) \right.$$
$$\left. + \dot{x}_i \frac{\mathrm{d}}{\mathrm{d}t} \left(\frac{\partial x_i}{\partial q_k} \right) + \dot{y}_i \frac{\mathrm{d}}{\mathrm{d}t} \left(\frac{\partial y_i}{\partial q_k} \right) + \dot{z}_i \frac{\mathrm{d}}{\mathrm{d}t} \left(\frac{\partial z_i}{\partial q_k} \right) \right]$$
$$= \sum_{i=1}^{n} \left[\left(\frac{\partial U}{\partial x_i} \frac{\partial x_i}{\partial q_k} + \frac{\partial U}{\partial y_i} \frac{\partial y_i}{\partial q_k} + \frac{\partial U}{\partial z_i} \frac{\partial z_i}{\partial q_k} \right) + m_i \left(\dot{x}_i \frac{\partial \dot{x}_i}{\partial q_k} + \dot{y}_i \frac{\partial \dot{y}_i}{\partial q_k} + \dot{z}_i \frac{\partial \dot{z}_i}{\partial q_k} \right) \right].$$

The first bracket on the right-hand side is $\partial U/\partial q_k$; the second is $\partial T/\partial q_k$. We thus have

$$\frac{\mathrm{d}}{\mathrm{d}t} \left(\frac{\partial T}{\partial \dot{q}_k} \right) = \frac{\partial U}{\partial q_k} + \frac{\partial T}{\partial q_k}. \qquad (6.48)$$

But U does not contain $\dot{q}_k$. We may therefore, by defining a function L as $L = T + U$, write

$$\frac{\mathrm{d}}{\mathrm{d}t} \left(\frac{\partial L}{\partial \dot{q}_k} \right) = \frac{\partial L}{\partial q_k} \qquad (k = 1, 2 \ldots 3n) \qquad (6.49)$$

which is the standard form of Lagrange's equations. The function L, often called the *kinetic potential* or *the Lagrangian*, is a function of the q, $\dot{q}$ and t.

The *momentum* corresponding to the generalized coordinate q_k is $\partial L/\partial \dot{q}_k$; if L does not contain q_k explicitly, then q_k is termed an *ignorable coordinate* and we see that by equation (6.49) that $\partial L/\partial \dot{q}_k =$ constant. It is also readily seen that if L does not contain t explicitly the Lagrange equations possess an energy integral. In this case

$$\frac{\mathrm{d}L}{\mathrm{d}t} = \sum_{j=1}^{3n} \left(\frac{\partial L}{\partial q_j} \dot{q}_j + \frac{\partial L}{\partial \dot{q}_j} \ddot{q}_j \right)$$
$$= \sum_{j=1}^{3n} \left[\frac{\mathrm{d}}{\mathrm{d}t} \left(\frac{\partial L}{\partial \dot{q}_j} \right) \dot{q}_j + \frac{\partial L}{\partial \dot{q}_j} \ddot{q}_j \right]$$
$$= \frac{\mathrm{d}}{\mathrm{d}t} \left(\sum_{j=1}^{3r} \frac{\partial L}{\partial \dot{q}_j} \dot{q}_j \right).$$

Hence

$$\sum_{j=1}^{3n} \frac{\partial L}{\partial \dot{q}_j} \dot{q}_j - L = C.$$

Now $L = T + U$ and T is of homogeneous quadratic form in $\dot{q}_j$ while U is not a function of $\dot{q}_j$. Hence by Euler's theorem

$$C = 2T - (T + U) = T - U$$

so that C is the total energy in the system.

196

As an illustration of the above ideas, consider a planet moving in an undisturbed heliocentric orbit with rectangular ecliptic coordinates (x, y, z). Suppose we wish to obtain Lagrange's equations of motion of the planet using the generalized coordinates (r, β, λ) where r is the planet's radius vector, β is its ecliptic latitude and λ is its ecliptic longitude. Then

$$x = r \cos \lambda \cos \beta$$

$$y = r \sin \lambda \cos \beta$$

$$z = r \sin \beta.$$

Forming $(\dot{x}, \dot{y}, \dot{z})$ from these transformation equations, the kinetic energy T is found to be

$$T = \frac{1}{2}(\dot{r}^2 + r^2\dot{\beta}^2 + r^2\dot{\lambda}^2 \cos^2 \beta).$$

Also

$$U = U(r, \beta, \lambda) = \frac{\mu}{r}$$

where $\mu = G(M + m)$ and in this case U is a function of r alone.
The equations of motion then follow from equation (6.49) using

$$L = \frac{1}{2}(\dot{r}^2 + r^2\dot{\beta}^2 + r^2\dot{\lambda}^2 \cos^2 \beta) + \frac{\mu}{r}.$$

For the first coordinate r, we have

$$\frac{d}{dt}(\dot{r}) = r\dot{\beta}^2 + r\dot{\lambda}^2 \cos^2 \beta - \frac{\mu}{r^2}$$

or

$$\ddot{r} - r\dot{\beta}^2 - r\dot{\lambda}^2 \cos^2 \beta + \frac{\mu}{r^2} = 0. \tag{6.50}$$

For the second coordinate β, we have

$$\frac{d}{dt}(r^2\dot{\beta}) = -r^2\dot{\lambda}^2 \cos \beta \sin \beta. \tag{6.51}$$

For the third coordinate λ, we have

$$\frac{d}{dt}(r^2\dot{\lambda} \cos^2 \beta) = 0$$

which can be integrated immediately to give

$$r^2\dot{\lambda} \cos^2 \beta = h. \tag{6.52}$$

In addition, since L is not an explicit function of time we have $T - U = C$, or in other words

$$\frac{1}{2}(\dot{r}^2 + r^2\dot{\beta}^2 + r^2\dot{\lambda}^2 \cos^2 \beta) - \frac{\mu}{r} = C. \tag{6.53}$$

Integrals (6.52) and (6.53) are the integrals of angular momentum and energy respectively.

6.9 Hamilton's Canonic Equations

In many textbooks on celestial mechanics large sections are devoted to Hamilton's canonic equations, to the Hamilton–Jacobi method of tackling dynamical problems and to the theory of contact transformations. Their detailed study is beyond the scope of the present text but, because of their importance in dynamics, a very brief summary of the main procedure will be inserted here. For a more extended account the reader can consult the works by Smart (1953), Sterne (1960) or Plummer (1960) listed in the bibliography at the end of this chapter.

If we define a set of variables p_r by the equations

$$p_r = \frac{\partial L}{\partial \dot{q}_r} \quad (r = 1, 2 \ldots 3n) \tag{6.54}$$

then a variable p_k is the *momentum conjugate* to q_k. By Lagrange's equations,

$$\dot{p}_r = \frac{\partial L}{\partial q_r} \quad (r = 1, 2 \ldots 3n). \tag{6.55}$$

If a function H of the form $H(q_r, p_r, t)$ is introduced such that H is defined by the relation

$$H(q_r, p_r, t) = \sum_{r=1}^{3n} p_r \dot{q}_r - L \tag{6.56}$$

it can be shown that

$$\dot{q}_r = \frac{\partial H}{\partial p_r}, \quad \dot{p}_r = -\frac{\partial H}{\partial q_r}. \tag{6.57}$$

These $6n$ differential equations of the first order are *Hamilton's canonic equations*. The function H is called the Hamiltonian.

It is seen that if the Lagrangian L does not contain the time explicitly then neither does H; hence

$$\frac{dH}{dt} = \sum_{r=1}^{3n} \frac{\partial H}{\partial p_r} \dot{p}_r + \sum_{r=1}^{3n} \frac{\partial H}{\partial q_r} \dot{q}_r = 0$$

using the Hamiltonian equations. Then $H = \text{constant} = T - U$, which is the energy integral.

A dynamical problem, once set up in the form of Hamilton canonic equations, becomes the problem of solving them. In the two-body problem they can be solved exactly. In most other problems met with in celestial mechanics and astrodynamics they cannot be solved exactly, but can be used in a general perturbation manner to give solutions in series valid for a certain length of time. By certain transformation rules it is possible to obtain, in successive approximations to the complete solution, differential equations that are still canonic in form and whose variables are the so-called canonic constants of integration obtained in the previous approximation. The process can be carried on as far as one pleases.

Formally, it can be proved that the solution of the canonic equations (6.57) can be written down if a function S can be found, where S is any complete solution of the equation

$$\frac{\partial S}{\partial t} + H\left(q_r, \frac{\partial S}{\partial q_r}, t\right) = 0. \tag{6.58}$$

This is the Hamilton–Jacobi equation, the Hamiltonian of the problem being expressed as a function of the q_r, the time, and quantities $\partial S/\partial q_r$, where

$$\frac{\partial S}{\partial q_r} = p_r.$$

From the Hamilton–Jacobi equation, S is obtained as a function of the q, $3n$ constants α_r, and t. The equations

$$p_r = \frac{\partial S(q_i, \alpha_i, t)}{\partial q_r} \quad (i = 1, 2 \ldots 3n)$$

$$\beta_r = \frac{\partial S(q_i, \alpha_i, t)}{\partial \alpha_r} \quad (r = 1, 2 \ldots 3n) \tag{6.59}$$

where the β_r are independent constants, then contain the solutions of the Hamilton canonic equations. The $6n$ constants α_r and β_r are the canonic constants of integration arising in the solution.

Now suppose that it is not possible to solve equation (6.57) by this method but that a solution can be obtained when H is replaced in the canonic equations by H_0. It may then be shown that the solution of these canonic equations by the above method results in $6n$ canonic constants α_r and β_r which become canonic variables in the next approximation, their differential equations being

$$\dot{\alpha}_r = \frac{\partial H_1}{\partial \beta_r}, \quad \dot{\beta}_r = -\frac{\partial H_1}{\partial \alpha_r} \tag{6.60}$$

where $H_1 = H_0 - H$.

A convenient part of H_1 may then be taken as a new Hamiltonian and the solution of (6.60) carried out to give new canonic constants.

In the planetary case, the equation of relative motion of a disturbed planet was of the form (see section 6.3)

$$\ddot{\mathbf{r}} = \nabla U = \nabla(U_0 + R). \tag{6.61}$$

In this case it is obvious that the component equations of (6.42)

$$\ddot{x} = \frac{\partial U}{\partial x}$$

$$\ddot{y} = \frac{\partial U}{\partial y}$$

$$\ddot{z} = \frac{\partial U}{\partial z}$$

are Lagrange's equations of motion where the Lagrangian L is given by

$$L = \frac{1}{2}(\dot{x}^2 + \dot{y}^2 + \dot{z}^2) + U.$$

It should be noted that since U is a function of the time-dependent coordinates of the disturbing bodies, it cannot be considered to be a potential energy. Then

$$H = \frac{1}{2}(p_x{}^2 + p_y{}^2 + p_z{}^2) - U$$

where the equations

$$\dot{p}_x = -\frac{\partial H}{\partial x}, \qquad \dot{x} = \frac{\partial H}{\partial p_x}$$

with similar equations in y and z, are the Hamilton canonic equations.

The two-body problem can be solved exactly, and so the first step is to solve the equations

$$\dot{p}_x = -\frac{\partial H_0}{\partial x}, \qquad \dot{x} = \frac{\partial H_0}{\partial p_x}$$

where

$$H_0 = \frac{1}{2}(p_x{}^2 + p_y{}^2 + p_z{}^2) - \frac{\mu}{r}.$$

The six canonic equations give canonic constants α_1, α_2, α_3, β_1, β_2, β_3 which then become variables satisfying the canonic equations

$$\dot{\alpha}_r = \frac{\partial H_1}{\partial \beta_r}, \qquad \dot{\beta}_r = -\frac{\partial H_1}{\partial \alpha_r}$$

where $H_1 = H_0 - H$.

To finish this section we will illustrate the procedure by applying the Hamilton–Jacobi method to the two-body (i.e. undisturbed) problem. As in the previous section the generalized coordinates r, β and λ are used, so we have

$$T = \frac{1}{2}(\dot{r}^2 + r^2\dot{\beta}^2 + r^2\dot{\lambda}^2 \cos^2 \beta), \qquad U_0 = \frac{\mu}{r}.$$

Then as before,

$$L_0 = \frac{1}{2}(\dot{r}^2 + r^2\dot{\beta}^2 + r^2\dot{\lambda}^2 \cos^2 \beta) + \frac{\mu}{r}.$$

Using equation (6.54)

$$p_r = \dot{r}, \qquad p_\lambda = r^2 \cos^2 \beta \dot{\lambda}, \qquad p_\beta = r^2 \dot{\beta}$$

so that by equation (6.55),

$$H_0 = \frac{1}{2}\left(p_r{}^2 + \frac{p_\lambda{}^2}{r^2 \cos^2 \beta} + \frac{p_\beta{}^2}{r^2}\right) - \frac{\mu}{r}.$$

The Hamilton–Jacobi equation (6.58) becomes

$$\frac{\partial S}{\partial t}+\frac{1}{2}\left[\left(\frac{\partial S}{\partial r}\right)^2+\frac{1}{r^2\cos^2\beta}\left(\frac{\partial S}{\partial\lambda}\right)^2+\frac{1}{r^2}\left(\frac{\partial S}{\partial\beta}\right)^2\right]-\frac{\mu}{r}=0.\qquad(6.62)$$

Now t does not appear explicitly in H_0 so that

$$H=\text{constant}=\alpha_1\ (\text{say}).$$

Then by (6.62) we have

$$\left(\frac{\partial S}{\partial r}\right)^2+\frac{1}{r^2}\left(\frac{\partial S}{\partial\beta}\right)^2+\frac{1}{r^2\cos^2\beta}\left(\frac{\partial S}{\partial\lambda}\right)^2=2\alpha_1+\frac{2\mu}{r}.\qquad(6.63)$$

This equation is seen to be in a form suitable for separating the variables. We note first that λ is absent from H_0 so that

$$\frac{\partial H_0}{\partial\lambda}=0$$

giving

$$\dot p_\lambda=\frac{\partial H_0}{\partial\lambda}=0.$$

Hence

$$\frac{\partial S}{\partial\lambda}=p_\lambda=\alpha_3.$$

Equation (6.63) may then be written as

$$\left[\left(\frac{\partial S}{\partial\beta}\right)^2+\alpha_3{}^2\sec^2\beta\right]+r^2\left[\left(\frac{\partial S}{\partial r}\right)^2-\left(2\alpha_1+\frac{2\mu}{r}\right)\right]=0$$

or

$$f_1(\beta)+f_2(r)=0.$$

These are functions of independent variables and so we may put

$$\left.\begin{array}{l}\dfrac{\partial S}{\partial r}=\left(2\alpha_1+\dfrac{2\mu}{r}-\dfrac{\alpha_2{}^2}{r^2}\right)^{1/2}\\[2mm]\dfrac{\partial S}{\partial\lambda}=\alpha_3\\[2mm]\dfrac{\partial S}{\partial\beta}=(\alpha_2{}^2-\alpha_3{}^2\sec^2\beta)^{1/2}.\end{array}\right\}\qquad(6.64)$$

Hence

$$S=-\alpha_1 t+\alpha_3\lambda+S_1(r)+S_2(\beta)$$

where

$$S_1=\int_{r_1}^{r}(2\alpha_1 r^2+2\mu r-\alpha_2{}^2)^{1/2}\frac{dr}{r}$$

and

$$S_2=\int_{0}^{\beta}(\alpha_2{}^2-\alpha_3{}^2\sec^2\beta)^{1/2}\,d\beta.$$

The constant r_1 is defined to be the smaller of the two roots of the equation

$$2\alpha_1 r^2 + 2\mu r - \alpha_2{}^2 = 0$$

where we assume that both roots are real and positive.

By equations (6.59) and (6.64) the complete solution thus consists of

$$\left. \begin{aligned} \beta_1 &= -t + \frac{\partial S_1}{\partial \alpha_1} \\[2mm] \beta_2 &= \frac{\partial S_1}{\partial \alpha_2} + \frac{\partial S_2}{\partial \alpha_2} \\[2mm] \beta_3 &= \lambda + \frac{\partial S_2}{\partial \alpha_3} \end{aligned} \right\} \tag{6.65}$$

together with

$$p_r = \dot{r} = \left(2\alpha_1 + \frac{2\mu}{r} - \frac{\alpha_2{}^2}{r^2} \right)^{1/2}$$

$$p_\beta = r^2 \dot{\beta} = (\alpha_2{}^2 - \alpha_3{}^2 \sec^2 \beta)^{1/2}$$

$$p_\lambda = r^2 \dot{\lambda} \cos^2 \beta = \alpha_3.$$

When the right-hand sides of (6.65) are integrated (using hindsight and our knowledge of the properties of two-body motion!) the interpretation of the canonic constants in terms of the familiar elliptic elements is as follows:

$$\alpha_1 = -\frac{\mu}{2a}, \qquad\qquad \beta_1 = -\tau = \chi/n$$

$$\alpha_2 = [\mu a(1 - e^2)]^{1/2}, \qquad \beta_2 = \omega$$

$$\alpha_3 = [\mu a(1 - e^2)]^{1/2} \cos i, \quad \beta_3 = \Omega.$$

6.10 Derivation of Lagrange's Planetary Equations from Hamilton's Canonic Equations

The relationship between the classical elliptical elements a, e, i; Ω, ω, τ and the Hamilton canonic constants α_i, β_i, $i = 1, 2, 3$, obtained in the previous section enables the Lagrange planetary equations to be derived easily from the differential equations of the canonic constants (6.60) when a disturbing function is present. Although it was not done historically in this fashion the derivation is instructive.

Let the disturbing function be $R = H_1$. Then

$$\dot{\alpha}_r = \frac{\partial R}{\partial \beta_r}, \qquad \dot{\beta}_r = -\frac{\partial R}{\partial \alpha_r}$$

where the α and β are new canonic variables.

Then, using $\mu = n^2 a^3$,

$$\dot{\alpha}_1 = \frac{\partial R}{\partial \beta_1} = -\frac{\partial R}{\partial \tau} = -\frac{\partial R}{\partial \chi}\frac{\partial \chi}{\partial \tau} = n\frac{\partial R}{\partial \chi}.$$

But

$$\dot{\alpha}_1 = \frac{d}{dt}\left(\frac{-\mu}{2a}\right) = \frac{\mu}{2a^2}\dot{a} = \frac{n^2 a}{2}\dot{a}.$$

Hence

$$\dot{a} = \frac{2}{na}\frac{\partial R}{\partial \chi}. \tag{6.66}$$

Now

$$\frac{\partial R}{\partial \omega} = \frac{\partial R}{\partial \beta_2} = \dot{\alpha}_2 = \frac{d}{dt}[\sqrt{\mu a(1-e^2)}] = \frac{1}{2}na(1-e^2)^{1/2}\left(\dot{a} - \frac{2ae}{1-e^2}\dot{e}\right).$$

Hence

$$\dot{a} - \frac{2ae}{1-e^2}\dot{e} = \frac{2}{na\sqrt{1-e^2}}\frac{\partial R}{\partial \omega}. \tag{6.67}$$

Now

$$\frac{\partial R}{\partial \Omega} = \frac{\partial R}{\partial \beta_3} = \dot{\alpha}_3 = \frac{d}{dt}[\sqrt{\mu a(1-e^2)}\cos i]$$

$$= \frac{1}{2}na(1-e^2)^{1/2}\cos i\left(\dot{a} - \frac{2ae}{1-e^2}\dot{e} - 2a\tan i\frac{di}{dt}\right).$$

Hence

$$\dot{a} - \frac{2ae}{1-e^2}\dot{e} - 2a\tan i\frac{di}{dt} = \frac{2}{na\sqrt{1-e^2}\cos i}\frac{\partial R}{\partial \Omega}. \tag{6.68}$$

We have

$$\frac{\partial R}{\partial i} = \frac{\partial R}{\partial \alpha_3}\cdot\frac{\partial \alpha_3}{\partial i} = -na^2\sqrt{1-e^2}\sin i\,(-\dot{\beta}_3) = na^2\sqrt{1-e^2}\sin i\,\dot{\Omega}$$

or

$$\dot{\Omega} = \frac{1}{na^2\sqrt{1-e^2}\sin i}\frac{\partial R}{\partial i}. \tag{6.69}$$

Also

$$\frac{\partial R}{\partial e} = \frac{\partial R}{\partial \alpha_2}\frac{\partial \alpha_2}{\partial e} + \frac{\partial R}{\partial \alpha_3}\frac{\partial \alpha_3}{\partial e}$$

$$= -\left(\dot{\beta}_2\frac{\partial \alpha_2}{\partial e} + \dot{\beta}_3\frac{\partial \alpha_3}{\partial e}\right) = \frac{na^2 e}{\sqrt{1-e^2}}(\cos i\,\dot{\Omega} + \dot{\omega})$$

or

$$\dot{\Omega} \cos i + \dot{\omega} = \frac{\sqrt{1-e^2}}{na^2 e} \frac{\partial R}{\partial e} \qquad (6.70)$$

and

$$\frac{\partial R}{\partial a} = \frac{\partial R}{\partial \alpha_1} \frac{\partial \alpha_1}{\partial a} + \frac{\partial R}{\partial \alpha_2} \frac{\partial \alpha_2}{\partial a} + \frac{\partial R}{\partial \alpha_3} \frac{\partial \alpha_3}{\partial a}$$

$$= -\left(\dot{\beta}_1 \frac{\partial \alpha_1}{\partial a} + \dot{\beta}_2 \frac{\partial \alpha_2}{\partial a} + \dot{\beta}_3 \frac{\partial \alpha_3}{\partial a} \right)$$

$$= -\frac{1}{2} na\dot{\chi} - \frac{1}{2} na\sqrt{1-e^2}(\dot{\omega} + \dot{\Omega} \cos i)$$

or

$$\dot{\chi} + \sqrt{1-e^2}(\dot{\omega} + \dot{\Omega} \cos i) = -\frac{2}{na} \frac{\partial R}{\partial a}. \qquad (6.71)$$

Equations (6.66) to (6.71) can easily be solved to give Lagrange's planetary equations, as listed in the set of equations (6.29).

Problems

6.1 A particle of unit mass moves in a straight line according to the differential equation

$$\frac{d^2 x}{dt^2} = g(1 + \epsilon x)$$

where g and ϵ are constants and $0 < \epsilon x \ll 1$. Use the method of the variation of parameters to show that the particle's motion is given approximately by

$$x = at + b + \tfrac{1}{2}gt^2[1 + \epsilon(\tfrac{1}{12}gt^2 + \tfrac{1}{3}at + b)]$$

where a and b are the values of dx/dt and x at $t=0$.

6.2 Using the expression for the disturbing function R given by equations (6.31) and (6.32), obtain the first approximation to the solution of the equation for e in the set (6.30).

6.3 The gravitational potential due to Jupiter in its equatorial plane at a distance r from its centre is approximately

$$\frac{\mu}{r}\left[1 + \frac{\lambda}{3}\left(\frac{\rho}{r} \right)^2 \right]$$

where ρ is the radius of Jupiter, λ is a small constant and $\mu = GM$.

Prove that in the absence of other perturbations, the major axis of a Jovian satellite's orbit, whose plane is at zero inclination to Jupiter's equatorial plane, rotates with a mean angular velocity of approximately

$$\frac{2\pi\rho^2\lambda}{a^2 T}$$

where T is the orbital period and a is the semimajor axis of its orbit. (You may take the eccentricity to be small so that $f = M + 2e \sin M$.)

6.4 Show that there are no perturbations in the inclination and the longitude of the node of the orbit of a single planet of negligible mass moving about a spherical star that is slowly radiating away its mass at a constant rate. If the eccentricity of the osculating orbit is small at a given time, examine the perturbations to the first order in the other elements.

6.5 If a planet moves about a star within a resisting medium such that the only disturbing acceleration on the planet is D given by

$$D = kV^2r^{-2}$$

where k is a constant and V and r are the planet's velocity and radius vector respectively, show that da/dt is given by

$$\frac{da}{dt} = \frac{-2kn}{(1-e^2)^{7/2}}(1+e\cos f)^2 (1+e^2+2e\cos f)^{3/2}.$$

6.6 Two planets of masses m and m_1 revolve about the Sun in orbits of small inclination i and i_1. When the transformations

$$p = \tan i \sin \Omega, \qquad q = \tan i \cos \Omega$$

$$p_1 = \tan i_1 \sin \Omega_1, \qquad q_1 = \tan i_1 \cos \Omega_1$$

are made, the relevant aperiodic part of the disturbing function R for planet m disturbed by planet m_1 is R', given by

$$R' = -Gm_1 D(p^2+q^2+p_1^2+q_1^2-2pp_1-2qq_1)$$

where the factor D is a symmetrical function of a and a_1. The corresponding disturbing function for m_1 is R_1' where

$$\frac{R_1'}{m} = \frac{R'}{m_1}.$$

It is also found that

$$\dot{p} = \frac{1}{na^2}\frac{\partial R'}{\partial q}, \qquad \dot{q} = -\frac{1}{na^2}\frac{\partial R'}{\partial p}$$

with two corresponding equations in p_1 and q_1. Show that:

(i) $mna^2\gamma^2 + m_1n_1a_1^2\gamma_1^2 = $ constant,

(ii) $p = A \sin (ft+c_1) + B \sin c_2,$

where $\gamma = \tan i$, $\gamma_1 = \tan i_1$ and A, B, f, c_1 and c_2 are constants.

Bibliography

Brouwer D and Clemence G M 1961 *Methods of Celestial Mechanics* (New York and London: Academic)
Danby J M A 1967 *Fundamentals of Celestial Mechanics* (New York: Macmillan)
Herrick S 1971, 1972 *Astrodynamics* vols 1 and 2 (London: Reinhold, Van Nostrand)
Moulton F R 1914 *An Introduction to Celestial Mechanics* (New York: Macmillan)
Plummer A C 1918 *An Introductory Treatise on Dynamical Astronomy* (London: Cambridge University Press)
Poincaré H 1895 *Les Méthodes Nouvelles de la Mécanique Céleste* (Paris: Gauthier-Villars) (*NASA* 1967 **TTF-450**, Washington)
Ramsey A S 1949 *Newtonian Attraction* (London: Cambridge University Press)
Smart W M 1953 *Celestial Mechanics* (London, New York, Toronto: Longmans)
Sterne T E 1960 *An Introduction to Celestial Mechanics* (New York: Interscience)
Tisserand F 1889 *Traité de Mécanique Céleste* (Paris: Gauthier-Villars)

7 Special Perturbations

7.1 Introduction

In many orbital motion problems it is not possible to derive a general perturbation theory, but it is always possible to use special perturbations, the method of numerically integrating the equations of motion of the bodies in some form or other. Starting with the positions and velocities of the bodies at a given date, the effects of all the forces on them during a small time interval may be computed from the equations of motion by one of a variety of methods, so that new positions and velocities at the end of this time interval can be found. A new computation using these positions and velocities enables the process to be carried forward for another time interval. Each computation is called a step and in theory the numerical integration may be continued as far as one pleases. In practice, a feature called *rounding-off error* is bound up in any numerical process. Since the operator will be working to so many significant figures, he or the machine will be constantly performing rounding-off computations and in doing so errors are inevitably produced.

The process in general is cumulative; the greater the number of steps required, the greater the error. As a result, by the end of the calculation an error of several thousands in the last place may exist, so the last four figures may be meaningless.

Obviously one remedy is to work with more figures than one needs (or indeed has data for) so that by the time the calculation has been completed the rounding-off error still does not affect the last figure one wishes to be significant. Another remedy is to work with the largest possible time interval so that the number of steps is reduced to a minimum. These are only partial remedies however. In the first case, the integration may have to be carried out for so long a time that the number of decimals required may be too many for the machine to carry conveniently; in the second case, the size of interval is held below a certain value fixed by the numerical integration formula one is using.

An important study by Brouwer (1937) shows that in units of the last place the *probable error* of a double integral is $0 \cdot 1124\ n^{3/2}$, where n is the number of steps. After numerically integrating the second-order (x, y, z) equations of motion of a satellite through, for example, 100 steps, we should expect that there is an even chance that the rounding-off error is smaller than $112 \cdot 4$ in units of the last decimal. The study also shows that the mean errors of the osculating elements of a body obtained by integrating numerically the

Lagrange planetary equations (which are first-order equations), or by using the usual formulae to obtain them from the (x, y, z) and $(\dot{x}, \dot{y}, \dot{z})$ components will be proportional to $n^{1/2}$ except the mean orbital longitude whose mean error is again proportional to $n^{3/2}$. It will be remembered that this last quantity is the result of a double integration.

7.2 Factors in Special Perturbation Problems

R. H. Merson (1973) has systematized the problems of special perturbations under five headings: (i) the type of orbit, (ii) the operational requirements, (iii) the formulation of the equations of motion, (iv) the numerical integration procedure and (v) the available computing facilities. We consider each of these in turn, noting that they react closely with each other in practice.

7.2.1 The type of orbit

Roughly speaking, the orbit to be computed may be classified as almost circular, highly eccentric, or parabolic–hyperbolic. Examples are respectively the orbits of a planet, a comet and a spacecraft escaping from Earth. It can be however that during the computation the orbit may change from one class to another. In addition the orbit may be slightly, moderately or highly perturbed: for example a planet's orbit, a close artificial-satellite orbit, or the fly-by phase of an interplanetary probe's planetary encounter.

7.2.2 The operational requirements

Among the requirements will be the desired accuracy (i.e. number of significant figures) and the length of the computation (possibly one extended orbital calculation or many short computations of the problem from a variety of starting conditions).

7.2.3 The formulation of the equations of motion

In some methods the differential equations are first-order, in others they are second-order, and in others again they may be a mixture of first- and second-order. The Lagrange planetary equations are an example of a first-order set; the equations of relative motion in rectangular coordinates form a second-order set, while Hansen's method leads to a mixed set.

7.2.4 The numerical integration procedure

If a procedure does not use previously computed sets of values of the variables concerned in finding the next set of values (i.e. at the end of the current numerical integration step), it is usually called a single-step method. Single-step methods have the advantages that no special starting procedure is required and the step size can be changed easily during the computation when necessary, for example in approaching and receding from perihelion in the computation of a highly eccentric cometary orbit.

Multistep methods use previous sets of values. Their formulation is usually simple, so that little computation per step is required. Special starting and step-changing procedures are needed however; the former requirement is no great disadvantage, especially in an extended integration, but the latter can be a disadvantage in cases where step-size changing is frequent, as in a highly eccentric orbit.

7.2.5 The available computing facilities

From the days of logarithm tables through the mechanical desk calculator era to the modern régime of solid-state pocket computers and large electronic computers, the main considerations have been speed, number of available digits and capacity. In G Darwin's day logarithm tables came in different sizes, capacity was provided by paper and pencil and speed was dictated by the human computer's ability and stamina. Even today, however, when most computers have more than adequate storage and compiler facilities and speeds such that it would take only hours to reproduce the results of Darwin's years of hard labour, there exist orbital motion problems that are too big to be tackled. Others can be processed only by carefully choosing the appropriate formulation of the equations of motion and the most suitable numerical integration procedure, and utilizing a double-precision program on the computer to avoid the loss of too many significant figures. Finally it should be remembered that computing is not only a question of growth of rounding-off error and speed; it also costs money. In any orbital motion problem, careful consideration of the points outlined above will often reduce by an order of magnitude the computation time and money.

In the next section we look more closely at the points outlined in sections 7.2.3 and 7.2.4, considering the advantages and disadvantages of various formulations of the equations of motion, and then comparing some of the wide 'variety of numerical-integration procedures currently in use. The discussion that follows is by no means exhaustive but the references at the end of the chapter will help to deepen the reader's understanding of this important field of study.

7.3 Cowell's Method

Cowell and Crommelin (1908) published a paper in which they investigated the motion of Jupiter's eighth satellite by special perturbations. They formulated their equations in rectangular coordinates and integrated them numerically by means of a multistep algorithm. Since then great confusion has arisen when the term 'Cowell's method' is used. In numerical analysis texts (e.g. Henrici 1962), 'Cowell-type methods' refers to multistep algorithms resembling that used in the 1908 paper. This type of algorithm may be used to solve any suitable differential equation, whether or not it arises from celestial mechanics. On the other hand, within the literature of celestial mechanics, the term 'Cowell's method' is becoming widely used to refer to the formulation of the equations (i.e. the method which employs the differential equations in rectangular coordinates) and where no knowledge of the orbit's behaviour is used to speed their solution. These equations may of course be integrated by any suitable numerical algorithm, for example by Runge–Kutta formulae.

It is a straightforward method of wide application since it makes no distinction between the disturbing function and the central (i.e., two-body) part of the acceleration. Its main disadvantage arises from this lack of distinction, since a large number of significant figures has to be carried due to the large central force term; consequently a smaller integration step has to be used. The development of high-speed electronic computers has removed much of the weight of this disadvantage and one of the first modern applications of Cowell's method was to calculate the rectangular coordinates of the five outer planets through a time interval of 400 years, using the IBM Selective Sequence Electronic Calculator.

Cowell's method has the advantage of being easy to formulate and to program. However, it is not without its pitfalls and disadvantages; for example difficulties arise when close encounters take place. The step size in such cases becomes so small that an inordinately large amount of machine time is used and much loss of accuracy accrues due to the growth of rounding-off error. In such circumstances it is customary to use other types of methods where we introduce some intermediate reference orbit. In the case of a highly eccentric cometary orbit, it is thus often advantageous to integrate the difference between the comet's path and the path of a hypothetical comet following an undisturbed Keplerian orbit. The greater amount of computing work per step is more than compensated for by the far larger step size which may be taken, especially when the eccentricity of the orbit is large. The above method is known as Encke's method. In recent years some authors have modified Encke's original method, and in the next sections we shall describe the original method and several recent variations.

7.4 Encke's Method

This method makes use of the fact that to a first approximation the orbit is a conic section. The integration gives the difference between the real coordinates and the conic-section coordinates. The conic-section orbit is an osculating one, so at the epoch of osculation the differences vanish. As time goes on the differences grow, until it becomes necessary to derive a new osculating orbit. This process is called *rectification of the orbit*. The main advantage of Encke's method is that a larger integration interval than is possible in Cowell's method can be adopted, since near the osculation epoch the differences are small and capable of being expressed by a few significant figures. On the other hand, rather more work is involved in an Encke integration step than in a Cowell one.

The following device, introduced by Encke, renders his method practical. Letting suffix e denote positions given by the two-body equation of motion, we have

$$\ddot{\mathbf{r}}_e + \mu \frac{\mathbf{r}_e}{r_e^3} = 0. \tag{7.1}$$

For the actual motion we have

$$\ddot{\mathbf{r}} + \mu \frac{\mathbf{r}}{r^3} = \mathbf{F}$$

where $\mathbf{F}$ is due to the attractions of other bodies, drag by atmosphere, and so forth.

Let $\boldsymbol{\rho}$ be defined by $\boldsymbol{\rho} = \mathbf{r} - \mathbf{r}_e$. Then

$$\ddot{\boldsymbol{\rho}} = \mathbf{F} + \mu \left(\frac{\mathbf{r}_e}{r_e^3} - \frac{\mathbf{r}}{r^3} \right). \tag{7.2}$$

The osculating orbit for some epoch is the solution of (7.1) and is known, so that the rectangular coordinates x_e, y_e, z_e and r_e can be computed for any time after this epoch.

The quantity

$$\frac{\mathbf{r}_e}{r_e^3} - \frac{\mathbf{r}}{r^3}$$

is the difference of two nearly equal vectors, since for some time after the epoch the true and the conic-section orbit are not much different. This would cause an increase in the number of significant figures required. To avoid this, Encke put

$$\frac{r^2}{r_e^2} = 1 + 2q$$

and

$$\left(\frac{r_e}{r} \right)^3 = (1 + 2q)^{-3/2} = 1 - fq.$$

The function f of q (which is a small quantity) is tabulated in Planetary Coordinates (1960–80). Then

$$\mu \left(\frac{\mathbf{r}_e}{r_e^3} - \frac{\mathbf{r}}{r^3} \right) = \frac{\mu}{r_e^3} (fq\mathbf{r} - \boldsymbol{\rho})$$

where

$$fq = \frac{3q}{1} - \frac{3 \times 5}{1 \times 2} q^2 + \frac{3 \times 5 \times 7}{1 \times 2 \times 3} q^3 - \frac{3 \times 5 \times 7 \times 9}{1 \times 2 \times 3 \times 4} q^4 + \cdots$$

and

$$q = \frac{1}{r_e^2} \left[\left(x_e + \frac{1}{2}\xi \right)\xi + \left(y_e + \frac{1}{2}\eta \right)\eta + \left(z_e + \frac{1}{2}\zeta \right)\zeta \right]$$

the vector $\boldsymbol{\rho}$ being related to ξ, η, ζ by the relation

$$\boldsymbol{\rho} = \mathbf{i}\xi + \mathbf{j}\eta + \mathbf{k}\zeta$$

210

where $\mathbf{i}$, $\mathbf{j}$ and $\mathbf{k}$ are unit vectors along the x-, y- and z-axes. Equation (7.2) then becomes

$$\ddot{\boldsymbol{\rho}} = \mathbf{F} + \frac{\mu}{r_e{}^3}(fq\mathbf{r} - \boldsymbol{\rho})$$

or

$$\ddot{\boldsymbol{\rho}} = \mathbf{F} + hfq\mathbf{r} - h\boldsymbol{\rho}$$

where

$$h = \mu/r_e{}^3.$$

An alternative device which avoids the use of the series for f is derived as follows:

$$\frac{\mathbf{r}_e}{r_e{}^3} - \frac{\mathbf{r}}{r^3} = \mathbf{r}_e\left(\frac{1}{r_e{}^3} - \frac{1}{r^3}\right) - \frac{\boldsymbol{\rho}}{r^3}.$$

Now

$$\boldsymbol{\rho} \cdot (\mathbf{r} + \mathbf{r}_e) = (\mathbf{r} - \mathbf{r}_e) \cdot (\mathbf{r} + \mathbf{r}_e) = r^2 - r_e{}^2$$

and so

$$\frac{1}{r_e{}^3} - \frac{1}{r^3} = \left(\frac{r - r_e}{r^3 r_e{}^3}\right)(r^2 + r r_e + r_e{}^2)$$

$$= \left(\frac{r^2 - r_e{}^2}{r^3 r_e{}^3}\right)\left(\frac{r^2 + r r_e + r_e{}^2}{r + r_e}\right)$$

so that

$$\frac{1}{r_e{}^3} - \frac{1}{r^3} = \left(\frac{\rho^2 + 2\boldsymbol{\rho} \cdot \mathbf{r}_e}{r^3 r_e{}^3}\right)\left(\frac{r^2 + r r_e + r_e{}^2}{r + r_e}\right).$$

Equation (7.2) then becomes

$$\ddot{\boldsymbol{\rho}} = \mathbf{F} + \mu\left(Q\mathbf{r}_e - \frac{\boldsymbol{\rho}}{r^3}\right)$$

where

$$Q = \left(\frac{r^2 + r r_e + r_e{}^2}{r + r_e}\right)\left(\frac{\rho^2 + 2\boldsymbol{\rho} \cdot \mathbf{r}_e}{r^3 r_e{}^3}\right).$$

Encke's method has had wide applications, not only in cometary orbit work but also in computing orbits in Earth–Moon space where the Moon is taken to be the perturbing body. It has also been used in calculating orbits that differ only slightly from some standard orbit because of slightly different initial conditions, as in investigations into the sensitivity of orbits to error.

There have been some recent efforts to improve Encke's method by the use of a better reference orbit. Kyner and Bennet (1966) have shown that when integrating the equations of motion of a near-Earth satellite, the Encke method is greatly improved when the first-order effects of the Earth's oblateness are included in the reference orbit. This improvement in the reference orbit not only greatly increased the interval before the rectification of the reference orbit became necessary; it also produced a considerable increase in the accuracy of the integration compared with that achieved by the classical Encke method and Cowell's method.

Stumpff and Weiss (1967) have shown that for the integration of the equations of motion when four or more bodies are involved, the execution time required for the Encke method can be one tenth of the time for the *classical* Encke method, when the reference orbit is taken to be a combination of several Keplerian orbits.

The philosophy behind the Encke approach is thus to find a reference orbit that is known and which remains very near to the real evolving orbit for a considerable time. The differential equations of the differences between the real-orbit variables and the corresponding quantities in the reference orbit are set up and integrated numerically. It should be noted that these reference quantities may not be constant. If the choice of reference orbit is good, the integration steps should be much larger than they would otherwise be if the original differential equations of the real orbit were integrated, the size of the step thereby more than offsetting the additional number of operations per step. It should also be noted that there is no necessity that the position and velocity in the reference orbit at any desired time be calculated from analytical expressions.

7.5 The Use of Perturbational Equations

The Lagrange planetary equations (6.29), (6.30) and (6.41) may be integrated numerically instead of analytically. This may be done step by step, the new elements at the end of each step being used in the computation of the next step. Another method in use with these equations is to insert the osculating elements into the right-hand sides and then integrate numerically over an extended period of time. By this procedure the first-order perturbations in the elements are obtained. The new perturbed elements are now inserted into the right-hand sides of the equations and the equations are integrated once more throughout the required length of time to give elements that include second-order perturbations, and so on. This method is analogous to the analytical method described in section 6.2.3.

As mentioned in section 6.7.4, equations (6.41), where the expressions on the right-hand sides are given in terms of the components S, T and W, are often used in special perturbations.

Ever since Lagrange introduced his planetary equations (where the rates of change of the osculating elements of a planet's orbit are given in terms of the elements of that planet and of the planets disturbing its heliocentric orbit), various authors have described methods which attempt to remove some of the serious disadvantages of a method that generally appears to offer a number of advantages in special perturbations. Among the advantages are:

(i) It is strictly a perturbation method and as such it bypasses the central-body acceleration.
(ii) For moderate perturbations, the differentials of the elements are small, and so a much larger step size can be used than is possible in a rectangular

coordinate method (such as Cowell's method) that calculates at each step the central-body acceleration.

(iii) The integration immediately exhibits the behaviour of the elements.

Among the disadvantages are:

(i) the more complicated nature of the right-hand sides of the equations compared to those of the rectangular coordinate equations,
(ii) the presence of sines and cosines of a number of angles,
(iii) the break-down of the equations when either the orbital eccentricity becomes zero or unity, or the orbital inclination goes to.zero,
(iv) the fact that the equations are usually given in terms of elliptic elements and as such are inapplicable to parabolic, hyperbolic or rectilinear orbits, and
(v) the necessity to solve Kepler's equation.

With respect to disadvantages (i) and (ii) the saving in machine time obtained by using a larger step than is possible with Cowell's method is reduced by the larger number of manipulations which are required at each step to evaluate the right-hand sides of the Lagrange equations. This reduction is further emphasized by the need to form the sines and cosines of as many as six different angles.

Disadvantage (iii) was quickly appreciated in work with cometary orbits of high eccentricity, and also in working with planetary orbits, because these are mainly of small eccentricity and are little inclined to the usual reference planes of the ecliptic or the solar system's invariable plane. As the eccentricity is decreased the position of the apse becomes indeterminate, and as the inclination goes to zero the longitude of the ascending node becomes impossible to compute accurately. The usual transformation to avoid this disadvantage consists of substituting the variables h, k, p, q for the offending elements e, ϖ, i and Ω where

$$h = e \sin \varpi, \qquad k = e \cos \varpi$$
$$p = \sin i \sin \Omega, \qquad q = \sin i \cos \Omega.$$

Other transformations avoided the difficulty of the eccentricity approaching unity.

Disadvantage (iv) was not so serious before the era of artificial satellites and interplanetary probes except where comet work was concerned, but is serious when for example the escape of a probe from earth into a heliocentric transfer orbit takes place essentially along a hyperbolic path, and when the circumnavigation of the moon involves a hyperbolic lunar encounter and a highly eccentric *cis*-lunar transfer.

The fifth disadvantage is more apparent than real, since a method such as the Newton–Raphson method (Henrici 1964) of successive approximations converges so quickly that it occupies very little machine time. It has been pointed out by various workers, however, that the solution of Kepler's equation can be avoided by changing the independent variable from the time to the true or eccentric anomaly. For example, the eccentric anomaly was first used by Oppolzer (1870) in computing the perturbations of Comet Pons–Winnecke through nine revolutions from 1819 to 1869.

Some authors have avoided a number of the disadvantages outlined above by using various combinations of the vectors

$$\mathbf{a} = e\mathbf{P}, \qquad \mathbf{b} = e\sqrt{p}\,\mathbf{Q}$$

$$\mathbf{c} = \sqrt{p}\,\mathbf{R}, \qquad \mathbf{g} = \frac{e}{\sqrt{p}}\,\mathbf{Q}$$

where e = eccentricity, p = semilatus rectum = $a\,(1 - e^2)$ (a being the semimajor axis), $\mathbf{P}$ and $\mathbf{R}$ are unit vectors directed respectively from the central body to pericentre and along the normal to the orbital plane, and $\mathbf{Q} = \mathbf{R} \times \mathbf{P}$. For example, Herrick (1953) used $\mathbf{a}$ and $\mathbf{b}$, Milankovic (1939) used $\mathbf{a}$ and $\mathbf{c}$, and the vector $\mathbf{g}$ is used implicitly in Hansen's theory. These pairs give only five independent scalars, so a sixth is required. This has usually been the mean anomaly, the time of pericentre passage, the mean anomaly at the epoch or the modified mean anomaly at the epoch. Merton (1949) describes a method using the mean anomaly as the independent variable.

Allan (1961) and Allan and Ward (1963) have used the vectors $\mathbf{h}$ and $\mathbf{e}$, where $\mathbf{h}$ is the osculating angular momentum and $\mathbf{e}$ is a vector of magnitude e directed along the major axis towards perihelion. The resulting equations are more concise and avoid the generation of most of the trigonometrical terms used in Lagrange's planetary equations, although they are still cumbersome to evaluate. Because of the definitions of these vectors a number of checks are provided;

$$\mathbf{a} \cdot \mathbf{b} = \mathbf{h} \cdot \mathbf{e} = \mathbf{r} \cdot \mathbf{h} = \dot{\mathbf{r}} \cdot \mathbf{h} = 0.$$

Musen (1954) has made use of the vectors $\mathbf{c}$ and $\mathbf{g}$ in formulating a set of differential equations for special perturbations; he points out that in Herrick's method the appearance of e in the vectors $\mathbf{a}$ and $\mathbf{b}$ is troublesome when e is small, though Herrick (1953) suggested replacing the mean anomaly by the mean longitude and the use of $\mathbf{c}$ instead of $\mathbf{b}$. Herget (1962) describes a set of equations which removes the singularity of zero eccentricity from the equations of Musen's method. The resulting equations, however, are very cumbersome.

Different approaches have been made by Garafalo (1960), Cohen and Hubbard (1962) and Pines (1961). To avoid the singularities $e = 0$, $e = 1$ and $i = 0$, Garafalo (1960) introduced a set of variables of which five are obtained by the integration of expressions that have the perturbing mass as a factor. However, the sixth expression

$$\dot{\theta} = \frac{h}{r^2}$$

(where θ is the true anomaly) is of zero order and, as Garafalo points out, requires a smaller interval in the integration.

Cohen and Hubbard (1962) provide a transformation of the elliptic orbital elements that eliminates the singularities $e = 0$, $i = 0$ and $i = 180°$. They also

214

mention that the use of the true longitude as the independent variable avoids the solution of Kepler's equation. As discussed above, however, the solution of Kepler's equation is not difficult or time-consuming when a method such as the Newton–Raphson method is used. The resulting set of equations, which are expressed in terms of the radial, transverse and normal components of the disturbing acceleration, are not of extreme simplicity and break down when $h=0$.

Pines (1961) also avoids the difficulties experienced when the eccentricity and inclination are small, and the need for an additional equation for the integration of the mean motion. His method uses as parameters a set of initial position and velocity vectors in the osculating orbital plane; but the resulting differential equations are complicated.

In the remainder of this section we present a set of perturbational equations that minimize the disadvantages listed above while still retaining all the advantages of the Lagrange equations. They also hold for all approximate conic-section orbits.

7.5.1 Derivation of the perturbation equations (case $h \neq 0$)

The equations of motion of a body P of mass m disturbed in its Keplerian orbit about a body S of mass M are given by

$$\frac{d^2\mathbf{r}}{dt^2} = \mathbf{F} - \mu\frac{\mathbf{r}}{r^3} \qquad (7.3)$$

where $\mathbf{r}$ is the radius vector from S to P, G is the gravitational constant, $\mathbf{F}$ is the disturbing acceleration and $\mu = G(M+m)$. Let E be the Keplerian energy, defined by

$$E = \tfrac{1}{2}(\dot{\mathbf{r}} \cdot \dot{\mathbf{r}}) - \frac{\mu}{r} \qquad (7.4)$$

and let $\mathbf{h}$ and $\mathbf{e}$ be respectively the osculating angular momentum and a vector with a magnitude of the osculating eccentricity drawn from S towards pericentre.

Define a vector $\boldsymbol{\epsilon}$ by

$$\boldsymbol{\epsilon} = \mu\mathbf{e}. \qquad (7.5)$$

Let λ be the true longitude of P, defined in the usual way by

$$\lambda = \Omega + \omega + \theta \qquad (7.6)$$

where Ω is the longitude of the ascending node, ω is the argument of pericentre and θ is the true anomaly. Then by the above definitions and using equation (7.3) with

$$\frac{d\mathbf{h}}{dt} = \mathbf{r} \times \ddot{\mathbf{r}} \qquad (7.7)$$

and Hamilton's integral, given by Milne (1948) as

$$\boldsymbol{\epsilon} = \dot{\mathbf{r}} \times \mathbf{h} - \mu \frac{\mathbf{r}}{r} \tag{7.8}$$

it may easily be shown that

$$\frac{d\mathbf{h}}{dt} = \mathbf{r} \times \mathbf{F} \tag{7.9}$$

$$\frac{d\boldsymbol{\epsilon}}{dt} = (\mathbf{F} \times \mathbf{h}) + (\dot{\mathbf{r}} \times \dot{\mathbf{h}}) \tag{7.10}$$

$$\frac{dE}{dt} = \dot{\mathbf{r}} \cdot \mathbf{F}. \tag{7.11}$$

The derivation of the time derivative for the true longitude λ involves considerably more work, the final expression for $(d\lambda/dt)$ being

$$\frac{d\lambda}{dt} = \frac{h}{r^2} + \frac{h_x h_y - h_y h_x}{h(h + h_z)} \tag{7.12}$$

where

$$\mathbf{h} = \mathbf{i} h_x + \mathbf{j} h_y + \mathbf{k} h_z$$

$$h = (h_x^2 + h_y^2 + h_z^2)^{1/2} \tag{7.13}$$

$$\mathbf{r} = \mathbf{i} x + \mathbf{j} y + \mathbf{k} z \tag{7.14}$$

$$\boldsymbol{\epsilon} = \mathbf{i} \epsilon_x + \mathbf{j} \epsilon_y + \mathbf{k} \epsilon_z \tag{7.15}$$

and $\mathbf{i}$, $\mathbf{j}$, $\mathbf{k}$ are orthogonal unit vectors such that $\mathbf{i}$ and $\mathbf{j}$ lie in the reference plane, $\mathbf{k}$ is normal to it and the vectors define the x, y and z axes respectively, Ω being measured along the reference plane from the x axis.

In the absence of any perturbation the osculating orbit would be undisturbed and Keplerian, its properties at any time being given by the usual conic-section two-body relations.

Letting the suffix k denote an undisturbed quantity and using Kepler's second law, we have

$$\frac{d\lambda_k}{dt} = \frac{h_k}{r_k^2} \tag{7.16}$$

by virtue of the fact that

$$\frac{d\Omega_k}{dt} = \frac{d\omega_k}{dt} = 0. \tag{7.17}$$

Subtracting equation (7.16) from equation (7.12), we have

$$\frac{d\lambda_p}{dt} = \left(\frac{h}{r^2} - \frac{h_k}{r_k^2} \right) + \frac{h_x h_y - h_y h_x}{h(h + h_z)} \tag{7.18}$$

where λ_p, the perturbation in λ, is given by

$$\lambda = \lambda_k + \lambda_p. \tag{7.19}$$

The angles λ and λ_k need not be coplanar.

216

We may take equations (7.9), (7.10) and (7.18) as a set suitable for integration. It should be noted that they give only six independent quantities, since

$$\mathbf{h} \cdot \boldsymbol{\epsilon} = 0. \tag{7.20}$$

In addition we have the relation

$$\mathbf{r} \cdot \mathbf{h} = \dot{\mathbf{r}} \cdot \mathbf{h} = \mathbf{r} \cdot \dot{\mathbf{h}} = 0. \tag{7.21}$$

which provides, with equation (7.20), useful integration checks.

Collecting equations (7.9), (7.10) and (7.18) below, we have

$$\frac{d\mathbf{h}}{dt} = \mathbf{r} \times \mathbf{F} \tag{7.9}$$

$$\frac{d\boldsymbol{\epsilon}}{dt} = (\mathbf{F} \times \mathbf{h}) + (\dot{\mathbf{r}} \times \dot{\mathbf{h}}) \tag{7.10}$$

$$\frac{d\lambda_p}{dt} = \left(\frac{h}{r^2} - \frac{h_k}{r_k^2}\right) + \frac{h_x h_y - h_y h_x}{h(h + h_z)}. \tag{7.18}$$

Although equation (7.11) may seem simpler than equation (7.10), so that (7.11) and (7.20) might be used to eliminate two of the three scalar equations given by (7.10), it is found in practice that equation (7.10) is usually concise in form and that (7.20) can cause trouble, any component being capable of becoming zero. Thus equation (7.20) is best retained as a check while E can be obtained from the relation

$$E = \frac{\epsilon^2 - \mu^2}{2h^2}. \tag{7.22}$$

It may be noted in passing that the use of expression (7.18) is similar to Encke's method, and so rectification of the orbit is required when the quantity in the bracket becomes too large. This question will be discussed later.

7.5.2 The relations between the perturbation variables, the rectangular coordinates and velocity components, and the usual conic-section elements.

We have by definition

$$h_x = y\dot{z} - z\dot{y}, \quad h_y = z\dot{x} - x\dot{z}, \quad h_z = x\dot{y} - y\dot{x} \tag{7.23}$$

with

$$h = (h_x^2 + h_y^2 + h_z^2)^{1/2}. \tag{7.24}$$

We also have

$$\epsilon_x = Dx - K\dot{x}, \quad \epsilon_y = Dy - K\dot{y}, \quad \epsilon_z = Dz - K\dot{z} \tag{7.25}$$

with

$$\epsilon = (\epsilon_x^2 + \epsilon_y^2 + \epsilon_z^2)^{1/2} \tag{7.26}$$

where

$$\left. \begin{array}{c} K = r\dot{r}, \quad D = V^2 - \dfrac{\mu}{r} \\[2mm] V^2 = \dot{x}^2 + \dot{y}^2 + \dot{z}^2. \end{array} \right\} \tag{7.27}$$

217

Also

$$\left.\begin{array}{l} \cos \lambda = \dfrac{x}{r} - \dfrac{z}{r} \dfrac{h_x}{(h+h_z)} \\[3mm] \sin \lambda = \dfrac{y}{r} - \dfrac{z}{r} \dfrac{h_y}{(h+h_z)}. \end{array}\right\}$$ (7.28)

Conversely, we have

$$\left.\begin{array}{l} c \equiv \epsilon \cos \theta = \gamma \cos \lambda + \delta \sin \lambda \\[2mm] s \equiv \epsilon \sin \theta = \gamma \sin \lambda - \delta \cos \lambda \end{array}\right\}$$ (7.29)

where

$$\left.\begin{array}{l} \gamma = \epsilon_x - \dfrac{\epsilon_z h_x}{h+h_z} \\[4mm] \delta = \epsilon_y - \dfrac{\epsilon_z h_y}{h+h_z}. \end{array}\right\}$$ (7.30)

Then

$$r = \frac{h^2}{\mu + c}, \qquad \dot{r} = \frac{s}{h}, \qquad \epsilon = (c^2 + s^2)^{1/2}.$$ (7.31)

The coordinates x and y are obtained from

$$x = r \cos \lambda + \frac{zh_x}{h+h_z}$$ (7.32)

$$y = r \sin \lambda + \frac{zh_y}{h+h_z}$$ (7.33)

when z has been computed from

$$z = -\frac{r}{h} (h_x \cos \lambda + h_y \sin \lambda).$$ (7.34)

The velocity components can then be calculated from

$$\dot{x} = \frac{1}{r^2} (zh_y - yh_z + r\dot{r}x)$$ (7.35)

$$\dot{y} = \frac{1}{r^2} (xh_z - zh_x + r\dot{r}y)$$ (7.36)

$$\dot{z} = \frac{1}{r^2} (yh_x - xh_y + r\dot{r}z).$$ (7.37)

Also, if ψ is the angle between the radius vector and the velocity vector, so that

$$\mathbf{r} \cdot \dot{\mathbf{r}} = rV \cos \psi$$ (7.38)

218

we have

$$\cos \psi = \frac{s}{hV}$$

$$\left. \sin \psi = \frac{h}{rV} = \frac{\mu + c}{hV} \right\}$$
(7.39)

where

$$V^2 = \frac{1}{h^2}(\mu^2 + \epsilon^2 + 2\mu c).$$
(7.40)

The usual angular elements Ω, ω, i are given by

$$\cos i = \frac{h_z}{h}, \quad \sin i = \frac{H}{h}$$
(7.41)

$$\cos \Omega = -\frac{h_y}{H}, \quad \sin \Omega = \frac{h_x}{H}$$
(7.42)

$$\cos \omega = \frac{(\epsilon_y h_x - \epsilon_x h_y)}{\epsilon H}, \quad \sin \omega = \frac{\epsilon_z h}{\epsilon H}$$
(7.43)

where

$$H = (h_x^2 + h_y^2)^{1/2}.$$

If i is zero, we may use $\varpi = \Omega + \omega$ and then

$$\cos \varpi = \frac{\epsilon_x}{\epsilon}, \quad \sin \varpi = \frac{\epsilon_y}{\epsilon}.$$
(7.44)

For the other three osculating elements, the osculating conic has to be considered as an ellipse, a parabola or hyperbola. We have three possibilities:

(i) $0 \leqslant \epsilon < \mu$; the orbit is elliptic.
(ii) $\epsilon = \mu$; the orbit is parabolic.
(iii) $\epsilon > \mu$; the orbit is hyperbolic.

These are discussed in detail below.

(i) The eccentricity $e = \epsilon/\mu$, the semimajor axis $a = h^2\mu/(\mu^2 - \epsilon^2)$, and the time of pericentre passage τ is given by

$$\tau = t - \frac{E - e \sin E}{n}$$

where

$$n = \mu^{1/2}a^{-3/2}$$

and

$$\tan \frac{E}{2} = \left(\frac{\mu - \epsilon}{\mu + \epsilon}\right)^{1/2} \tan \frac{\theta}{2}$$

E being the eccentric anomaly.

(ii) The eccentricity is equal to 1. The pericentre distance q is given by

$$q = h^2/2\mu$$

and Barker's equation gives us

$$\tau = t - \frac{h^3}{2\mu^2} \left[\tan\left(\frac{\theta}{2}\right) + \frac{1}{3}\tan^3\left(\frac{\theta}{2}\right) \right].$$

(iii) The hyperbolic analogue to the elliptic eccentric anomaly is F, given by

$$F = \ln \left[\frac{\mu \cos\theta + \epsilon + (\epsilon^2 - \mu^2)^{1/2} \sin\theta}{\mu + \epsilon \cos\theta} \right]. \tag{7.45}$$

Then

$$\tau = t - \frac{h^3 \mu}{(\epsilon^2 - \mu^2)^{3/2}} \left\{ \frac{\epsilon}{2\mu}(e^F - e^{-F}) - F \right\}. \tag{7.46}$$

Expressions (7.45) and (7.46) avoid the use of hyperbolic functions.

It should be noted that when r is obtained from (7.31) from a knowledge of c, accuracy will be lost when $\mu \sim c$, the latter being negative. This occurs when the orbit is almost parabolic with the true anomaly approaching 180°. It is best in such cases to switch to the régime of section 7.5.4 since the orbit is then approximately rectilinear.

7.5.3 Numerical integration procedure

We suppose that at time $t = t_0$, the values of x, y, z; $\dot{x}$, $\dot{y}$, $\dot{z}$ are known, and also that the values of the components F_x, F_y, F_z of the disturbing acceleration vector $\mathbf{F}$ are known. By relations (7.23)–(7.28), the information required to calculate the right-hand sides of equations (7.9), (7.10) and (7.18) is thereby obtained. With respect to equation (7.18) it should be noted that if the intermediate orbit coincides with the real orbit at $t = t_0$, the quantity I, given by

$$I = \frac{h}{r^2} - \frac{h_k}{r_k^2}$$

is zero at that time. It will not remain zero since, not only does h vary while h_k remains constant, but r and r_k are obtained at any time from

$$r = \frac{h^2}{\mu + \epsilon \cos\theta}, \qquad r_k = \frac{h_k^2}{\mu + \epsilon_k \cos\theta_k}.$$

By a numerical integration method, values of h_x, h_y, h_z; ϵ_x, ϵ_y, ϵ_z; λ_p at the end of the interval are then computed.

There remains the task of calculating the value of λ_k (the Keplerian true longitude) at the end of the step ($t = t_1$) in order to obtain the perturbed true longitude $\lambda = \lambda_k + \lambda_p$. Let the notation (0) and (1) denote values of a quantity at $t = t_0$ and $t = t_1$ respectively.

Now

$$\lambda_k = \varpi_k + \theta_k$$

so that

$$\lambda_k(1) - \lambda_k(0) = \theta_k(1) - \theta_k(0)$$

since ϖ_k for the intermediate orbit is constant.

Hence the change in λ_k during a step is the change in the Keplerian (i.e. undisturbed) true anomaly during the interval. To compute this the standard elliptic, parabolic or hyperbolic formulae are used as follows:

At $t=t_0$, the value of ϵ_k is known. Then if

(i) $0 \leqslant \epsilon_x < \mu$, the intermediate orbit is elliptic during the step,
(ii) $\epsilon_k = \mu$, the intermediate orbit is parabolic during the step,
(iii) $\epsilon_k > \mu$, the intermediate orbit is hyperbolic during the step.

The treatment of each of these three cases is as follows:

(i) Use the relation

$$\tan \frac{E}{2} = \left(\frac{\mu - \epsilon}{\mu + \epsilon}\right)^{1/2} \tan \frac{\theta}{2} \tag{7.47}$$

with the values of the Keplerian ϵ and θ at $t=t_0$ to obtain E_0, which is the value of the Keplerian eccentric anomaly at $t=0$. At the end of the step $(t=t_1)$, the value of E is E_1, given by

$$E_1 - \frac{\epsilon}{\mu} \sin E_1 = M_1 \tag{7.48}$$

where

$$M_1 = n(t_1 - t_0) + \left(E_0 - \frac{\epsilon}{\mu} \sin E_0\right)$$

and

$$n = (\mu^2 - \epsilon^2)^{3/2}/h^3 \mu.$$

Kepler's equation (7.48) may be solved by the usual Newton–Raphson method.

We can then use E_1 in the equations

$$\left.\begin{array}{l} \cos \theta_k = \dfrac{\mu \cos E - \epsilon}{\mu - \epsilon \cos E} \\[3mm] \sin \theta_k = \dfrac{(\mu^2 - \epsilon^2)^{1/2} \sin E}{\mu - \epsilon \cos E} \end{array}\right\} \tag{7.49}$$

to give θ_k at $t=t_1$.

(ii) Let $J = \tan \theta/2$. By Barker's equation, if J_0, J_1 are the values of J at t_0 and t_1, we then have

$$J_1 + \tfrac{1}{3}J_1^3 = M_1' \tag{7.50}$$

where

$$M_1' = 2\bar{n}(t_1 - t_0) + (J_0 + \tfrac{1}{3}J_0^3)$$

and

$$\bar{n} = \mu^2/h^3.$$

Barker's equation may be solved by the method of section 4.6.

(iii) The hyperbolic 'eccentric anomaly' is F, where

$$F = \ln \left\{\frac{\mu \cos \theta + \epsilon + (\epsilon^2 - \mu^2)^{1/2} \sin \theta}{\mu + \epsilon \cos \theta}\right\}. \tag{7.51}$$

221

The value of F at $t=t_1$ is then F_1, given by

$$\frac{\epsilon}{\mu}(\sinh F_1) - F_1 = M_1'' \tag{7.52}$$

$$M_1'' = \nu(t_1 - t_0) + M_0$$

$$\nu = (\epsilon^2 - \mu^2)^{3/2}/h^3\mu$$

and

$$M_0 = \frac{\epsilon}{\mu}(\sinh F_0) - F_0 = \frac{\epsilon}{2\mu}[(\exp F_0) - (\exp - F_0)] - F_0.$$

Equation (7.52) may be solved by the method of section 4.7.2.

Having found F_1 (the value of F for $t=t_1$) the Gudermannian function of F (namely q) is obtained, where

$$\tan(45° + \tfrac{1}{2}q) = \exp F.$$

Then θ_k is finally calculated for $t=t_1$ from

$$\left.\begin{array}{l} \sin\theta_k = \dfrac{(\epsilon^2 - \mu^2)^{1/2}\sin q}{\epsilon - \mu\cos q} \\[3mm] \cos\theta_k = \dfrac{\epsilon\cos q - \mu}{\epsilon - \mu\cos q}. \end{array}\right\} \tag{7.53}$$

We have

$$\lambda_k(1) = \lambda_k(0) + \theta_k(1) - \theta_k(0)$$

giving

$$\lambda_1 = \lambda_k(1) + \lambda_p(1).$$

Equations (7.29) to (7.40) give the values of x, y, z; $\dot{x}$, $\dot{y}$, $\dot{z}$; r, $\dot{r}$ and V at $t=t_1$. If desired, values of the new osculating elements a (or q), e, τ; Ω, ω and i may be computed from equations (7.41) to (7.43) and the appropriate elliptic, parabolic or hyperbolic set, selected by the value of ϵ at $t=t_1$. They are not, of course, necessary to continue the integration.

The decision as to whether or not to rectify (i.e. update) the intermediate orbit is now taken. Factors involved in this decision include the work involved, the size of the term

$$\frac{h}{r^2} - \frac{h_k}{r_k^2}$$

and the type of numerical integration procedure adopted.

The work involved is certainly less than that involved in updating the reference orbit in Encke's method. The old values of h_k, r_k, ϵ_k and θ_k existing at the end of the step are simply replaced by the values of h, r, ϵ, θ that have already been calculated.

7.5.4 Rectilinear or almost rectilinear orbits

The method described above holds in principle for all values of the eccentricity and inclination; it breaks down for rectilinear or almost rectilinear orbits

222

since h appears in the denominator of a number of the relations involved, in particular equation (7.18). It is therefore necessary to change to a new set of variables when all three components of h become smaller than a certain size, to avoid loss of accuracy. These new variables may be taken to be the polar coordinates (r, α, β) where

$$
\left.
\begin{aligned}
x &= r \cos \alpha \cos \beta \\
y &= r \sin \alpha \cos \beta \\
z &= r \sin \beta.
\end{aligned}
\right\}
\tag{7.54}
$$

α and β are therefore the ecliptic longitude and latitude (or right ascension and declination) respectively.

We also have

$$
\left.
\begin{aligned}
\dot{x} &= \frac{x}{r} \dot{r} - z \dot{\beta} \cos \alpha - y \dot{\alpha} \\
\dot{y} &= \frac{y}{r} \dot{r} - z \dot{\beta} \sin \alpha + x \dot{\alpha} \\
\dot{z} &= \frac{z}{r} \dot{r} + r \dot{\beta} \cos \beta.
\end{aligned}
\right\}
\tag{7.55}
$$

Hence given $r, \alpha, \beta; \dot{r}, \dot{\alpha}, \dot{\beta}$, we can compute $x, y, z; \dot{x}, \dot{y}, \dot{z}$ from (7.54) and (7.55).

Conversely

$$
\left.
\begin{aligned}
\sin \beta &= \frac{z}{r} \\
\tan \alpha &= \frac{y}{x} \\
r &= (x^2 + y^2 + z^2)^{1/2}.
\end{aligned}
\right\}
\tag{7.56}
$$

It should be noted that $\sin \beta$ takes the sign of z, while the relations

$$
\sin \alpha = \frac{y}{r \cos \beta}
$$

$$
\cos \alpha = \frac{x}{r \cos \beta}
$$

give the correct value of α.

Also

$$
\left.
\begin{aligned}
\dot{r} &= \frac{x\dot{x} + y\dot{y} + z\dot{z}}{r} \\
\dot{\alpha} &= h_z/(x^2 + y^2) \\
\dot{\beta} &= \frac{\dot{z} - \dot{r} \sin \beta}{r \cos \beta} = \frac{r\dot{z} - z\dot{r}}{r^2 \cos \beta}
\end{aligned}
\right\}
\tag{7.57}
$$

so that from $x, y, z; \dot{x}, \dot{y}, \dot{z}$ we can obtain $r, \alpha, \beta; \dot{r}, \dot{\alpha}, \dot{\beta}$.

223

Differentiating the first equation of (7.57) and using equation (7.3), we obtain

$$\ddot{r}=\frac{1}{r}\left[\left(\frac{h}{r}\right)^2-\frac{\mu}{r}+\mathbf{r}\cdot\mathbf{F}\right]$$

where

$$h^2=r^2(V^2-\dot{r}^2).$$

It should be noted that h is zero for an exactly rectilinear orbit.

If suffix k again denotes the Keplerian undisturbed quantity, we have

$$\ddot{r}_k=\frac{1}{r_k}\left[\left(\frac{h_k}{r_k}\right)^2-\frac{\mu}{r_k}\right].$$

The perturbation in the magnitude of the radius vector is then r_p, given by the relation $r=r_k+r_p$. By differentiating to give $\dot{r}=\dot{r}_k+\dot{r}_p$, we see that

$$\ddot{r}_p=\left\{\frac{1}{r}\left[\left(\frac{h}{r}\right)^2-\frac{\mu}{r}\right]-\frac{1}{r_k}\left[\left(\frac{h_k}{r_k}\right)^2-\frac{\mu}{r_k}\right]\right\}+\frac{\mathbf{r}\cdot\mathbf{F}}{r}. \tag{7.58}$$

Again, differentiating the second and third of (7.57) and letting the perturbations in α and β be α_p and β_p, where

$$\alpha=\alpha_k+\alpha_p$$

$$\beta=\beta_k+\beta_p$$

we have

$$\ddot{\alpha}_p=\frac{h_z}{x^2+y^2}-\frac{2h_z(x\dot{x}+y\dot{y})}{(x^2+y^2)^2}-\left[\frac{h_z}{x^2+y^2}-\frac{2h_z(x\dot{x}+y\dot{y})}{(x^2+y^2)^2}\right]_k \tag{7.59}$$

and

$$\ddot{\beta}_p=\left\{\left[\left(F_z-\frac{z}{r^2}\mathbf{r}\cdot\mathbf{F}\right)-\frac{z}{r^2}\left(\frac{h}{r}\right)^2+\dot{\beta}(z\dot{\beta}-2\dot{r}\cos\beta)\right]\Big/r\cos\beta\right\}$$

$$+\left\{\left[\frac{z}{r^2}\left(\frac{h}{r}\right)^2-\dot{\beta}(z\dot{\beta}-2\dot{r}\cos\beta)\right]\Big/r\cos\beta\right\}_k \tag{7.60}$$

where k denotes that the quantities inside the brackets have the Keplerian undisturbed values.

We assume that the reference orbit *is always exactly rectilinear*, so that unlike the former ($h\neq0$) it is not obtained from an osculating orbit if the true orbit is itself not exactly rectilinear. The reference orbit being rectilinear, we have

$$\left.\begin{array}{l}h_k=\dot{h}_k=0\\\dot{\alpha}_k=\dot{\beta}_k=0\end{array}\right\} \tag{7.61}$$

and

$$\frac{\dot{x}_k}{x_k}=\frac{\dot{y}_k}{y_k}=\frac{\dot{z}_k}{z_k}=\frac{\dot{r}_k}{r_k}$$

so that equations (7.58), (7.59) and (7.60) become respectively

$$\ddot{r}_p = R - \mu\left(\frac{1}{r^2} - \frac{1}{r_k^2}\right) \tag{7.62}$$

$$\ddot{\alpha}_p = \frac{h_z}{x^2+y^2} - \frac{2h_z(x\dot{x}+y\dot{y})}{(x^2+y^2)^2} \tag{7.63}$$

$$\ddot{\beta}_p = \left[F_z - \left(\frac{z}{r}\right)R + \dot{\beta}(z\dot{\beta} - 2\dot{r}\cos\beta)\right]\bigg/ r\cos\beta \tag{7.64}$$

where

$$R = \frac{1}{r}\left[\left(\frac{h}{r}\right)^2 + \mathbf{r}\cdot\mathbf{F}\right].$$

The procedure resembles that used in the non-rectilinear case.

At $t=t_0$, from values of x, y, z and $\dot{x}$, $\dot{y}$, $\dot{z}$ we form r, α, β; $\dot{r}$, $\dot{\alpha}$, $\dot{\beta}$; h_z, h_z, h and compute the right-hand sides of equations (7.62) to (7.64). Again, if rectification of the reference orbit is made at the end of each step we have $r_k = r$, so that $\ddot{r}_p = R$ at the beginning of the step. Within a step, of course, r departs in general from r_k. We now integrate through a step to obtain the value of $\dot{r}_p$, $\dot{\alpha}_p$, $\dot{\beta}_p$ and r_p, α_p, β_p at $t=t_1$.

To obtain $\dot{r}_k$, $\dot{\alpha}_k$, $\dot{\beta}_k$ and r_k, α_k, β_k at $t=t_1$, we remember that the reference orbit is always exactly rectilinear so that $\dot{\alpha}_k = \dot{\beta}_k = 0$ and the values of α_k and β_k are therefore what they were at $t=t_0$. The remaining quantities r_k and $\dot{r}_k$ change during the step and are obtained from the appropriate set of relations for the rectilinear ellipse, parabola or hyperbola (see section 4.8).

The choice of rectilinear ellipse, parabola or hyperbola as reference orbit during the step is dictated by whether the energy E, given by

$$E = \tfrac{1}{2}\dot{r}^2 - \frac{\mu}{r}$$

is negative, zero or positive at the beginning of the step.

The quantities r, α, β and $\dot{r}$, $\dot{\alpha}$, $\dot{\beta}$ at $t=t_1$ can now be computed since $r = r_p + r_k$, etc. By equations (7.54) and (7.55), the values of x, y, z and $\dot{x}$, $\dot{y}$, $\dot{z}$ at $t=t_1$ can thus be found.

The reference orbit can now be updated. If the osculating orbit is still almost rectilinear, the new reference orbit is again taken to be rectilinear. Its parameters are then given at the beginning of the new step by

$$x_k = x, \qquad \dot{x}_k = x_k\dot{r}_k/r_k$$

$$y_k = y, \qquad \dot{y}_k = y_k\dot{r}_k/r_k$$

$$z_k = z, \qquad \dot{z}_k = z_k\dot{r}_k/r_k$$

$$r_k = r, \qquad \dot{r}_k = \dot{r}$$

$$\alpha_k = \alpha, \qquad \dot{\alpha}_k = 0$$

$$\beta_k = \beta, \qquad \dot{\beta}_k = 0.$$

The new energy, given by

$$E_k = \tfrac{1}{2}\dot{r}_k{}^2 - \frac{\mu}{r_k}$$

dictates whether the new rectilinear reference orbit is elliptic, parabolic or hyperbolic. When the angular momentum $\mathbf{h}$ becomes large enough, the method for nonzero h can be adopted.

7.6 Regularization Methods

An important feature of the Newtonian law of gravitation is that the force acting between particles approaches infinity as the distance between them approaches zero. Of course, the concept of a 'point mass' is entirely mathematical and in practice the singularities are never reached, since the surfaces of the colliding bodies will touch before this happens. However, in numerical work point masses can be manipulated and the singularities are of great importance. Further, as one body approaches closely to another (for example at pericentre in a highly eccentric orbit), the relative velocity increases greatly. This necessarily causes a considerable decrease in the step size which can be used in a numerical integration procedure.

Multistep integration methods are used most efficiently when the problem requires only a minimum in the rate at which halving and doubling of step size takes place during the numerical integration.

The singularities occurring at collisions are not of an essential character and can be eliminated by the proper choice of independent variable. This process is known as regularization. The problems attendant upon regularization have been extensively investigated and Szebehely (1967) gives an excellent bibliography on the subject as well as treating the regularization of the restricted three-body problem. A full treatment of the linearization of the equations of motion as well as their regularization is to be found in a book by Stiefel and Scheifele (1971). The usual approach is to replace the physical time t by a fictitious time s where $dt = r^k ds$ for some k. Here r is the radial distance between the attracting centres. When $k=1$, s is equivalent to the eccentric anomaly; when $k=2$, s is equivalent to the true anomaly.

This process has been called 'analytical step regulation' by Stiefel and Scheifele.

Stiefel (1970) used $k=1$ and linearized the equations of motion for the two-body problem. By comparing this to the normal formulation he found that an increase in accuracy of about 30 times could be achieved by regularization with no corresponding loss in speed. This and other recent work show that the concept of regularization is of the foremost importance in the numerical solution of problems in celestial mechanics.

Heggie (1971) has described a regularization using the potential or kinetic energy as a time regularization function, which is suitable for systems of two or more bodies. For straightforward two-body encounters it is not as useful as the Kustaanheimo–Stiefel regularization which is described by Peters (1968) but is more powerful in more complex situations. The use of the regularized equations with this regularization has yielded a reduction in computing time of 50% from that required by the unregularized equations when integrating the IAU 25-body problem which is described by Lecar (1968).

Kustaanheimo and Stiefel proposed and developed a regularization method (now usually called the KS transformation) in which the three-dimensional differential equations of motion, for example the equation

$$\frac{d^2\mathbf{r}}{dt^2} + \mu\frac{\mathbf{r}}{r^3} = 0$$

in the two-body problem, are regularized by transforming the three-vector $\mathbf{r}$ into a four-vector $\mathbf{u}$, the independent variable t being changed to the variable s by the relation $dt/ds = r$. The two-body motion is then represented by four second-order simple harmonic linear differential equations of the form

$$\frac{d^2\mathbf{u}}{ds^2} + \omega^2\mathbf{u} = 0$$

where ω is a constant. Stiefel and Scheifele developed the application of the KS transformation to problems of perturbed motion, producing a perturbational equations version.

Regularization is especially important in high-precision numerical studies of many-body systems in stellar dynamics, where many close encounters between pairs of particles can take place. Without a suitable regularization technique (or some alternative procedure) each close encounter can be both time-consuming and productive of a sharp rise in rounding-off error.

7.7 Numerical Integration Methods

In order to illustrate the essential difference between single and multistep numerical integration methods let us consider the numerical integration of the second-order equation

$$\frac{d^2x}{dt^2} = -x \tag{7.65}$$

where we are told that at $t = t_0$, $x = x_0$ and $dx/dt = \dot{x}_0$.

Consider the Taylor series

$$x = \sum_{n=0}^{\infty} \overset{(n)}{x_0}(t-t_0)^n/n! \tag{7.66}$$

where

$$\overset{(n)}{x_0} = \left(\frac{d^n x}{dt^n}\right)_0$$

the zero suffix denoting the value of the derivative at $t = t_0$.

Let

$$t_j = t_0 + jh, \qquad t_{-j} = t_0 - jh \text{ etc}$$

where h is the step size (assumed constant), and j is an integer.

Then for example, the value of x at time $t_1 = t_0 + h$ is x_1, where

$$x_1 = x_0 + \dot{x}_0 h + \frac{\ddot{x}_0 h^2}{2!} + \frac{\dddot{x}_0 h^3}{3!} + \frac{\ddddot{x}_0 h^4}{4!} + \ldots \tag{7.67}$$

while at $t_1 = t_0 - h$,

$$x_{-1} = x_0 - \dot{x}_0 h + \frac{\ddot{x}_0 h^2}{2!} - \frac{\dddot{x}_0 h^3}{3!} + \frac{\ddddot{x}_0 h^4}{4!} \ldots \tag{7.68}$$

where x_{-1} is the value of x at time $t_{-1} = t_0 - h$.

Now we know the values of x_0 and $\dot{x}_0$ at t_0. Furthermore, by equation (7.65) we have

$$\overset{(n)}{x} = \frac{\mathrm{d}^n x}{\mathrm{d}t^n} = -\frac{\mathrm{d}^{n-2} x}{\mathrm{d}t^{n-2}} = -\overset{(n-2)}{x} \tag{7.69}$$

so that in principle we may compute $\ddot{x}_0$, $\dddot{x}_0$, $\ddddot{x}_0$ etc as far as we please. By then evaluating equation (7.67) we may calculate x_1 to the desired accuracy, since $h = t_1 - t_0$ and is known.

In similar fashion, using the Taylor series

$$\dot{x}_1 = \dot{x}_0 + \ddot{x}_0 h + \frac{\dddot{x}_0 h^2}{2!} + \frac{\ddddot{x}_0 h^3}{3!} + \ldots \tag{7.70}$$

we can obtain $\dot{x}_1$, the value of $\mathrm{d}x/\mathrm{d}t$ at t_1.

By using equation (7.68) and the corresponding series for $\dot{x}_{-1}$, we may also calculate x_{-1} and $\dot{x}_{-1}$. Obviously this procedure may be extended to compute x_2, $\dot{x}_2$ for example, by now using the equations

$$x_2 = x_1 + \dot{x}_1 h + \frac{\ddot{x}_1 h^2}{2!} + \frac{\dddot{x}_1 h^3}{3!} + \ldots$$

$$\dot{x}_2 = \dot{x}_1 + \ddot{x}_1 h + \frac{\dddot{x}_1 h^2}{2!} + \frac{\ddddot{x}_1 h^3}{3!} + \ldots$$

and equation (7.69).

At this stage it may be remarked that:

(i) this is a single-step procedure, only data from the beginning of the current step being used in the calculation of the variable values at the end of the step,
(ii) it is self-starting,
(iii) halving and doubling the interval or step would obviously cause no difficulty if some error criterion dictated this change in step size,
(iv) the easy calculation of the higher derivatives is an essential requirement if such a straightforward Taylor series procedure is used. If however the equation or equations are nonlinear, then it may be more and more cumbersome and time-consuming to compute the higher derivatives.

Let us now transform the method into a multistep procedure. Adding equations (7.67 and 7.68) we obtain

$$x_1 = 2x_0 - x_{-1} + h^2 \ddot{x}_0 + \frac{2h^4 \ddddot{x}_0}{4!} + \ldots \tag{7.71}$$

Also, by adding (7.70) and the corresponding equation for $\dot{x}_{-1}$, we obtain

$$\dot{x}_1 = 2\dot{x}_0 - \dot{x}_{-1} + h^2 \dddot{x}_0 + \frac{2h^4 \ddddot{x}_0}{4!} + \dots \qquad (7.72)$$

To calculate x_1 and $\dot{x}_1$ we now require data from the beginning of the previous step (i.e. x_{-1}, $\dot{x}_{-1}$) as well as from the beginning of the present step. The main advantage is not self-evident yet. It may be shown however that by a suitable combination of sets of Taylor series, it is possible to avoid the calculation of derivatives beyond the second if a sufficient number of data from previous steps are involved. We may thus write in general

$$x_1 = 2x_0 - x_{-1} + h^2 \sum_{i=0}^{k} a_i \ddot{x}_i \qquad (7.73)$$

where h is the step size as before, k is a positive integer, the a_i are numerical coefficients and the $\ddot{x}_i$ are the values of the second derivatives at the beginning of the present step and the k previous ones. The numerical values taken by the a_i depend upon the value of k. For example, if $k=0$, we have the simple formula

$$x_1 = 2x_0 - x_{-1} + h^2 \ddot{x}_0$$

so that $a_0 = 1$. We may note that:

(i) this formula is correct to the order of h^3, since in producing it the first term neglected in the Taylor series is $(2h^4 \ddddot{x}_0 / 4!)$. A formula such as equation (7.73) is therefore said to be correct to order h^{2k+3}; that is, the first term neglected is

$$q h^{(2k+4)} \overset{(2k+4)}{x_0}$$

where q is some numerical factor.

(ii) In general, the higher the order neglected, the larger the step size that can be taken. However, not only does the law of diminishing returns set in, but stability considerations usually make it advisable to keep the order below double figures.

(iii) A multistep procedure obviously involves fewer computations than a single-step method correct to the same order. It is therefore much faster, subject to the constraint that it is not self-starting, and subject also to the fact that it requires special procedures for halving and doubling the step size. It is therefore best applied to situations where step changes are kept to a minimum (e.g. almost circular orbits, or when the equations have been regularized).

We now consider several single-step methods.

7.7.1 Recurrence relations

The use of recurrence-relation methods has already been discussed in section 4.13. It is sufficient to remark here that by their use the task of numerically calculating the higher derivatives in a Taylor series single-step method is greatly speeded up when, as is usually the case in orbital motion problems, the differential equations are nonlinear. Reference may be made to the series of papers (Roy *et al* 1972, Moran 1973, Roy and Moran 1973, Moran *et al* 1973, Emslie and Walker 1979) for a thorough exposition of such topics as well as a comparison of speeds and accuracies obtained by the adoption of various sets of auxiliary variables, accuracy criteria and step adjustment procedures.

7.7.2 Runge–Kutta four

This is a single-step procedure, with *truncation error* (i.e. the order of the first term neglected) of the order of h^5. Consider the first-order differential equation

$$\frac{dx}{dt} = f(x, t)$$

where $x = x_0$ at $t = t_0$. The value of x at $t = t_1 = t_0 + h$ is then denoted x_1, where

$$x_1 = x_0 + \tfrac{1}{6}(k_1 + 2k_2 + 2k_3 + k_4)$$

and

$$k_1 = hf(x_0, t_0)$$

$$k_2 = hf[(x_0 + \tfrac{1}{2}k_1), (t_0 + \tfrac{1}{2}h)]$$

$$k_3 = hf[(x_0 + \tfrac{1}{2}k_2), (t_0 + \tfrac{1}{2}h)]$$

$$k_4 = hf[(x_0 + k_3), (t_0 + h)].$$

The Runge–Kutta Four (RK4) is very popular. Most computer libraries contain an RK4 routine; it has all the advantages of single-step procedures and can be simply extended to second-order equations and to sets of equations. For example, the equation

$$\frac{d^2x}{dt^2} = f(x, t)$$

becomes

$$\frac{dx}{dt} = v, \quad \frac{dv}{dt} = f(x, t).$$

It has the disadvantages of being far slower and less accurate than a high-order Taylor series with recurrence relations, or a multistep method. It may be anything up to 50 times slower! It also necessitates the calculation of the function f four times each step. Various workers have attempted to remove or moderate these difficulties. Shanks (1966) and Butcher (1965) have developed higher-order RK-type formulae. Fehlberg (1968, 1972) has given an eighth-order process requiring only nine function evaluations (usually known as a Runge–Kutta–Fehlberg procedure).

7.7.3 Multistep methods

We have seen that multistep methods are simple and fast. On the other hand, ill-chosen high-order multistep methods tend to be unstable in the sense that any errors committed will propagate to future steps rather than be damped out (Lapidus and Seinfeld 1971). However, much work has been done to correct this instability and one feels that if a fixed step can be chosen (or the number

230

of changes in step size kept to a minimum), a high-order multistep algorithm is both accurate and fast. Merson (1973), in his study of a wide variety of special perturbation methods, concludes that for second-order equations the *Gauss–Jackson* eighth-order method applied to the Cowell equations (with analytical step regulation if required) is probably the optimum combination. Herrick (1971) also judged the Gauss–Jackson method (alternatively called the *Gaussian 'second-sum' formula* or *procedure*) to be the method most preferred. To understand the terms involved, we illustrate below some basic ideas in finite-difference theory as used in numerical integration.

7.7.4 Numerical methods

Suppose a function x of t is tabulated at equal intervals of time (the interval being h) so that t_p is given by $t_p = t_0 + ph$ where t_0 is some epoch at which x has the value x_0.

A table such as Table 7.1 may be made up.

Table 7.1

Argument	Sums		Function	Differences			
	2nd	1st		1st	2nd	3rd	4th
t_{-2}	$\Sigma^2 x_{-2}$		x_{-2}		$\delta^2 x_{-2}$		$\delta^4 x_{-2}$
		$\Sigma^1 x_{-1\frac{1}{2}}$		$\delta x_{-1\frac{1}{2}}$		$\delta^3 x_{-1\frac{1}{2}}$	
t_{-1}	$\Sigma^2 x_{-1}$		x_{-1}		$\delta^2 x_{-1}$		$\delta^4 x_{-1}$
		$\Sigma^1 x_{-\frac{1}{2}}$		$\delta x_{-\frac{1}{2}}$		$\delta^3 x_{-\frac{1}{2}}$	
t_0	$\Sigma^2 x_0$		x_0		$\delta^2 x_0$		$\delta^4 x_0$
		$\Sigma^1 x_{\frac{1}{2}}$		$\delta x_{\frac{1}{2}}$		$\delta^3 x_{\frac{1}{2}}$	
t_1	$\Sigma^2 x_1$		x_1		$\delta^2 x_1$		$\delta^4 x_1$
		$\Sigma^1 x_{1\frac{1}{2}}$		$\delta x_{1\frac{1}{2}}$		$\delta^3 x_{1\frac{1}{2}}$	
t_2	$\Sigma^2 x_2$		x_2		$\delta^2 x_2$		$\delta^4 x_2$

The *first difference* $\delta x_{p+1/2}$ is obtained by subtracting x_p from x_{p+1}, the *second difference* $\delta^2 x_p$ by subtracting $\delta x_{p-1/2}$ from $\delta x_{p+1/2}$, and so on.

Again, the *first sum* $\Sigma^1 x_{p+1/2}$ is got from the formula

$$\Sigma^1 x_{p+1/2} = \Sigma^1 x_{p-1/2} + x_p, \tag{7.74}$$

while the *second sum* $\Sigma^2 x_{p+1}$ is obtained from the formula

$$\Sigma^2 x_{p+1} = \Sigma^2 x_p + \Sigma^1 x_{p+1/2}. \tag{7.75}$$

Half-differences are often introduced into the blank spaces on the line in the odd difference and summation columns and on the half-lines in the even difference and summation columns according to the formulae

$$\mu \Sigma^1 x_p = \frac{1}{2}(\Sigma^1 x_{p-1/2} + \Sigma^1 x_{p+1/2}) \tag{7.76}$$

$$\mu \delta x_p = \frac{1}{2}(\delta x_{p-1/2} + \delta x_{p+1/2}). \tag{7.77}$$

These half-differences are distinguished by preceding them by the letter μ.

It is possible to interpolate using such a table (i.e. to obtain the value of x for any value of the independent variable t, even when that value of t is not given by an integral value of p) as long as t falls within the table's range.

Various formulae using the quantities tabulated exist for this purpose. For example, Bessel's formula is

$$x_p = x_0 + p\delta x_{1/2} + B_2(\delta^2 x_0 + \delta^2 x_1) + B_3 \delta^3 x_{1/2} + B_4(\delta^4 x_0 + \delta^4 x_1) + \ldots$$

where B are *Bessel's interpolation coefficients*. These are functions of p and are given in many works.

Again, Everett's formula is

$$x_p = x_0 + p\delta x_{1/2} + E_2^0 \delta^2 x_0 + E_2^1 \delta^2 x_1 + E_4^0 \delta^4 x_0 + E_4^1 \delta^4 x_1 + \ldots$$

where the *Everett coefficients* (functions of p) are also tabulated in a number of references, for example the Interpolation and Allied Tables (1956).

The successive orders of the differences in a table such as table 7.1 are related to the successive derivatives of the function x with respect to t, and formulae have been derived with which to perform numerical differentiation. Thus Bessel's formula for numerical differentiation is

$$h\left(\frac{dx}{dt}\right)_0 = \delta x_{1/2} + \tfrac{1}{4}(2p-1)(\delta^2 x_0 + \delta^2 x_1)$$
$$+ B_3'\delta^3 x_{1/2} + B_4'(\delta^4 x_0 + \delta^4 x_1) + \ldots \quad (7.78)$$

where the B' are tabulated.

In many problems the values of the derivatives are wanted only at tabular or half-way points. If this is so, equation (7.78) becomes

$$h\left(\frac{dx}{dt}\right)_0 = \mu\delta x_0 - \frac{1}{6}\mu\delta^3 x_0 + \frac{1}{30}\mu\delta^5 x_0 - \frac{1}{140}\mu\delta^7 x_0 + \ldots$$

We now consider the numerical integration of a differential equation that cannot be integrated analytically. Let the equation be

$$\frac{dx}{dt} = F$$

where F is some function of x and t.

Suppose we insert a series

$$x = \sum_{n=0}^{\infty} a_n(t - t_0)^n$$

into the equation and obtain the first few constants a_n in terms of the initial condition $x = x_0$ when $t = t_0$. The series will then enable values of x and F for a small range of values of $(t - t_0)$ to be calculated. A table for x and one for F after the manner of table 7.1 can then be set up within this range for certain values of t (namely $t_p = t_0 + ph$, where p is a positive or negative integer and h is a suitable tabular interval).

232

It is usual to include the factor h in the values computed for the function F so that in fact we take

$$X = hF = h\frac{dx}{dt}$$

as the function for which we wish to make a table.

Then we have

$$X_p = h\left(\frac{dx}{dt}\right)_p$$

and it may be shown that

$$\Sigma^1 X_{1/2} = x_0 + \frac{1}{2} X_0 + \frac{1}{12} \mu\delta X_0 - \frac{11}{720} \mu\delta^3 X_0 + \frac{191}{60480} \mu\delta^5 X_0 - \ldots \quad (7.79)$$

Also, at a subsequent tabular epoch,

$$x = \mu\Sigma^1 X - \frac{1}{12} \mu\delta X + \frac{11}{720} \mu\delta^3 X - \frac{191}{60480} \mu\delta^5 X + \ldots \quad (7.80)$$

Where necessary the differences are estimated from a knowledge of the way in which they are running in the table.

There are also formulae for extrapolation, for example

$$x_2 = x_0 + 2X_1 + \frac{1}{3}\delta^2 X_1 - \frac{1}{90}\delta^4 X_1 + \frac{1}{756}\delta^6 X_1 - \ldots \quad (7.81)$$

$$= x_{-2} + 4X_0 + \frac{8}{3}\delta^2 X_0 + \frac{14}{45}\delta^4 X_0 - \frac{8}{945}\delta^6 X_0 + \ldots \quad (7.82)$$

In orbital motion the differential equations to be solved are usually simultaneous second-order nonlinear equations. If the equation

$$\frac{d^2x}{dt^2} = F \quad (7.83)$$

symbolizes one of the equations of the set, we may then write

$$X_p = (h^2 F)_p = h^2 \left(\frac{d^2x}{dt^2}\right)_p. \quad (7.84)$$

The starting procedure can be the same as in the first-order case in that a series solution of the set of equations of the form (7.83), valid for a short time, could be used to set up a sum and difference table. In practice it is customary to use the undisturbed two-body orbit to provide a table for a given interval of time from which to start.

For this second-order case, if x_0 and $(dx/dt)_0 = x_0'$ are the values of x and dx/dt at $t = t_0$,

$$\Sigma^1 X_{1/2} = hx_0' + \frac{1}{2} X_0 + \frac{1}{12} \mu\delta X_0 - \frac{11}{720} \mu\delta^3 X_0 + \ldots \quad (7.85)$$

$$\Sigma^2 X_0 = x_0 - \frac{1}{12} X_0 + \frac{1}{240}\delta^2 X_0 - \frac{31}{60480}\delta^4 X_0 + \ldots \quad (7.86)$$

233

At a subsequent tabular date,

$$hx^1 = \mu\Sigma^1 X - \frac{1}{12}\mu\delta X + \frac{11}{720}\mu\delta^3 X - \frac{191}{60\,480}\mu\delta^5 X + \dots \tag{7.87}$$

$$x = \Sigma^2 X + \frac{1}{12}X - \frac{1}{240}\delta^2 X + \frac{31}{60\,480}\delta^4 X - \dots . \tag{7.88}$$

The differences may again be estimated, and in practice the estimate can be made so accurately that, after x has been calculated and (with y and z) used to calculate the value of X from equation (7.84), it is often found that a further iteration is not required for that step.

Equation (7.88) can be used to provide an extrapolated value of x by estimating values of X and the same line differences. Alternatively, one may use

$$x_2 = x_{-1} + 3\Sigma^1 X_{1/2} + \frac{5}{4}\delta X_{1/2} + \frac{17}{240}\delta^3 X_{1/2} - \dots . \tag{7.89}$$

It sometimes happens that as a body (for example a comet) nears perihelion, it becomes necessary to halve the tabular interval. After perihelion passage the interval may be doubled again.

To illustrate some of the above ideas we take as an example the numerical integration of the second-order equation

$$\frac{d^2 x}{dt^2} = -x \tag{7.90}$$

where we are told that at $t = 1\cdot10\,512$, $x = 0\cdot21\,856$ and $dx/dt = 0\cdot48\,273$.

The substitution of the series

$$x = \sum_{n=0}^{\infty} a_n(t-t_0)^n$$

and equating coefficients of the powers of $(t-t_0)$ yields the series

$$x = x_0\left[1 - \frac{(t-t_0)^2}{2!} + \frac{(t-t_0)^4}{4!} - \frac{(t-t_0)^6}{6!} + \dots\right]$$

$$+ \dot{x}_0\left[(t-t_0) - \frac{(t-t_0)^3}{3!} + \frac{(t-t_0)^5}{5!} - \dots\right] \tag{7.91}$$

where $x_0 = 0\cdot21\,856$, $\dot{x}_0 = 0\cdot48\,273$, $t_0 = 1\cdot10\,512$.

Take the tabular interval h to be $0\cdot1$. The series (7.91) can then be used to calculate the values of x in column 2 of table 7.2.

Values for the function X can then be inserted into column 3 of the table.

We can now set up a table for function X, putting in the differences δX, $\delta^2 X$, $\delta^3 X$, $\delta^4 X$ where available, and also the half-differences $\mu\delta X_0$ and $\mu\delta^3 X$.

Using (7.85) and (7.86), values of $\Sigma^1 X_{1/2}$ and $\Sigma^2 X_0$ are calculated and entered in the table. Succeeding values of the first and second sums are

234

Table 7.2

t_p	x_p	$X_p = h^2\left(\dfrac{d^2x}{dt^2}\right)_p = -h^2 x_p$
t_{-2}	0·90512	0·11830	−0·0011830
t_{-1}	1·00512	0·16928	−0·0016928
t_0	1·10512	0·21856	−0·0021856
t_1	1·20512	0·26566	−0·0026566
t_2	1·30512	0·31010	−0·0031010

obtained by using (7.74) and (7.75). The new table, as far as it has gone, is shown in table 7.3 above the staggered line.

To obtain the value of x at t_3 we can estimate values of the differences required in (7.88). Guessing that $\delta^4 X$ is zero, we can write values for $\delta^3 X_{1\frac{1}{2}}$, $\delta^2 X_2$, $\delta^1 X_{2\frac{1}{2}}$ and X_3. These are respectively 0·0000048, 0·0000314, −0·0004130 and −0·0035140. If we further suppose that $\delta^4 X_2$ is zero we can also write values for $\delta^3 X_{2\frac{1}{2}}$ and $\delta^2 X_3$. These are respectively 0·0000048 and 0·0000362. We know $\Sigma^2 X_3$, so that a first approximation to x_3 can then be calculated from (7.88), giving $x_3 = 0·35145$. From this value, new values of the differences can be written down. It is found that this results in a change of only 5 in the last place of these differences, so a new value for x_3 need not be computed from (7.88) and the new differences can now be confirmed. They are shown in table 7.3 below the staggered line. The first and second sums $\Sigma^1 X_{3\frac{1}{2}}$ and $\Sigma^2 X_4$ can now be entered, and the next step (the calculation of x_4) can be begun.

Alternatively, equation (7.89) in the form

$$x_3 = x_0 + 3\Sigma^1 X_{1\frac{1}{2}} + \frac{5}{4}\delta X_{1\frac{1}{2}} + \frac{17}{240}\delta^3 X_{1\frac{1}{2}} - \dots$$

could have been used to provide a first approximation to x_3, only $\delta^3 X_{1\frac{1}{2}}$ requiring estimation.

If $\dot{x}$ is to be found, the successive half-differences must also be entered in table 7.3 to enable (7.87) to be used.

There is a vast literature on numerical procedures; many mathematicians such as Newton, Gauss, Lagrange, Bessel, Stirling and others have contributed elegant methods of tackling such subjects as interpolation, numerical differentiation and integration, the solution of differential equations, the fitting of data and so forth.

As previously mentioned, the Gauss–Jackson method is among the best for use in the numerical integration of the second-order differential equations most commonly used in orbital motion problems. In the nomenclature given above, the equation

$$x = \Sigma^2 X + \frac{1}{12}X - \frac{1}{240}\delta^2 X + \frac{31}{60480}\delta^4 X - \frac{289}{3628800}\delta^6 X \dots$$

235

Table 7.3

	Sums		Function	Differences			
Argument	2nd	1st		1st	2nd	3rd	4th
t_{-2} 0·90512			−0·0011830				
				−0·0005098			
t_{-1} 1·00512			−0·0016928		0·0000170		
		0·047140		−0·0004928		0·0000048	
t_0 1·10512	0·21874		−0·0021856	μ −0·0004819	0·0000218	μ0·0000048	0·0000000
		0·044483		−0·0004710		0·0000048	
t_1 1·20512	0·26588		0·0026566		0·0000266		−0·0000005
		0·041382		−0·0004444		0·0000043	
t_2 1·30512	0·31036		−0·0031010		0·0000309		
		0·037867		−0·0004135			
t_3 1·40512	0·35174		−0·0035145				
t_4 1·50512	0·38961						

is used for double integration, while for single integration the equation

$$h\dot{x}=\mu\Sigma^1 X-\frac{1}{12}\mu\delta X+\frac{11}{720}\mu\delta^3 X-\frac{191}{60480}\mu\delta^5 X+\frac{2497}{3628800}\mu\delta^7 X\ldots$$

is used.

The equation is used as a predictor; that is to say, a first approximation to the value of x is calculated from it, having estimated values of the differences 'below' the line as previously described in forming their first approximations. If the step size or tabular interval has been chosen judiciously a corrector cycle will be unnecessary, but can be included for the human computer's peace of mind. It will utilize the equation giving X from x.

Problems

7.1 Form a difference table for the equation

$$x=1+t+t^2+t^3$$

taking the step size to be $t=1$; that is $h=0$, $t_1=1$, $t_{-1}=-1$, etc. Why is the fourth difference zero?

7.2 If the step size is doubled or halved, what effect will it have on the differences in a table of differences? Check your result by doubling and halving the interval or step size in the table obtained in problem 7.1.

7.3 In problem 6.1, take $g=9\cdot81$ m s^{-2}, $\epsilon=0\cdot01$; at $t=0$, $x=0$, $dx/dt=0\cdot56$ m s^{-1}. Use the Gauss–Jackson method of numerical integration to obtain the value of x at $t=1\cdot00$ s. Check your answer by the approximate formula given in problem 6.1.

7.4 In the example of section 7.4, obtain the value of x by numerical integration and use the series (7.91) to check your answer.

References

Allan R R 1961 *Nature* **190** (No. 4776) 117
Allan R R and Ward G N 1963 *Proc. Camb. Phil. Soc.* **59** 669
Brouwer D 1937 *Astron. J.* **46** 199
Butcher J C 1965 *Math. Comput.* **19** 408
Cohen C J and Hubbard E C 1962 *Astron. J.* **67** 10
Cowell P H and Crommelin A D 1908 *Mon. Not. R. Astron. Soc.* **68** 576
Emslie A E and Walker I W 1979 *Cel. Mech.* **19** 147
Fehlberg E 1968 *NASA Tech. Rep.* R-248
—— 1972 *NASA Tech. Rep.* R-381
Garafalo A M 1960 *Astron. J.* **65** 117
Heggie D C 1971 *Astrophys. Space Sci.* **14** 35
Henrici P 1962 *Discrete Variable Methods in Ordinary Differential Equations* (New York: Wiley)
—— 1964 *Elements of Numerical Analysis* (New York: Wiley)
Herget P 1962 *Astron. J.* **67** 16
Herrick S 1953 *Astron. J.* **58** 156
—— 1971 *Astrodynamics* (London: Van Nostrand Reinhold)
Interpolation and Allied Tables 1956 (London: HMSO)
Kyner W T and Bennet M M 1966 *Astron. J.* **71** 579
Lapidus L and Seinfeld J H 1971 *Numerical Solution of Ordinary Differential Equations* (New York: Academic)

Lecar M 1968 *Bull. Astron.* **3** 91
Merson R H 1973 *Numerical Integration of the Differential Equations of Celestial Mechanics* (Farnborough: Royal Aircraft Establishment)
Merton G 1949 *Mon. Not. R. Astron. Soc.* **109** 421
Milankovic M 1939 *Acad. Serbe. Bull. Acad. Sci. Mat. Nat. A* (No. 6)
Moran P E 1973 *Cel. Mech.* **7** 122
Moran P E, Roy A E and Black W 1973 *Cel. Mech.* **8** 405
Musen P 1954 *Astron. J.* **59** 262
Oppolzer T R 1870 *Sitzungsberichte der Wiener Akad.* (Math. Classe)
Peters C F 1968 *Bull. Astron.* **3** 167
Pines S 1961 *Astron. J.* **66** 5
Roy A E and Moran P E 1973 *Cel. Mech.* **7** 236
Roy A E, Moran P E and Black W 1972 *Cel. Mech.* **6** 468
Shanks E B 1966 *Math. Comput.* **20** 21
Stiefel E L 1970 *Cel. Mech.* **2** 274
Stiefel E L and Scheifele G 1971 *Linear and Regular Celestial Mechanics* (Berlin: Springer-Verlag)
Stumpff S and Weiss E H 1967 *NASA Tech. Note* D-4470
Szebehely V 1967 *Theory of Orbits* (New York: Academic)

Bibliography

Brouwer D and Clemence G M 1961 *Methods of Celestial Mechanics* (New York and London: Academic)
Buckingham R A 1957 *Numerical Methods* (London: Pitman)
Herget P 1948 *The Computation of Orbits* (University of Cincinnati)
Herrick S 1971, 1972 *Astrodynamics* vols 1 and 2 (London: Van Nostrand Reinhold)
Jackson J 1924 *Mon. Not. R. Astron. Soc.* **84** 602
Khabaza I M 1965 *Numerical Analysis* (London: Pergamon)
Milne E A 1948 *Vectorial Mechanics* (London: Methuen)
Moore R E 1966 *Interval Analysis* (New York)
Stumpff K 1959 *Himmelsmechanic* vol 1 (Berlin: VEB Deutscher Verlag der Wissenschaften)

8 The Stability and Evolution of the Solar System

8.1 Introduction

We return now to a consideration of some important problems in Solar System dynamics. These problems are concerned with questions of evolution and stability. When we observe the members of the Sun's family we see planets moving about the Sun in well spaced orbits, which are gradually altering in ways given precisely by the theories of general perturbations. Most satellites behave likewise, though suspicion arises that the retrograde moons of Jupiter and Saturn are captured asteroids. The abundance of near-commensurabilities in mean motions is a notable feature, as is the seeming avoidance of certain commensurabilities in the asteroid belt and in the ring structure of Saturn. It is also a matter of record that on occasion comets may have their orbits suddenly and drastically altered by close planetary encounters.

Some of the questions we would like to answer may be formulated along the following lines:

(i) How old is the Solar System?
(ii) Does the distribution of planetary orbits alter appreciably in an astronomically long time?
(iii) If so, do the orbits alter slowly; or can sudden far-reaching changes occur in one or more of the planetary orbits, even to the extent of planets changing their order from the Sun or colliding?
(iv) If the Solar System is stable and only slowly evolving, is this due to its present set-up with almost circular orbits, low inclinations and near-commensurabilities in mean motion?

These questions have been tackled in one way or another by many researchers. The first question is one to which geophysics, lunar sample dating and solar astrophysics suggest agreeably close answers. Radioactive dating of terrestrial and lunar rocks give figures of the order of $4 \cdot 5 \times 10^9$ years as the minimum ages of the Earth and the Moon. The theory of stellar structure and energy generation applied to the Sun estimates its age as $5 \cdot 0 \times 10^9$ years. It therefore seems unlikely that the Solar System is any younger. This length of time, of the order of 5×10^9 revolutions of the Earth about the Sun, makes us suspect that the answer to question (ii) is 'probably not'. This view is strengthened when the geological record of fossils is examined, and we find that complicated life forms have inhabited the Earth for at least the past 2×10^9 years. During that time at least, the Sun's radiation output cannot have altered to any major extent; nor can the major and minor distances of Earth from Sun have strayed far from their present values. Certain marine-life studies even give us data on how slow the evolution of the Earth–Moon system has been under tidal action.

It is humiliating to acknowledge that, even today, celestial mechanics is not capable of making such confident statements on the age, stability and evolution of the Solar System. This is not to say that progress has not been made in recent years. A great deal of progress has undoubtedly been made in many parts of the general problem and as a result we now understand more clearly the gravitational mechanisms running some of its subsystems. We consider some of these topics in the following sections:

8.2 Planetary Ephemerides

Another aspect of the problem of Solar System dynamics is the production of the various national ephemerides. Most of the tables published today are based on various theories of the Sun, Moon and planets which were carefully and laboriously computed by many eminent astronomers. These are for the most part analytical theories based on general perturbation methods. However, in the last few years several projects have been under way which approach the problem of the compilation of ephemerides from different directions. One project of this nature is described in Oesterwinter and Cohen (1972). Others are being carried out at the Massachusetts Institute of Technology and the Jet Propulsion Laboratory, Pasadena. Since an important factor in such work is the accurate numerical integration of the Solar System, it is appropriate to outline Oesterwinter's and Cohen's approach.

These workers point out that the major defects in the classical theories arise from the fact that these theories were done for the most part by hand. Due to the limited amount of algebra which a man can do during a lifetime the series had to be truncated somewhere. It is difficult to find all of the terms which are greater than a certain threshold. It is also very laborious to fit these theories directly to the observations, and generally some previous investigator's residuals were used instead. As a consequence some of the published places are in considerable error. For example, Pluto's published places are now approximately $10''$ in error. These considerations led Oesterwinter and Cohen to attempt a global solution of the Solar System, simultaneously determining the elements of the planets and the Moon in such a way as to give a least-squares fit to the observations over a large time span. They used an n-body program and simply treated the Moon as another planet in orbit about the Sun. This treatment can cause much difficulty during the integration, since the Moon's highly perturbed heliocentric orbit dictates the use of a very small step size. A choice of 0·4 days as a step size for the whole system was in fact made. The model incorporated many features of interest, including an estimate of the Earth–Moon tidal coupling and an extrapolation of atomic time back to 1912, and is probably one of the best sources of planetary positions available today.

8.3 Commensurabilities in Mean Motion

Much research in celestial mechanics has been devoted to attempts to understand the presence of so many near-commensurabilities in mean motions, or resonances, found between pairs of planets such as Pluto and Neptune, or pairs of satellites such as Mimas and Tethys, or even trios of satellites such as Io, Europa and Ganymede. Roy and Ovenden (1954) showed that too many occurred to be due to chance. Attempts have also been made to explain why some commensurabilities are avoided, as in the Kirkwood gaps in the asteroid region or Cassini's division in Saturn's rings.

Several questions may be asked. Was the mode of formation of the planetary system and the satellite systems such that it gave rise to near-commensurabilities? If so, were there more in the past, the mutual gravitational forces tending to destroy them? Or are they particularly stable arrangements against such perturbations, so that objects pursuing noncommensurable orbits in the Solar System have had their orbits altered, even to the point of collision or escape? Can a pair of bodies, not in a commensurable relationship, drift under the action of forces operating into such a relationship and thence remain in it? We now consider some research that has shed light on such questions.

8.4 The Asteroids

One problem in the orbital motions of the asteroids is the overall distribution of these objects; namely the way in which asteroid numbers vary with mean heliocentric distance, or more relevantly with mean motion. A related problem involves the avoidance by asteroids of certain mean motions (the Kirkwood gaps) and their preference for certain other mean motions (for example, the Hilda group and the Trojans). Figure 8.1 (Brouwer 1963) plots the distribution of asteroids with respect to their mean motions about the Sun in seconds of arc per day (q denotes order of commensurability). The gaps corresponding to mean motions which would be commensurable with that of Jupiter ($n_J = 299 \cdot 13''$ per day) are evident. The positions of the commensurabilities involving the smallest integers are also given. The sharp cut-off beyond 2/1 (the so-called Hecuba gap) is obvious, as is the clustering about 3/2 (the Hilda group) and 1/1 (the Trojan group).

Asteroid problems have been attacked by both analytical and numerical methods. The mass of any asteroid is so tiny compared with the masses of Sun and Jupiter that many of the problems may be considered as practical examples of the elliptic or circular restricted three-body problem. Tisserand.

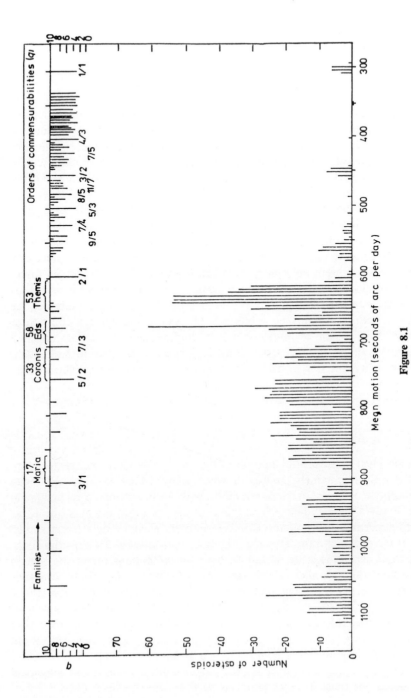

Figure 8.1

242

Poincaré, Andoyer, Hirayama, Brouwer, Farinella, Cl and Ch Froeschlé, Ferraz-Mello, Hadjidemetriou, Kozai, Scholl and Message are only a few who have developed and used analytical methods applicable to the cases of asteroids where their mean motions are commensurable with that of Jupiter. Ordinary general perturbation theory is useful, even in the cases of pairs of planets the ratio of whose mean motions approximates to a whole-number ratio. In such cases, so-called critical terms in the disturbing function produce terms in the perturbations which have small divisors, giving rise to the inequalities characterized by the Jupiter–Saturn 'great inequality' of 900 years (section 6.7.3). When the commensurability is very close however, as in certain satellite pairs, or in the case of Pluto and Neptune, or in some of the asteroid–Jupiter cases, different perturbation methods have to be created.

It is found both by the application of such methods and by numerical integration that the gaps and concentrations at commensurabilities are indeed due to Jupiter's perturbing effect. We have already dealt in chapter 5 with the Trojans as a practical case of the Lagrange equilateral-triangle solution of the three-body problem, which is stable in that the Trojan asteroids merely oscillate (or *librate*) about the equilateral points.

For noncommensurable orbits the perturbations in the mean motions of the asteroids are proportional to the ratio of Jupiter's mass to that of the Sun. For commensurable orbits there will be critical terms giving rise to large long-period librations in the mean motion and in the other orbital elements, the result being that the asteroid's mean motion is rarely observed at its small-integer commensurable value. This is analogous to taking randomly timed flashlight photographs in darkness of a pendulum swinging. Most of the snapshots would show the pendulum away from its vertical position. If we therefore take a distribution of mean motions right across a small-integer commensurability, we would expect to observe fewer minor planets with osculating mean motions in the immediate vicinity of the commensurability, even though the commensurability is stable. Both Brouwer (1963) and Message (1966), using different arguments, have put forward evidence supporting their view. On such a view therefore, the gaps are not regions of instability. Some work by Schubart (1966) indicates that the 3/2 commensurability (the Hilda group) is a region where stable librations about periodic orbits can exist. There are about 40 members of this group.

Research by Hunter (1967) and more recently by Lecar and Franklin (1974) shows the relationship of asteroids lying not only between Mars and Jupiter but also between Jupiter and Saturn, with the capture of satellites by Jupiter and their escape.

Figure 8.1 shows the sharp cut-off of asteroid numbers beyond the 2/1 Hecuba gap, leaving a zone essentially devoid of asteroids apart from the Hilda group at 3/2 and the Trojans at 1/1. From a study of the escape of Jovian satellites under solar perturbations, Hunter found that such hypothetical satellites would become asteroids in that empty zone or go into solar orbit as asteroids in the region between the orbits of Jupiter and Saturn. Since orbits are traversed in the opposite direction if time is reversed, the implication is that Jupiter could deplete any original distribution of asteroids in the now empty zone, even sending them after a close encounter or a temporary existence as satellites of itself into the Saturn–Jupiter region or back into the asteroid zone again. The Hilda group is stable against such a process.

Lecar and Franklin examined the effect of Jupiter on an initially uniform distribution of asteroids extending from Mars to Jupiter. By numerical integration they showed that, after as short a period of time as 2400 years, most of the asteroids in the region extending from the 3/2 commensurability to Jupiter were ejected, with the exception of the stable librators (the Hilda group). Between the 2/1 and 3/2 commensurabilities however, the depletion was small. Lecar and Franklin concluded that far longer times would have to elapse before this region was emptied by Jovian perturbations, or that some other mechanism would have to be invoked to sweep the region clear of asteroids.

With respect to the region between Jupiter and Saturn, they found that the perturbing effects of these two massive planets on an initially uniform distribution of asteroids would remove at least 85% of them in only 6000 years, leaving two bands at distances 1·30 and 1·45 Jupiter units from the Sun (6·8 and 7·5 AU). Asteroids at such distances are at least temporarily stable, and it is interesting to note that the '1·30' distance gives commensurabilities in mean motion with respect to Jupiter and Saturn's of close to 3/2 and 3/5, while the '1·45' distance gives commensurabilities close to 7/4 and 7/10 respectively. Whether or not such orbits are stable over much longer periods is still unknown. The important implication of such work is that even if asteroids had existed between Jupiter and Saturn, and had had masses as large as those of the Earth, Venus or Mars, the vast majority of them would have been slung into other parts of the Solar System in a few thousand years at the most.

Ch Froeschlé and H Scholl (1988) have shown that interesting effects occur in the evolution of asteroid orbits when they are located in secular resonances. Such a resonance will produce strong secular perturbations on the asteroid orbit when the precession rate of the perihelion longitude ϖ or nodal longitude $\dot{\Omega}$ is nearly equal to the corresponding rate $\dot{\varpi}$ or $\dot{\Omega}$ of a planetary orbit. In the asteroid region, three strong resonances ν_5, ν_6 and ν_{16} can occur, where

$$\dot{\nu}_5 \sim \dot{\varpi}_{\mathrm{J}}, \qquad \dot{\nu}_6 \sim \dot{\varpi}_{\mathrm{S}}, \qquad \dot{\nu}_{16} \sim \dot{\Omega}_{\mathrm{J}} \sim \dot{\Omega}_{\mathrm{S}},$$

J and S referring to Jupiter and Saturn.

Such asteroids have their eccentricities increased to such an extent that their orbits may cross those of Mars, Earth and even Venus. Obvious consequences of such an evolution include the possibility of a further dramatic transformation of the orbit by a close encounter of the asteroid with one of these planets or even a collision.

8.5 Rings, Shepherds, Tadpoles, Horseshoes and Co-orbitals

The title of this section, reminiscent more of biology and bucolic pastimes than celestial mechanics, acknowledges a class of dynamical problems brought into prominence in recent years by ground-based and spacecraft-based discoveries in the outer Solar System. It raises a number of interesting questions of stability.

8.5.1 Ring systems

Prior to 1977 only Saturn had been found to have a ring system. Apart from a few Earth-based observers such as Bernard Lyot and Andouin Dollfus who had, under momentary conditions of good seeing, detected fine structure in the rings, it was thought that the rings comprised three in number; the bright outermost ring A was separated from the bright middle ring B by a dark space called Cassini's division. Ring C (a hazy, transparent ring—the so-called crepe ring), was situated just inside ring B. Theoretical investigations by, among others, Clerk Maxwell (1859) and spectroscopic observations by Keeler (1895) involving Doppler measurements, showed that the rings were neither solid nor liquid but had to consist of numerous small solid particles in orbit about the planet. It could be shown also that their individual orbits were perturbed by Saturn's innermost three moons: Mimas, Enceladus and Tethys. Cassini's division contained distances where the mean motion of hypothetical particles would be twice that of Mimas and three and four times those of Enceladus and Tethys, while the boundary between rings B and C lay at a distance where the mean motion would be three times that of Mimas.

The challenging picture presented by the Voyager encounters with Saturn is much more complicated. Not only do rings A, B and C consist of many hundreds of ringlets but numerous distinct ringlets exist in the Cassini division. The F ring, discovered by Pioneer 11, is itself composed of a number of separate ringlets. Rings D and E also exist.

The detection of a ring system about Uranus in 1977, from anomalous occultations of starlight, was only the second discovery of a ring system in 350 years. The third discovery took place in 1979, just two years after the second. On Voyager pictures taken during the fly-pasts of Jupiter, a single narrow (7000 km wide) bright ring of radius 1.81 Jovian radii appeared. The outer edge is sharp, the inner is fuzzy and may extend all the way to Jupiter.

Whether Neptune has rings is as yet unknown; the question will have to wait for the Hubble Space Telescope to be flown or, more likely, the Voyager fly-past of Neptune in 1989.

The nine Uranian rings are quite unlike the Saturnian ones. Their dimensions are given in table 8.1.

The data in table 8.1, from Elliot *et al* (1981), are derived from occultation observations and a kinematic model in which the rings are taken to be

Table 8.1

Ring	Semimajor axis (km)	Eccentricity e $(\times 10^3)$	Precession rate from fitted J_2 and J_4 (deg/day)^2
6	$41\,863\cdot8 \pm 32\cdot6$	$1\cdot36 \pm 0\cdot07$	$2\cdot7600$
5	$42\,270\cdot3 \pm 32\cdot6$	$1\cdot77 \pm 0\cdot06$	$2\cdot6678$
4	$42\,598\cdot3 \pm 32\cdot7$	$1\cdot24 \pm 0\cdot09$	$2\cdot5963$
α	$44\,750\cdot5 \pm 32\cdot8$	$0\cdot72 \pm 0\cdot03$	$2\cdot1832$
β	$45\,693\cdot8 \pm 32\cdot8$	$0\cdot45 \pm 0\cdot03$	$2\cdot0288$
η	$47\,207\cdot1 \pm 32\cdot9$	$(0\cdot03 \pm 0\cdot04)$	$1\cdot8094$
γ	$47\,655\cdot4 \pm 32\cdot9$	$(0\cdot04 \pm 0\cdot04)$	$1\cdot7503$
δ	$48\,332\cdot0 \pm 33\cdot0$	$0\cdot054 \pm 0\cdot035$	$1\cdot6657$
ϵ	$51\,179\cdot7 \pm 33\cdot8$	$7\cdot92 \pm 0\cdot04$	$1\cdot3625$

Values of harmonic coefficients of Uranian gravitational potential:

$$J_2 = (3\cdot352 \pm 0\cdot006) \times 10^{-3}; \quad J_4 = (-2\cdot9 \pm 1\cdot3) \times 10^{-5}.$$

coplanar ellipses of zero inclination, precessing because of Uranus' gravitational potential's zonal harmonics.

In fact, more recent observations have shown that some of the rings are inclined to the equatorial plane of Uranus by a few hundredths of a degree.

8.5.2 Small satellites of Jupiter and Saturn

In 1974, the discovery of Jupiter XIII (Leda) brought the number of known natural satellites in the Solar System to 33. By the end of 1987, the number had grown to 44. Those recently discovered are, not surprisingly, small objects but some of them exhibit dynamical cases of great interest. Data for them are provided in table 8.2.

The satellite Adrastea (1979 J1) moves just outside Jupiter's ring; its sharp outer edge would appear to be controlled by the satellite.

In the Saturnian system, 1980S13 (Telesto) and 1980S25 (Calypso) librate about the Lagrangian L4 and L5 equilateral positions in the Saturn–Tethys system while 1980S6 librates about the L4 position in the Saturn–Dione system. These three satellites are therefore analogous in their positions to the Trojan asteroids in the Sun–Jupiter system. The ratios of the masses of Tethys and Dione to that of Saturn are so far below Routh's value of 0·0385 (see Section 5.10.4) that the satellites 1980S13, 1980S25 and 1980S6 are able to perform stable periodic oscillations about the Lagrange points.

When we come to consider the other small satellites of interest, it is instructive to move from the restricted three-body problem where two of the bodies have finite mass and one has infinitesimal mass to a special case of the general three-body problem where all three bodies have finite mass but two have masses of comparable size, both being small in comparison to the third.

Consider now the system Saturn and its two small satellites 1980S1 (Janus) and 1980S3 (Epimetheus). Let their masses be respectively M, m_1 and m_3.

Table 8.2

Temporary designation	Permanent number	Name	Semimajor axis a (km)	Mean motion n (deg/day)	Eccentricity e	Inclination i (deg)	Radius (km)	Mass† relative to planet's mass
	JUPITER							
1979J3	XVI	Metis	127 960	1221·249 ± 0·002	$<4 \times 10^{-3}$	~0	20	
1979J1	XV	Adrastea	128 980	1206·995 ± 0·002	~0	~0	12	
1979J2	XIV	Thebe	221 895	533·700 ± 0·013	0·015 ± 0·006	0·8 ± 0·2	40	
	SATURN							
1980S28	XV	Atlas	137 640	598·306	~0	~0	20 × 10	$2·2 \times 10^{-11}$
1980S27	—	—	139 353	587·2890 ± 0·0005	0·0024 ± 0·0006	0·0 ± 0·1	70 × 50 × 40	$1·0 \times 10^{-9}$
1980S26	—	—	141 700	572·7891 ± 0·0005	0·0042 ± 0·0006	0·0 ± 0·1	55 × 45 × 35	$6·4 \times 10^{-10}$
1980S3	XI	Epimetheus	151 422	518·490 ± 0·01	0·009 ± 0·002	0·34 ± 0·05	70 × 60 × 50	$1·5 \times 10^{-9}$
1980S1	X	Janus	151 472	518·236 ± 0·01	0·007 ± 0·002	0·14 ± 0·05	110 × 100 × 80	$6·5 \times 10^{-9}$
1980S13	XIII	Telesto	294 660	190·70	~0	~0	17 × 14 × 13	$2·2 \times 10^{-11}$
1980S25	XIV	Calypso	294 660	190·70	~0	~0	17 × 11 × 11	$1·5 \times 10^{-11}$
1980S6	XII	—	377 400	131·536	0·005	0·2	18 × 16 × 15	$3·1 \times 10^{-11}$

† Assuming mean densities of 1 g cm^{-3} for Saturn's satellites.
Data mostly from Synnott (1984), Synnott et al (1981, 1983).

Then m_1 and m_3 are given by $m_1 = \epsilon_1 M$ and $m_3 = \epsilon_3 M$ where ϵ_1 and ϵ_3 are $6 \cdot 5 \times 10^{-9}$ and $1 \cdot 5 \times 10^{-9}$, respectively.

Yoder *et al* (1983) described their behaviour according to the following model.

If 1980S3 had infinitesimal mass, it could librate (figure 8.2, orbit a) about either L4 or L5 in the Saturn–1980S1 system with 1980S1 performing a circular orbit about Saturn.

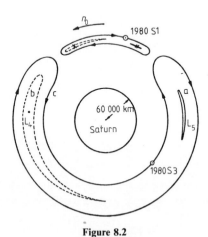

Figure 8.2

Now if the libration were enlarged its shape would resemble that of a tadpole and the limiting tadpole orbit would be similar to orbit b of figure 8.2. Any enlargement of the libration orbit would produce a horseshoe orbit c (figure 8.2). Satellite 1980S3, however, is not infinitesimal in mass compared with the mass of satellite 1980S1 and so the picture has to be modified in the manner shown in figure 8.2. 1980S1 is perturbed by 1980S3 and consequently performs its own oscillations. Both bodies therefore pursue horseshoe shaped orbits about their mean positions which of course rotate about Saturn. If A_1 and A_3 are the amplitudes of the oscillations, it can be shown that $A_1/A_3 = m_3/m_1$. The widths of the horseshoes in figure 8.2 are grossly exaggerated, being 700 times narrower than shown and being proportional to the cube root of the perturbing satellite's mass.

Harrington and Seidelmann used numerical integration to show that the libration period was about 3000 days, the saturnocentric radius vectors of the two satellites 1980S1 and 1980S3 never approaching within 6° over the integration duration of 100 years. Changes in initial conditions did not cause instability, nor did the effects of Saturn's oblateness and the perturbations of the eight major satellites. Colombo (1982) has calculated the librational amplitudes as 60° for S1 and 285° for S3. Other investigators such as Dermott and Murray (1981a, b) have also studied the possible tadpole and horseshoe orbits of 1980S1 and 1980S3.

The satellites 1980S26 and 1980S27, unlike the co-orbital pair 1980S1 and S3 which exchange inner and outer orbits at close encounter, merely overtake each other every 25 days. Nevertheless they are of great interest lying as they are with one just inside the F ring of Saturn and the other just outside. Because of their role in confining and maintaining the F ring, they have been christened the 'shepherd' satellites.

8.5.3 Spirig and Waldvogel's analysis

In a paper by Spirig and Waldvogel (1985) the authors study the three-body problem with one large central mass M; the other two bodies are satellites of comparable masses ϵm_1 and ϵm_2, with $\epsilon \ll 1$. Using the techniques of perturbation theory, the satellites' motion is described by an 'outer' and an 'inner' approximation, the former being valid when the satellites are far apart, the latter when they are close together. In the outer solution the satellites are found to pursue independent Keplerian motions about the central mass; the inner solution satisfies Hill's lunar equation. Spirig and Waldvogel's elegant achievement is to show that the discussion of Hill's problem with appropriate boundary conditions at infinity predicts that the co-orbiting satellites of Saturn, 1980S1 (Janus) and 1980S3 (Epimetheus), exchange orbits at the close encounter every 4 years whereas the shepherds 1980S26 and 1980S27 do not. In what follows we follow closely Spirig and Waldvogel's analysis.

Firstly, we set up the equations of motion of the problem.

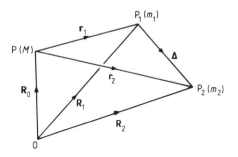

Figure 8.3

Let O be an origin in an inertial frame with vectors $\mathbf{R}_0$, $\mathbf{R}_1$, $\mathbf{R}_2$ denoting the positions of P(mass M), P_1(mass m_1) and P_2(mass m_2).

Then, as shown in figure 8.3, we can define relative positions

$$\mathbf{r}_i = \mathbf{R}_i - \mathbf{R}_0, \qquad (i = 1, 2), \qquad \mathbf{\Delta} = \mathbf{R}_2 - \mathbf{R}_1.$$

The force function U and kinetic energy T can then be written as

$$U = \frac{Mm_1}{|\mathbf{r}_1|} + \frac{Mm_2}{|\mathbf{r}_2|} + \frac{m_1 m_2}{|\mathbf{\Delta}|}, \qquad T = \frac{1}{2}\left(M|\dot{\mathbf{R}}_0|^2 + \sum_{i=1}^{2} m_i|\dot{\mathbf{R}}_i|^2 \right).$$

We can eliminate the centre-of-mass integral by introducing the relative coordinates $\mathbf{r}_1$ and $\mathbf{r}_2$ into the Lagrangian $L = T + U$. The kinetic energy T is now given by

$$T = \frac{1}{2(M + m_1 + m_2)}[m_1(M + m_2)\dot{\mathbf{r}}_1 \cdot \dot{\mathbf{r}}_1 + m_2(M + m_1)\dot{\mathbf{r}}_2 \cdot \dot{\mathbf{r}}_2 - 2m_1 m_2 \dot{\mathbf{r}}_1 \cdot \dot{\mathbf{r}}_2].$$

The Lagrangian equations of motion are then

$$\frac{\mathrm{d}}{\mathrm{d}t}\left(\frac{\partial L}{\partial \dot{\mathbf{r}}_k}\right) = \frac{\partial L}{\partial \mathbf{r}_k}, \qquad (k = 1, 2), \qquad \boldsymbol{\Delta} = \mathbf{r}_2 - \mathbf{r}_1. \tag{8.1}$$

Let M be the mass of the central body and m_1, m_2 the masses of the small satellites and let

$$\epsilon = \frac{m_1 + m_2}{M}, \qquad \mu_k = \frac{m_k}{m_1 + m_2}, \qquad (k = 1, 2).$$

Note that, whereas ϵ is small, of order 10^{-9}, the μ_k are of order unity. Take the central body's mass as unit mass. Then equations (8.1) become the perturbation equations

$$\ddot{\mathbf{r}}_1 + (1 + \epsilon\mu_1)\frac{\mathbf{r}_1}{r_1^3} = \epsilon\mu_2\left(-\frac{\mathbf{r}_2}{r_2^3} + \frac{\boldsymbol{\Delta}}{\Delta^3}\right)$$

$$\ddot{\mathbf{r}}_2 + (1 + \epsilon\mu_2)\frac{\mathbf{r}_2}{r_2^3} = -\epsilon\mu_1\left(\frac{\mathbf{r}_1}{r_1^3} + \frac{\boldsymbol{\Delta}}{\Delta^3}\right). \tag{8.2}$$

We now expand $\mathbf{r}_k(t)$ in a Taylor series with respect to ϵ, namely

$$\mathbf{r}_k(t) = \sum_{l=0}^{\infty} \mathbf{r}_{kl}(t)\epsilon^l. \tag{8.3}$$

Now as $\varepsilon \to 0$, the equations (8.2) reduce to

$$\ddot{\mathbf{r}}_1 + \frac{\mathbf{r}_1}{r_1^3} = 0, \qquad \ddot{\mathbf{r}}_2 + \frac{\mathbf{r}_2}{r_2^3} = 0$$

which are the familiar two-body equations of motion giving as solution two independent Kepler motions $\mathbf{r}_k(t)$ with appropriate initial conditions. This solution, in singular perturbation theory, is referred to as the *outer solution*. As long as the distance between the satellites is large, the outer solution will approximate to the solution of the system (8.2) even when $\epsilon \neq 0$.

To obtain an approximate solution when $|\boldsymbol{\Delta}|$ is small we replace $\mathbf{r}_1$ and $\mathbf{r}_2$ by $\boldsymbol{\Delta}$ and $\mathbf{R}$ as variables and introduce Jacobian coordinates, where $\mathbf{R}$ is the position of the centre of mass of the satellites with respect to the central body. Then

$$\left.\begin{aligned} \mathbf{R} &= \mu_1\mathbf{r}_1 + \mu_2\mathbf{r}_2 \\ \mathbf{r}_1 &= \mathbf{R} - \mu_2\boldsymbol{\Delta}, \qquad \mathbf{r}_2 = \mathbf{R} + \mu_1\boldsymbol{\Delta}. \end{aligned}\right\} \tag{8.4}$$

The force function U and kinetic energy T become

$$\left. \begin{array}{l} U = M\left(\dfrac{1}{\mu_2 |\mathbf{r}_1|} + \dfrac{1}{\mu_1 |\mathbf{r}_2|} + \dfrac{\epsilon}{|\Delta|}\right) \\[3mm] T = \dfrac{1}{2} M\left(\dfrac{\epsilon}{1+\epsilon} \dot{\mathbf{R}} \cdot \dot{\mathbf{R}} + \epsilon \mu_1 \mu_2 \dot{\Delta} \cdot \dot{\Delta}\right) \end{array} \right\} \qquad (8.5)$$

while the equations of motion may be written as

$$\left. \begin{array}{l} \ddot{\mathbf{R}} = (1+\epsilon)\mu_1\mu_2 \dfrac{\partial U}{\partial \mathbf{R}} = -M(1+\epsilon)\left(\dfrac{\mu_1 \mathbf{r}_1}{r_1^3} + \mu_2 \dfrac{\mathbf{r}_2}{r_2^3}\right) \\[3mm] \ddot{\Delta} = \dfrac{\partial U}{\partial \Delta} = M\left(\dfrac{\mathbf{r}_1}{r_1^3} - \dfrac{\mathbf{r}_2}{r_2^3} - \epsilon \dfrac{\Delta}{\Delta^3}\right). \end{array} \right\} \qquad (8.6)$$

Because we are dealing with a close encounter, we magnify the neighbourhood by using ϵ^α where α is chosen so that in the limit $\epsilon \to 0$ a maximum number of terms remains. We therefore introduce

$$\Delta = \epsilon^{1/3} \mathbf{r}$$

into equations (8.6) and obtain

$$\ddot{\mathbf{R}} = -\frac{\mathbf{R}}{R^3} + O(\epsilon^{2/3}) \qquad (8.7)$$

$$\ddot{\mathbf{r}} = -\left(\frac{1}{r^2} + \frac{1}{R^3}\right)\mathbf{r} + 3\frac{(\mathbf{r} \cdot \mathbf{R})\mathbf{R}}{R^5} + O(\epsilon^{1/3}) \qquad (8.8)$$

where we have expanded the RHS with respect to E. As $\epsilon \to 0$ a new perturbation problem is obtained, the solution of the reduced problem ($\epsilon = 0$) being the *inner solution*, valid only in a 'boundary layer' near the close encounter.

Now let R, r denote complex numbers with $|R|$, $|r|$ as their amplitudes. Then (8.7) and (8.8), $\epsilon \to 0$, become

$$\ddot{R} = -\frac{R}{|R|^3}, \qquad \ddot{r} = -\left(\frac{1}{|r|^3} + \frac{1}{|R|^3}\right)r + 3\frac{rR}{|R|^5}R.$$

Take rotating, pulsating coordinates where the scale is continually varied so that the distance $|R|$ is constant and the x axis always lies along R. Introduce the complex number $z = r/R$. Then the equation for $\ddot{r}$ becomes

$$\ddot{z}R + 2\dot{z}\dot{R} + z\ddot{R} = -\left(\frac{1}{|Rz|^3} + \frac{1}{|R|^3}\right)zR + \frac{3(zR)R}{|R|^5}R$$

or

$$\ddot{z} + 2\frac{\dot{R}}{R}\dot{z} = \frac{3(zR)R}{|R|^5} - \frac{z}{|Rz|^3} - \left(\frac{\ddot{R}}{R} + \frac{1}{|R|^3}\right)z.$$

251

Using the equation for $\ddot{R}$ and noting that the projection of r on R is the real part of $z(\textbf{Re } Z)$, we arrive at

$$\ddot{z} + 2\left(\frac{\dot{R}}{R}\right)\dot{z} = \frac{1}{|R|^3}\left(3 \textbf{ Re } z - \frac{z}{|z|^3}\right). \tag{8.9}$$

We now change the independent variable to s, the true anomaly, given by

$$|R| = \frac{p}{1 + e \cos s}, \qquad |R|^2 \frac{ds}{dt} = \sqrt{p}$$

where we note that

$$R = \frac{p \exp (is)}{1 + e \cos s} = |R| \exp is. \tag{8.10}$$

Now

$$\frac{d}{dt} = \frac{\sqrt{p}}{|R|^2}\frac{d}{ds}, \qquad \frac{d^2}{dt^2} = p\left(\frac{1}{|R|^2}\frac{d}{ds}\left(\frac{1}{|R|^2}\frac{d}{ds}\right)\right)$$

hence

$$\frac{\dot{R}}{R} = \frac{1}{|R| e^{is}}\frac{d}{ds}(|R| e^{is}) = i + |R|'$$

where the prime denotes d/ds. Also

$$\dot{Z} = \frac{p}{|R|^2}\left(\frac{1}{|R|^2}z'' - \frac{2}{|R|^3}|R|'Z'\right)$$

giving, after a little reduction,

$$z'' + 2iz' = (1 + e \cos s)^{-1}(3 \textbf{ Re } z - z/|z|^3) \tag{8.11}$$

where primes denote derivation by s. This equation is known as Hill's elliptic lunar problem.

In the near circular case

$$e = O(\epsilon^{1/3}), \qquad ds/dt = p^{-2/3} + O(\epsilon^{1/3})$$

the eccentricities of the orbits being of the order of the cube root of the mass ratio. The inner system (5.117) to zeroth order thus becomes

$$x'' - 2y' = 3x - \frac{x}{(x^2 + y^2)^{3/2}}$$

$$y'' + 2x' = -\frac{y}{(x^2 + y^2)^{3/2}} \tag{8.12}$$

where real notation with $z = x + iy$ has been used.

If we multiply the first equation of (8.12) by x', the second by y' and add, we can integrate to give the Jacobi integral

$$\frac{1}{2}(x'^2 + y'^2) - \frac{3}{2}x^2 - (x^2 + y^2)^{-1/2} = h. \tag{8.13}$$

We now require to match the inner and outer solutions. The procedure can be limited to those cases where the constant $h < 0$. Let the variables $\bar{x}$, $\bar{y}$, $\bar{s}$ be related to x, y, s by the introduction of $\bar{c} > 0$, related to h by $h = -\frac{3}{8}\bar{c}^2$, so that

$$x = \bar{c}\bar{x}, \qquad y = \bar{y}/\bar{c}^2, \qquad s = \bar{s}/\bar{c}^3. \qquad (8.14)$$

In the limit $\bar{c} = 0$, the first equation of (8.12) and equation (8.14) become

$$\frac{d\bar{y}}{d\bar{s}} = -\frac{3}{2}\bar{x}, \qquad \left(\frac{d\bar{y}}{d\bar{s}}\right)^2 = 3\bar{x}^2 + \frac{2}{\bar{y}} - \frac{3}{4}. \qquad (8.15)$$

On eliminating $d\bar{y}/d\bar{s}$ between these and returning to variables x, y and s we obtain

$$y = \frac{8/3}{\bar{c}^2 - x^2}, \qquad (|x| < \bar{c}). \qquad (8.16)$$

This represents a U-shaped orbit travelled from right to left and is an extremely accurate approximation to non-oscillating solutions of (8.12) if $\bar{c} < 0{\cdot}7$. Spirig and Waldvogel refer to orbits leading from the first to the second quadrant of the x-y plane as E-orbits (or *exchange* orbits).

If $\bar{c}$ takes large values the scaling transformation of $x = \bar{c}\bar{x}$, $y = \bar{c}\bar{y}$ and passing to the limit $\bar{c} \to \infty$ leads to the linear problem

$$\bar{x}'' - 2\bar{y}' - 3\bar{x} = 0, \qquad \bar{y}'' + 2\bar{x}' = 0 \qquad (8.17)$$

which has a solution

$$\bar{x} = c + a \cos(s - s_1), \qquad \bar{y} = -\frac{3}{2}c(s - s^*) - 2a \sin(s - s_1). \quad (8.18)$$

The quantities c, a, s^*, s_1 are constants of integration, the Jacobi integral giving the relation $c^2 = \frac{4}{3}a^2 + 1$. Spirig and Waldvogel designate orbits leading from the first to the fourth quadrant such as that given by (8.18) as P-orbits (or *passing* orbits).

It is possible to give the asymptotic expansion for $|s| \to \infty$ of a four-parameter family of solutions and it may be shown that the pair of series

$$x = \sum_{j=0}^{\infty} s^{-j} \sum_{k=0}^{j} a_{jk}(s)l^k, \qquad y = \sum_{j=0}^{\infty} a^{1-j} \sum_{k=0}^{j} b_{jk}(s)l^k \qquad (8.19)$$

with

$$l = \log s/s_0 \qquad (8.20)$$

is a formal solution of (8.12) if the coefficients $a_{jk}(s)$, $b_{yk}(s)$ are chosen as appropriate trigonometric polynomials (including constants) in s. Thus

$$a_{00} = c + a \cos(s - s_1), \qquad a_{10} = \frac{8}{9}c^{-2}, \qquad a_{11} = 0$$

$$a_{20} = -\frac{32}{81}c^{-5} + \frac{14}{27}c^{-3}a \sin(s - s_1), \qquad a_{21} = -\frac{64}{81}c^{-5}$$

253

$$b_{00} = -\frac{3}{2}c, \qquad b_{10} = -2a \sin (s - s_1), \qquad b_{11} = -\frac{4}{3}c^{-2}$$

$$b_{30} = \frac{14}{27} \frac{a^2}{c^4} - \frac{44}{27} c^{-2} - \frac{64}{243} c^{-8} + \frac{28}{27} c^{-3} a \cos (s - s_1)$$

$$b_{31} = b_{32} = \frac{128}{243} c^{-8}, \qquad a_{jj} = b_{jj} = 0, \qquad j > 1.$$

The coefficients a_{4k}, b_{4k} can, with considerable effort, also be calculated In these expressions c, a, s_0, s_1 are the four integration constants. The Jacobi constant h of this family of solutions is given by

$$h = \frac{a^2}{2} - \frac{3}{8}c^2. \tag{8.21}$$

We note that if $a = 0$, all the periodic coefficients are constants, hence $x \to c$ in a non-oscillating way as $s \to \infty$. In this case $c = \bar{c}$ so that (8.21) is an approximation for this type of solution. If however $a \neq 0$, no limit of x exists as $s \to \infty$; in contrast, the solution asymptotically shows oscillating behaviour.

Spirig and Waldvogel note that in the matter of matching solutions of (8.12) with circular outer solutions, the relevant orbits are those whose asymptotic behaviour for $s \to -\infty$ is given by (8.19) with $a = 0$. They obtained solutions of this type by numerical integration with initial conditions sufficiently close to $s = -\infty$. Examples of the family of solutions for various values of the parameter c are shown in figure 8.4.

It turns out that for $c < 0.7$ the orbit is almost perfectly symmetrical with respect to the y axis. It also closely resembles (8.6). The solution is still an E-orbit but the outgoing branch shows noticeable oscillations ($a \neq 0$). As c approaches 1.33 a close encounter with O occurs. E- and P-orbits mix chaotically in the range $1.33 < c < 1.72$ providing an arbitrary number of revolutions around 0 (possibly involving close encounters). For $c > 1.72$ only P-orbits occur. If $c > 2$ they quickly assume an almost straight line.

Matching the outer and inner solutions is done by expressing both of them in the same set of variables.

The two Kepler motions are defined by their longitudes ϕ_k of pericentre, their eccentricities e_k and their latus rectums p_k, $k = 1, 2$. Take the vicinity of an aligned configuration where the true anomalies s_k are equal. We can take $s_k = 0$ at $t = 0$.

Assuming the eccentricities to be small and applying the laws of Keplerian motion, also

$$p_k = p(1 + q_k e_k), \qquad q_k = O(1), \qquad (k = 1, 2) \tag{8.22}$$

we obtain, in complex notation

$$r_k = p \exp (is)\{1 + e_k[q_k(1 - \tfrac{3}{2}is) - \cos (s - \phi_k)$$
$$+ 2i \sin (s - \phi_k) + 2i \sin \phi_k]\} + O(e_k^2), \qquad s = t/p^{3/2}.$$

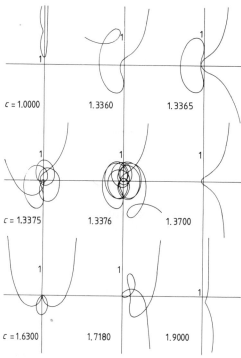

$c = 1.0000$	1.3360	1.3365
$c = 1.3375$	1.3376	1.3700
$c = 1.6300$	1.7180	1.9000

Figure 8.4

The equations defining the inner coordinates x, y are

$$\mathbf{R} = \mu_1\mathbf{r}_1 + \mu_2\mathbf{r}_2, \qquad \mathbf{\Delta} = \mathbf{r}_2 - \mathbf{r}_1 = \epsilon^{1/3}R(x+iy).$$

Scaling $e_k = \epsilon^{1/3}\bar{e}_k$ with $\bar{e}_k = O(1)$, we obtain

$$\left.\begin{aligned} R &= p \exp(is) + O(\epsilon^{1/3}) \\ x &= c + a\cos(s - s_1), \qquad y = -\frac{3}{2}cs - 2a\sin(s - s_1) + a\sin s_1 \end{aligned}\right\} \quad (8.23)$$

where c, a and s_1 are defined by

$$\left.\begin{aligned} pc\epsilon^{1/3} &= p_2 - p_1 \\ a\exp(is_1) &= \bar{e}_1\exp(i\phi_1) - \bar{e}_2\exp(i\phi_2) \end{aligned}\right\} \quad (8.24)$$

in view of equation (8.19). The second relation may be interpreted by means of the eccentricity vectors of the two Kepler motions.

Thus to a given outer solution the asymptotic initial values c, a, s_1, s_0 of the inner solution can be calculated by equation (8.24) and a similar relation for s_0. The inner solution then describes the motion of m_2 relative to m_1 during their interaction. A rather more complicated procedure is required to return from inner to outer solution when $s = +\infty$.

If the orbits in the outer system happen to be circular, i.e. $e_k = 0$, the matching procedure may proceed as follows.

255

We assume, instead of (8.22)

$$p_k = p(1 + \delta_k), \qquad p = \sum \mu_k p_k, \qquad \delta_k = O(\epsilon^{1/3}).$$

Then c, found from (8.24), is the only essential parameter of the inner solution: its value will govern the behaviour of the close encounter (exchange, collision, overtaking).

Expanding as before the outer solution in inner variables we find to first order that

$$\left. \begin{aligned} x_{\text{out}} &= c - \frac{9}{8}\epsilon^{1/3}(\mu_1 - \mu_2)c^2 s^2 + O(\epsilon^{2/3}) \\ y_{\text{out}} &= s[-\tfrac{3}{2}c + \tfrac{3}{8}\epsilon^{1/3}(\mu_1 - \mu_2)c^2 + O(\epsilon^{2/3})]. \end{aligned} \right\} \tag{8.25}$$

The expansion of the inner solution yields

$$x_{\text{in}} = c + \frac{8}{9}c^{-2}s^{-1} + O(s^{\theta-2})$$

$$y_{\text{in}} = s\left(-\frac{3}{2}c - \frac{4}{3}c^{-1}s^{-1}\log\frac{c}{s_0} + O(s^{\theta-2}) \right) \tag{8.26}$$

for any $\theta > 0$.

The two expansions match if there exist transformations

$$s = \epsilon^{\alpha}\bar{s}, \qquad \alpha < 0, \qquad \bar{s} < 0$$

such that for fixed $\bar{s}$ the non-matching terms in (8.25) and (8.26) tend to 0 as $\epsilon \to 0$. This is so if $-\frac{1}{6} < \alpha < -\theta$.

The 'upper' boundary of the matching region, $s = O(\epsilon^{-1/6})$, denotes the onset of the two satellites' strong interaction: the crossing of the outer satellite over the tangent to the inner satellite's orbit. Spirig and Waldvogel give an expression for the time the outer satellite—in a non-interacting circular co-orbital pair—spends on the outer side of the tangent, namely

$$t = \frac{4}{3}\sqrt{\frac{2}{M\rho}}p^2 = \frac{4}{3}\sqrt{\frac{2p^3}{Mc}}\epsilon^{-1/6}$$

where $\rho = p_2 - p_1$.

In table 8.3, data and numerical results are given for Saturn's co-orbital satellites 1980S1 (Janus), 1980S3 (Epimetheus) and the F-ring shepherds 1980S26 and 1980S27. Spirig and Waldvogel's calculations are based on a coplanar circular model. $\Delta_{\min}$ is the minimum distance between two exchanging co-orbital satellites, T_{syn} is the synodic period of the pair, and γ is the angle within which the region of strong interaction is seen from M.

It is seen that the c-values for the satellite pairs lie well within the regions of E-orbits or of P-orbits so that the corresponding inner solutions are almost perfectly symmetric or straight, respectively. The outer solution after a close encounter is therefore again almost perfectly circular so that the long-term stability of these two pairs of satellites is assured.

Table 8.3

Parameter	Epimetheus 1980S3, $k=1$	Janus 1980S1, $k=2$	F-ring shepherds	
			1980S27, $k=1$	1980S26, $k=2$
$p_k = \|r_k\|$	151 422 km	151 472 km	139 353 km	141 700 km
$T_k = 2\pi\sqrt{p_k^3/m_0}$	16·682 804 h	16·691 067 h	14·728 55 h	15·102 20 h
m_k/m_0	$1\cdot5\times10^{-9}$	$6\cdot5\times10^{-9}$	$1\cdot0\times10^{-9}$	$0\cdot64\times10^{-9}$
$\mu_k = m_k/(m_1+m_2)$	0·1875	0·8125	0·61	0·39
$\epsilon = (m_1+m_2)/m_0$	$8\cdot0\times10^{-9}$		$1\cdot64\times10^{-9}$	
$\epsilon^{1/3}$	0·002 00		0·001 18	
$\rho = p_2 - p_1$	50 km		2347 km	
$p = \Sigma\mu_k p_k$	151 463 km		140 269 km	
$c = \rho/(p\epsilon^{1/3})$	0·165		14·2	
$\Delta_{min} = (8/3)\rho/c^3$	29 650 km			
$T_{syn} = T_1 T_2/(T_2 - T_1)$	1404 d		24·8 d	
$t^* = (4\sqrt{2}/3)p^2\sqrt{m_0\Delta}$	275·7 h = 16·52 revol. m_1		34·5 h = 2·34 revol. m_1	
$\gamma = 2\sqrt{2\rho/p}$	2·94°		20·96°	

8.5.4 Satellite–ring interactions

We now consider some problems raised by the presence of the rings in the Jovian, Saturnian and Uranian systems and their possible interactions with neighbouring satellites. The simple picture painted in subsection 8.5.1 of the classical rings of Saturn acted upon resonantly by Saturn's inner satellites, especially Mimas, was complicated by the Voyager finding of a multitude of fine ring detail, by the discovery of Uranus' nine discretely separated elliptic rings, by the discovery of Jupiter's ring, as well as the discoveries of the shepherd satellites 1980S26, 1980S27 and 1980S28 and the Jovian satellite 1979J1 (Adrastea).

In the case of the Uranian rings it would seem (Goldreich and Tremaine 1979) that they could be disrupted by particle collision, radiation drag (the Poynting–Robertson effect) and differential precession because of the oblateness of Uranus, leading to destruction of a ring in less than 10^8 years. These authors suggested, however, that stability of the rings could be provided by a series of small satellites orbiting within the ring system and that self-gravity within a ring also provided a ring-maintaining mechanism. A variation of the small satellite theory by Dermott et al (1979) proposed that a small satellite resided in each Uranian ring which kept the ring particles pursuing horseshoe orbits about the Lagrange equilibrium points L_4 and L_5 (see figure 5.2).

In 1983 Borderies et al considered the problem of how a nearby satellite in a coplanar orbit can affect the eccentricity and precession rate of a ring particle's orbit, a problem stimulated by the discovery of the shepherd satellites 1980S26 and 1980S27 which orbit Saturn, the former just outside Saturn's F ring, the latter just inside.

Imagine that the satellite's mass is distributed evenly along its orbit to form a wire of linear density $\rho_s = M_s/2\pi a$, where a is the satellite's orbital semimajor axis. Let a ring particle be radially distant from the wire by a small distance $\Delta r = r_r - r_s$ where r_r and r_s are the radius vectors of the ring particle and the satellite. The particle will experience a gravitational force F_r which is nearly equal to that produced by an infinitely long straight wire, so that

$$F_r = \frac{2G\rho_s}{\Delta r} = -\frac{GM_s}{\pi a \Delta r}.$$

The Gaussian form of Lagrange's planetary equations (see chapter 6, subsection 6.7.4) provides a means of calculating the rates of change of e_r and ϖ_r, the eccentricity and longitude of the apse of the particle orbit because of the radial force F_r. If we neglect the square of the eccentricity we can write

$$\frac{de_r}{dt} = \frac{F_r}{na}\sin f_r, \qquad \frac{d\varpi_r}{dt} = -\frac{F_r}{nae_r}\cos f_r, \qquad n^2 a^3 = GM$$

where M is the mass of the planet. There is also the perturbation in these quantities and in the satellite's orbit caused by the planet's oblateness, principally due to the J_2 term.

The differential precession is then given by

$$\frac{d}{dt}\Delta\varpi_{J_2} = \frac{d}{dt}(\varpi_r - \varpi_s)_{J_2} = -\frac{21}{4}nJ_2R^2\frac{\Delta a}{a^3} \qquad (8.27)$$

where R denotes the planet's radius.

Further development of these equations by Borderies *et al* has been used to study the F ring and the two shepherd satellites. They concluded that only the inner shepherd 1980S27 has an appreciable effect on the ring. The precessional period with respect to the satellite is $2\pi/\Delta\dot\varpi_{J_2}$, or 18 years, a figure derived from equation (8.27). The satellite will reduce the minimum distance Δr from 133 km to 50 km. At this time $\Delta\varpi = \varpi_r - \varpi_s = \pi$, $f_r = 0$ and $f_s = \pi$, a situation where the satellite is at aposaturnium, when that point is collinear with the ring's perisaturnium and Saturn (see figure 8.5).

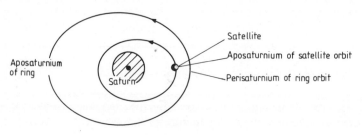

Figure 8.5

Now the satellite's longest axis is aligned towards Saturn and is of length 140 km. At such a time therefore the satellite actually ploughs into the ring, an event that must disturb it and may produce ripples in it. The last time this happened was around 1975.

The problem of how an eccentric ring can maintain its apse alignment against the disruptive force of the planet's oblateness has been investigated by a number of researchers. The problem arose because of the observed eccentricity of at least six of the rings. In addition the F ring and many of the other Saturnian rings are elliptical.

Consider a ring to be bounded by two aligned ellipses with elements $a = a_0 \pm \delta a/2$ and $e = e_0 \pm \delta e/2$.

In the 1979 treatment by Goldreich and Tremaine they treat an extended ring as a set of N elliptical wires or streamlines. We can designate each ring wire by its semimajor axis a, eccentricity e and mass m. Let the jth wire's parameters be a_j, e_j and m_j so that

$$a_j = a_{in} + (j - \tfrac{1}{2})\delta a/N$$

$$m_j = h_j M_r, \qquad \sum_{j=1}^{M} h_j = 1$$

where a_{in} is the inner radius of the ring assembly of width δa and mass M_r, and h_j, $j = 1, 2, \ldots, N$ are constants defining the density profile of the ring. The rate of precession of wire j due to the primary planet's oblateness is then $(d\varpi_j/dt)_{J_2}$, given by

$$\frac{d\varpi_j}{dt} = \frac{3}{2}J_2 R^2 (GM)^{1/2} a_j^{-7/2} \sim \text{const} - \frac{21}{4}J_2 \left(\frac{R}{a_0}\right)^2 \frac{n_0}{N} j \frac{\delta a}{a_0}.$$

The density ρ_k along wire k will be inversely proportional to the particle speed. Hence

$$\rho_k = \frac{m_k}{2\pi a_k}(1 - e_k \cos f_k) + O(e_k^2).$$

The force exerted by wire k on a particle in wire j can be calculated as in the satellite wire case, except that there will now be a tangential as well as a radial component of force. We insert these forces into the variational equation for $d\varpi_j/dt$ and average the equation over one revolution. Summing over all wires, we obtain the total precession rate of wire j due to all the other wires, namely

$$\frac{d\varpi_j}{dt} = -\frac{n_0 a_0 N M_r}{\pi e_0 \delta a M} \sum_{k \neq j} \left(\frac{h_k}{k-j}\right) \frac{\tan \psi_{jk}}{\cos 2\psi_{jk}}$$

provided that

$$\sin 2\psi_{jk} = \frac{N a_e}{\delta a}\left(\frac{e_k - e_j}{k-j}\right) \gg e_0.$$

Goldreich and Tremaine then make the following condition hold so that the ring wires precess together as one whole ring:

$$\frac{d\varpi_j}{dt} + \left(\frac{d\varpi_j}{dt}\right)_{J_2} = \text{constant}. \tag{8.28}$$

Condition (8.28) may be written as

$$j\frac{C}{N} + N \sum_{k \neq j} \left(\frac{h_k}{k-j}\right) \frac{\tan \psi_{jk}}{\cos 2\psi_{jk}} = D = \text{constant}, \qquad (j = 1, 2, \ldots, N)$$

and

$$C = \frac{21}{4} \pi e_0 J_2 \frac{M}{M_r} \left(\frac{R}{a_0}\right)^2 \left(\frac{\delta a}{a_0}\right)^2.$$

Now e_2 and e_n are the observed eccentricities of the ring borders so that we can write down N equations for the N unknowns $C, D, e_3, \ldots, e_{N-1}$. To apply these equations a density distribution h_k across the ring has to be assumed, for example based on measurements of optical depth.

Two important results follow.

(1) The total mass M_r of the ring can be deduced from the value of C.
(2) Theory requires that the ring eccentricity increases outwards, as is indeed found by observation, since $e_N - e_1 = \delta e > 0$.

Borderies *et al* (1983) extended this theory to inclined rings and predicted from it that the inclination would increase from the inner to the outer edge of the ring. Thus $\delta i / \delta a > 0$. They further found that if $\delta i / i_0$, $\delta e / e_0$, $a_0 \delta i / \delta a$ and $a_0 \delta e / \delta a$ are all much less than unity, then $\delta i / i_0 = \delta e / e_0$.

The cause of ring eccentricities was considered by Goldreich and Tremaine in a paper published in 1981. They showed that if gaps in the ring develop at dominant resonance with an external satellite of mass M_s, orbital radius a_s, and mean motion n_s, the ring eccentricity e_r will tend to increase as

$$\frac{d(e_r^2)}{dt} = 1 \cdot 52 \left(\frac{M_s}{M}\right)^2 n_s \left|\frac{a_s}{a_s - a_r}\right|^5 e_r^2.$$

This equation is valid if $e_r \ll |(a_s - a_r)/a_s|$. It gives an explanation for the existence of elliptical rings. If no ring gaps exist, the factor $1 \cdot 52$ changes to $-0 \cdot 148$ which causes a decrease in the eccentricity.

A formidable amount of work has evidently been inspired by our recently acquired knowledge of the Solar System's stock of rings and small satellites: there must be a possibility that the Voyager fly-past of Neptune will provide a few more surprises for celestial mechanicians.

260

8.6 Near-Commensurable Satellite Orbits

In Saturn's satellite system there are three pairs of satellites with closely commensurable mean motions, the closeness of the mean motion ratios to small integer vulgar fractions giving rise to stable resonant behaviour; the longitude of the conjunction line of each satellite pair thus librates about a specific direction.

(i) The mean motions of Titan (Saturn's most massive satellite) and Hyperion, nearly in the ratio 4:3, are such that the satellites' conjunction line librates about the moving aposaturnium of Hyperion with an amplitude of 36°.

If λ, λ' and ϖ are the longitudes of Hyperion, Titan and the perisaturnium of Hyperion, it is found that a quantity θ called the *critical argument* can be defined by

$$\theta = 4\lambda - 3\lambda' - \varpi \sim 180°$$

while

$$4n - 3n' - \dot{\varpi} \sim 0°$$

where n, n' and $\dot{\varpi}$ are the mean motions of Hyperion, Titan and the apse line of Hyperion. The critical argument θ, as stated above, has an amplitude of 36°. The value of $\dot{\varpi}$ is about $-20.3°$ per annum. In fact the Saturn–Titan–Hyperion system is quite close to being a periodic solution of Poincaré's second kind in the restricted problem of three bodies.

(ii) The mean motion ratio of Enceladus and Dione is close to 2:1 and the conjunction line oscillates about the perisaturnium of Enceladus with an amplitude of 1.5°. In this case,

$$\theta = 2\lambda - \lambda' - \varpi' \sim 180°$$

while

$$2n - n' - \dot{\varpi}' \sim 0°$$

the primed and unprimed quantities referring to Enceladus and Dione respectively. Again, the motion of the Saturn–Enceladus–Dione system resembles Poincaré's second kind of periodic solution.

(iii) The mean motion ratio of Mimas and Tethys is also close to 2:1, but in this case the satellites' conjunction line librates about the midpoint of their ascending nodes on Saturn's equatorial plane. The amplitude of the oscillation is 48.5°. The Saturn–Mimas–Tethys system exhibits a critical argument θ, where

$$\theta = 4\lambda - 2\lambda' - \Omega - \Omega' \sim 0°$$

where Ω is the longitude of the ascending node, and the primed and unprimed quantities refer to Mimas and Tethys respectively. This system resembles a Poincaré periodic solution of the third sort in the restricted three-body problem.

All three systems are stable; in fact all three systems are in configurations which ensure that their major perturbations are quickly reversed, because each pair passes frequently through near-mirror configurations (see section 5.6).

261

In recent years much attention has been paid to the question of the origin of such resonant systems. It has been shown by Goldreich (1965) that a remarkable transfer mechanism exists, whereby a pair of satellites under the influence of their planet's tidal forces will change their semimajor axes so that, even if their mean motion ratio was noncommensurable, it will not only have a chance to become commensurable but, having done so, will thereafter maintain that situation *even while both orbits continue to evolve.* Goldreich was the first to suggest this mechanism. He showed that, given certain assumptions, if two satellites P_1 and P_2 are in orbits of semimajor axes a_1 and a_2 $(a_1 < a_2)$, tidal forces will act upon P_1 to cause it to spiral outwards faster than P_2. Once it has reached a near-commensurable relationship with P_2 of the type found in nature, it will be able to feed angular momentum into the orbit of P_2 at just the correct rate needed to maintain the relationship.

This then is possibly the origin of the Mimas–Tethys and Dione–Enceladus resonances. Colombo and Franklin (1973) have argued that even if the Goldreich tidal mechanism is not the cause of the Titan–Hyperion resonance it could have arisen naturally. In other words, it is possible that Titan and Hyperion were formed at that resonance and, because the resonance is stable, have remained there.

The newly-discovered Saturnian satellite 1980S6 (Dione B) forms, with Dione, a new practical example of the Lagrange equilateral triangle solution of the three-body problem (section 5.8). Dione B orbits Saturn $60°$ ahead of Dione thus forming with Dione and Saturn an equilateral triangle. Since the value of the ratio of the mass of Dione to the sum of the masses of Saturn and Dione is very much below Routh's value of $0 \cdot 0385$ (sections 5.10.4 and 5.10.7), the orbit of Dione B would appear to be linearly stable to small perturbations of the kind it will suffer.

We conclude this section by considering the Galilean satellites of Jupiter: Io (J1), Europa (J2), Ganymede (J3) and Callisto (J4). In the usual notation (the mean motions are in degrees per day) we have

$$n_1 = 203 \cdot 488\,992\,435$$

$$n_2 = 101 \cdot 374\,761\,672$$

$$n_3 = 50 \cdot 317\,646\,290$$

$$n_4 = 21 \cdot 971\,109\,630.$$

Then

$$n_1 - 2n_2 = 0 \cdot 739\,469\,091$$

$$n_2 - 2n_3 = 0 \cdot 739\,469\,092$$

$$3n_3 - 7n_4 = -0 \cdot 044\,828\,540.$$

We also note that $n_1 - 3n_2 + 2n_3 = 0$ to the limit of observational accuracy. In the mean longitudes of the first three satellites we have

$$l_1 - 3l_2 + 2l_3 = 180°.$$

These relationships are obviously stable; the four satellites have been observed for more than 350 years (which corresponds to about 10^5 revolutions of these bodies) corresponding to a period of the order of 10^5 years for the inner planets of the Solar System. Their motions, however, are difficult to analyse. The pairs J1–J2, J2–J3 and J3–J4, according to Goldreich and Griffin, probably consist of two-body stable commensurabilities involving the eccentricities and the apses, since the four orbits are essentially coplanar. Laplace also showed that the relations involving the mean motions and longitudes of J1, J2 and J3 are stable.

The seventh resonant relationship in the Solar System (between Pluto and Neptune) also appears to be stable. It will be discussed later.

8.7 Large-Scale Numerical Integrations

Since their advent about thirty years ago, high-speed digital computers have been used by astronomers for the solution of a large variety of problems. Included among these problems is the large-scale integration of dynamical systems. The word 'large' might here be confusing, since its significance has changed considerably over the last 15 years. A problem which would then have consumed many hours of computer time can now be solved in a matter of minutes. Our use of the word 'large' will generally signify problems which tax the available resources of the computer system to a fairly high degree, typically requiring many hours of computer time. We will restrict our attention to problems typical of the Solar System rather than general dynamical systems, which include star cluster n-body problems where $n > 10$ (e.g. Lecar 1970), see also Section 15.10, and the dynamics of continuous media (e.g. Dormand and Woolfson 1971).

8.7.1 The outer planets for 120 000 years

One of the first attempts to integrate numerically a Solar System type of problem was made by Cohen and Hubbard (1965). This was prompted by a desire to extend backwards the 400 year ephemeris of Eckert *et al* (1951). The orbits of the five outer planets were numerically integrated backwards for 120 000 years from the present. Their computations were carried out on the Naval Ordnance Research Calculator, a 13 place binary-coded decimal computer. The numerical method used was Cowell's method, using the ninth differences and employing a fixed step size of 40 days. The total machine time required was of the order of 80 hours.

One of the principal purposes of their work was to monitor the distance between Pluto and Neptune. Since the perihelion distance of Pluto is less than that of Neptune, the suggestion had been made that Pluto might make a close approach to Neptune. In fact the authors discovered that the angle $\theta_N = 3\lambda_P - 2\lambda_N - \varpi_P$ (where λ is the mean longitude, ϖ is the longitude of perihelion and P and N refer to Pluto and Neptune respectively) librates about

180° with an amplitude of 76° and a period of 19670 years. As a consequence of this libration, the two planets can never approach one another in the vicinity of Pluto's perihelion, and in fact the closest approach found was 18 AU occurring at aphelion.

In a further and more accurate study, Cohen *et al* (1967) improved the elements for Pluto's orbit and performed a 300000 year integration with the new elements. The results showed that θ_N librated with an amplitude of 80° and a period of 19440 years.

8.7.2 Element plots for 1000000 years

Motivated by a desire to obtain a more complete picture of the motions of the outer planets resulting from their interactions, Cohen *et al* (1972) again calculated the orbits for a total of 1 000 000 years centred at the epoch 1941 Jan 6·0. This time a more powerful machine was used (the IBM 7030-Stretch) and the Cowell predictor increased to 12th order, again with a fixed step size of 40 days. The Stretch computer uses a 48 bit mantissa, giving a precision of about 14 decimal places. The total time taken for their integrations was somewhat less than 20 hours. Their results were presented in the form of element plots for each of the five planets and were accompanied by an extensive discussion of the various periodic modulations apparent in the plots.

In the Jupiter–Saturn system for example, the famous 900 year oscillation due to the 2:5 near-commensurability in the period of these planets is prominent in all of the plots. This fundamental frequency, when viewed over the full million-year span, appears to be modulated with a signal of a period of about 54000 years. The modulation appears in the semimajor axis and eccentricity plots for both planets. When the plots of the motion of the two perihelia are studied, it is seen that that of Jupiter completes one revolution in 300000 years and that the period of Saturn's perihelion is 46000 years. These mean motions of perihelia lead to a synodic period of the perihelia of 54000 years, which appears as a signal on Jupiter's perihelion plot as well as on the semimajor axis and eccentricity plots. A further interesting feature of the Jupiter–Saturn system becomes apparent on inspecting the plots of the inclination and the longitude of the nodes. It appears that the inclinations of the two planets oscillate with almost identical amplitudes, but 180° out of phase. Hence the two orbital planes move almost like a rigid body with the common 50000 year period of the nodes.

Due to the former results on the motion of Pluto, it is of interest to view its motion over the extended period. As we might expect, the plots of Pluto's elements show a strong signal with a period of about 19500 years, due to the Neptune–Pluto libration already discussed. However, in the inclination and eccentricity plots there is an apparently secular variation over the 1000000 years. One might expect this to be simply part of a periodic variation with a period much greater than 1000000 years, but no help in deciding this question can be obtained from the plots. As a final comment on the motion of Pluto, the authors extrapolated the longitude of perihelion and deduced a possible period of the order of 4 000 000 years. In actual fact Brouwer (1966) pointed out that the high inclination of Pluto's orbit should give rise to another angle similar to θ_N (viz. $\theta'_N = 3\lambda_P - 2\lambda_N - \Omega_P$ where Ω is the longitude of the node), and that this angle should librate as did θ_N. He therefore proposed that the argument of perihelion $\omega = \theta'_N - \theta_N$ might librate rather than circulate. However, a plot by Cohen *et al* over 1 000 000 years could not resolve this question.

8.7.3 Does Pluto's perihelion librate or circulate?

Hori and Giacaglia (1967) carried out a study which concluded that the argument of perihelion for Pluto should circulate with a period of 30 million years. However, Williams and Benson (1971) believed that the results of Cohen *et al* hinted at libration, and so they embarked on a 4 500 000 year integration of Pluto's orbit. In contrast to the simultaneous integration of the rectangular coordinates of the five outer planets by Cohen *et al*, Williams and Benson numerically integrated the planetary equations for Pluto only. The orbits of the other four planets were considered to be completely known and unaffected by Pluto. Pluto's motion was integrated as though it were a point-mass. The secular variation of the elements of Neptune, Uranus, Saturn and Jupiter were mainly modelled according to the calculations of Brouwer and Van Woerkom (1950). Furthermore, Williams and Benson did not integrate the planetary equations as they stood since the integration step size would have been held down by the short-period terms. In order to eliminate these terms they employed the device of Gauss for isolating the secular terms. Here the disturbing function R_j is averaged over the mean anomalies of the disturbed and disturbing bodies M and M_j, while the other elements are held constant. R_j is then replaced by

$$\langle R_j \rangle = \frac{1}{4\pi^2} \int_0^{2\pi} \int_0^{2\pi} R_j \, \mathrm{d}M_j \, \mathrm{d}M.$$

Using this simplified model they employed a fourth-order Runge–Kutta algorithm with a step size of 500 years to integrate backwards to 2·1 million BC and forwards to 2·4 million AD, the time required being 1 minute on an IBM 360/91 computer.

Their results were presented as plots of ω, e and i for Pluto, with the 19 500 year Neptune–Pluto libration averaged out. They showed that ω (the argument of perihelion) librates about $90°$ with an amplitude of approximately $24°$ and with a period 3 955 000 years. The authors referred to a discussion of librating ω given by Hori and Giacaglia (1967), who stated that for a given value of the semimajor axis, libration of ω would be expected if $I = (1 - e^2) \cos^2 i$ is less than a critical value, while circulation of ω is to be expected if I is above this value. They argued that if Pluto were close to this critical value the amplitude of libration would be near to $90°$. Since it is only $24°$ however, Pluto must lie well within the libration region. Williams and Benson believed that the reason for their results being in conflict with those of Hori and Giacaglia was due to an erroneous value of Neptune's mass used by the latter. They also quoted the interesting conclusion that Neptune tries to make ω regress while the other three planets try to make ω progress. This results in a near-cancellation which is sometimes positive and sometimes negative. In their simple model Hori and Giacaglia ignored the effects of the planets other than Neptune. It is of interest to note that only one other natural body in the Solar System has an argument of pericentre which librates rather than circulates; this is the asteroid 1373 Cincinnati (Marsden 1970).

8.7.4 The outer planets for 10^8 years

So fast has been the progress in computer development and data-handling techniques in the past ten years that recently two numerical integrations of the outer planets for 10^8 years and 210 million years have been computed. One, by the LONGSTOP consortium (Milani 1988, Roy *et al* 1988), used an Encke-type procedure and a CRAY-1S computer to compute forward and back in time over a total of 10^8 years; the other, by Applegate *et al* (1986) used a specially designed computer called the Digital Orrery to complete a 210 million year integration. In addition, in 1983, Kinoshita and Nakai performed a 5 million year integration of the five outer planets on a Fujitsu FACOM 380R. This last computation took 4 h CPU time and was post-processed (partially) by Kinoshita and Nakai and also by Milani and Nobili. Among the resulting insights into the dynamical behaviour of the outer Solar System over timescales of millions of years there exists a secular resonance locking the perihelion of Uranus and the aphelion of Jupiter (the libration period being 1 100 000 years) which turns out to be the major mechanism controlling the stability of the outer Solar System over these timescales.

In carrying out numerical integrations of such magnitude in machine terms and in length of time two entirely different sets of problems have to be assessed. Set 1 arises from the consideration of precisely what dynamical model should be adopted for integration together with considerations such as

(1) possible relativistic effects
(2) tidal effects
(3) stellar and galactic central bulge perturbations
(4) decreasing mass of the Sun by radiation
(5) perturbations by the inner four planets
(6) satellite masses
(7) changing masses of the planets by accretion
(8) drag and radiation pressure
(9) possible unknown planets
(10) quadrupole moment of the Sun.

One or more of these effects, able to be neglected in the Cohen–Hubbard–Oesterwinter study, could possibly affect a study over 10^8 years.

The second set of problems are directly related to the numerical integration, namely

(1) boundary values: to match starting values to the best available ephemeris
(2) choice of numerical integration techniques
    (a) multi-step or single step: Encke-type or Cowell-type procedure
    (b) computing speed and efficiency
    (c) numerical errors, round-off, truncation
    (d) software for array processing
(3) data handling
    (a) data storage
    (b) data plotting and presentation
    (c) smoothing of data
(4) investigation of stability of solutions
    (a) effect of variations in starting values
    (b) monitoring for possible close encounters.

Because of the potential enormous output of a very long numerical integration of the orbits of the outer planets, it becomes a main purpose to derive a synthetic secular perturbation theory, namely to reduce the huge amount of the numerical output to a few tens of numbers—the frequencies, amplitudes and phases of the main periodicities—and so represent as accurately as possible the long-term dynamical structure of the real system, the spectrum of frequencies of the problem, the frequencies being the 'lines'. The word 'synthetic' is used because the theory is obtained from the numerical output, hence opposed to analytic. Once constructed the synthetic theory can be compared with analytic theories or with other synthetic theories obtained independently.

Lack of space prevents details of the method of constructing a synthetic theory to be given here and the reader may consult Milani (1988). Essentially the enormous computer output is filtered to delete all short-period terms up to 4900 years in length, a process that removes the quasi-resonances in mean motion between Jupiter and Saturn and between Uranus and Neptune but retains the Neptune–Pluto 3:2 resonance in mean motion of period 19 900 years. The process then gives the secular periods of the perihelia and the nodes, together with their harmonics and combinations. The final data also include the semimajor axes, eccentricities and inclinations which have their own dynamical behaviour.

The examination of the spectrum of frequencies or 'lines' obtained from the 10^8 year integration reveals indications, however, that there may be a limit to this kind of investigation. The spectrum reveals a bewildering multitude of lines in the form of multiplets of lines of comparable amplitude which very often cannot be identified with theoretically allowed combinations of fundamental frequencies. Milani has deduced that there is therefore a possibility that the solution to the planetary equations of motion might not be quasi-periodic and that it may therefore not be possible to predict the motions of the outer planets for an arbitrarily long span of time, no matter how good the computer and the numerical algorithm used to propagate the orbits are. This view is in accordance with the concept of the *predictability horizon*, recently introduced by Sir James Lighthill (1986). Whenever new and analytical or computational tools become available the horizon may be pushed forward but if the dynamical system is essentially unstable, the predictability horizon is reduced and if chaos is involved the predictability may go to zero.

But what about Pluto, the one planet among the five that might be expected to misbehave itself?

Williams and Benson's 1971 conclusion that Pluto's argument of perihelion ω librates with a period of 3·955 million years (3·955 Myr) has been confirmed, the value obtained from the long interpretation of Applegate *et al* being 3·798 Myr. This libration is modulated in amplitude with a period near 34 Myr. Williams and Benson's additional conclusion that the variation in ω is locked to the variations of e and i is also confirmed, all three elements being found to be modulated with a 35 Myr period. There are signs of a still longer period which Applegate *et al* suggest might have dangers for Pluto's stability over a 10^9 year time span. Finally, although the argument of perihelion ω does librate, it should be noted that the longitude of perihelion $\varpi (= \Omega + \omega)$ does circulate with a period of 3·69 Myr, the period of circulation of the longitude of the ascending node Ω.

8.7.5 The analytical approach against the numerical approach

To conclude the discussion on the study of the evolution of the outer planets, we compare the different approaches in its solution. The first approach is that of an analytical theory. Typical of this is the secular theory of Brouwer and Van Woerkom (1950). The term 'secular' indicates that a solution is sought which is valid for a long period of time. Short-period terms are of no interest and are immediately eliminated from the disturbing function. This is then expanded to some order in the disturbing masses, the eccentricities and the inclinations. It is of course important to include critical terms due to any near-resonances which occur. In fact Cohen *et al* (1973) plotted the elements derived from this theory in a similar way to their numerical integration results. They truncated the disturbing function after the first order in the disturbing masses and after the second order in the eccentricities and inclinations. Excellent agreement was obtained in the case of Jupiter and Saturn, since the terms associated with the near-resonance of about 900 years were included. However, in the other cases significant differences were apparent, although the general pattern was reproduced. Cohen *et al* suggested that this was due to their neglecting the great inequality terms for Uranus-Neptune.

It is of course possible now to program a computer to do literal algebra, employing the machine's large capacity, high speed and 'untiring dedication' to carry out analytical expansions to powers much higher than any achieved in former days by human beings. At worst the machine can re-capture and check in a fraction of the time the planetary and lunar theories achieved in former times by celestial mechanicians: hopefully, the computer will provide theories valid for far longer time intervals than hitherto reached. Nevertheless, the possibility revealed by the synthetic theory constructed from the long numerical integrations that there could be a predictability horizon must be kept in mind.

The second possible approach is that of special perturbations rather than general perturbations, again exemplified in the work of Cohen *et al* and their successors. In this case accurate values of the positions and velocities will be obtained for as large a time of integration as the truncation and rounding-off errors of the method permit. Another important factor here is the machine time required, although at the present time the speed of machines is sufficient to allow any integration we might require. To combat the growth of rounding-off and truncation errors, we require a method of formulating the problem so that large integration steps may be taken with no corresponding increase in the accumulated errors. If, however, we are already effectively pushing our predictability horizon to nearly its effective limit, then the discovery of new concepts may be our only hope. In that context the work of Williams and Benson (1971), already discussed, is relevant. They clearly chose the simplifications involved in their method with skill and insight and were rewarded by very good agreement with the results of the long-term integrations.

8.8 Ovenden's Principle

Hills (1970) gave numerical integrations of eleven hypothetical coplanar solar systems (a large primary mass and several small secondary masses) for different initial conditions and mass values. He found that after vigorous interactions all the systems tended towards quasistationary states, each state showing a tendency for the periods of adjacent orbits to be near-commensurable in small integer fractions. In further work along the same lines, Ovenden (1973) confirmed this tendency, concluding that the ratio of any pair of planets' mean motions was (in the course of the numerical integration) most often found close to a small whole number rational fraction. It should be noted that neither Hills nor Ovenden used tidal action in their simulations, only Newtonian point-mass gravitational forces being modelled.

Ovenden concluded that a general characteristic stood out from such studies. A system spent only a short time with the planets close together and interacting violently; most of its time the planets were distributed far apart and interacting mildly. He went on to establish his *principle of least interaction action* which states that 'a satellite system of n point-masses will spend most of its time close to a configuration for which the time-mean of the action associated with the mutual interactions is an overall minimum'. This condition is shown to be equivalent to finding the overall minimum of the time-mean

$$\bar{R} = \sum_i \sum_j \frac{m_i m_j}{r_{ij}} \quad (j \neq i)$$

where m_i and m_j are the masses of the ith and jth satellites and r_{ij} is their instantaneous distance apart. This principle was essentially anticipated by Bass (1958) some years before as a purely abstract theorem, unapplied at the time to the Solar System. Ovenden showed that if the principle is applied to two-satellite cases it leads to the resonant situations of Saturn's satellite pairs (described in section 8.6). This stems from the theorem of Poincaré (1895) which states that a resonant gravitational system has the property that the disturbing function, averaged with respect to the critical or resonant parameter, has a local minimum (Hori 1960, Giacaglia and Nacozy 1969).

Ovenden then applied his principle to systems containing more than two satellite bodies, obtaining several interesting results:

(i) The distribution of the five satellites of Uranus is within 5% of the calculated minimum interaction action distribution.
(ii) The triplets U5, U1, U2 and J1, J2, J3 show the Laplace relationships between their mean motions:

$$\text{Jupiter: } n_1 - 3n_2 + 2n_3 = 2 \times 10^{-7} \simeq 0$$

$$\text{Uranus: } n_5 - 3n_1 + 2n_2 = 3 \times 10^{-4} \simeq 0.$$

(iii) The time taken to approach within the observed precision of resonance is 2×10^9 years and 6×10^9 years for the Jovian and Uranian systems respectively. These correlate well with the best determined age of the solar system of $4 \cdot 5 \times 10^9$ years.
(iv) The system Jupiter–Saturn–Uranus–Neptune is not in a state of minimum interaction action.

Ovenden then postulated that a planet of mass M_A used to exist at the distance of the present asteroid belt. He calculated the family of minimum interaction action configurations as a function of the one parameter M_A. He also calculated the family of configurations of M_A–Jupiter–Saturn–Uranus–Neptune as a function of the one parameter, time. He found that, for $M_A = 90$ Earth masses and a time of $1 \cdot 6 \times 10^7$ years, the system M_A–Jupiter–Saturn–Uranus–Neptune was at a minimum interaction action configuration, and that if the mass M_A had by some process been dissipated at that time, the remaining system would now be moving to its own minimum but would not yet have reached it. Note that by varying only two parameters, four separate quantities were matched to about $0 \cdot 3 \%$. To support his hypothesis further, he quoted the evidence obtained from achrondite meteorites that their ages are of the order of 10^7 years, suggesting that their origin was from within M_A.

270

These extremely controversial results, in spite of their mutual consistency, were derived by a relatively crude form of integration procedure. Ovenden assumed that the system evolved through a series of quasicircular orbits, driven by a secular term (proportional to em^3 where m is the disturbing mass and e the orbital eccentricity) in the series expansion for the semimajor axis. The existence of such a secular term is still the subject of great debate, and Ovenden's calculations leave much to be desired as far as reliability is concerned. However, it is of interest that recent dramatic advances in our ability to integrate accurately planetary systems for time intervals of the order of 2×10^8 years holds out hope that the day is not distant when we will be able to confirm or deny Ovenden's hypothesis.

8.9 Empirical Stability Criteria

The Solar System is obviously a many-body hierarchical dynamical system. The planetary orbits may be ordered in their sizes; likewise the satellite systems have orbits that may be said to be ordered in size if we are prepared to group the outermost ten satellites of Jupiter in clusters about three 'spectral lines' distinct in size.

The Jacobian coordinate system (section 5.12), applied to an n-body hierarchical dynamical system $(n > 3)$, has been the starting point of a number of studies of Solar System dynamics, starting with the expression of the planetary equations of motion in Jacobian coordinates, equation (5.98), namely

$$\frac{m_i M_{i-1}}{M_i} \ddot{\rho}_i = \nabla_i U \tag{8.29}$$

where

$$U = \tfrac{1}{2} G \sum_{k=1}^{n} \sum_{l=1}^{n} \frac{m_k m_l}{r_{kl}} \qquad l \neq k$$

is the force function. In these equations m_i denotes the ith mass, $i = 0, 1, 2, \ldots, n$; $(m_0 = 0)$,

$$M_i = \sum_{j=0}^{i} m_j, \qquad \rho_i = \mathbf{R}_i - \bar{\mathbf{R}}_{i-1}$$

$\mathbf{R}_i$ and $\bar{\mathbf{R}}_i$ being the position vectors of m_i and the mass-centre of $(m_1, m_2,\ldots, m_i)$ respectively in an inertial system

$$\nabla_i \equiv \frac{\partial}{\partial \mathbf{r}_i} = \mathbf{i}\frac{\partial}{\partial x_i} + \mathbf{j}\frac{\partial}{\partial y_i} + \mathbf{k}\frac{\partial}{\partial z_i}$$

$$\mathbf{r}_{kl} = \mathbf{R}_l - \mathbf{R}_k, \qquad r_{kl} = |\mathbf{r}_{kl}|.$$

$\mathbf{i}, \mathbf{j}$ and $\mathbf{k}$ are unit vectors.

In the planetary system, m_1 is the Sun's mass, m_2 is Mercury's and so on. Essentially, each body's radius vector is taken from the centre of mass of the

271

bodies lower down in the hierarchy. Thus Jupiter's radius vector is drawn from the centre of mass of the Sun, Mercury, Venus, Earth, Mars and the asteroids.

By equation (5.99)

$$\mathbf{r}_{kl} = \rho_l - \rho_k + \sum_{j=k}^{l-1} \frac{m_j}{M_j} \rho_j.$$

Applying this relationship to the expansion of U in equation (8.1), the following expression, correct to the second order, may be obtained,

$$\ddot{\rho}_i = GM_i \, \nabla_i \left[\frac{1}{\rho_i} \left(1 + \sum_{k=1}^{i-1} \epsilon^{ki} P_2(C_{ki}) + \sum_{l=i+1}^{n} \epsilon_{li} P_2(C_{il}) \right) \right] \qquad (8.30)$$

where

$$\epsilon^{ki} = \frac{m_k M_{k-1}}{M_k M_{i-1}} \alpha_{ki}^2$$

$$\epsilon_{li} = \frac{m_l}{M_i} \alpha_{il}^3 \qquad i = 1, 2, \ldots, n. \qquad (8.31)$$

In these expressions

$$\alpha_{ij} = \rho_i / \rho_j < 1 \qquad C_{ij} = \frac{\rho_i \cdot \rho_j}{\rho_i \rho_j} \qquad \rho_i = |\rho_i|$$

while P_2 is the Legendre polynomial of order 2 in the parameter C.

On examination it is seen that the first term of the right-hand side of equation (8.30) represents the undisturbed elliptic motion of the ith mass about the mass-centre of the subsystem of masses $m_1, \ldots, m_{i-1}$, while the ϵ^{ki}, ϵ_{li} provide a measure of the disturbance of the elliptic motion by the remaining masses, i.e. masses other than the ith. It may be noted that a superscripted ϵ denotes the disturbance of a body by an inferior body (smaller orbit) while a subscript denotes the disturbance of a body by a superior body.

If $n = 3$, equations (8.30) and (8.31) reduce to

$$\ddot{\rho}_2 = GM_2 \nabla_2 \left[(1 + \epsilon_{32} P_2(C_{23})) / \rho_2 \right] \qquad (8.32)$$

$$\ddot{\rho}_3 = GM_3 \nabla_3 \left[(1 + \epsilon^{23} P_2(C_{23})) / \rho_3 \right] \qquad (8.33)$$

with

$$\epsilon^{23} = \frac{m_2 M_1}{M_2^2} \alpha_{23}^2 = \frac{m_1 m_2}{(m_1 + m_2)^2} \alpha_{23}^2$$

$$\epsilon_{32} = \frac{m_3}{M_2} \alpha_{23}^3 = \frac{m_3}{m_1 + m_2} \alpha_{23}^3. \qquad (8.34)$$

Thus ϵ_{32} is a measure of the ratio of the disturbance by P_3 on the orbit of P_2 about P_1, to the central two-body force between P_2 and P_1. Likewise ϵ^{23} is a measure of the ratio of the disturbance by P_1 and P_2 on the orbit of P_3 about

272

the centre of mass of P_1 and P_2, to the central two-body force between P_3 on the one hand and P_1 and P_2, assumed to lie at their mass-centre.

If we introduce μ and μ_3 by the relations

$$\mu = m_2/(m_1 + m_2) \qquad \mu_3 = m_3/(m_1 + m_2) \qquad (8.35)$$

then

$$\epsilon^{23} = \mu(1-\mu)\alpha_{23}^2 \qquad \epsilon_{32} = \mu_3 \alpha_{23}^3. \qquad (8.36)$$

We now examine this picture in the light of the hierarchical three-body stability criterion (section 5.13) based on the quantity $S = |\mathbf{C}|^2 E$, where $\mathbf{C}$ and E are the constants appearing respectively in the angular momentum and energy integrals of the general three-body problem. If the three-body system was a hierarchical one (a binary plus a third body in a large orbit about the binary's mass-centre), and $S \leqslant S_{cr}$, the binary could never be broken up. The critical stability value S_{cr} was derived from the collinear solution of the general three-body problem. To obtain S_{cr}, the ratio X must be found, where X was the solution of Lagrange's quintic equation (equation (5.37)). In its turn $\alpha_{cr} = (\rho_2/\rho_3)_{cr}$ is related to X through S_{cr}.

The initial value of the quantity $\alpha = \rho_2/\rho_3$, however, is independent of μ and μ_3 as is S, both being fixed in value by the initial setting-up of the hierarchical three-body problem. It we assume that the three-body system is initially set off in circular, coplanar orbits (P_2 about P_1; P_2 about the mass-centre of P_1 and P_2), then to the stability criterion, namely $S \leqslant S_{cr}$, there corresponds the stability criterion $\alpha \leqslant \alpha_{cr}$, for a given $\alpha = \rho_2/\rho_3 = \alpha_2/\alpha_3$ (the radii of the initially circular orbits) and a given μ, μ_3. Note that the value of α_{cr} is dictated solely by μ, μ_3 and the solution of Lagrange's quintic equation (equation (5.37)) in μ, μ_3 and X.

Thus for all pairs of possible values of μ and μ_3, plotted on the μ–μ_3 plane, a surface of values of α_{cr} exists above it in the third dimension α. Therefore for a hierarchical three-body problem with initially circular, coplanar orbits, α is known, as is μ and μ_3 The point μ, μ_3, α can therefore be plotted. If it lies below or on the point μ, μ_3, α_{cr}, the system is stable in the sense that the binary P_1–P_2 cannot be broken up.

From relations (8.36), it is obvious that a system may be expressed not only as a set of values μ, μ_3, α but also as a set of values ϵ^{23}, ϵ_{32}, α. Calculating α_{cr} from μ, μ_3 and the Lagrange quintic equation gives, by substitution in (8.8), values $(\epsilon^{23})_{cr}$, $(\epsilon_{32})_{cr}$. It is thus possible to use the criterion in relation to the ϵ parameters as well as the μ parameters.

The Solar System and satellite systems can now be broken into hierarchical three-body subsets. Examples might be Sun–Jupiter–Saturn, Earth–Moon–Sun, Jupiter–Io–Europa, Sun–Earth–Uranus, and so on, the first two in each set forming the binary, the third being looked upon as being in orbit about the mass-centre of the first two. If this is done and the relevant ε parameters are computed so that the criterion of stability may be applied, it is found that, with certain exceptions, the criterion is well satisfied with the real alphas all much smaller than the α_{cr} values for these systems.

Several comments are necessary.

The exceptions include the retrograde satellites of Jupiter, which is satisfactory since they are possibly captured asteroids and could well escape again.

Eccentricities and inclinations were neglected in the above study. The Solar System is so flat, however, that inclusion of actual inclinations would probably leave the results essentially unchanged. A more recent study by Valsecchi *et al* (1984) has included eccentricities of satellite and planetary orbits. For pairs of planets or pairs of major satellites disturbing each other, the previous results are unaltered. For the case of triple systems of the type planet–satellite–Sun, however, the surprising result emerges that for all Solar System satellites except Triton, the $S = |C|^2 E$ criterion of stability is not satisfied.

This does not mean that satellites are unstable against solar perturbations. The indications are that the major satellite orbits, though disturbed by solar perturbations, are hierarchically stable: if so it merely indicates that the $|C|^2 E$ criterion is far too strict and that while it is desirable for a satellite to have that guarantee of stability, orbital existence about a planet may continue for an astronomically long time without it.

The discussion in section 5.10.3 of the surfaces of zero velocity in the circular restricted three-body problem is illuminating in this respect. For a certain value of the Jacobi constant C (say C_2) the inner ovals that bound the particle to the vicinity of one *or* other of the massive objects met at the double point L_2. A slight change in value of C from C_2 to C_3 caused the ovals to coalesce into a dumbbell-shaped figure allowing the particle to wander from the vicinity of one mass through the narrow neck to the vicinity of the other. The guarantee of Hill stability was now broken. Nevertheless, the time it would take the particle to 'find' the neck and follow a trajectory through it could range from a tiny to an astronomically long duration depending strongly upon its initial conditions of position and velocity. A new version of stability can then be introduced which may be called 'statistical' or 'empirical' which has nothing absolute about it, providing as it does estimates of the time it will take for half the members of a family of particles with similar starting conditions to escape through the neck.

In the general three-body problem Walker and Roy (1982) have demonstrated by numerical integration of a large number of three-body hierarchical dynamical systems that the $|C|^2 E$ criterion is unnecessarily restrictive, a zone of empirical stability existing.

In fact, within the Solar System, no triple subset is totally isolated gravitationally from other members of the Solar System. The Sun, Jupiter and Saturn have often been spoken of as essentially making up the Solar System with a little bit of debris left over, such as Earth, Venus and so on, but even the triple subset of Sun–Jupiter–Saturn is to some measure disturbed. The important question from the point of view of stability is therefore: what effect in the long term will these additional perturbations produce?

Although the subset may satisfy the criterion at present, with its alpha value lying a good way below α_{cr}, the system is being disturbed. The alpha height at which its point in the $\epsilon_{32} - \epsilon_{23} - \alpha$ space lies will move in a pseudo-random or pseudo-periodic fashion because of the smaller disturbances by the other bodies. As long as the point lies below α_{cr}, the subset is stable in that the orbits of Jupiter and Saturn will not intersect. However, if the point wanders in a sort of random walk so that it ultimately reaches a situation where $\alpha > \alpha_{cr}$, then the subset *may* become unstable. The same argument applies to other triple subsets. Equations (8.2) show that the epsilons may well be the crucial parameters in a consideration of the stability of the Solar System. They are a measure of the disturbances that each body produces on the others' orbits.

In an attempt to treat the disturbing effect of a fourth body on a triple subset, Milani and Nobili (1983) sought for a general perturbation theory relating the hierarchical stability lifetime of a four-body hierarchical dynamical system to the rate of change of the absolute stability criteria $|C|^2 E$ of each of its three-body subsets as they are disturbed by the fourth body. Since the critical value of the function, $|C|^2 E_{crit}$, is a function only of the three masses, it is constant. While $|C|^2 E \leqslant |C|^2 E_{crit}$, for a given three-body subset, the subset remains hierarchically stable. Milani and Nobili's approach, which in its analytical development used the Roy–Walker empirical stability parameters, provided a means of calculating the minimum time perturbations would take to increase $|C|^2 E$ to $|C|^2 E_{crit}$ for a given three-body subset. In applying their method to the four-body system Sun–Mercury–Venus–Jupiter, they concluded that the hierarchy of the subset Sun–Mercury–Venus was stable for at least $1 \cdot 1 \times 10^8$ years while the subset Sun–Venus–Jupiter was stable for at least 3×10^9 years.

The empirical stability studies by Roy and his co-workers (Roy 1979, Walker *et al* 1980, Walker and Roy 1981, 1982, 1983, Walker 1982, Roy *et al* 1985) have been extended from three-body to four-body systems. In the case $n = 4$, studies were made to establish how different initial sets of starting conditions (the ϵ, α and μ values) govern the time it takes the hierarchy of the system to be violated. From such experiments it is becoming clear that it should ultimately be possible from an examination of the 'starting conditions' in an n-body system to provide a statistical estimate of its stability—the dynamical equivalent of the lifetime of a planetary atmosphere.

The kinetic theory of gases enables a half-life T (the time it will take half the molecules in the atmosphere to escape into space) to be calculated from x, the ratio of the mean molecular velocity to the velocity of escape from the planet.

For $x = 1$, the value of T is very small indeed. As x decreases, T grows slowly at first and is measured in minutes, hours, weeks. But quite soon a region of x is reached where T shoots up to durations of astronomical length.

It is possible that the stability of the Solar System may have to be treated like this. If we begin with a large number of hierarchical dynamical systems (solar systems) where they all have epsilon and alpha values within certain ranges, we may be able to state that the statistical *status quo* lifetime of these systems is of such and such a duration in the sense that such a lifetime will have to elapse before half the systems will have suffered any change in the *status quo* of their ordered orbits.

If this is so, then with the exception of the 'hard' commensurabilities in the Solar System (see sections 8.6 and 8.7) there would appear to be nothing remarkably esoteric about the distribution of Solar System orbits or the values of the elements that describe these orbits. In their distribution, near-circularity and near-coplanarity, they merely reflect the sizes of the epsilons and alphas that have reduced the orbits' pseudo-random walks to such small strolls, enabling the Solar System's *status quo* to be maintained over a long time, perhaps an astronomically long time.

8.10 Conclusions

Can we now go some way towards answering the questions asked in section 8.1? We have seen that geophysical, selenophysical and solar astrophysical evidence agree that the Earth, Moon and Sun are roughly $4 \cdot 5 \times 10^9$ years old, while the fossil record of complex life forms on our planet suggest that the Earth's orbit has not been drastically altered in at least 2×10^9 years. But what does celestial mechanics have to say? It is not nearly so confident as it once was in making dogmatic statements about the stability and good behaviour of the planetary orbits.

In 1773 Laplace published a theorem, later improved by Poisson to the second order in the disturbing masses, supposedly showing that the Solar System was stable in the sense that each planet was permanently restricted to the inside of its own spherical annulus, no two planetary annuli ever intersecting. In other words, the changes in the semimajor axes were purely periodic. In 1784, by using Lagrange's planetary equations, Laplace further stated that the inclinations and eccentricities of the planetary orbits must always remain small. He achieved these results by neglecting everything but the first and second orders in these small quantities. The American astronomer Simon Newcomb (1876) showed that if all but one of a number of point-masses are small with respect to a large mass and they are in orbit round it in orbits of small eccentricity and inclination, there exists a multiply-periodic, trigonometric infinite series solution of that *n*-body problem. The crucial question of convergency or divergency of Newcomb's series, however, remained. If convergent, the actual planetary motions would be quasiperiodic. If divergent, nothing could be said about the long-term behaviour of the planetary orbits.

Poincaré in 1899 proved rigorously that in general Newcomb's series are divergent. This effectively dismissed the Laplace–Poisson–Lagrange theorems. Nevertheless, this seeming defeat was the beginning of the theory of asymptotic expressions, which has been applied so fruitfully in fluid dynamics.

In recent years, mathematical work done by Siegel and Moser (1971) has shown that some of the classical series expansions in celestial mechanics are convergent and give rise to a rigorous description of solutions of the *n*-body problem valid for all time. This work has clarified the status of Newcomb's series where most planetary-type motions are concerned. As Bass (1975) concisely puts it; 'For all very nonresonant initial states, Newcomb's series converge (nonuniformly), and so these motions are quasiperiodic; but they are not orbitally stable, and so arbitrarily small perturbations in the initial conditions can (so far as we know) yield wild motions. For resonant or nearly resonant motions the series can converge uniformly (orbitally stable quasi-periodic motion), or converge nonuniformly (orbitally unstable quasiperiodic motion) or diverge (wild motion).'

We have seen that the work of J G Hills and M W Ovenden demonstrated that, after a short period of wild behaviour, a planetary system could settle down into a distribution of orbits very similar to a commensurable Bode-type configuration. Such a configuration would, under the action of other forces such as tidal friction, nudge the system into a neighbouring truly stable configuration, which on inspection might be thought to have been the system's state for a very long time. Indeed numerical integrations backward in time could take the system, still well behaved, through the episode of wild behaviour as if it had never been.

As far as the major planetary bodies in the Solar System are concerned, the long-term numerical integrations have shown that over 2×10^8 years there is definite stability in the hierarchical sense. Roy and Walker's work leading to statistical or 'half-life' concepts of stability, again in the hierarchical sense, also suggest that the planetary system and the major satellite systems have been stable for a time which is a considerable fraction of the putative age of the Solar System. But even today, as far as celestial mechanics is concerned, it would be a bold person who made that fraction approach unity.

In Scotland we have a third verdict in the courts in addition to 'guilty' and 'not guilty'. If someone is accused of a crime and the evidence is less than sufficient to convict him, but we suspect he has done it, we can bring in a verdict of 'not proven'. It seems that at the time of writing the verdict on the Solar System, accused of long-term stability, is from the testimony of dynamical astronomy 'not proven'.

References

Aksnes K 1985 *Stability of the Solar System and its Minor Natural and Artificial Bodies* ed V Szebehely (Dordrecht: Reidel)

Applegate J H, Douglas M R, Gursel Y, Sussman G J and Wisdom J 1986 *Astron. J.* **92** 176

Bass R W 1958 *Solution of the N-body Problem* part 3 (Martin Marietta Corporation)

—— 1975 *Can Worlds Collide?* (June 13, *Pensée*)

Borderies N, Goldreich P and Tremaine S 1983a *Icarus* **53** 84

—— 1983b *Astron. J.* **88** 226

—— 1983c *Astron. J.* **88** 1560

Brouwer D 1963 *Astron. J.* **68** 152

—— 1966 *The Theory of Orbits in the Solar System and in Stellar Systems* ed G Contopoulos (New York: Academic) p 227

Brouwer D and Van Woerkom A J J 1950 *Astron. Pap. Am. Ephemeris* **13**

Carusi A, Roy A E and Valsecchi G B 1986 *Astron. Astrophys.* **162** 312

Cohen C J and Hubbard E C 1965 *Astron. J.* **70** 10

Cohen C J, Hubbard E C and Oesterwinter C 1967 *Astron. J.* **72** 973

—— 1972 *Astron. Pap. Am. Ephemeris* **13**

—— 1973 *Cel. Mech.* **7 438**

Colombo G 1982 *Applications of Modern Dynamics to Celestial Mechanics and Astrodynamics* ed V Szebehely (Dordrecht: Reidel)

Colombo G and Franklin F A 1973 *Recent Advances in Dynamical Astronomy* ed B Tapley and V Szebehely (Dordrecht: Reidel)

Dermott S F, Gold T and Sinclair A T 1979 *Astron. J.* **84** 1225

Dermott S F and Murray D 1981a *Icarus* **48** 1

—— 1981b *Icarus* **48** 12

Dormand J R and Woolfson M M 1971 *Mon. Not. R. Astron. Soc.* **151** 307

Eckert W J, Brouwer D and Clemence G M 1951 *Astron. Pap. Am. Ephemeris* **12**

Elliot J L, French R G, Frogel J A, Elias J H, Mink D J and Liller W 1981 *Astron. J.* **86** 464

Elliot J L and Kerr R 1984 *Rings* (Cambridge, MA: MIT Press)

Froeschlé Ch and Scholl H 1988 *Long-Term Dynamical Behaviour of Natural and Artificial N-Body Systems* ed A E Roy (Dordrecht: Reidel)

Giacaglia G E O and Nacozy P E 1969 *Periodic Orbits, Stability and Resonances* ed G E O Giacaglia (Dordrecht: Reidel) p 96

Goldreich P 1965 *Mon. Not. R. Astron. Soc.* **130** 159

Goldreich P and Tremaine S 1979 *Astron. J.* **84** 1638

Harrington R S and Seidelmann P K 1981 *Icarus* **47** 97

Hills J G 1970 *Nature* **225** 840

Hori G 1960 *Astrophys. J.* **74** 1254

Hori G and Giacaglia G E O 1967 *Research in Celestial Mechanics and Differential Equations* (University of Sao Paulo)

Hunter R B 1967 *Mon. Not. R. Astron. Soc.* **136** 245, 267

Keeler J E 1895 *Astrophys. J.* **1** 416

Lecar M (ed) 1970 *IAU Colloquium No. 10* (Dordrecht: Reidel)

Lecar M and Franklin F A 1974 *IAU Symposium No. 62* ed T Kozai (Dordrecht: Reidel)

Lighthill J 1986 *Proc. R. Soc.* **A407** 35

Marsden B G 1970 *Astron. J.* **75** 206

Maxwell J C 1859 *On the Stability of the Motions of Saturn's Rings* (London: Macmillan)

Message P J 1966 *IAU Symposium No. 25* (New York: Academic) p 197

Milani A 1988 *Long-Term Dynamical Behaviour of Natural and Artificial N-Body Systems* ed A E Roy (Dordrecht: Reidel)

Milani A and Nobili A 1983 *Cel. Mech.* **31** 241

Newcomb S 1876 *Smithsonian Contribution to Knowledge* **21**

Oesterwinter C and Cohen C J 1972 *Naval Weapons Laboratory Technical Report* TR-2693 (Virginia, USA)

Ovenden M W 1973a *Recent Advances in Dynamical Astronomy* ed B Tapley and V Szebehely (Dordrecht: Reidel)

—— 1973b *Vistas in Astronomy* **16** (Oxford: Pergamon)

Ovenden M W, Feagin T and Graf O 1974 *Cel. Mech.* **8** 465

Poincaré H 1895 *Les Méthodes Nouvelles de la Mécanique Céleste* (Paris: Gauthier-Villars) (1967 *NASA Technical Translation* TTF-450–2 (Washington))

Roy A E 1979 *Instabilities in Dynamical Systems* ed V Szebehely (Dordrecht: Reidel)

—— 1982 *Applications of Modern Dynamics to Celestial Mechanics* ed V Szebehely (Dordrecht: Reidel)

Roy A E, Carusi A, Valsecchi G B and Walker I W 1984 *Astron. Astrophys.* **141** 25

Roy A E and Ovenden M W 1954 *Mon. Not. R. Astron. Soc.* **114** 232

Roy A E, Walker I W and McDonald A J C 1985 *Stability of the Solar System and its Minor Natural and Artificial Bodies* ed V Szebehely (Dordrecht: Reidel)

278

Roy A E, Walker I W, McDonald A J, Williams I P, Fox K, Murray C D, Milani A, Nobili A, Message P J, Sinclair A T and Carpino M 1988 *Vistas in Astronomy* in press

Schubart J 1966 *IAU Symposium No. 25* ed G Contopoulos (New York: Academic) p 187

Siegel C L and Moser J K 1971 *Lectures on Celestial Mechanics* (Berlin: Springer-Verlag)

Spirig F and Waldvogel J 1985 *Stability of the Solar System and its Minor Natural and Artificial Bodies* ed V Szebehely (Dordrecht: Reidel)

Synnott S P 1984 *Icarus* **58** 178

Synnott P, Peters C F, Smith B A and Morabito L A 1981 *Science* **212** 192

Synnott S P, Terrile R J, Jacobson R A and Smith B A 1983 *Icarus* **53** 156

Valsecchi G B, Carusi A and Roy A E 1984 *Cel. Mech.* **32** 217

Walker I W 1983 *Cel. Mech.* **29** 149

Walker I W, Emslie A G and Roy A E 1980 *Cel. Mech.* **22** 371

Walker I W and Roy A E 1981 *Cel. Mech.* **24** 195

—— 1983 *Cel. Mech.* **29** 117, 267

Williams J G and Benson G S 1971 *Astron. J.* **76** 167

Yoder C F, Colombo G, Synnott S P and Yoder K A 1983 *Icarus* **53** 431

Bibliography

Gehrels T (ed) 1979 *Asteroids* (Tucson: University of Arizona Press)

Message P J 1958 *Astron. J.* **63** 443

Moser J K 1973 *Stable and Random Motions in Dynamical Systems, with Special Emphasis on Celestial Mechanics* (Princeton University Press)

—— 1974 *IAU Symposium No. 62* ed T Kozai (Dordrecht: Reidel) p 1

Roy A E (ed) 1988 *Long-Term Dynamical Behaviour of Natural and Artificial N-Body Systems* (Dordrecht: Reidel)

Szebehely (ed) 1982 *Applications of Modern Dynamics to Celestial Mechanics* (Dordrecht: Reidel)

—— 1985 *Stability of the Solar System and its Minor Natural and Artificial Bodies* (Dordrecht: Reidel)

9 Lunar Theory

9.1 Introduction

Lunar theory is concerned in general with the orbital motion of a satellite about a planet; in particular it has largely been devoted to the case of the motion of the Moon about the Earth. In what follows we shall be principally concerned with the Earth–Moon case but much of what is said applies to any lunar problem. Indeed Delaunay's lunar theory, developed for the Earth–Moon case, can be applied to other similar satellite problems.

As a starting point we set down the basic facts of the Earth–Moon system.

9.2 The Earth–Moon System

The Moon moves in an approximately elliptic orbit inclined at about five degrees to the plane of the ecliptic. The mean values of the semimajor axis a, the eccentricity e and the inclination i are given below

$$a = 384\,400 \text{ km}$$

$$e = 0.05490$$

$$i = 5°\,09'.$$

Because of solar perturbations, all three elements are subject to periodic variations about these values. In particular, the eccentricity varies from 0.044 to 0.067 while the inclination oscillates between 4° 58' and 5° 19'.

Various periods of revolution of the Moon in its orbit may be defined, namely the *sidereal* (the time required by the Moon to move through 360°), the *synodic* (the time between successive similar configurations with the Sun), the *nodical* (the time between successive passages through the ascending node), the *anomalistic* (the time between successive passages through perigee) and the *tropical* (the time between successive conjunctions with Aries). Their mean values are given in table 9.1.

Although in any revolution of the Moon in its orbit these months may differ by a few hours from the mean values given above, the mean values remain steady over many centuries to within one second.

The other three elements of the Moon's orbit, namely the longitude of the ascending node Ω, the longitude of perigee ϖ and the time of perigee passage τ suffer secular as well as periodic changes, due predominantly to the action of the Sun's gravitational pull. The line of nodes regresses in the

Table 9.1

Length of month	(Days)	d	h	m	s
Synodic	29·53059	29	12	44	03
Sidereal	27·32166	27	07	43	12
Anomalistic	27·55455	27	13	18	33
Nodical	27·21222	27	05	05	36
Tropical	27·32158	27	07	43	05

plane of the ecliptic, making one revolution in 6798·3 days (about 18·6 years) while the line joining perigee to apogee (the line of apses) advances, making one revolution in 3232·6 days (8·85 years).

The planets have small but not negligible effects on the Moon's orbit, and the shape of Earth and Moon themselves contribute to the perturbations. An idea of the relative orders of size of the various perturbations due to the Sun, planets, figures of Moon and Earth and so on is given in table 9.2, taken from Brown's lunar theory displaying the secular components of the movements of perigee and node.

Table 9.2

Mean annual motion of the:	Perigee	Node
Principal solar action	+146426·92″	− 69672·04″
Mass of the Earth	− 0·68	+ 0·19
Direct planetary action	+ 2·69	− 1·42
Indirect planetary action	− 0·16	+ 0·05
Figure of the Earth	+ 6·41	− 6·00
Figure of the Moon	+ 0·03	− 0·14
	+ 146435·21	− 69679·36

The construction of a complete lunar theory which not only includes the effects of Earth, Sun, planets and the figures of Earth and Moon but can also be compared with observations is one of the most difficult in astronomy. Newton, Euler, Clairaut, Hansen, Delaunay, Hill and Brown, to name but a few, have worked on the problem using many different approaches. Brown's lunar theory and his 'Tables of the Moon' are the most exhaustive treatment of the lunar problem. His theory includes 1500 separate terms, of which the so-called equation of the centre, the evection and the variation (see below) are the main ones. The theory is still used in preparing the lunar ephemeris. The first few terms in the expression for the Moon's longitude λ are given approximately by

$$\lambda = L + (377 \sin l)' + (13 \sin 2l)' + [74 \sin (2D - l)]'$$
$$+ (40 \sin 2D)' - (11 \sin l')' + \ldots$$

281

where L is the Moon's mean longitude, l is the angular distance of a fictitious mean moon from the mean perigee, D is its distance from the mean sun and l' is the mean sun's distance from the perigee point of the Sun's apparent orbit about the Earth. Essentially similar series give the Moon's latitude and parallax (the angle subtended at the Moon by an equatorial radius of the Earth).

The terms in l and $2l$ are ordinary elliptic two-body terms. The term in $(2D-l)$ is the *evection* and is due to the variation in the eccentricity of the orbit caused by the Sun's gravitational pull. Its period is 31·8 days. The term in $2D$ is the *variation*, an inequality in the Moon's motion due to a variation in the magnitude of the solar perturbing force during a synodic month. The other main inequality, the *annual equation*, given by the term in l', has a period of one anomalistic year and is due to the annual variation of the Earth's distance from the Sun.

There are other major inequalities of the Moon's motion caused by the Sun's gravitational pull. The parallactic inequality is a variation in the longitude with an amplitude containing the expression

$$\frac{E-M}{E+M}\left(\frac{a}{a_1}\right)$$

as a factor, where E and M are the masses of the Earth and the Moon respectively while a and a_1 are the mean geocentric distances of Moon and Sun respectively. It has an amplitude of just over 2′ and a period of one synodic month. In addition, the main inequality in the inclination has an amplitude of about 9′ and a period of half a nodical year.

The evection was noticed and discussed by Ptolemy in the Almagest. The variation, with a period of one-half of a synodic month, was described by Tycho Brahe who also discovered the annual equation. He also seems to have been the first to observe the variation in inclination, noting that i is at its maximum of 5° 18′ at first and third quarters and at its minimum of 4° 58′ at new and full moon. This oscillation is bound up with the regression of the nodes and, as mentioned above, has a period of half a nodical year; not one synodic month.

9.3 The Saros

There is one further property of the Earth–Moon–Sun system that has been known for at least 2500 years. The Saros, known to the ancient Chaldeans, is a period of time of approximately 18 years and 10 or 11 days (depending upon the number of leap years in the interval). At the end of a Saros, the geometry in the Earth–Moon–Sun system is repeated to a close enough extent that solar and lunar eclipses can be predicted from the occurrence of past eclipses at the Saros' beginning. Table 9.3 shows, for example, the values of the semi-diameters of Moon and Sun during four eclipses, each set of four occurring in the years 1898, 1916, 1934, 1952 and 1970. The eclipses were:

(i) partial eclipse of the Moon (February 21, 1970),
(ii) total eclipse of the Sun (March 7, 1970),
(iii) partial eclipse of the Moon (August 17, 1970),
(iv) annular eclipse of the Sun (August 31–September 1, 1970).

Table 9.3

Semi-diameter of Sun and Moon during eclipse

Year	1898	1916	1934	1952	1970
Date	Jan. 7	Jan. 19	Jan. 30	Feb. 10–11	Feb. 21
Moon	14′ 52·00″	14′ 49·8″	14′ 48·5″	14′ 47·3″	14′ 46·8″
Sun	16′ 15·87″	16′ 15·3″	16′ 14·1″	16′ 12·4″	16′ 10·3″
Date	Jan. 21	Feb. 3	Feb. 13–14	Feb. 25	Mar. 7
Moon	16′ 24·30″	16′ 25·4″	16′ 27·3″	16′ 29·2″	16′ 31·6″
Sun	16′ 14·83″	16′ 13·5″	16′ 11·6″	16′ 09·4″	16′ 06·8″
Date	Jul. 3	Jul. 14	Jul. 26	Aug. 5	Aug. 17
Moon	16′ 43·32″	16′ 42·9″	16′ 43·1″	16′ 43·2″	16′ 43·9″
Sun	15′ 43·86″	15′ 44·1″	15′ 44·9″	15′ 46·2″	15′ 47·9″
Date	Jul. 18	Jul. 29	Aug. 10	Aug. 20	Aug. 31–Sep. 1
Moon	14′ 45·87″	14′ 44·0″	14′ 43·2″	14′ 42·5″	14′ 42·6″
Sun	15′ 44·36″	15′ 45·3″	15′ 46·8″	15′ 48·6″	15′ 50·8″

The characteristics of all four eclipses were unchanged in the five years in which they occurred. In comparing the values of the lunar semi-diameter (and therefore its geocentric distance) from Saros to Saros it is seen how little it varies. The same is true of the Sun's semi-diameter even though the ranges within which both lunar and solar semi-diameters can vary are large (Sun, 15′ 45″–16′ 18″; Moon, 14′ 42″–16′ 44″). If we also take additional eclipse data from the respective *Nautical Almanacs* and the 1970 *Astronomical Ephemeris* concerning solar and lunar ecliptic longitudes λ and latitudes β, and also the rates of change of these quantities, we find that their values at the beginning of a Saros are very nearly repeated at the end of the Saros. Thus in table 9.4, data for the partial lunar eclipses of 1952 (February 10–11) and 1970 (February 21) are compared. In the table the differences between the Sun and Moon's geocentric ecliptic coordinates during eclipse are tabulated for each eclipse. Suffixes M and S refer to Moon and Sun respectively, the dots denote daily rates of change and σ stands for semi-diameter.

One more example, not at an eclipse but taken at random in the lunar ephemeris, is illustrated in Table 9.5. Again it is seen how accurately the relative positions and velocities of Sun and Moon are repeated after one Saros. The reason is of course the interesting set of near-commensurabilities existing among the Moon's synodic period, its anomalistic period and its

283

Table 9.4

Data	$\lambda_S - \lambda_M$	$\beta_S - \beta_M$	σ_M	σ_S	λ_M	β_M	$\dot\sigma_M$	$\dot\lambda_S$	$\dot\beta_S$	$\dot\sigma_S$
1952 Feb. 11·02729^d	179° 54·9'	−0° 50·84'	14' 47·3"	16' 12·4"	11·955°	−65·345'	3·84"	1·0114°	−0·15"	−0·18"
1970 Feb. 21·35467^d	179° 54·7'	−0° 51·69'	14' 46·8"	16' 10·3"	11·922°	−65·305'	3·49"	1·0069°	−0·28"	−0·22"

Table 9.5

Data	$\lambda_S - \lambda_M$	$\beta_S - \beta_M$	σ_M	σ_S	λ_M	β_M	$\dot\sigma_M$	$\dot\lambda_S$	$\dot\beta_S$	$\dot\sigma_S$
1952 Mar. 18·0	104° 21·1'	5° 06·2'	15' 54·7"	16' 05·6"	13·801°	16·567'	8·01"	0·9945°	0·01"	−0·28"
1970 Mar. 29·347	104°21·9'	5° 05·8'	15' 52·9"	16' 02·6"	13·984°	16·006'	8·52"	0·9883°	−0·14"	−0·28"

nodical period. From the *Astronomical Ephemeris* (1970) their mean values are:

$$\text{Synodic (S)} = 29 \cdot 530\,589^d$$
$$\text{Anomalistic (L)} = 27 \cdot 554\,551^d$$
$$\text{Nodical (D)} = 27 \cdot 212\,220^d.$$

Then, as is well known,

$$223\,S = 6585 \cdot 3213^d$$
$$239\,L = 6585 \cdot 5377^d$$
$$242\,D = 6585 \cdot 3572^d.$$

The close agreement ensures that the geometry of the Earth–Moon–Sun system at any epoch is almost exactly repeated one Saros later. When the Moon's elongation is repeated at the end of the Saros its argument of perigee and true anomaly also have very nearly the same values as before. In addition, because the Saros length is only ten days longer than 18 years, the Sun is almost back to its original true anomaly and length of radius vector. The closeness of the fit is thus not only in position but also in velocities.

It should also be noted that, within any Saros, the perturbations of the Sun on the Earth–Moon system almost completely cancel themselves out, in particular the large disturbances in semimajor axis, eccentricity and inclination. It is perhaps easiest to see this if we take the situation at the beginning of a Saros to be such that full Moon occurs when the Moon and the Sun are at perigee, the Moon's latitude being zero. The velocity vectors of the Sun and Moon are then perpendicular to both the radius vectors. This is a mirror condition, and by the mirror theorem (Roy and Ovenden 1955) the history of the system after that time is a mirror image of its history prior to that time.

But nine years and approximately five days later, a new mirror condition very nearly occurs—a new Moon, Sun within 6° of perigee, Moon at apogee, Moon's latitude zero. The velocity vectors of Sun and Moon are very nearly perpendicular to both the radius vectors. If this second mirror configuration were exact, the Moon's orbit would be exactly periodic, returning at the end of the Saros to a repeat of the first mirror configuration so that the perturbations built up in the first half of the Saros would have been cancelled completely in the second, the only result being that the sidereal position of the line of nodes of the Moon's orbital plane would have regressed approximately 11°. As it is, the Moon's orbit under solar perturbation is very nearly periodic with a period of one Saros, the close repetition of the geometrical properties of solar and lunar eclipses being the outward manifestation of how closely the Earth–Moon–Sun system approximates to a purely periodic motion. All other perturbations (planetary, tidal, figures of Earth and Moon) are very small indeed.

9.4 Measurement of the Moon's Distance, Mass and Size

The semimajor axis of the Moon's orbit has been determined in a wide variety of ways. The trigonometric method involved the use of two observatories widely separated in latitude to provide a long enough baseline, from the ends of which the Moon's sidereal positions could be measured. A knowledge of the size of the Earth, the observatories' coordinates and the observations and the times at which they were made, provided sufficient information from which to calculate the Moon's orbital semimajor axis.

The use of short-wavelength radar also enables the Moon's distance to be found, while the range and range-rate tracking of artificial lunar satellites has also provided observational data from which the mean Earth-centre to Moon-centre distance may be determined. The most modern and most accurate method involves the use of laser beams reflected from the banks of corner reflectors left on the Moon's surface by the Apollo astronauts. The error in such measurements is probably less than 0·2 m. The size of the Moon is then found by measuring its angular diameter and using its known distance. A value of 3476 km for the linear diameter is obtained.

A direct method of measuring the mass of the Moon is to use the apparent monthly oscillations in the directions of external bodies (such as the Sun and asteroids) produced by the elliptical movement of the Earth's centre about the centre of mass of the Earth–Moon system. For the Sun the amplitude is of the order of 6″, but for an asteroid that makes a close approach to the Earth the amplitude may be several times this amount. From this method, a value for the Moon's mass of 1/81·27 times that of the Earth is deduced.

A second method makes use of one of the perturbations in the Moon's motion caused by the solar attraction, namely the *parallactic inequality*.

The observational value is (according to Brouwer) 124·97″, while by lunar theory it is (in seconds) given by the expression

$$49\,853\cdot2''\,\frac{E-M}{E+M}\left(\frac{a}{a_1}\right).$$

These may therefore be equated and a knowledge of M/E obtained if a and a_1 are known. A value of 1/81·22 is obtained for M/E in this way. Our modern and much more accurate measurements of the Moon's mass are derived from observations of the orbits of artificial lunar satellites, in essence by the use of Newton's form of Kepler's third law.

Knowing the mass and linear diameter of the Moon, its mean density may be calculated immediately. It is found to be about 3·33 times that of water, very close to that of the basic rocks under the thin surface crust of the Earth.

9.5 The Moon's Rotation

The rotation of the Moon about its centre of mass is described by three empirical laws stated by Cassini in 1721. They are:

First law: The Moon rotates eastward about an axis fixed within it, with constant angular velocity in a period of rotation equal to the mean sidereal period of revolution of the Moon about the Earth.

Second law: The inclination of the mean plane of the lunar equator to the plane of the ecliptic is constant.

Third law: The poles of the lunar equator, the ecliptic, and the Moon's orbital plane all lie in one great circle in the above order; that is, the line of intersection of the mean lunar equatorial plane with the ecliptic is also the line of nodes of the Moon's orbit, the descending node of the equator being at the ascending node of the orbit (see figure 9.1).

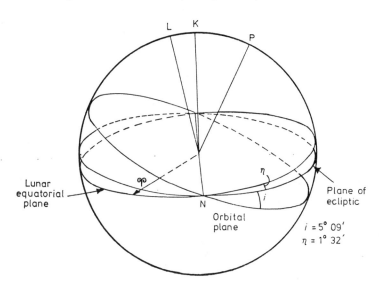

Figure 9.1

In figure 9.1, which represents a selenocentric celestial sphere, the great circles made by the intersections of these planes with the sphere are shown.

Cassini's laws are valid to a high degree of approximation; departures in the Moon's rotation from them consist of small oscillations called the *physical libration* made up of a free oscillation and forced oscillations. The causes of these slight wobbles are the shape of the Moon (which is approximately a triaxial ellipsoid with the longest axis always pointing in the general direction of the Earth) and the attraction of the Earth on this protuberance. Because the Moon in its elliptic orbit obeys Kepler's second law while the Moon rotates uniformly on its axis, the long axis of the Moon oscillates about the line joining the centres of Earth and Moon as shown in figure 9.2, the amplitude EM̂A of this oscillation being about 8°. The Earth thus tends to swing the Moon in various directions giving rise to the forced oscillations. The maximum amplitude of the *physical* libration is about 3·5′.

Because of Cassini's laws and Kepler's second law, the so-called *geometrical librations* (or *optical librations*) in longitude and latitude are observed. The libration in longitude, resulting from Cassini's first law and Kepler's second law, means that objects on the lunar surface are displaced in longitude by ±7·9°, as measured from the Moon's centre. The latitude libration is a

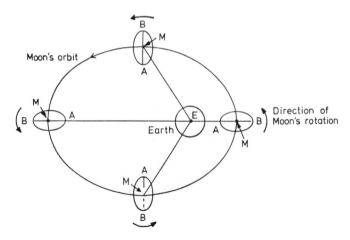

Figure 9.2

consequence of Cassini's second law so that lunar objects are displaced in latitude by $\pm 6 \cdot 7°$, again as measured from the lunar centre.

There is a third geometrical libration called the *diurnal* or *parallactic libration* arising from the position of the observer on a finite-sized, rotating Earth, enabling him to see about $1°$ around the edges of the Moon's visible face.

These geometrical librations allowed maps to be constructed of 59% of the lunar surface even before Lunik III photographed the other side of the Moon in 1959.

9.6 Selenographic Coordinates

In order to take account of the lunar geometrical and physical librations, astronomers have adopted the system known as the *selenographic coordinate system*. The origin of this system is the Moon's centre. When the Moon is at the mean ascending node of its orbit at a time when the node coincides with either the mean perigee or mean apogee, the point where the line joining Earth centre to lunar centre cuts the surface of the Moon is defined to be the mean centre of the apparent disk. This point, like Greenwich on the Earth, defines a prime lunar meridian from which *selenographic longitudes* λ of places on the Moon may be measured, the positive direction being towards Mare Crisium (i.e. towards the west on a geocentric celestial sphere). The *selenographic latitude* β is measured from the lunar equator along a meridian and is taken to be positive when the latitude is of a place in the northern hemisphere of the Moon (i.e. in the hemisphere containing Mare Serenitatis).

At any time, according to the phases of the geometrical and physical librations, the line joining the centres of Earth and Moon will intersect the Moon's surface at a point possessing a certain selenographic latitude and longitude. These latitudes and longitudes are tabulated for every day of the

288

year in the *Astronomical Almanac* as the Earth's selenographic latitude and longitude. They are the sums of the geocentric optical and physical librations. Also tabulated is the position angle of the axis, namely the angle that the lunar meridian through the centre of the Moon's visible disc makes with the declination circle passing through that central point.

9.7 The Moon's Figure

The Moon's figure is approximately triaxial and so it possesses three moments of inertia A, B and C about three unequal mutually perpendicular axes. The longest axis (Ox) points approximately towards the Earth, while the shortest (Oz) is nearly perpendicular to the plane of the orbit (O being the Moon's centre of mass). The moment of inertia A about the longest axis is thus the least, while the moment of inertia C about the shortest axis is the greatest. From a study of the dynamics of the Earth–Moon system, it may be shown that the above relationship among the moments of inertia must hold (i.e. that $A < B < C$) if Cassini's laws are to be obeyed, leading to small stable oscillations about the steady motion.

The best method of obtaining an accurate description of the Moon's figure is by studying the perturbations its gravitational field produces in the orbits of artificial lunar satellites. Such satellites are also attracted by the Sun and the Earth, so their orbits are subject to perturbations produced by those bodies. It is possible, however, to separate the effects produced by the Moon's gravitational potential's departure from that of a point-mass from those caused by solar and terrestrial attractions. In the next chapter we consider in some detail how artificial Earth satellite theories may be constructed and used to obtain values of the harmonic constants describing the Earth's figure. We content ourselves here by saying that essentially similar theories may be constructed for the lunar-satellite problem. Lists of values for the Moon's gravitational potential have been published (Michael *et al* 1970).

In terms of the dynamical ellipticities, we have

$$\frac{C-A}{C} = 0.000\,629, \qquad \frac{B-A}{C} = 0.000\,225.$$

If M is the mass of the Moon, and r_{m} is its mean radius, we also have

$$\frac{C}{M r_{\mathrm{m}}{}^2} = 0.401.$$

It would appear that the difference in length between longest and shortest axes is about 1·1 km, while that between longest and shortest equatorial axes is about 0·3 km.

9.8 The Main Lunar Problem

Before qualitatively considering the various approaches of investigators to the problem of lunar orbital motion it is instructive to set up the equations of motion of the main lunar problem, where the Earth, Moon and Sun attract each other according to Newton's law of gravitation, all three bodies being taken to be point-masses. Everything else: the finite sizes of Earth and Moon, tidal effects, the attractions of the planets, etc. may be taken to be small (table 9.2) and can be added later.

In the planetary problem, bodies moved about the Sun at roughly comparable distances and disturbed each other's heliocentric orbit, so that the most convenient form of the equations of motion is one where the origin lies at the Sun's centre. It is also most convenient to use the ratio of the mass of a disturbing planet to that of the Sun as a small quantity, expanding the disturbing function in successive powers of this. In addition, auxiliary expansions in powers and products of the eccentricities and inclinations are involved.

In the lunar problem both Moon and Earth are at almost the same distance from the Sun, but this distance is always a large multiple of their separation; in addition the mass of the disturbing body (the Sun) is of the order of 330 000 times the mass of Earth and Moon combined. A convenient small quantity is the ratio of the Earth–Moon mean distance to the Earth–Sun mean distance, which is of the order of $1/400$. A set of equations that demonstrates a useful property of the lunar problem may be set up as follows.

In section 5.11.3 we saw that by using Jacobian coordinates the general three-body equations of motion could be expressed by relations (5.90) and (5.91). If the force function U was defined by

$$U = G \left(\frac{m_1 m_2}{r} + \frac{m_2 m_3}{r_{23}} + \frac{m_3 m_1}{r_{31}} \right) = U(r, \rho) \tag{9.1}$$

then the equations of motion took the form

$$\ddot{\mathbf{r}} = \frac{1}{g_1} \frac{\partial U}{\partial \mathbf{r}} \tag{9.2}$$

$$\ddot{\boldsymbol{\rho}} = \frac{1}{g_2} \frac{\partial U}{\partial \boldsymbol{\rho}}$$

where

$$g_1 = \frac{m_1 m_2}{\mu} \quad \text{and} \quad g_2 = \frac{m_3 \mu}{M} \tag{9.3}$$

remembering that $\mu = m_1 + m_2$ and $M = m_1 + m_2 + m_3$.

Let us now write the function U as

$$U = G \left(\frac{m_1 m_2}{r} + F \right)$$

where

$$F = \left(\frac{m_2 m_3}{r_{23}} + \frac{m_3 m_1}{r_{31}} \right). \tag{9.4}$$

Remembering that

$$\frac{\partial}{\partial \rho} \left(\frac{1}{r} \right) = 0,$$

we can now rewrite equation (9.2) in the form

$$\ddot{\mathbf{r}} + G(m_1 + m_2)\frac{\mathbf{r}}{r^3} = \frac{G(m_1 + m_2)}{m_1 m_2}\nabla F \tag{9.5}$$

and

$$\ddot{\boldsymbol{\rho}} = \frac{GM}{m_3(m_1 + m_2)}\nabla_c F. \tag{9.6}$$

We now identify m_1, m_2 and m_3 as the masses of Earth, Moon and Sun respectively and denote them E, M and S (figure 9.3). The equations of motion in the main lunar problem in Jacobian coordinates are then

$$\ddot{\mathbf{r}} + G(E + M)\frac{\mathbf{r}}{r^3} = \frac{GS(E + M)}{EM}\nabla F \tag{9.7}$$

$$\ddot{\boldsymbol{\rho}} = \frac{G(S + E + M)}{E + M}\nabla_c F \tag{9.8}$$

where

$$F \equiv \frac{E}{r_{ES}} + \frac{M}{r_{MS}}. \tag{9.9}$$

It is to be noted that so far no approximations in this problem have been made. We now consider what these equations tell us about the orbit of the Sun.

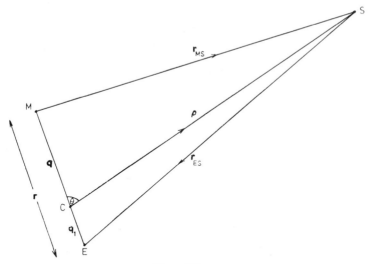

Figure 9.3

9.9 The Sun's Orbit in the Main Lunar Problem

To begin with we expand the function F, given by

$$F = \frac{E}{r_{ES}} + \frac{M}{r_{MS}} \tag{9.10}$$

291

in much the same way that the denominator in the expression for the potential of a body of arbitrary shape was treated (i.e. by introducing Legendre polynomials). Let $M\hat{C}S = \theta$, and $c = \cos \theta$. Take the vectors $CM = \mathbf{q}$, $EC = \mathbf{q_1}$. Then

$$\mathbf{q_1} = \frac{M\mathbf{r}}{E+M}$$

$$\mathbf{q} = \frac{E\mathbf{r}}{E+M}.$$

From triangle CMS, we have

$$r_{MS}{}^2 = \rho^2 + q^2 - 2\rho q \cos \theta$$

or

$$\frac{\rho}{r_{MS}} = \left[1 + \left(\frac{q}{\rho} \right)^2 - 2 \left(\frac{q}{\rho} \right) \cos \theta \right]^{-1/2}$$

in other words

$$\frac{\rho}{r_{MS}} = 1 + \sum_{i=1}^{\infty} \alpha^i P_i(c) \tag{9.11}$$

where $\alpha = q/\rho$ and the $P_i(c)$ are Legendre polynomials.

Similarly from triangle ECS, putting $\alpha_1 = q_1/\rho$ and noting that $E\hat{C}S = \pi - \theta$, we may write

$$\frac{\rho}{r_{ES}} = 1 + \sum_{i=1}^{\infty} (-\alpha_1)^i P_i(c). \tag{9.12}$$

Hence, by writing equation (9.10) in the form

$$F = \frac{1}{\rho} \left[E \frac{\rho}{r_{ES}} + M \frac{\rho}{r_{MS}} \right]$$

and substituting expressions (9.11) and (9.12) in it, we obtain

$$F = \frac{E+M}{\rho} + \frac{ME}{E+M} \left[\frac{r^2}{\rho^3} P_2 + \left(\frac{E-M}{E+M} \right) \frac{r^3}{\rho^4} P_3 + \frac{(E^2 - EM + M^2)}{(E+M)} \frac{r^4}{\rho^5} P_4 + \dots \right]. \tag{9.13}$$

We can now use this expression to investigate the Sun's orbit. By equation (9.8),

$$\ddot{\boldsymbol{\rho}} = E(S+E+M)\nabla_c \left[\frac{1}{\rho} + \frac{ME}{(E+M)^2} \frac{r^2}{\rho^3} P_2 + \dots \right].$$

The second term within the bracket divided by the first is of size

$$\frac{ME}{(E+M)^2} \frac{r^2}{\rho^2} = \left(\frac{M}{E} \right) \frac{1}{(1+M/E)^2} \left(\frac{r}{\rho} \right)^2 \sim 8 \times 10^{-8}.$$

Hence the second and following terms, to a high degree of approximation, may be neglected. The equation of motion of the Sun about the centre of mass of the Earth–Moon system is therefore of the form

$$\ddot{\boldsymbol{\rho}} + G(S+E+M) \frac{\boldsymbol{\rho}}{\rho^3} = 0.$$

This is the familiar two-body equation of motion, which shows that the Sun very nearly follows a fixed Keplerian elliptic orbit. The Sun's coordinates are therefore given by the usual analytical expressions and its orbital elements are constant. To this extent the lunar problem is simpler than the planetary problem where the disturbing bodies are themselves sensibly disturbed. It is, however, the only bonus we get!

9.10 The Orbit of the Moon

By equation (9.7) it is seen that the disturbing function for the Moon is R, given by

$$R = \frac{GS(E+M)F}{EM}. \tag{9.14}$$

Inspecting equation (9.13) it is seen that the first term in F has ρ as a variable in it. But $\nabla \rho^{-1} = 0$ and therefore we may neglect this term. If we let the mean motions of the Sun and the Moon be n_1 and n respectively, and define a parameter m by

$$m = n_1/n \tag{9.15}$$

then by Kepler's third law

$$n_1{}^2 a_1{}^3 = G(S+E+M) = GS \tag{9.16}$$

accurate to 3×10^{-6}, where a_1 is the Sun's orbital semimajor axis.

By equations (9.13), (9.14), (9.15) and (9.16), we therefore have

$$R = m^2 n^2 r^2 \left(\frac{a_1}{\rho}\right)^3 \left[P_2 + \frac{E-M}{E+M}\left(\frac{r}{\rho}\right)P_3 + \frac{E^2 - EM + M^2}{E+M}\left(\frac{r}{\rho}\right)^2 P_4 + \ldots\right].$$

The disturbing function is now arranged in ascending orders of the small quantity $r/\rho \sim 1/400$. Further progress lies in expressing R in the elements and then in expanding the subsidiary small quantities provided by the eccentricities of the lunar and solar orbits, the inclination of the lunar orbits to the ecliptic and the ratio m of the mean motions. A straightforward but incredibly time-consuming approach (if carried out by a human operator) would be to set up the Lagrange planetary equations in the Moon's orbital elements, expand the lunar disturbing function in powers of these auxiliary small quantities and then solve the equations by the method of successive approximations. This approach was attempted by Poisson. Having solved the main lunar problem, the other perturbations due to the figures of Moon and Earth etc. can be included.

9.11 Lunar Theories

From Newton's time, many mathematical astronomers have attempted to create lunar theories. Apart from the natural desires to produce an analytical theory capable of furnishing predictions as accurate as the best observed positions of the Moon, to study the evolution of the lunar orbit and to check how completely Newton's law of gravitation explained the satellite's motion, there were other reasons for creating theories. The lack of accurate clocks (before Harrison produced his chronometer in 1761) made it impossible to provide a solution to the important practical problem of determining longitude at sea. Galileo had thought of determining time by comparing observations of the moons of Jupiter with tables of their positions. Newton's preference was that the Moon be used. In the first century of the search for a lunar theory therefore, there were military–exploratory–mercantile pressures urging it on. The removal of these pressures did not stop the search. There were always enough people interested in the problem for its own sake for research to continue. Furthermore, as observational methods became more precise, older theories became inadequate or were found to possess errors (for example Damoiseau's extension of Laplace's lunar theory) and so became superseded. More recently, researches in geophysics and tidal theory (in addition to the advent of lunar laser-ranging methods) have necessitated the improvement of our means of computing lunar ephemerides.

Newton found the lunar problem so difficult that he complained, 'it made his head ache and kept him awake so often that he would think of it no more.' But he did show that the known inequalities in the Moon's orbital motion were due to the Sun; he also computed the motion of perigee to within 8% of the observed value by taking second-order terms into account.

Important contributors to lunar theory have included Newton, Euler, Clairaut, Poisson, Laplace, Damoiseau, Hansen, Delaunay, Hill, Brown and Deprit. All of their theories have two common features—the large number of terms they contain and the need for selecting a zero-order intermediate orbit. The number of terms required is dictated not only by the required accuracy but also by the choice of intermediate orbit and method of development. Most theories began with the equations of motion expressed in terms of polar coordinates or functions of the orbital elements, though Euler's theory of 1772 used rectangular coordinates, the x and y axes rotating with the Moon's mean angular motion. De Pontécoulant's theory published in 1846 was based on polar coordinates. Hill's theory utilized rotating rectangular coordinates but with the x axis restrained to point at the Sun's mean position. A fixed Keplerian ellipse, a rotating ellipse of fixed shape, and a periodic orbit more complicated than either have all been used at various times as intermediate orbits. For example, Hill chose a periodic orbit which was a particular solution of two second-order differential equations in u and s, where

$$u = X + iY, \; s = X - iY \quad (i = \sqrt{-1})$$

X and Y being the Moon's geocentric ecliptic coordinates; the X axis always points to the Sun's mean geocentric position. The independent variable ζ was defined by

$$\zeta = \exp [i(n - n_1)(t - t_0)]$$

where n_1 is the mean motion of the Sun about the Earth, t is time and t_0 and n are undetermined constants at that stage.

Hill obtained these equations by neglecting the solar eccentricity, the solar parallax and the Moon's latitude and eccentricity. The solution used by Hill as his intermediate orbit was expressed in a Fourier series of $(n-n_1)t$. It is an oval, symmetrical about the axes with the longer axis of the oval perpendicular to the Sun's direction. This figure is known as Hill's variational curve. The deviations of the real lunar orbit from this intermediate orbit were then developed analytically by Hill and Brown. Brown later provided tables of the Hill–Brown lunar theory for use in computing the lunar ephemerides. In recent years however, with the advent of electronic computers, the more accurate theory has been used to compute improved lunar positions. Further improvements have also been made.

Two additional features of the development of lunar theory must be considered:

(i) The theories themselves have fallen into three classes: analytic, analytic–numerical and numerical. Delaunay's lunar theory is the supreme example of the purely analytic approach. The disturbing function was completely developed to the seventh order in small parameters. Over 500 canonical transformations were applied to reduce it, term by term, finally producing the ecliptic latitude and longitude and the sine parallax of the Moon. The work took twenty years. Because of its completely analytical nature it can be applied to any three-body problem.

The analytic–numerical approach was begun by Laplace. While retaining the two eccentricities and $\sin (i/2)$ as undetermined parameters (i being the inclination), he gave a numerical value to $m = n_1/n$. The Hill–Brown theory strictly falls into this class.

Sir George Airy proposed a purely numerical approach to the problem of improving Delaunay's theory. It showed great promise but Airy's own attempt at it, published in 1886, was faulty. Recently Eckert has applied Airy's technique to Brown's theory of the main lunar problem.

The drift from analytical theories to purely numerical ones was due to a realization that for a specific lunar theory, the goal of the desired accuracy was more quickly achieved with far less work if a numerical approach was chosen. The advent of high-speed, large-capacity electronic computers has changed this view. As Herget and Musen showed as far back as 1959, computers can be programmed to carry out the literal developments so often used in celestial mechanics. Using a computer in this way is not easy; it may take a year to write, test and debug a program for a particular task; but when it is done, the computer will produce a purely analytical print-out. Delaunay's development of the lunar disturbing function is a typical example. But instead of taking years to do so the machine time is measured in hours and the analytical development can be taken to a far higher order. Symbolic manipulation by computer opens a new era in orbital motion studies.

An analytical lunar ephemeris by Deprit has been produced that goes far beyond Brown's lunar theory in accuracy, where the main lunar problem is concerned. Table 9.6 (from Deprit) compares the number of trigonometric arguments in the ecliptic longitude, latitude and sine parallax appearing in Brown's theory, in Eckert's revision of the improved Lunar Ephemeris (ILE) and in the computer-produced analytical Lunar Ephemeris (ALE).

Table 9.6

	Longitude	Latitude	Sine parallax
Brown	552	487	304
ILE	703	623	577
ALE	2427	2355	2013

(ii) The second feature is the considerable improvement in observational accuracy and the change in order of importance of the measured quantities. Until the advent of radar, lunar theory was primarily concerned with the ecliptic longitude and latitude of the Moon, while the distance (or the related quantity, the sine parallax) came third. This order or priority was dictated by the observational astronomers' optical measurements of lunar positions on the celestial sphere. Radar, directly concerned with distance, enhanced the importance of sine parallax. The establishment of laser-ranging corner reflectors on the Moon confirmed the prime importance of the sine parallax series in the lunar theory. In addition, the potential accuracy of being able to measure at any time by laser the Earth–Moon distance with an error of the order of 25 cm makes it necessary that that series is established in lunar theory to an equivalent accuracy; the series for the other two coordinates must likewise be improved since all three are interdependent. Only a computer-generated literal lunar theory such as Deprit's has this capability.

For an up-to-date account of the history of lunar theories and a presentation of modern developments the reader is referred to Cook (1988).

9.12 The Secular Acceleration of the Moon

So far it has been assumed that the Moon's mean distance suffers only periodic variations. It should consequently be expected through Kepler's third law that the Moon's mean motion would behave likewise. The expression for the Moon's mean longitude l should therefore be given by

$$l = l_0 + n_0 t + P$$

where l_0 and n_0 are constants and P denotes the value of periodic inequalities at time t. In fact, by a study of ancient eclipses described in Ptolemy's Almagest and of a number of eclipses observed by Arabian astronomers in the ninth century AD, Halley in 1693 demonstrated that the expression for l is of the form

$$l = l_0 + n_0 t + \sigma \left(\frac{t}{100} \right)^2 + P.$$

In this expression t is measured in Julian years; σ is the *coefficient of the secular acceleration* and has a value of about 11 seconds of arc.

Laplace gave an explanation for this acceleration by pointing out that planetary perturbations on the Earth's orbit are changing its eccentricity. The change is in fact periodic, the main period being of the order of 24 000 years; for much shorter intervals it can be treated as a secular change. Through the appearance of the Earth's orbital eccentricity in the Lagrange planetary equation for ϵ in the lunar theory, it turns out that ϵ behaves as

$$\epsilon = \epsilon_0 + \epsilon_1 t + \epsilon_2 t^2$$

ϵ being the mean longitude at the epoch. The Moon's mean longitude l is given by

$$l = \int n \, dt + \epsilon$$

so that it is seen to include the acceleration term $\epsilon_2 t^2$.

Subsequent refinements of the theory by J C Adams in 1880 showed that the value of σ is less than 6 seconds of arc (i.e. just over half the observed value of 11 seconds of arc). The discrepancy is now believed to be due to tidal friction. The Earth, rotating once every sidereal day, tries to carry the tidal bulges produced by the Moon's gravitational pull round with it; the Moon holds them back since it revolves about the Earth in the much longer period of the sidereal month (27·22 days). The consequence is that angular momentum is lost by the Earth by tidal friction, principally in the shallower seas, so that the Earth's period of rotation increases. The transfer of angular momentum to the Moon causes it to recede from the Earth, increasing the length of the month. Calculations indicate that the Moon appears to accelerate in its orbit at a rate making up the observed discrepancy. This process will continue until the Moon spirals out to a distance where the length of the period of rotation of the Earth (the day) equals the Moon's period of revolution (the month), an interval of time about 40 times our present mean solar day. The lunar tide effect then ceases. Tidal friction due to solar gravitation must still operate; this will decrease the angular momentum of the Earth–Moon system since solar tidal drag tries to slow down the system's rotation. As a consequence, the Moon will approach the Earth once more, spiralling in slowly. It is not without interest that in the astronomical long run, an effect that is tiny compared with the major and obvious perturbations of the Moon's orbit should be the principal agent in shaping the Moon's orbital history.

References

Astronomical Ephemeris 1970
Cook A 1988 *The Motion of the Moon* (Bristol: Adam Hilger)
Michael W H Jr, Blackshear W T and Gapcynski L P 1970 *Dynamics of Satellites* ed Bruno Morando (Berlin: Springer-Verlag)

Bibliography

Brouwer D and Clemence G M 1961 *Methods of Celestial Mechanics* (New York and London: Academic)

Brown E W 1896 *An Introductory Treatise on the Lunar Theory* (London: Cambridge University Press)

—— 1919 *Tables of the Motion of the Moon* (New Haven: Yale University Press)

Danby J M A 1962 *Fundamentals of Celestial Mechanics* (New York: Macmillan)

Deprit A 1971 *ELDO/ESRO Scientific and Technical Review* **3** (**No. 1**) 77

Herget P and Musen P 1959 *Astron. J.* **64** 11

Moulton F R 1914 *An Introduction to Celestial Mechanics* (New York: Macmillan)

Plummer H C 1918 *An Introductory Treatise on Dynamical Astronomy* (London: Cambridge University Press); 1960 paperback edition (New York: Dover Publications)

Smart W M 1953 *Celestial Mechanics* (London, New York, Toronto: Longmans)

10 Artificial Satellites

10.1 Introduction

In this chapter an account is given of the dynamics of artificial satellites. Most of our attention will be given to artificial Earth satellite orbits but many of their properties may be taken over unchanged to the study of artificial satellites of other planets. To understand and compare the magnitudes of the different forces acting upon an artificial Earth satellite, the Earth and its environment require study. In what follows we first of all consider the Earth as a planet then describe briefly its structure, atmosphere, and magnetic field. From there we proceed to the orbit of a satellite under the action of the major forces involved.

10.2 The Earth as a Planet

The Earth's orbit, lying between the orbits of Venus and Mars, is to a high degree of approximation an ellipse of small eccentricity. The elements of this orbit suffer changes of the nature described in chapter 6, the changes being measured with respect to some fixed reference plane and direction such as the position of the ecliptic and vernal equinox at a given epoch. These changes are caused by the attractions of the planets; in addition the Moon, because of its proximity, also affects the Earth's orbit. We have seen that it is the centre of mass of the Earth–Moon system that revolves in a disturbed ellipse about the Sun while the Earth and the Moon revolve about this centre. Because the Moon's mass is only 1/81 that of the Earth, and its geocentric distance is some 60 Earth radii, the centre of mass lies about 1600 km below the Earth's surface.

Astronomers have found it convenient to use data connected with the Earth's orbit and the Sun as their units of time, distance, and mass. Taking the solar mass, the mean solar day and the Earth's mean distance from the Sun as the units of mass, time and distance respectively, the precise statement of Kepler's third law for a planet of mass m_2 revolving about the Sun of mass m_1, which is given by

$$k^2(m_1+m_2)T^2=4\pi^2a^3$$

becomes

$$k^2(1+m_2)T^2=4\pi^2a^3$$

where k^2 is written for G (the gravitational constant), and m_2, T and a are in the units defined above. The quantity k is called the *Gaussian constant of gravitation*.

If (as was done by Gauss) the planet is taken to be the Earth and T given the value of 365·2563835 mean solar days (the length of the sidereal year adopted by Gauss) while m_2 is taken to be 1/354710 solar masses, k is found to have the value 0·0172020989 5 (the value of a being of course unity).

Since then, these quantities have from time to time been determined more accurately; but to avoid having to recompute k every time, astronomers have adopted Newcomb's practice and retained the original value of k as absolutely correct. This means that the Earth is treated like any other planet. The unit of time is now the ephemeris day. The Earth's mean distance from the Sun is now 1·00000003 astronomical units while the Earth–Moon system's mass is 1/329390 solar masses.

We may note then that the definition of the astronomical unit is given by Kepler's third law

$$k^2(1+m)T^2 = 4\pi^2 a^3$$

with the Sun's mass taken to be unity, $k = 0·0172020989 5$ and the unit of time taken to be one ephemeris day. It is the radius of a circular orbit in which a body of negligible mass, free from perturbations, will revolve about the Sun in one Gaussian year of $2\pi/k$ ephemeris days.

In feasibility studies it is often accurate enough when working in years and astronomical units to take $GM = 4\pi^2$, where M is the Sun's mass and G is the constant of gravitation, since for any planet and any probe we have the relation

$$GMT^2 = 4\pi^2 a^3.$$

Hence for a body in a heliocentric orbit of period T_1 years and semimajor axis a_1 (measured in AU), we have

$$T_1 = a_1^{3/2}.$$

At this point it may be mentioned that for satellite motion about the Earth, the ephemeris minute, mass and radius of the Earth can be conveniently taken as the units of time, mass and distance respectively. If then m_E is the Earth's mass and G is the constant of gravitation, we can introduce k_E^2 by setting

$$k_E^2 = Gm_E. \tag{10.1}$$

This quantity can be determined accurately.

As before, k_E may be taken to be absolutely accurate and defines a unit of distance, namely the radius of an equatorial circular orbit in which a particle of negligible mass (free of perturbations) will revolve about the Earth in a period of $2\pi/k_E$ ephemeris minutes. For $k_E = 0·07436574$, we have $2\pi/k_E = 84·49032$ and the unit of distance is 6378·270 km. The use of k_E^2 defined by equation (10.1) bypasses the poor knowledge we have of the values of G and m_E.

300

Any distance within the Solar System may be expressed in astronomical units to a high degree of accuracy, since only angular and temporal measurements need be made. But to obtain the astronomical unit in kilometres, or in other words to obtain the scale of the Solar System, other methods must be adopted. The quantity called the *solar parallax*, defined as the angle subtended by the equatorial radius of the Earth at a distance of one astronomical unit, connects the astronomical unit with the size of the Earth. Its value is about 8·80″.

Many methods have been devised for measuring this important quantity directly or indirectly. Some, such as the use of transits of Venus across the Sun's disc, are of purely historical interest and could not give answers of high accuracy. Until recently the most reliable methods used observations of the asteroid Eros, which occasionally approaches to within 23 000 000 km of the Earth. In one such method the geocentric distance of Eros was found essentially by a triangulation method under the direction of Spencer Jones. The solar parallax could then be computed. A second method, carried out by Rabe, used the dynamics of the problem, taking into account the perturbations of the planets on Eros's orbit.

The most modern method uses radar. The distance between Venus and the Earth can now be measured with very high accuracy by transmitting radio pulses to the planet, the times of transmission and reception of the echo being measured. The time interval (or travel time) and the known velocity of electromagnetic radiation enables the distance to be found. Various corrections must be applied to derive the distance of Venus-centre to Earth-centre. From values obtained, the solar parallax P can be calculated. The value is $P = 8·794″$.

10.2.1 The Earth's shape

The Earth's shape is roughly that of an oblate spheroid. A consequence of the Earth's departure from a sphere is the luni–solar precession (section 3.4) due to the attractions of the Sun and Moon on the equatorial bulge of the rotating Earth. Some understanding of the processes involved may be obtained from the following simple picture.

It has been seen in chapter 6 that if two planets are mutually perturbing each other's orbit, their orbital planes regress. If now the Moon and a close satellite moving in a circular orbit in the Earth's equatorial plane are substituted for the planets (a spherical Earth taking the place of the Sun), the mutual perturbations of the two satellites will cause their orbital planes to regress, since the orbital plane of the Moon's orbit and the Earth's equatorial plane are not coplanar. If the satellite is attached to the rotating spherical Earth, and if there are indeed many such attached 'satellites' of the Earth spread round its equator to simulate the equatorial bulge, it is readily seen that the Moon's perturbing effect on the Earth will cause a regression of the Earth's equatorial plane. The Sun, taken as a satellite of the Earth, adds its effect to that of the Moon. The period of precession is about 26 000 years.

Although Clairaut and others had worked out in broad outline the theory of the Earth's figure by the eighteenth century, it is only within the present century (and especially since the advent of artificial Earth satellites) that most of our knowledge of our planet has been gathered. The figure of the Earth itself may be found by geodetic measurements, the constant of precession and the motions of the Moon and artificial satellites.

Geodetic triangulation enables the shape and dimensions of the Earth to be determined by measuring the separation of places whose latitudes and longitudes are known. The basic method is to measure very accurately the distance between two points defining a baseline. A third point is then observed by theodolite from each end of the baseline, the two angles and the length of the baseline enabling the position of the third point to be calculated. The theodolite is then used to extend the survey to a fourth point by using one of the two original points and the third point as the ends of a new baseline. In this way a net of triangulation points is obtained. Since errors in measuring are in general cumulative, more than one measured baseline is used, and at various points in the triangulation (known as Laplace stations) astronomical observations are made to obtain their longitudes and latitudes. In the United States, geodetic surveys made in this way have established a net which is supposed to give an internal accuracy of one part in 200 000. Similar surveys have been carried out in Europe and Africa.

The triangulation measurements must be referred to a suitable spheroid of reference. The International Ellipsoid of 1924 is one such convenient mathematical model for the Earth's surface. This is the Hayford Ellipsoid of 1909 with a polar radius of 6 356 912 metres and an equatorial radius of 6 378 388 meters, giving an ellipticity of exactly 1/297·0. Other models such as the Clarke ellipsoid of 1880 exist, and their differences are of the order of 200 metres. Satellites specially designed for geodetic work have been put into orbit in recent years. Observations of the satellite direction and range from a number of stations in Europe and the United States enable the North American Datum to be tied in to the European Datum.

The concept of the *geoid* may be mentioned here. It is the equipotential surface that coincides on average with mean sea level in the oceans and is everywhere perpendicular to a plumb-line, since gravity is always normal to its surface. The geoid is more nearly an ellipsoid than the Earth. The landmasses have attractions that make the figure of the geoid slightly irregular, though the surfaces of ellipsoid and geoid are never more than 100 meters from each other.

10.2.2 Clairaut's formula

We now consider briefly the type of reasoning that leads to the conclusion that the figure of the Earth approximates to that of an oblate spheroid. To do so we derive Clairaut's formula for gravity.

Let U be the potential of the Earth's gravitational field and let ω be the angular velocity of the Earth's rotation about its polar axis. If the surface is an equipotential surface and in equilibrium, then a quantity U', defined by the equation

$$U' = U + \tfrac{1}{2}\omega^2 r^2 \cos^2 \phi,$$

will be constant over the surface (r and ϕ are respectively the radius vector and the angle which the radius vector makes with the equatorial plane).

Now we have seen (section 6.5) that the gravitational potential may be written as

$$U = \frac{GM}{r} - \frac{G(C-A)}{2r^3} (3 \sin^2 \phi - 1) + \text{higher-order terms}$$

so that, neglecting those higher-order terms, we have

$$U' = \frac{GM}{r} - \frac{G(C-A)}{2r^3} (3 \sin^2 \phi - 1) + \tfrac{1}{2}\omega^2 r^2 \cos^2 \phi. \tag{10.2}$$

The quantity

$$\tfrac{1}{2}\omega^2 r^2 \cos^2 \phi$$

may be taken to be a disturbing potential due to the Earth's rotation.

If we now set

$$r = R(1-\eta) \tag{10.3}$$

where η is a small quantity and R is the Earth's equatorial radius, then on substitution into (10.2) we obtain

$$\frac{GM}{R(1-\eta)} - \frac{G(C-A)}{2R^3(1-\eta)^3} (3 \sin^2 \phi - 1) + \tfrac{1}{2}\omega^2 R^2 (1-\eta)^2 \cos^2 \phi = \text{constant}.$$

Putting $(1 - \sin^2 \phi)$ for $\cos^2 \phi$ and expanding by the binomial theorem we obtain, on neglecting powers of η higher than the first, the equation

$$\frac{GM}{R} (1+\eta) - \frac{G(C-A)}{2R^3} (1+3\eta)(3 \sin^2 \phi - 1)$$

$$+ \tfrac{1}{2}\omega^2 R^2 (1-2\eta)(1 - \sin^2 \phi) = \text{constant}.$$

If cross-products of small quantities such as η and $G(C-A)/R^3$ are neglected, then we must have

$$\eta = \left[\frac{3(C-A)}{2MR^2} + \frac{\omega^2 R^3}{2GM} \right] \sin^2 \phi. \tag{10.4}$$

Defining a quantity m as $(\omega^2 R^3/GM)$, it is seen that m is the ratio of the centrifugal force at the equator to gravity at the equator.

Then

$$\eta = \left[\frac{3(C-A)}{2MR^2} + \tfrac{1}{2}m \right] \sin^2 \phi. \tag{10.5}$$

Now the equation of an oblate spheroid is

$$r^2 = \frac{a^2(1-e^2)}{\sin^2\phi + (1-e^2)\cos^2\phi} \qquad (10.6)$$

where a and e are the semimajor axis and eccentricity of an elliptic cross-section containing the polar axis. The ellipticity (or flattening) ϵ is given by

$$\epsilon = \frac{a-b}{a} = 1 - \sqrt{1-e^2}$$

or

$$(1-\epsilon)^2 = 1 - e^2.$$

Hence equation (10.6) may be written as

$$r^2 = \frac{a^2(1-\epsilon)^2}{\sin^2\phi + (1-\epsilon)^2\cos^2\phi}.$$

Expanding by the binomial theorem and retaining terms of the order of ϵ^2, we obtain

$$r = a\left[1 - \left(\epsilon + \frac{3\epsilon^2}{2}\right)\sin^2\phi + \frac{3}{2}\epsilon^2\sin^4\phi\right]$$

or

$$r = a\left(1 - \epsilon\sin^2\phi - \frac{3}{8}\epsilon^2\sin^2 2\phi\right). \qquad (10.7)$$

Comparing equations (10.3), (10.5) and (10.7), it is seen that to the first order in ϵ the equilibrium surface is that of an oblate spheroid given by

$$r = R(1 - \epsilon\sin^2\phi) \qquad (10.8)$$

where

$$\epsilon = \frac{3(C-A)}{2MR^2} + \tfrac{1}{2}m. \qquad (10.9)$$

If ϵ can be measured, m being known, then the difference between the polar and equatorial moments of inertia can be found. The flattening ϵ is derived from gravity measurements and from the motions of artificial satellites.

If we now form $-\partial U'/\partial r$ we obtain the acceleration due to gravity. To the order of small quantities to which we are working,

$$g = \frac{GM}{r^2} - \frac{3G(C-A)}{2r^4}(3\sin^2\phi - 1) - \omega^2 r(1-\sin^2\phi). \qquad (10.10)$$

Using equation (10.8) and eliminating $(C-A)$, it is found after a little reduction that

$$g = \frac{GM}{R^2}\left(1 + \epsilon - \frac{3}{2}m\right) + \left(\frac{5}{2}R\omega^2 - \frac{\epsilon GM}{R^2}\right)\sin^2\phi$$

304

or

$$g = g_0 \left[1 + \left(\frac{5}{2} m - \epsilon \right) \sin^2 \phi \right] \qquad (10.11)$$

where g_0 (the value of gravity at the equator) is given by

$$g_0 = \frac{GM}{R^2} \left(1 + \epsilon - \frac{3}{2} m \right).$$

The relation (10.11) is Clairaut's equation, and it shows that to a first approximation the value of gravity increases proportionally as the square of the sine of the latitude. It should be noted that no assumption is made about the internal constitution of the Earth.

Observations of the precession of the equinoxes give information about the quantity $(C-A)/C$, called the *mechanical ellipticity* of the Earth. Using equation (10.9) it is then possible to obtain a value for C/MR^2.

Airy, Callandreau, and others have developed Clairaut's theory to the second order. When this is done the formula for g becomes

$$g = g_0 (1 + 0.005\,288\,4 \sin^2 \phi'' - 0.000\,005\,9 \sin^2 2\phi'')$$

where $g_0 = 978.049$ cm s^{-2} and ϕ'' is the geodetic or geographic latitude. It goes without saying that g_0 is the value of gravity at the equator, uncorrected for the effect of the equatorial rotation. If corrected, the value of g_0 becomes 981.43 cm s^{-2}.

The difference between geographic latitude ϕ'' and geocentric latitude ϕ' is given by the formula

$$\phi'' - \phi' = 695.66'' \sin 2\phi'' - 1.17'' \sin 4\phi''.$$

With respect to geographic latitude, equation (10.7) can be easily shown to become

$$r = R \left(1 - \epsilon \sin^2 \phi'' + \frac{5}{8} \epsilon^2 \sin^2 2\phi'' \right).$$

If the International Ellipsoid is used,

$$r = 6\,378\,388\,(1 - 0.003\,3670 \sin^2 \phi'' + 0.000\,0071 \sin^2 2\phi'') \text{ metres.}$$

Finally we can now introduce the modification in Kepler's third law for a satellite in a circular orbit about an oblate planet in the plane of the planet's equator. The gravitational acceleration on the satellite is obtained from equation (10.10), omitting the ω^2 term and setting ϕ equal to zero. Then

$$g = \frac{GM}{r^2} + \frac{3G(C-A)}{2r^4}$$

where r is the planetocentric distance of the satellite. Using equation (10.9)

$$g = \frac{GM}{r^2} \left[1 + \left(\frac{R}{r} \right)^2 \left(\epsilon - \frac{m}{2} \right) \right].$$

305

Then, instead of the simple relation for two point-masses m_1 and m_2, which is

$$T = 2\pi \sqrt{\frac{r^3}{G(m_1 + m_2)}}$$

we replace $G(m_1 + m_2)$ by

$$GM \left[1 + \left(\frac{R}{r}\right)^2 \left(\epsilon - \frac{m}{2}\right) \right]$$

giving

$$T = 2\pi \left\{ \frac{r^3}{GM} \left[1 - \left(\frac{R}{r}\right)^2 \left(\epsilon - \frac{m}{2}\right) \right] \right\}^{1/2}$$

neglecting the mass of the satellite and remembering that

$$\left(\frac{R}{r}\right)^2 \left(\epsilon - \frac{m}{2}\right) \ll 1.$$

10.2.3 The Earth's interior

Information about the interior of the Earth is obtained indirectly from the motions of satellites, the study of earthquake waves and the physics and chemistry of matter under high temperatures and pressures. The measured value (~ 0.98) of the ratio ϵ/m indicates that there is an increase of density towards the Earth's centre. The refraction, reflection and diffraction of earthquake waves show the presence of a core with a diameter of more than 6400 km. Its density is from ten to twelve times that of water. Above it is a shell (the mantle) with a mean density about four times that of water, possibly made up of heavy basic rocks, while above this shell is a thin lighter granite layer less than 80 km thick.

There seems no doubt that the core is fluid, though according to Bullen the presence of a smaller solid inner core is possible. Where the central temperature and the constitution of the Earth's interior are concerned, we are on more speculative ground. Many theories have been put forward, including the older theory that the core is almost entirely molten iron. Ramsey has shown that this view contains serious difficulties.

10.2.4 The Earth's magnetic field

To a first approximation the Earth's magnetic field simulates that of a simple dipole embedded within and near the centre of the Earth at an angle of about $11.4°$ to the Earth's axis of rotation. In fact, the line connecting the two geomagnetic poles misses the centre by some hundreds of kilometres. The vertical field strength at the geomagnetic poles is 0·63 gauss; at the equator it is 0·31 gauss.

More accurately, it is found that the field departs from a simple dipole field at various places due to the presence of magnetic materials in the crust. In addition, fluctuations of short period occur, caused by solar activities. At a point on the Earth's surface the magnetic field changes slowly, such a change being called the *secular variation*. Much information about the extent and strength of the field to distances of many Earth radii from the surface has been gathered in recent years by using artificial satellites.

The source of the Earth's magnetic field almost certainly lies in the Earth's core, possibly in a self-acting dynamo action set up by motions in the electrically conducting fluid core. Thermal convection provides a satisfactory mechanism for such motions.

10.2.5 The Earth's atmosphere

The International Union of Geodesy and Geophysics at its 1951 Brussels meeting recommended the nomenclature summarized in figure 10.1 for classifying the structure of the Earth's atmosphere.

The troposphere, stratosphere, mesosphere, thermosphere and exosphere are classified on a thermal basis; the layers dividing them are named by substituting the suffix 'pause' for the suffix 'sphere'. If the classification is by chemical composition, the main regions are the homosphere and heterosphere. The structure of the atmosphere can in addition be classified from a number of other viewpoints such as its degree of ionization.

In the last few years, work with rockets, satellites and other instruments of atmospheric research has enormously increased our knowledge of the

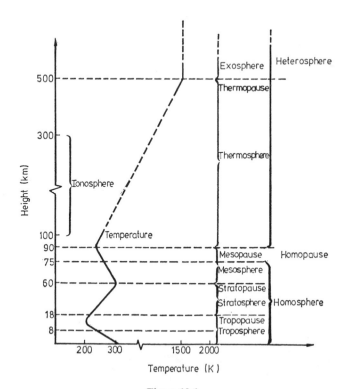

Figure 10.1

constitution and extent of the atmosphere, which is now very well known up to an altitude of about 30 km; above this region there exists a shell of low density reaching as far as 700 km, finally merging into the interplanetary medium.

Up to a height of 70 km, the composition is unchanging. By volume the principal constituents are molecular nitrogen (78 per cent) and molecular oxygen (21%), with argon, water vapour and carbon dioxide taking up most of the remaining 1%. In addition, other permanent gases such as neon are present in very small quantities. Ozone (O_3) appears in a layer some 25 km up as a result of the dissociation of molecular oxygen by ultraviolet radiation, the atomic oxygen then combining with oxygen molecules.

At the homopause (see figure 10.1) the composition begins to change, and within the heterosphere a number of processes such as diffusion, mixing and photodissociation are at work, changing the make-up of this tenuous region.

The ionosphere is a region of ions and electrons created by the Sun's short-wave radiation and by cosmic rays. This region is usually divided into several layers called the D, E, F_1 and F_2 layers in order of ascending height. The ionosphere is extremely variable, the number of electrified particles depending on sunspots, season, latitude and the change from day to night.

In attempts to obtain insight into the relations between pressure, density and temperature throughout the atmosphere, model atmospheres have been constructed mathematically and their predictions compared with data derived from vertical rocket flights and observations of atmospheric drag on artificial satellites. Such models use the equation of hydrostatic equilibrium

$$dp = -g\rho \, dh$$

where g is the acceleration due to gravity at a given height h, ρ is the density at that height and p is the pressure. The equation gives the slight decrease in pressure when the height is increased slightly from h to $h+dh$.

The ideal-gas law is also used,

$$p = \frac{\rho \mathcal{R} T}{\mu}$$

where $\mathcal{R}$ is the universal gas constant, μ is the mean molecular weight and T is the temperature. As more and more data have accumulated, revisions of such model atmospheres as the Air Research and Development Command (ARDC) Model Atmosphere of 1956 have been made.

From the changes in satellite orbits due to atmospheric drag, the following diagram (figure 10.2) was constructed (King-Hele 1974).

These figures are not invariant with time, but give an indication of the order of magnitude of the density at various heights. It has also been found that seasonal, diurnal and latitude variations in density take place. Solar

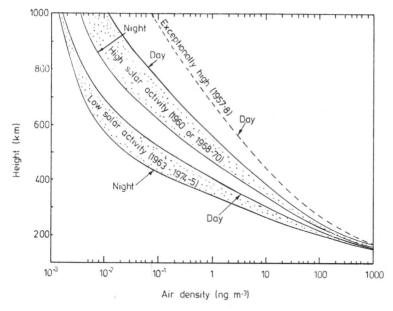

Figure 10.2

activity is a major cause of atmospheric density variations at a given height and latitude.

From an astrodynamical viewpoint, any Earth satellite in an orbit below 160 km suffers enough atmospheric drag to destroy it within a few revolutions, while a satellite in an orbit higher than 500 km is acted upon by too small a drag to bring it back to Earth within a period measured in years.

10.2.6 Solar–terrestrial relationships

The correlation of such terrestrial events as auroral displays and magnetic storms with solar activity reveals an intimate relationship between the output of electromagnetic and corpuscular radiation from the Sun and changes in the Earth's atmospheric density, magnetic field and atmospheric electrical activity. The Van Allen radiation belts surrounding the Earth above the atmosphere owe their existence to solar activity and to the Earth's magnetic field.

In addition to the fluctuations in air density caused by solar radiation, streams of charged particles (especially at times of solar flare outbursts) impinge on the atmosphere causing violent magnetic storms, changes in air density and auroral displays. Such streams also contribute to the numbers of charged particles in the radiation belts. It should be noted that in this context the term 'radiation' really refers to particles. The particles (protons and electrons) are trapped in the Earth's magnetic field and spiral along the lines of magnetic force. The pitch of the spiral becomes smaller as the particle approaches the Earth until it reverses its direction and roughly retraces

309

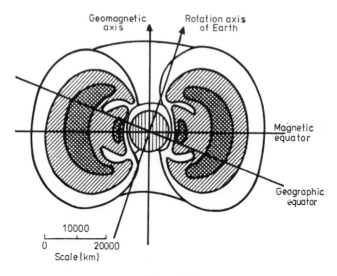

Figure 10.3

its path. There is also a drift in longitude so that an injection of charged particles at a point above the atmosphere quickly results in a spread about the Earth. The radiation zones and the process are sketched roughly in figures 10.3 and 10.4.

There are two belts or regions of maximum concentration of such particles: one about 4000 km above the Earth's surface, the other about 16000 km up. The regions of maximum intensity are shown cross-hatched in figure 10.3. The orbits of the particles are quasistable in that irregularities in the Earth's field and collisions with air molecules ultimately reduce the numbers

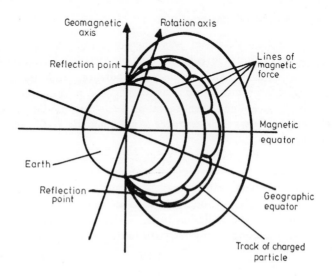

Figure 10.4

in the belts; but solar outbursts are continually replenishing the supply. The processes involved are complicated and are not well understood even now. A further ring current of electrons at a distance of some 56000 km circles the Earth.

Finally, the solar wind (protons and electrons ejected by the Sun in a steady flow) pushes in the Earth's magnetic field on the sunward side of the planet and stretches it out on the opposite side. The term *magnetosphere* has been given to the resulting tear-drop shaped region about the Earth in which the Earth's magnetic field is dominant.

10.3 Forces Acting on an Artificial Earth Satellite

We are now in a position to list and compare the forces on an artificial satellite in orbit about the Earth. In general forces due to the following causes will affect its orbit:

(i) the Earth's gravitational field,
(ii) the gravitational attractions of Sun, Moon and planets,
(iii) the Earth's atmosphere,
(iv) the Earth's magnetic field,
(v) solar radiation, and
(vi) charged and uncharged particles.

We examine these in turn.

(i) The Earth's gravitational field is the major controller of the orbit of an Earth satellite. It has been seen that the potential is of the form

$$U = \frac{GM}{r} - \frac{G(C-A)}{2r^3} (3 \sin^2 \phi - 1) + \text{higher-order terms}$$

so that to a first approximation the orbit of the satellite is given by the two-body formulae, both bodies being point-masses. The second- and higher-order terms perturb this orbit.

(ii) For a satellite in an orbit of less than 1600 km in altitude, the effects of Sun and Moon on the orbit are very small, though not negligible if information about the higher harmonics in the Earth's potential is sought from observations of satellites. Kozai (1959a), among others, has set up the expression for the disturbing function R due to the attractions of Sun and Moon and obtained by the method of the variation of parameters the changes in the Keplerian elements of the satellite orbit. There is no secular change in the semimajor axis. The planets have no appreciable effect on an Earth satellite.

(iii) The Earth's atmosphere gives rise to a drag on the satellite. Such a drag force is due to the continual collision of air molecules, atoms and ions with the satellite. The magnitude of the force depends upon a number of factors that vary with time, such as altitude, longitude and, unless the satellite is spherical, its attitude. Unless the satellite is below an altitude of 150 km, the drag force can be treated as a perturbing force. Fortunately, the perturbations due to drag are different in their effects from those due to the harmonics in the Earth's gravitational field.

(iv) If the satellite has metal in its construction the Earth's magnetic field induces eddy currents in the satellite. In addition, a slight retardation acts on the satellite. The changes in the orbit due to this are very small.

(v) Solar radiation can produce marked effects on a satellite orbit if the mean density of the satellite is small, as in the case of balloon satellites. For example, an oscillation in perigee height of about 500 km was produced in the orbit of Echo I, the period of the cycle (about 10 months) being the synodic period of the perigee point; that is, the time it took to make one rotation of the Earth relative to the Sun. These changes however, even for balloon satellites, can be treated by perturbation techniques.

(vi) Uncharged particles (such as neutral atoms or meteoritic dust) encountered by a satellite must have a braking effect upon it similar to that of the atmosphere; but the magnitude of this effect is negligible.

The drag due to charged particles either of direct solar origin or contained within the atmosphere is difficult to calculate accurately, since the electrostatic potential on the satellite surface and also the characteristics of the charged material surrounding the satellite must be known. Order-of-magnitude calculations, however, make it clear that any drag due to this cause can be safely neglected.

It is therefore seen that for almost all Earth satellites the major perturbations of the two-body Keplerian orbit are caused by the Earth's oblateness and atmospheric drag. In the rest of this chapter, this main artificial satellite problem will be treated; included is a sketch of the use of Hamilton–Jacobi theory as it has been applied to the problem by Sterne, Garfinkel and others.

10.4 The Orbit of a Satellite About an Oblate Planet

In this section we study the satellite orbit under the gravitational influence of the Earth, neglecting the effect of atmospheric drag. Many authors in recent years have treated this problem, among them Kozai (1959b), Merson (1960), Brouwer (1959), Sterne (1958), Garfinkel (1958, 1959) and King-Hele (1958).

In the treatment below we follow Kozai (1959b). In figure 10.5 the position S of the satellite in its orbit at time t has coordinates r, δ, λ as shown, where the axes (nonrotating) have the Earth's centre of mass as origin; they are given by OX (in the direction of the First Point of Aries), OY (90° along the equator from OX in the direction of increasing right ascension) and OZ (along the Earth's axis of rotation).

Then, letting the projection of S upon the celestial sphere be S′ and drawing the arc of the great circle ZS′Q through S′, we have

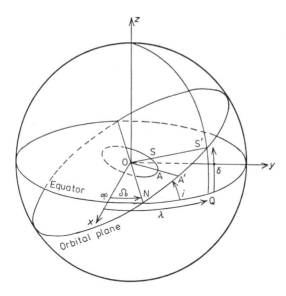

Figure 10.5

$$r = OS = \text{radius vector of S}$$

$$\delta = Q\hat{O}S' = \text{declination of S}$$

and

$$\lambda = X\hat{O}Q = \text{right ascension of S}.$$

The osculating orbit is defined by the six elements a, e, i, Ω, ω and M where a is the semimajor axis, e is the eccentricity, i is the inclination of the orbital plane to the equator, Ω is the right ascension of the ascending node, ω is the argument of perigee (the arc NA′) and M is the mean anomaly. The radius r and the declination δ are then related to the elements and to the true anomaly f by the relations

$$\left. \begin{array}{c} r = \dfrac{a(1-e^2)}{1+e \cos f} \\[2mm] \sin \delta = \sin i \sin (f+\omega). \end{array} \right\} \qquad (10.12)$$

Now the equation of motion of the satellite is

$$\ddot{\mathbf{r}} = \nabla U$$

where U is the Earth's potential. For a body possessing axial symmetry, its potential (see section 6.5) at a point external to it may be written as

$$U = \frac{Gm}{r} \left[1 - \sum_{n=2}^{\infty} J_n \left(\frac{R}{r} \right)^n P_n (\sin \phi) \right]$$

where r is the distance of the point from the body's centre of mass, the J_n are constants, R is the body's equatorial radius, m is the mass of the body,

313

ϕ is the angle between the body's equator and the radius to the point and P_n (sin ϕ) is the Legendre polynomial of order n in sin ϕ. Then, since $\delta \equiv \phi$, and writing μ for Gm, we have

$$U = \frac{\mu}{r}\left[1 - \sum_{n=2}^{\infty} J_n \left(\frac{R}{r}\right)^n P_n (\sin \delta)\right].$$

In using this expression for the Earth's gravitational potential we are assuming that no effects due to an ellipticity of the equator are present, though we are allowing for effects due to an asymmetry between northern and southern hemispheres. The disturbing potential F is then given by

$$F = U - U_0 = U - \frac{\mu}{r} = -\frac{\mu}{r}\sum_{n=2}^{\infty} J_n \left(\frac{R}{r}\right)^n P_n (\sin \delta).$$

Now for the Earth, J_2 is of the order of 10^{-3}, while J_3, J_4 ... are of the order of 10^{-6} or less. Since J_4, J_5 ... do not contribute anything fundamentally new to the effects due to J_2 and J_3, we will confine our study to the second and third harmonics.

Then

$$F = -\frac{\mu}{r}\left[J_2 \left(\frac{R}{r}\right)^2 \left(\frac{3}{2}\sin^2 \delta - \frac{1}{2}\right) + J_3 \left(\frac{R}{r}\right)^3 \left(\frac{5}{2}\sin^2 \delta - \frac{3}{2}\right) \sin \delta\right]$$

which is deduced by using the relations

$$P_2 (\sin \delta) = \tfrac{1}{2}(3\sin^2 \delta - 1)$$

and

$$P_3 (\sin \delta) = \tfrac{1}{2}(5\sin^3 \delta - 3\sin \delta).$$

Applying the second of equations (10.12), F becomes

$$F = \mu\left\{\frac{3}{2}\frac{J_2 R^2}{a^3}\left(\frac{a}{r}\right)^3 \left[\frac{1}{3} - \frac{1}{2}\sin^2 i + \frac{1}{2}\sin^2 i \cos 2(f+\omega)\right]\right.$$

$$\left. - \frac{J_3 R^3}{a^4}\left(\frac{a}{r}\right)^4 \left[\left(\frac{15}{8}\sin^2 i - \frac{3}{2}\right)\sin (f+\omega) - \frac{5}{8}\sin^2 i \sin 3(f+\omega)\right]\sin i\right\}.$$

The true anomaly f is easily transformed to the mean anomaly M, which is a linear function of time in unperturbed motion, by the relation

$$\frac{df}{dM} = \frac{a^2}{r^2}(1-e^2)^{1/2}.$$

The quantities r/a and f in the disturbing function F are then functions of e and M only and are periodic with respect to M. Terms in F depending neither on M nor on ω are secular; terms depending on ω but not on M are long-period, while those depending on M are short-period.

Now the long-period perturbations will arise from terms of the second order in F, and so secular terms and long-period terms must be retained up to the second order. For short-period terms, on the other hand, only terms of the first order need be considered.

314

In order to sort out such terms we remember that short-period perturbations result from the variation of M around the orbit, while the long-period perturbations arise from the secular variation of ω. With this in mind we take the mean value of the disturbing function F with respect to M to obtain the long-period perturbations. To obtain the secular perturbations we likewise average with respect to M those parts of the disturbing function which are dependent neither on M nor ω.

To carry out these operations the quantities are integrated between zero and 2π so that, if Q is any term treated in this way, we obtain

$$\bar{Q} = \frac{1}{2\pi} \int_0^{2\pi} Q \, dM.$$

The required relations, given by Tisserand (1889) are

$$\overline{\left(\frac{a}{r}\right)^3} = \frac{1}{2\pi} \int_0^{2\pi} \left(\frac{a}{r}\right)^3 dM = (1-e^2)^{-3/2}$$

$$\overline{\left(\frac{a}{r}\right)^3 \sin 2f} = \overline{\left(\frac{a}{r}\right)^3 \cos 2f} = 0$$

$$\overline{\left(\frac{a}{r}\right)^4 \cos f} = e(1-e^2)^{-5/2}$$

$$\overline{\left(\frac{a}{r}\right)^4 \sin f} = \overline{\left(\frac{a}{r}\right)^4 \cos 3f} = \overline{\left(\frac{a}{r}\right)^4 \sin 3f} = 0.$$

The relevant parts of the disturbing function F are then

$$\left. \begin{array}{l}
F_1 = \dfrac{3}{2} \dfrac{\mu J_2 R^2}{a^3} \left(\dfrac{1}{3} - \dfrac{1}{2}\sin^2 i\right)(1-e^2)^{-3/2} \\[4mm]
F_2 = 0 \\[4mm]
F_3 = \dfrac{3}{2} \dfrac{\mu J_3 R^3}{a^4} \sin i \left(1 - \dfrac{5}{4}\sin^2 i\right) e(1-e^2)^{-5/2} \sin \omega \\[4mm]
F_4 = \dfrac{3}{2} \dfrac{\mu J_2 R^2}{a^3} \left(\dfrac{a}{r}\right)^3 \left\{\left(\dfrac{1}{3} - \dfrac{1}{2}\sin^2 i\right)\left[1 - \left(\dfrac{r}{a}\right)^3(1-e^2)^{-3/2}\right]\right. \\[4mm]
\hspace{3.5cm} \left. + \dfrac{1}{2}\sin^2 i \cos 2(f+\omega)\right\}
\end{array} \right\} \quad (10.13)$$

where F_1, F_2, F_3 and F_4 are the first-order secular, second-order secular, long-period and short-period parts respectively of the disturbing function.

10.4.1 The short-period perturbations of the first order

The differential equations of the elements used are

$$
\left.
\begin{aligned}
\frac{da}{dt} &= \frac{2}{na}\frac{\partial F}{\partial M} \\[2mm]
\frac{de}{dt} &= \frac{(1-e^2)}{na^2e}\frac{\partial F}{\partial M} - \frac{\sqrt{1-e^2}}{na^2e}\frac{\partial F}{\partial \omega} \\[2mm]
\frac{di}{dt} &= \frac{\cos i}{na^2\sqrt{1-e^2}\sin i}\frac{\partial F}{\partial \omega} - \frac{1}{na^2\sqrt{1-e^2}\sin i}\frac{\partial F}{\partial \Omega} \\[2mm]
\frac{d\Omega}{dt} &= \frac{1}{na^2\sqrt{1-e^2}\sin i}\frac{\partial F}{\partial i} \\[2mm]
\frac{d\omega}{dt} &= -\frac{\cos i}{na^2\sqrt{1-e^2}\sin i}\frac{\partial F}{\partial i} + \frac{\sqrt{1-e^2}}{na^2e}\frac{\partial F}{\partial e} \\[2mm]
\frac{dM}{dt} &= n - \frac{1-e^2}{na^2e}\frac{\partial F}{\partial e} - \frac{2}{na}\frac{\partial F}{\partial a}
\end{aligned}
\right\}
\tag{10.14}
$$

where n is given by the relation

$$
n^2 a^3 = \mu.
\tag{10.15}
$$

The set of equations (10.14) are a version of the Lagrange planetary equations (6.29), where the mean anomaly M has been substituted in place of χ using the relation

$$
M = n(t - \tau) = nt + \chi.
$$

To derive the first-order short-period perturbations, the disturbing function is replaced in (10.14) by F_4, and we note that to this order the quantities a, n, e, i and ω on the right-hand sides of the resulting equations may be taken to be constants, except that where n appears in the last equation in the first term without a factor it must be regarded as variable, even in a first-order treatment. The variable n is, however (by means of equation (10.15)), a known function of time once the expression for the semimajor axis has been obtained.

The independent variable is now transformed from t to f by using the relation

$$
dt = \frac{dt}{dM}\,dM = \frac{1}{n}\left(\frac{r}{a}\right)^2 \frac{1}{\sqrt{1-e^2}}\,df.
$$

If the inclination is taken as an example, we have

$$
\Delta i_p = \frac{\cos i}{n^2 a^2 (1-e^2)\sin i} \int \left(\frac{r}{a}\right)^2 \frac{\partial F_4}{\partial \omega}\,df
$$

where the suffix p denotes the short-period perturbation.

Substituting for F_4, it is found that the integrand may be expressed as a

316

finite trigonometric series capable of being integrated. The resulting expressions for the six elements are:

$$\Delta a_p = \frac{J_2 R^2}{a}\left\{\left(1-\frac{3}{2}\sin^2 i\right)\left[\left(\frac{a}{r}\right)^3-(1-e^2)^{-3/2}\right]\right.$$

$$\left.+\frac{3}{2}\left(\frac{a}{r}\right)^3\sin^2 i\cos 2(f+\omega)\right\}$$

$$\Delta e_p = \frac{3}{2}\frac{J_2 R^2\,(1-e^2)}{a^2}\frac{1}{e}\left\{\frac{1}{3}\left(1-\frac{3}{2}\sin^2 i\right)\left[\left(\frac{a}{r}\right)^3-(1-e^2)^{-3/2}\right]\right.$$

$$+\frac{1}{2}\left(\frac{a}{r}\right)^3\sin^2 i\cos 2(f+\omega)\right\}$$

$$-\frac{3}{4}\frac{J_2 R^2}{ap}\frac{\sin^2 i}{e}\left[\cos 2(f+\omega)+e\cos (f+2\omega)+\frac{1}{3}e\cos (3f+2\omega)\right]$$

$$\Delta i_p = \frac{3}{8}\frac{J_2 R^2}{p^2}\sin 2i\left[\cos 2(f+\omega)+e\cos (f+2\omega)+\frac{e}{3}\cos (3f+2\omega)\right]$$

$$\Delta \Omega_p = -\frac{3}{2}\frac{J_2 R^2}{p^2}\cos i\left[f-M+e\sin f-\frac{1}{2}\sin 2(f+\omega)\right.$$

$$\left.-\frac{e}{2}\sin (f+2\omega)-\frac{e}{6}\sin (3f+2\omega)\right]$$

$$\Delta \omega_p = \frac{3}{2}\frac{J_2 R^2}{p^2}\left\{\left(2-\frac{5}{2}\sin^2 i\right)(f-M+e\sin f)\right.$$

$$+\left(1-\frac{3}{2}\sin^2 i\right)\left[\frac{1}{e}\left(1-\frac{1}{4}e^2\right)\sin f+\frac{1}{2}\sin 2f+\frac{e}{12}\sin 3f\right]$$

$$-\frac{1}{e}\left[\frac{1}{4}\sin^2 i+\left(\frac{1}{2}-\frac{15}{16}\sin^2 i\right)e^2\right]\sin (f+2\omega)$$

$$+\frac{e}{16}\sin^2 i\sin (f-2\omega)-\frac{1}{2}\left(1-\frac{5}{2}\sin^2 i\right)\sin 2(f+\omega)$$

$$+\frac{1}{e}\left[\frac{7}{12}\sin^2 i-\frac{1}{6}\left(1-\frac{19}{8}\sin^2 i\right)e^2\right]\sin (3f+2\omega)$$

$$\left.+\frac{3}{8}\sin^2 i\sin (4f+2\omega)+\frac{e}{16}\sin^2 i\sin (5f+2\omega)\right\}$$

$$e\Delta M_p = \frac{3}{2}\frac{J_2 R^2}{p^2}\sqrt{1-e^2}\left\{-\left(1-\frac{3}{2}\sin^2 i\right)\right.$$

$$\times\left[\left(1-\frac{e^2}{4}\right)\sin f+\frac{e}{2}\sin 2f+\frac{e^2}{12}\sin 3f\right]$$

$$+\sin^2 i\left[\frac{1}{4}\left(1+\frac{5}{4}e^2\right)\sin(f+2\omega)-\frac{e^2}{16}\sin(f-2\omega)\right.$$

$$-\frac{7}{12}\left(1-\frac{e^2}{28}\right)\sin(3f+2\omega)-\frac{3}{8}e\sin(4f+2\omega)$$

$$\left.\left.-\frac{e^2}{16}\sin(5f+2\omega)\right]\right\}$$

where

$$p = a(1-e^2).$$

Now the mean value of $\cos jf$ $(j=1, 2 \ldots)$ with respect to M does not vanish. In fact,

$$\overline{\cos jf} = \left(\frac{-e}{1+\sqrt{1-e^2}}\right)^j(1+j\sqrt{1-e^2}). \tag{10.16}$$

The mean values of the above perturbations are not zero, with the exception of those of a. Their mean values may in fact (with respect to M) be shown to be

$$\overline{\Delta a_p}=0$$

$$\overline{\Delta e_p}=\frac{1}{4}\frac{J_2 R^2}{p^2}\sin^2 i\left(\frac{1-e^2}{e}\right)\overline{\cos 2f}\cos 2\omega$$

$$\overline{\Delta i_p}=-\frac{1}{8}\frac{J_2 R^2}{p^2}\sin 2i\,\overline{\cos 2f}\cos 2\omega$$

$$\overline{\Delta\Omega_p}=-\frac{1}{4}\frac{J_2 R^2}{p^2}\cos i\,\overline{\cos 2f}\sin 2\omega$$

$$\overline{\Delta\omega_p}=\frac{3}{2}\frac{J_2 R^2}{p^2}\left[\sin^2 i\left(\frac{1}{8}+\frac{(1-e^2)}{6e^2}\overline{\cos 2f}\right)+\frac{1}{6}\cos^2 i\,\overline{\cos 2f}\right]\sin 2\omega$$

$$\overline{\Delta M_p}=-\frac{3}{2}\frac{J_2 R^2}{p^2}\sqrt{1-e^2}\sin^2 i\left(\frac{1}{8}+\frac{1+e^2/2}{6e^2}\overline{\cos 2f}\right)\sin 2\omega$$

where $\overline{\cos 2f}$ is given by (10.16) with $j=2$. The short-period perturbations whose mean values with respect to the mean anomaly are zero are therefore $(\Delta e_p-\overline{\Delta e_p})$, $(\Delta\omega_p-\overline{\Delta\omega_p})$ and so on.

318

10.4.2 The secular perturbations of the first order

These are obtained by putting $F = F_1$ in (10.14) and are

$$
\left.
\begin{aligned}
\bar{a} &= a_0 \\[4pt]
\bar{e} &= e_0 \\[4pt]
\bar{\imath} &= i_0 \\[4pt]
\bar{\omega} &= \omega_0 + \frac{3}{2}\frac{J_2 R^2}{p^2}\,\bar{n}\!\left(2 - \frac{5}{2}\sin^2 i\right) t \\[4pt]
\bar{\Omega} &= \Omega_0 - \frac{3}{2}\frac{J_2 R^2}{p^2}\,\bar{n}(\cos i)\,t \\[4pt]
\bar{M} &= M_0 + \bar{n} t \\[4pt]
\bar{n} &= n_0\left[1 + \frac{3}{2}\frac{J_2 R^2}{p^2}\left(1 - \frac{3}{2}\sin^2 i\right)(1 - e^2)^{1/2}\right]
\end{aligned}
\right\} \tag{10.17}
$$

where the zero-suffixed quantities are the mean values at the epoch, that is, the initial values from which periodic perturbations have been removed. In particular n_0 is the unperturbed mean motion, related to the unperturbed semimajor axis by

$$
n_0{}^2 a_0{}^3 = \mu.
$$

It is in fact more convenient to adopt as a mean value of the semimajor axis not a_0, but

$$
\bar{a} = a_0\left[1 - \frac{3}{2}\frac{J_2 R^2}{p^2}\left(1 - \frac{3}{2}\sin^2 i\right)(1 - e^2)^{1/2}\right]
$$

with

$$
\bar{n}^2 \bar{a}^3 = \mu\left[1 - \frac{3}{2}\frac{J_2 R^2}{p^2}\left(1 - \frac{3}{2}\sin^2 i\right)(1 - e^2)^{1/2}\right].
$$

Summing up at this stage, it is seen that while all the elements are subject to periodic perturbations Ω, ω and M are also changed secularly. In particular, to the order to which we are working, the orbital plane precesses unless $i = 90°$ (the condition for a polar orbit) when $\Delta\Omega_p = 0$.

The perigee advances in the orbital plane if $i < 63° \, 26'$ or regresses within the orbital plane if $i > 63° \, 26'$. This critical inclination is got by setting the term $[1 - (5/4)\sin^2 i]$ equal to zero. If the inclination is moderate however, a close Earth satellite's orbit will exhibit secular movements in Ω and ω of the order of 4°/day.

It is also seen that the perturbation in M will cause the actual period to vary. This may be allowed for by averaging over many revolutions to get rid of the short-period perturbations and by adopting a perturbed value of n (namely $\bar{n}$) given above.

319

10.4.3 Long-period perturbations from the third harmonic

The third harmonic J_3 contributes to F_3 in equation (10.13) and will give rise to various periodic perturbations. Now J_3 is of the order of $10^{-3} J_2$ for the Earth, so that the amplitudes of the short-period perturbations will be very small. On the other hand, amplitudes of the long-period perturbations, which depend on the secular variation of ω, may be much larger. To illustrate such long-period perturbations we consider the variation of the inclination under the effect of the third harmonic. Collecting the relevant equations, we have

$$\frac{di}{dt} = \frac{\cos i}{na^2 (1-e^2)^{1/2} \sin i} \left(\frac{\partial F_3}{\partial \omega} \right) \tag{10.18}$$

$$F_3 = \frac{3}{2} \frac{\mu J_3 R^3}{a^4} \sin i \left(1 - \frac{5}{4} \sin^2 i \right) e(1-e^2)^{-5/2} \sin \omega \tag{10.19}$$

$$\frac{d\omega}{dt} = \frac{3 J_2 R^2 n}{p^2} \left(1 - \frac{5}{4} \sin^2 i \right) \tag{10.20}$$

where equation (10.20), since we are interested in the secular part of the variation in ω, is obtained from the fourth equation in (10.17). Then substituting for F_3 from (10.19) in (10.18), differentiating with respect to ω and using the relation

$$n^2 a^3 = \mu$$

we have

$$\frac{di}{dt} = \frac{3}{2} \frac{n J_3 R^3}{a^3} \frac{e}{(1-e^2)^3} \cos i \left(1 - \frac{5}{4} \sin^2 i \right) \cos \omega.$$

Now

$$\frac{di}{dt} = \frac{di}{d\omega} \frac{d\omega}{dt}$$

or

$$\frac{di}{d\omega} = \frac{di}{dt} \Big/ \frac{d\omega}{dt} = \frac{1}{2} \frac{J_3}{J_2} \left(\frac{R}{a} \right) \frac{e}{(1-e^2)} \cos i \cos \omega.$$

Integrating, we obtain the long-period perturbation in i, denoted $\Delta_3 i$, due to the third harmonic:

$$\Delta_3 i = \frac{1}{2} \frac{J_3}{J_2} \left(\frac{R}{a} \right) \frac{e}{1-e^2} \cos i \sin \omega.$$

A long-period perturbation in the eccentricity due to J_3, of the form

$$\Delta_3 e = -\frac{1}{2} \frac{J_3}{J_2} \left(\frac{R}{a} \right) \sin i \sin \omega$$

has been used to measure the size of J_3 (Kozai 1961) since it does not give rise to secular terms capable of being utilized for this purpose.

320

10.4.4 Secular perturbations of the second-order and long-period perturbations

The derivation of these perturbations in the elements is based essentially on a process akin to the one sketched in section 6.7.2 for the solution of the Lagrange planetary equations where the functions of the elements on the right-hand sides of the equations are expanded in a Taylor series.

Thus, if σ_i is any one of the six orbital elements, so that its variational equation is $(\mathrm{d}\sigma_i/\mathrm{d}t) = \phi_i$, we may write

$$\frac{\mathrm{d}\sigma_i}{\mathrm{d}t} = (\phi_i)_0 + \sum_j \left(\frac{\partial \phi_i}{\partial \sigma_j}\right)_0 \Delta\sigma_j + \ldots$$

where the brackets and zero suffix denote that after differentiation the mean values of the elements at the epoch (taken to be constant) are used (Kozai 1959b).

On examining the resulting expressions it is found that a factor $(4-5\sin^2 i)$ enters the denominator of some of the perturbations, showing that the theory breaks down near the critical inclination of $63° \, 26'$. Various authors have since shown that other methods of development can be adopted to provide theories valid around the critical inclination.

10.5 The Use of Hamilton–Jacobi Theory in the Artificial Satellite Problem

The application of Hamilton–Jacobi theory to the many-body problem has been outlined in section 6.9. It was seen that in the first approximation a Hamiltonian function H_0 was taken with a potential of μ/r, so that the unperturbed solution, arising from a knowledge of the solution S of the Hamilton–Jacobi equation, gave an ordinary Keplerian ellipse. The disturbing Hamiltonian H_1 then entered the new canonic equations of the changes with time of the former canonic constants obtained in the first approximation.

The same unperturbed Hamiltonian H_0 may be used in the solution of the artificial satellite problem, where the disturbing Hamiltonian H_1 would arise from the second, third etc. harmonics omitted from the unperturbed solution. It has however been shown by Sterne (1958) and Garfinkel (1958, 1959) that it is possible to use an unperturbed Hamiltonian H_0 that contains the major part of the oblateness effects and leads to a Hamilton–Jacobi equation that is separable (i.e. capable of being solved).

Sterne and Garfinkel use different H_0 functions; but in both cases the perturbing Hamiltonian H_1, consisting of the remainder of the second harmonic and higher harmonics, contains no first-order secular perturbations. For lack of space we do no more than sketch Sterne's treatment.

Sterne's Hamiltonian function for which an exact canonical solution may be obtained is

$$H = \frac{1}{2}\left(p_r^2 + \frac{p_\lambda^2}{r^2 \cos^2 \delta} + \frac{p_\delta^2}{r^2}\right) + U_1(r) + \frac{1}{r^2} U_2(\delta) \qquad (10.21)$$

321

where r, δ and λ are defined as in figure 10.4; p_r, p_λ and p_δ are the conjugate momenta to r, λ and δ; and U_1 and U_2 are any functions of the radius vector and of the declination respectively.

It may be easily verified that the Hamilton–Jacobi equation using (10.21) is separable, giving as a solution

$$S = \int_{r_0}^r \frac{L}{r}\, dr + \int_0^\delta M\, d\delta + \alpha_3 \lambda - \alpha_1 t \tag{10.22}$$

where

$$L = [2r^2\alpha_1 - 2r^2 U_1(r) - \alpha_2{}^2]^{1/2}$$

$$M = [\alpha_2{}^2 - \alpha_3{}^2 \sec^2 \delta - 2U_2(\delta)]^{1/2}$$

and r_0 is the perigee distance.

The canonic constants α_1, α_2, α_3 have the respective meanings (all per unit mass of particle) of total energy, a quantity that would be the orbital angular momentum if U_2 were zero, and the axial component of orbital angular momentum.

The canonic solution is then (see section 6.9)

$$
\left.
\begin{aligned}
t + \beta_1 &= \frac{\partial S}{\partial \alpha_1} = \int_{r_0}^r \frac{r}{L}\, dr \\[2mm]
\beta_2 &= \frac{\partial S}{\partial \alpha_2} = \int_0^\delta \frac{\alpha_2}{M}\, d\delta - \int_{r_0}^r \frac{\alpha_2}{rL}\, dr \\[2mm]
\beta_3 &= \lambda + \frac{\partial S}{\partial \alpha_3} = \lambda - \int_0^\delta \frac{\alpha_3}{M} \sec^2 \delta\, d\delta \\[2mm]
p_r &= \dot{r} = \frac{L}{r} \\[2mm]
p_\delta &= r^2 \dot{\delta} = M \\[2mm]
p_\lambda &= r^2\, \dot{\lambda} \cos^2 \delta = \alpha_3
\end{aligned}
\right\}
\tag{10.23}
$$

where r_0 is the perigee distance and where the canonic constants β_1, β_2 and β_3 are respectively the negative of the time of some particular perigee passage, the argument of the declination of that perigee if U_2 were zero, and the right ascension of a particular ascending equatorial node.

Now it has been seen that

$$U = \frac{\mu}{r}\left[1 - \frac{1}{2} J_2 \left(\frac{R}{r}\right)^2 (3 \sin^2 \delta - 1) \right]$$

is a close approximation to the actual potential of the Earth, since the terms omitted (J_3, J_4 etc.) are of the order of 10^3 times smaller than the J_2 term.

322

Sterne then chooses as his unperturbed Hamiltonian H_0 the function

$$H_0 = \frac{1}{2}\left(p_r^2 + \frac{p_\lambda^2}{r^2\cos^2\delta} + \frac{p_\delta^2}{r^2}\right) - \frac{\mu}{r}\left[1 + \frac{J_2R^2}{2r^2}\left(1 - \frac{3}{2}\sin^2 i\right)\right.$$

$$\left. - \frac{3}{2}\frac{J_2R^2}{ra(1-e^2)}\left(\sin^2\delta - \frac{1}{2}\sin^2 i\right)\right] \quad (10.24)$$

which is of the same form as equation (10.21).

In equation (10.24) the constant i is the maximum declination of the particle, while the constant $a(1-e^2)$ is twice the product of the apogee and the perigee distances divided by their sum. The perturbing Hamiltonian H_1 is then given by

$$H_1 = H_0 - H$$

and becomes

$$H_1 = \frac{3}{2}\mu J_2 R^2\left(\sin^2\delta - \frac{1}{2}\sin^2 i\right)\left(\frac{1}{r^3} - \frac{1}{ar^2(1-e^2)}\right)$$

entering the canonic equations of the former canonic constants α_1, α_2, α_3, β_1, β_2, β_3, namely

$$\dot{\alpha}_i = \frac{\partial H_1}{\partial \beta_i}, \quad \dot{\beta}_i = -\frac{\partial H_1}{\partial \alpha_i} \quad (i = 1, 2, 3).$$

It should be noted that H_1 can contain any other harmonics so far neglected, but when partially differentiating H_1 all its terms must be regarded as functions of the canonic constants and the time. The exceptions are a, e and i introduced in equation (10.24) as constants.

The next step is the evaluation of the four integrals appearing in equation (10.23). It is found that they are elliptic integrals and are best treated by first expanding them in series, and then integrating them term by term (Sterne 1958).

The unperturbed solution obtained in this way, with slight adjustments in two of its canonic constants, is of the same order of accuracy as that of a conventional Keplerian elliptical orbit plus its first-order perturbations. Indeed, when Sterne's solution has first-order perturbations added, it is found that it is competitive in all respects with a conventional treatment plus first- and second-order perturbations. This work of Sterne's, and also the similar treatment by Garfinkel of the same problem, shows the power of Hamilton–Jacobi theory when applied to this type of problem.

10.6 The Effect of Atmospheric Drag on an Artificial Satellite

For most Earth satellites, drag changes the orbit secularly and is usually the force that finally removes the satellite's energy, causing it to spiral inwards to Earth. In a practical case, although the secular perturbations produced by atmospheric drag affect elements (namely a and e) that are not changed secularly by the harmonics of the Earth's gravitational field, the use of two separate theories (one for drag and one for gravitational field perturbations) is not a solution to the problem. A theory embodying both oblateness and drag effects must be constructed. At the same time, to keep the picture clear we will neglect oblateness effects in this section and suppose that we are dealing with a non-rotating spherical planet possessing an atmosphere. The gravitational potential function is then simply $U = \mu/r$ and the drag force acts as a perturbing force on the resulting Keplerian elliptic orbit of the satellite.

The shape of the satellite is a parameter of importance, as is its mean density. In general a satellite of arbitrary shape moving with some velocity v in an atmosphere of density ρ is subject to lift as well as drag. Both types of force will vary with time if the satellite is spinning and tumbling in its orbit as well as passing with varying velocity through regions of varying density. In the absence of precise knowledge of the satellite's attitude and of the atmospheric density as any instant, it is pot possible to predict exactly the changes in the satellite orbit.

For practical purposes however, it is sufficient to assume that the lift forces average out to zero, since the satellite's attitude is changing, and to assume an average cross-sectional area for the satellite when computing the drag. If indeed the satellite is spherical, the cross-sectional area is constant. The law of density change with altitude is sometimes taken to be a simple exponential fall-off of density with height, or is based on some model atmosphere with parameters determined empirically from satellite observations.

In what follows we consider that a satellite of mass m (negligible with respect to the Earth's mass) suffers a drag force per unit mass of magnitude F acting in the reverse direction to the satellite's geocentric velocity V. This force is given by

$$F = \frac{1}{2m} C_D A \rho V^2$$

where C_D is the aerodynamic drag coefficient, A is the average cross-sectional area of the satellite and ρ is the air density.

The coefficient C_D has a value between 1 and 2. It takes a value near 1 when the mean free path of the atmospheric molecules is small compared with the satellite size, and takes a value close to 2 when the mean free path is large compared with the size of the satellite. The density ρ is a function of height above the Earth's surface, and therefore of the distance from the Earth's centre.

Equations (6.41) gave the rates of change of the osculating elements of an orbit in terms of the components S, T and W of the disturbing acceleration; S, T and W being the radial, transverse and orthogonal components respectively, as shown in figure 10.6, where in this case E is the Earth's centre and P is the satellite position.

324

Equations (6.42) gave the relations between S, T, N and T', namely

$$T = \frac{(1+e \cos f)T'}{\sqrt{1+e^2+2e \cos f}} + \frac{(e \sin f)N}{\sqrt{1+e^2+2e \cos f}}$$

$$S = \frac{(e \sin f)T'}{\sqrt{1+e^2+2e \cos f}} - \frac{(1+e \cos f)N}{\sqrt{1+e^2+2e \cos f}}$$

where T' was the component of the perturbing acceleration tangential to the orbit in the direction of motion, while N was the component perpendicular to the tangent, positive when directed inwards (see figure 10.5). Then the drag $F = -T'$, while $N = W = 0$.

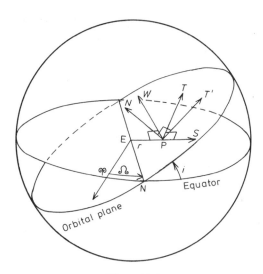

Figure 10.6

Using the elliptic orbit relationship

$$\cos E = \frac{\cos f + e}{1 + e \cos f}$$

equations (6.41) become

$$\frac{da}{dt} = -\left(\frac{A}{m}\right) \frac{C_D \rho V^2}{n(1-e^2)^{1/2}} (1+e^2+2e \cos f)^{1/2} \tag{10.25}$$

$$\frac{de}{dt} = -\left(\frac{A}{m}\right) \frac{C_D \rho V^2 (1-e^2)^{1/2}}{na} \frac{\cos f + e}{(1+e^2+2e \cos f)^{1/2}} \tag{10.26}$$

$$\frac{di}{dt} = 0 \tag{10.27}$$

$$\frac{d\Omega}{dt} = 0 \tag{10.28}$$

325

$$\frac{d\varpi}{dt} = -\left(\frac{A}{m}\right)\frac{C_D\rho V^2(1-e^2)^{1/2}}{nae}\frac{\sin f}{(1+e^2+2e\cos f)^{1/2}} \tag{10.29}$$

$$\frac{d\epsilon}{dt} = -\left(\frac{A}{m}\right)\frac{C_D\rho V^2}{na}\frac{e(1-e^2)\sin f}{\sqrt{1+e^2+2e\cos f}}\left\{\frac{1}{1-e^2+\sqrt{1-e^2}}\right.$$
$$\left.-\frac{1}{1+e\cos f}\right\} \tag{10.30}$$

Examining equations (10.27) and (10.28) it is seen that (as expected) neither the right ascension of the ascending node nor the inclination of the orbital plane is affected by drag. In addition, we note that the nonzero right-hand sides have the factor A/m, showing that a high ratio of cross-sectional area to mass produces the greatest drag effects. Ideally, a satellite designed for studying the outer atmosphere should be spherical and have a high A/m ratio.

In the remaining four equations we may replace V^2 and transform from t to f as the independent variable by using the two-body relationships

$$r = \frac{a(1-e^2)}{1+e\cos f}$$

$$r^2\dot{f} = h = \sqrt{\mu a(1-e^2)}$$

$$n^2 a^3 = \mu.$$

Hence

$$V^2 = \dot{r}^2 + r^2\dot{f}^2 = \frac{n^2 a^2}{(1-e^2)}(1+e^2+2e\cos f)$$

and

$$\frac{dt}{df} = \frac{(1-e^2)^{3/2}}{n(1+e\cos f)^2}.$$

The four equations then become

$$\frac{da}{df} = -\left(\frac{A}{m}\right)C_D\rho a^2\frac{(1+e^2+2e\cos f)^{3/2}}{(1+e\cos f)^2}$$

$$\frac{de}{df} = -\left(\frac{A}{m}\right)C_D\rho a(1-e^2)\frac{(1+e^2+2e\cos f)^{1/2}}{(1+e\cos f)^2}(\cos f+e)$$

$$\frac{d\varpi}{df} = -\left(\frac{A}{m}\right)C_D\rho\frac{a(1-e^2)}{e}\frac{(1+e^2+2e\cos f)^{1/2}}{(1+e\cos f)^2}\sin f.$$

$$\frac{d\epsilon}{df} = -\left(\frac{A}{m}\right)C_D\rho ae(1-e^2)^{3/2}\frac{(1+e^2+2e\cos f)^{1/2}}{(1+e\cos f)^2}$$
$$\times\left(\frac{1}{1-e^2+\sqrt{1-e^2}}-\frac{1}{1+e\cos f}\right)\sin f.$$

The density ρ is an even function of f and r. Examining the right-hand sides of the four equations with this in mind, it is seen that the equations for a and e are such that on integration, keeping a and e constant on the right-hand sides for a first approximation, a secular term appears, indicating that a and e decrease secularly with f and consequently with time. On the other hand, the presence of the $\sin f$ term in the other two equations ensures that both ϖ and ϵ are periodic functions of the time, the oscillations in general having small amplitudes because of the smallness of the coefficient $(A/m)C_D\rho$. These latter two elements are omitted from further consideration.

To solve the equations in a and e it is found useful to change the independent variable again, this time to the eccentric anomaly, using the relations

$$\cos f = \frac{\cos E - e}{1 - e \cos E}$$

and

$$\sin f = \frac{\sqrt{1 - e^2}\, \sin E}{1 - e \cos E}.$$

When this is done, we obtain

$$\frac{da}{dE} = -\left(\frac{A}{m}\right) C_D \rho a^2 \frac{(1 + e \cos E)^{3/2}}{(1 - e \cos E)^{1/2}}$$

$$\frac{de}{dE} = -\left(\frac{A}{m}\right) C_D \rho a \frac{(1 + e \cos E)^{1/2}}{(1 - e \cos E)^{1/2}} (1 - e^2) \cos E.$$

If Δa and Δe are the perturbations in a and e over one revolution of the satellite in its orbit, we have

$$\Delta a = -\left(\frac{A}{m}\right) C_D a^2 \int_0^{2\pi} \frac{\rho(1 + e \cos E)^{3/2}}{(1 - e \cos E)^{1/2}}\, dE$$

$$\Delta e = -\left(\frac{A}{m}\right) C_D a \int_0^{2\pi} \frac{\rho(1 + e \cos E)^{1/2}}{(1 - e \cos E)^{1/2}} (1 - e^2) \cos E\, dE$$

the integrations being carried out numerically.

The density ρ is an empirically determined function of r, although in a number of studies it is approximated by a simple exponential law

$$\rho = \rho_0 \exp\left[-(\eta - \eta_0)/H\right]$$

where η is the altitude, η_0 is some standard altitude (usually taken to be the altitude of perigee), ρ_0 is the density at the standard altitude and H is the scale height (assumed constant). The scale height is that vertical distance in which the density changes by a factor e and depends upon the altitude. H is about 6 km at sea level, reaching 40 km at a height of about 200 km.

Several further remarks may be made at this point. The perigee and apogee distances are $a(1-e)$ and $a(1+e)$ respectively. When the changes in these over one revolution are computed using the easily derived relations

$$\Delta[a(1-e)] = -\left(\frac{A}{m}\right) C_{\mathrm{D}} a^2 (1-e) \int_0^{2\pi} \rho(1-\cos E)\left(\frac{1+e\cos E}{1-e\cos E}\right)^{1/2} dE$$

$$\Delta[a(1+e)] = -\left(\frac{A}{m}\right) C_{\mathrm{D}} a^2 (1+e) \int_0^{2\pi} \rho(1+\cos E)\left(\frac{1+e\cos E}{1-e\cos E}\right)^{1/2} dE$$

it is found that unless the eccentricity is very small, the apogee change is much larger than the perigee change. Thus the change in a satellite orbit due to drag may be illustrated qualitatively, as in figure 10.7.

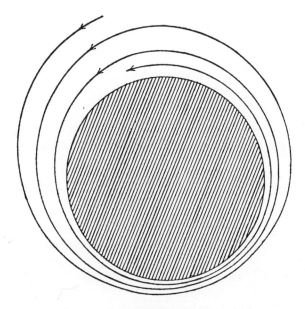

Figure 10.7

In the above discussion no account has been taken of the oblateness of the atmosphere over a nonspherical Earth, nor of the rotation of such an atmosphere. For a spheroidal planet, the density is a function of the vertical height along the normal to the surface of the planet while the difference between air speed and satellite speed is important (Sterne 1959, Roy 1963, Morando 1969, King-Hele 1964, 1987).

The temperature (and therefore density) of the upper atmosphere change because of diurnal and seasonal variations in the amount of radiation falling upon it; such changes have been studied by the effects they have produced in the orbits of Earth satellites. Solar activity such as the occurrence of flares also produces perturbations due to atmospheric heating, the density at heights of order 800 km increasing temporarily by factors of 3 to 7 on occasions.

10.7 Tesseral and Sectorial Harmonics in the Earth's Gravitational Field

So far it has been assumed that the Earth is symmetrical about its polar axis so that its potential U at a point P distant r from its centre of mass and with declination δ is given by

$$U = \frac{\mu}{r}\left[1 - \sum_{n=2}^{\infty} J_n\left(\frac{R}{r}\right)^n P_n(\sin \delta)\right] \qquad (10.31)$$

where $\mu = GM$, the J_n are constants and R and M are the earth's equatorial radius and mass respectively. In general, however, it appears that the Earth's potential departs slightly from that of a body having axial symmetry. The more general formula for the potential that includes such departures is

$$U = \frac{\mu}{r}\left[1 - \sum_{n=2}^{\infty} \sum_{m=0}^{n} J_n^{(m)}\left(\frac{R}{r}\right)^n \cos m(\lambda - \lambda_n^{(m)}) P_n^{(m)}(\sin \delta)\right]$$

or

$$U = \frac{\mu}{r}\left[1 - \sum_{n=2}^{\infty} \sum_{m=0}^{n} \left(\frac{R}{r}\right)^n \left(C_n^{(m)} \cos m\lambda + S_n^{(m)} \sin m\lambda\right) P_n^{(m)}(\sin \delta)\right].$$

The $P_n^{(m)}(\sin \delta)$ are the associated Legendre functions, given by

$$P_n^{(m)}(x) = \frac{1}{2^n n!}(1-x^2)^{m/2}\frac{d^{n+m}(x^2-1)^n}{dx^{n+m}}$$

where $x \equiv \sin \delta$, while the constants $C_n^{(m)}$ and $S_n^{(m)}$ are measures of the amplitudes of the various harmonics. The longitude λ enters the formula since the geoid cannot now be regarded as axially symmetrical. If $m=0$ (i.e. the geoid is axially symmetrical) the general formula reduces to equation (10.31) which consists of *zonal harmonics* only. In the general case however, the so-called tesseral and sectorial harmonics, depending not only on latitude but also on longitude, appear. These latter harmonics are of small amplitude and in the first order have no secular effects, causing only periodic perturbations in the elements of a satellite orbit. The long-period oscillations have been used by a number of workers to derive some of the values of some of the constants. In particular the ellipticity of the equator has been measured (see section 10.2.1). Determinations of tesseral and sectorial harmonics have been achieved from precisely reduced observations of artificial satellites (Morando 1969).

Problems

10.1 Show that:

(i)

$$\epsilon = \frac{1}{2}\left[3J_2 + \frac{\omega^2 R^3}{\mu}\right]$$

where R and ω are the Earth's radius and angular velocity of rotation respectively, ϵ is the flattening, J_2 is the second harmonic constant, and $\mu = GM$;

(ii) the period T of an artificial Earth satellite in a circular orbit of radius a in the plane of the Earth's equator is given approximately by

$$T=2\pi a^{3/2}\mu^{-1/2}\left[1-\frac{3}{4}\left(\frac{R}{a}\right)^2 J_2\right].$$

10.2 Prove that

(i)

$$\overline{\left(\frac{a}{r}\right)^3}\equiv\frac{1}{2\pi}\int_0^{2\pi}\left(\frac{a}{r}\right)^3 dM=(1-e^2)^{-3/2}$$

(ii)

$$\overline{\left(\frac{a}{r}\right)^3 \sin 2f}=0.$$

10.3 If the average is taken with respect to the true anomaly f, prove that

(i)

$$\overline{\left(\frac{r}{a}\right)}\equiv\frac{1}{2\pi}\int_0^{2\pi}\left(\frac{r}{a}\right) df=(1-e^2)^{1/2}$$

(ii)

$$\overline{\left(\frac{a}{r}\right)}=(1-e^2)^{-1}.$$

10.4 Prove that the J_4 terms in the disturbing function F of an Earth satellite are given by the expression

$$-\frac{\mu J_4 R^4}{r^5}\left(\frac{3}{8}+\frac{5}{8}\sin^2\delta-\frac{35}{32}\sin^2 2\delta\right).$$

Transform the expression into a function in terms of f, i and ω and hence show that the second-order secular part of the disturbing function F_2 is

$$-\frac{\mu J_4 R^4}{8a^5}\left[3-15\sin^2 i+\left(\frac{105}{8}\right)\sin^4 i\right](1-e^2)^{-7/2}\left(1+\frac{3e^2}{2}\right).$$

10.5 Show that, to the first order, there are two values of the inclination of an artificial satellite's orbit to the equator for which ϖ, given by

$$\varpi=\Omega+\omega$$

does not change secularly, and hence find their values.

10.6 Using the data of Appendix II, calculate the values of the first-order rates of change (in degrees per day) of the argument of perigee ω and the right ascension of the ascending node Ω of an artificial Earth satellite whose semimajor axis a and eccentricity e are given by

$$a=1\cdot30262R$$

$$e=0\cdot16561$$

$$i=32°\,52'$$

where R is the Earth's equatorial radius.

10.7 Verify that the function S, given by equation (10.22), is the solution of the Hamilton–Jacobi equation when the Hamiltonian is of the form given in equation (10.21).

References

Brouwer D 1959 *Astron. J.* **64** 378
Garfinkel B 1958 *Astron. J.* **63** 88
—— 1959 *Astron. J.* **64** 353
King-Hele D G 1958 *Proc. R. Soc.* **A247** 49
—— 1962 *Satellites and Scientific Research* (London: Routledge and Kegan Paul)
—— 1964 *Theory of Satellite Orbits in an Atmosphere* (London: Butterworths)
—— 1974 *A View of Earth and Air* (Royal Aircraft Establishment Tech. Memo. 212)
—— 1987 *Satellite Orbits in an Atmosphere: Theory and Applications* (Glasgow: Blackie)

Kozai Y 1959a *Smithsonian Institution Astrophysical Observatory; Special Report* 22
—— 1959b *Astron. J.* **64** 367
—— 1961 *Astron. J.* **66** 355
Merson R H 1960 *Geophysical Research* **4** 17
Morando B (ed) 1969 *Dynamics of Satellites* (Berlin: Springer-Verlag)
Roy M (ed) 1963 *Dynamics of Satellites* (Berlin: Springer-Verlag)
Sterne T E 1958 *Astron. J.* **63** 28
—— 1959 *J. Am. Rocket Soc.* **29** 777
Tisserand F 1889 *Traité de la Mécanique Céleste* (Paris: Gauthier-Villars)

Bibliography

Eckstein M C 1963 *Astron. J.* **68** 231
Izsak I G 1961a *Space Res.* **2** 352
—— 1961b *Astron. J.* **66** 226
Jeffreys H 1959 *The Earth* (4th edn) (London: Cambridge University Press)
Kaula W M 1961 *Space Res.* **2** 360
Kozai Y 1961 *Smithsonian Institution Astrophysical Observatory; Special Report* 72
Kuiper G (ed) 1954 *The Earth as a Planet* (Chicago: University of Chicago Press)
Muhleman D O *et al* 1962 *Astron. J.* **67** 191
Newton R R 1962 *J. Geophys. Res.* **67** 415
O'Keefe J A, Eckels A and Squires R K 1959 *Astron. J.* **64** 245
Pettengill G H *et al* 1962 *Astron. J.* **66** 226
Plummer H C 1960 *An Introductory Treatise on Dynamical Astronomy* (New York: Dover Publications)
Rabe E 1949 *Astron. J.* **55** 112

11 Rocket Dynamics and Transfer Orbits

11.1 Introduction

As far as present-day technology is concerned, space flight is practical only because the rocket (working by Newton's laws of motion) enables a vehicle, manned or unmanned, to transfer from the gravitational field of one Solar System body to that of another. An important part of orbital motion studies in the last twenty years has therefore been concerned with the dynamic behaviour of rockets in gravitational fields and their ability to effect such transfers.

In this chapter some basic principles of such motion are established. In the first part of the chapter the emphasis is on the rocket; in the second part applications of rocket motors in changing from one orbit to another are considered, and in the final part there is an elementary discussion of errors involved in such applications.

11.2 Motion of a Rocket

As an introduction let us consider a rocket moving in a vacuum in gravity-free space. Let its mass at time t be m and let its thrust, assumed constant, act continuously in one direction. The rocket works by ejecting part of its mass at a high velocity; in assuming its thrust to be constant we will also assume the mass ejected per second and the exhaust velocity v_e (measured with respect to the vehicle) to be constant. Then if the rocket's velocity in the opposite direction at time t is v, the momentum is mv.

If a mass dm is ejected, resulting in an increase of velocity dv, then by the law of conservation of momentum we may write

$$(m + dm)(v + dv) + (v_e - v)\, dm = mv.$$

Neglecting the product of dm and dv and cancelling out common terms, we obtain

$$m\, dv = -v_e\, dm \tag{11.1}$$

which may be immediately integrated to give

$$v - v_0 = v_e \ln (m_0/m) \tag{11.2}$$

where v_0 and m_0 are the initial velocity and mass of the rocket and m is the mass remaining when a velocity v has been attained.

The quantity m_0/m is called the *mass ratio*. If a velocity equal to the exhaust velocity is to be added to the original velocity, then a mass ratio of $e=2\cdot718\ldots$ has to be realized. Equation (11.2) is the fundamental equation of rocket flight. It also shows that, for a mass ratio greater than $e=2\cdot718$, the final velocity added to the rocket may exceed its exhaust velocity.

An important parameter in rocket design is the specific impulse I. The exhaust velocity of the rocket using chemical propellants depends upon the heat energy liberated per pound and on the molecular weight. For best results the former should be as large as possible, the latter as small as possible. The specific impulse I is then defined as

$$I = \frac{v_e}{g} = \frac{\text{thrust}}{\text{gravity} \times \text{fuel-mass flow rate}}$$

since $\text{thrust} = -v_e(dm/dt)$ and therefore has the dimensions of time.

For a liquid oxygen–alcohol motor (such as the wartime V-2), I has a value of about 240 s, while a fluorine–hydrogen motor has a specific impulse in the 300–380 s region.

11.2.1 Motion of a rocket in a gravitational field

Let the rocket be ascending in a straight line against a constant gravity g. The change in momentum in time dt due to the force g per unit mass is then $mg\,dt$ and equation (11.1) becomes

$$m\,dv = -v_e\,dm - mg\,dt$$

giving

$$dv = -v_e \frac{dm}{m} - g\,dt. \tag{11.3}$$

Integrating, we obtain

$$v - v_0 = v_e \ln (m_0/m) - gt. \tag{11.4}$$

If g varies with height,

$$v - v_0 = v_e \ln (m_0/m) - \int_0^t g(h)\,dt \tag{11.5}$$

where h is the rocket's height above some reference point.

If the gravity field is an inverse-square one, due to a planet of radius R with a surface acceleration due to gravity of value g_E, then the value of g at a distance r from the planet's centre is given by

$$g = g_E \left(\frac{R}{r}\right)^2.$$

This distance r is a function of time through the motion of the rocket.

Now in practice only a certain part of the rocket mass is fuel; so if m is the mass of the empty rocket, equations (11.4) and (11.5) give the maximum possible increase in velocity for a rocket having exhaust velocity v_e. If all the fuel is burnt by time t, the rocket will coast upwards under gravity, its maximum distance from the burn-out point being decided by the energy (the sum of potential and kinetic energy) it has acquired at burn-out. By sections 4.5 and 4.11, this depends upon its distance r from the centre of the gravitational field and the velocity v.

Equations (11.4) and (11.5) show that to increase v (and hence increase the maximum attainable distance) the mass ratio and/or the exhaust velocity should be increased. In addition a faster fuel consumption should be sought, since the longer the time spent under powered flight, the less will be the benefit from the fuel expenditure. The subject of gravity losses is highlighted by equation (11.3) if we assume that the fuel consumption rate dm/dt is so small and varies such that

$$\frac{dv}{dt} = -\frac{v_e}{m}\frac{dm}{dt} - g = 0.$$

Then $v = v_0$ and the rocket exhausts its fuel supply in maintaining its original position.

The distance s travelled by the rocket during the burning time t may be easily found.

If the rate of fuel consumption f is constant, then

$$\frac{dm}{dt} = -f$$

and hence

$$m = m_0 - ft.$$

Then

$$s = \int_0^t v\, dt = \int_0^t \left(v_0 - gt + v_e \ln \frac{m_0}{m}\right) dt$$

or

$$s = \int_0^t \left[v_0 - gt - v_e \ln \left(1 - \frac{ft}{m_0}\right)\right] dt$$

giving on integration

$$s = v_0 t + v_e \left[t - \left(t - \frac{m_0}{f}\right) \ln \left(1 - \frac{ft}{m_0}\right)\right] - \int_0^t gt\, dt$$

or

$$s = \frac{1}{f}\left[(v_0 + v_e)(m_0 - m) + v_e m \ln \frac{m}{m_0}\right] - \frac{1}{2}gt^2$$

having assumed g to be constant.

11.2.2 Motion of a rocket in an atmosphere

If the rocket is ascending through an atmosphere of density ρ, the density being some function of height, lift and drag forces will operate (see section 10.6). If the rocket is ascending vertically under power, lift forces may be neglected and the drag force per unit mass is F, given by

$$F = \frac{1}{2m} C_D A \rho V^2 \qquad (11.6)$$

where as before m is the rocket mass, V is its velocity, C_D is the drag coefficient and A is the cross-sectional area of the rocket. It should be noted that the drag coefficient depends on the rocket's shape and the speed, and can vary by a factor of two.

Examining equation (11.6) it is seen that the drag force is roughly proportional to the square of the velocity and the first power of the density, indicating that to minimize drag effects the rocket should ascend vertically through the atmosphere as slowly as possible. But this low speed is contrary to the policy of attaining as high a velocity as fast as possible to minimize gravitational losses. Drag losses, however, are far less important than those due to gravitation where ascending space vehicles are concerned, and so the problem of prime importance is to minimize gravitational losses.

The practical way of doing this is to adopt a flight path for the rocket that very quickly bends away from the vertical until a horizontal trajectory is followed. If at any instant the angle between thrust and horizon is θ, the gravitational component acting against the thrust is $g \sin \theta$. If an atmosphere is present, the bending must be delayed so that the rocket does not build up high speeds in the lower and denser atmospheric regions. There is a large literature on deflected powered trajectories which we have no space to consider here.

11.2.3 Step rockets

Typical values for the mass ratio R of a rocket and its exhaust velocity are 5 and 2·5 km s^{-1}. Substituting these figures into equation (11.2) it is found that

$$v - v_0 = 4 \cdot 02 \text{ km s}^{-1}.$$

Escape velocity from the Earth is 11·2 km s^{-1}, so a single rocket using a highly efficient design and powerful fuel does not provide the necessary velocity. If drag and gravitational losses are taken into account the picture is even gloomier.

It was recognized early in the history of space flight that only multistage rockets (or step rockets) possessed the ability of attaining velocities as great as or greater than escape velocity. To illustrate the principle of staging, which depends upon being able to jettison parts of the vehicle such as empty fuel tanks for which there is no further use, consider a two-stage rocket made up as shown in figure 11.1.

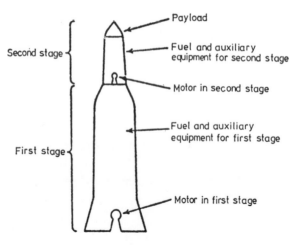

Payload

Fuel and auxiliary equipment for second stage

Motor in second stage

Fuel and auxiliary equipment for first stage

Motor in first stage

Second stage

First stage

Figure 11.1

Let

M_0 = total initial mass,

M_1 = mass of first stage (empty of fuel),

m_1 = mass of fuel in first stage,

M_2 = mass of second stage (empty of fuel),

m_2 = mass of fuel in second stage,

$(v_e)_1$ = exhaust velocity of first stage,

$(v_e)_2$ = exhaust velocity of second stage.

Then

$$M_0 = (M_1 + m_1) + (M_2 + m_2).$$

For simplicity we neglect drag and gravitational losses. The velocity increase achieved after all the fuel in the first stage has been burnt is, using equation (11.2),

$$v_1 = (v_e)_1 \ln \left(\frac{M_0}{M_0 - m_1} \right). \tag{11.7}$$

The empty first stage of mass M_1 is now jettisoned as the second-stage motor is ignited, and the new velocity increase provided by the fuel of the second stage is

$$v_2 = (v_e)_2 \ln \left[\frac{M_0 - (M_1 + m_1)}{M_0 - (M_1 + m_1 + m_2)} \right]. \tag{11.8}$$

The total increase in velocity of the second stage since take-off is then v, given by

$$v = v_1 + v_2. \tag{11.9}$$

We now introduce the permissible mass ratio R for a single-stage rocket, and also the fraction x of the mass of the first stage (including fuel) that the second stage represents, and we suppose that the mass ratio for both stages is R. Then the relations

$$\frac{M_1+m_1}{M_1}=\frac{M_2+m_2}{M_2}=R$$

$$(M_2+m_2)/(M_1+m_1)=x$$

and

$$(M_1+m_1)+(M_2+m_2)=M_0$$

coupled with equations (11.7) and (11.8) give us

$$v_1=(v_e)_1 \ln \left[\frac{R(1+x)}{1+Rx}\right] \qquad (11.10)$$

and

$$v_2=(v_e)_2 \ln R. \qquad (11.11)$$

In equation (11.10) the effective mass ratio R' is given by

$$R'=\frac{R(1+x)}{1+Rx}$$

which gives a maximum value of R when $x=0$ (since $R>1$). For the conditions:

$$x=0\cdot1, \qquad R'=1\cdot1R/(1+0\cdot1R)$$

$$x=1\cdot0, \qquad R'=2/(1+1/R)$$

$$x=10\cdot0, \qquad R'=11R/(1+10R)\sim1.$$

On this simple argument, the second-stage mass should therefore be much smaller than the first-stage mass. If we put $R=5$ as before, take $x=0\cdot1$ and set $(v_e)_1=(v_e)_2=2\cdot5$ km s^{-1}, the increase in velocity of the final stage is found to be $7\cdot27$ km s^{-1} which compares favourably with the $4\cdot02$ km s^{-1} obtained with a one-stage rocket.

The above picture is oversimplified. Apart from the omission of gravity and drag losses, we have not considered the additional structure made necessary by the complications of a second stage put on top of a first, nor have we considered the fact that in modern rockets the first-stage motor is usually a cluster of motors, delivering a thrust far greater than that of the second. But even when these complications have been added, there is no major change in the main conclusion that step rockets are essential for escape from the Earth, or even to put a satellite into Earth-orbit.

11.2.4 Alternative forms of rocket

At the time of writing, only chemical rockets are capable of providing thrusts large enough to lift themselves into orbit through a planet's gravitational field from the planetary surface or to land upon it. Other forms of rocket are being currently developed: of these only a nuclear reactor-powered type can compare in thrust with the chemical rocket. The other forms, such as the ion rocket, have very small thrusts and long 'burning' times and so will have to produce the energies required in moving from the neighbourhood of one planetary mass to that of another by building up these energy changes from sustained powered operations, possibly lasting for many days. Such power systems have a number of advantages over conventional chemical high-thrust systems: for example, in giving low mass ratios for interplanetary missions and appreciably shorter transfer time, especially with respect to flights to the outer planets of the Solar System.

Since all rocket motors depend for their drive effect upon the ejection of a fraction of their mass at a high velocity, the basic equation (11.2) holds for such low-thrust systems. The treatment of such systems when they operate for a long period of time in a gravitational field is nonetheless different from that of high-thrust systems where the thrust is so large that it may, with a high degree of accuracy, be considered to act for so short a time that only the vehicle's velocity vector is altered by it during operation. The scope of this book dictates that in the remainder of this chapter we consider only high-thrust systems, omitting the study of low-thrust manoeuvres (Ehricke 1961, 1962).

11.3 Transfer Between Orbits in a Single Central Force Field

If a vehicle is in an orbit about a massive spherical body, without perturbations by other masses, it moves in a central force field. If the motors are not being used the vehicle's orbit is a conic section, the properties of the orbit being described by the formulae of chapter 4. The firing of the motors will cause changes in the orbit, affecting in general all six elements. Since we are dealing with high-thrust systems we can assume that the thrust operates for so short a time that the impulse it provides instantaneously changes the vehicle's momentum vector but not its position. The attitude of the motor thrust to the tangent to the orbit determines the change in speed and direction. The fact that the change is effected without appreciable change in position ensures that no gravitational losses occur. In what follows we consider first the changes in the orbit due to various types of impulse, and we will then go on to study the requirements for a transfer from one orbit to another. Only the simplest cases will be treated.

11.3.1 Transfer between circular, coplanar orbits

Let us suppose that the vehicle is in one circular orbit, radius a_1 about a mass M, and it is desired to transfer to a larger circular orbit of radius a_2 as shown in figure 11.2. The most convenient way to treat the problem is to regard it as a problem in change of energy.

338

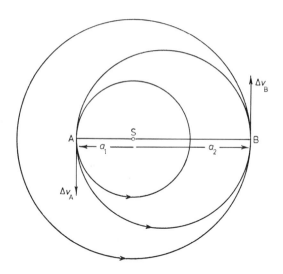

Figure 11.2

The vehicle's energy is (sections 4.5 and 4.11)

$$C_1 = \frac{1}{2} v_1{}^2 - \frac{\mu}{a_1} \qquad (11.12)$$

where $\mu = GM$ and $v_1 = (GM/a_1)^{1/2} =$ the velocity in the orbit.
The energy in the larger orbit is C_2, given by

$$C_2 = \frac{1}{2} v_2{}^2 - \frac{\mu}{a_2} \qquad (11.13)$$

where $v_2 = (GM/a_2)^{1/2}$ is the velocity in the larger orbit. Thus

$$C_1 = -\frac{1}{2} \frac{\mu}{a_1}, \qquad C_2 = -\frac{1}{2} \frac{\mu}{a_2}.$$

Then the energy required to effect the transfer is at least ΔC, where

$$\Delta C = C_2 - C_1 = \frac{\mu}{2} \left(\frac{1}{a_1} - \frac{1}{a_2} \right). \qquad (11.14)$$

If the transfer is effected by means of an elliptic transfer orbit cotangential to both circular orbits as shown in figure 11.2, then the operation requires two impulses, the first (taking place at A) putting the vehicle into the ellipse, the second (taking place at B) putting the vehicle into the larger circular orbit. These impulses are applied tangentially to the orbit by firing the rocket motor in the opposite direction.

If the impulse I does not act in the same direction in which the velocity vector lies but at some angle θ to it, producing a change in momentum $m\Delta v$, then the new velocity vector v' is given by adding Δv vectorially to the old velocity vector v as in figure 11.3.

339

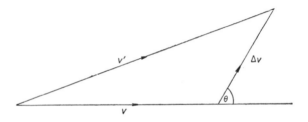

Figure 11.3

The increase in kinetic energy is given by the expression $\frac{1}{2}(v'^2 - v^2)$, which for a given impulse magnitude is a maximum for $\theta = 0$. Thus the tangentially applied impulse is the most economic in fuel for effecting a given change in kinetic energy.

Now the energy of the transfer ellipse C_T is given by

$$C_T = \frac{1}{2}v^2 - \frac{\mu}{r}.$$

But for elliptical motion,

$$v^2 = \mu\left(\frac{2}{r} - \frac{1}{a}\right)$$

and hence

$$C_T = -\frac{\mu}{2a_T} \tag{11.15}$$

a_T being the ellipse's semimajor axis. But

$$a_T = \frac{1}{2}(a_1 + a_2) \tag{11.16}$$

and therefore

$$C_T = \frac{-\mu}{a_1 + a_2}$$

and the required energy increment at A to place the vehicle into the correct transfer orbit is

$$\Delta C_A = C_T - C_1 = -\frac{\mu}{2a_1}\left(\frac{a_1 - a_2}{a_1 + a_2}\right). \tag{11.17}$$

Similarly the energy increment required at B is given by

$$\Delta C_B = C_2 - C_T = -\frac{\mu}{2a_2}\left(\frac{a_1 - a_2}{a_1 + a_2}\right). \tag{11.18}$$

The energy changes are due to changes in kinetic energy. Hence

$$\Delta C_A = \frac{1}{2}(v_A + \Delta v_A)^2 - \frac{1}{2}v_A^2 \tag{11.19}$$

and

$$\Delta C_B = \frac{1}{2}v_B^2 - \frac{1}{2}(v_B - \Delta v_B)^2 \tag{11.20}$$

340

where Δv_A and Δv_B are the necessary changes in velocity at A and B respectively. Equating (11.17) and (11.19) gives

$$\Delta v_A = \left(\frac{\mu}{a_1}\right)^{1/2} \left[\left(\frac{2a_2/a_1}{1+a_2/a_1}\right)^{1/2} - 1\right]. \qquad (11.21)$$

Similarly, equating (11.18) and (11.20) gives

$$\Delta v_B = \left(\frac{\mu}{a_2}\right)^{1/2} \left[1 - \left(\frac{2}{1+a_2/a_1}\right)^{1/2}\right]. \qquad (11.22)$$

By equation (11.2), applicable since there are no drag or gravity losses, we obtain the mass of fuel required for the impulses. For the first impulse,

$$\Delta v_A = v_e \ln \left(\frac{m_0}{m_A}\right)$$

giving $m_0 - m_A$ (the mass of fuel used) as a function of Δv_A, v_e (the exhaust velocity) and m_0 (the vehicle's mass before the operation).

For the second impulse,

$$\Delta v_B = v_e \ln \left(\frac{m_A}{m_B}\right)$$

giving $m_A - m_B$ (the mass of fuel used) as a function of Δv_B, v_e and m_A.

Combining Δv_A and Δv_B, the total velocity increment for transfer from one circular orbit to the other is given by

$$\Delta v = \Delta v_A + \Delta v_B = v_e \ln \left(\frac{m_0}{m_B}\right)$$

enabling the total fuel expenditure to be computed in one calculation.

The eccentricity e of the transfer orbit is obtained from

$$a_1 = a_T(1-e)$$
$$a_2 = a_T(1+e)$$

giving

$$e = \frac{a_2 - a_1}{a_2 + a_1}. \qquad (11.23)$$

The period of time t_T spent in making the transfer is half the period of revolution of a body in the transfer orbit T given by equation (4.26), namely

$$t_T = T/2 = \pi \left(\frac{a_T^3}{\mu}\right)^{1/2}. \qquad (11.24)$$

Positions and velocities of the vehicle in the transfer orbit at any other time may be computed using the formulae of chapter 4.

11.3.2 Parabolic and hyperbolic transfer orbits

Any circular orbit can be converted into a parabolic or hyperbolic orbit by increasing the velocity by applying a big enough impulse, tangentially or otherwise. To obtain a parabolic orbit from a circular one in which the velocity is $v_c = (GM/a)^{1/2}$, the velocity increment that must be added is

$$\Delta v = (\sqrt{2} - 1)\, v_c \qquad (11.25)$$

since parabolic velocity at a given distance is $\sqrt{2} \times$ circular velocity (see equation (4.81). Any velocity in excess of this parabolic velocity will convert the orbit to a hyperbolic path of eccentricity greater than unity.

Now at pericentre (the point in the orbit nearest the central mass), the hyperbolic velocity is

$$V_P^2 = \frac{\mu}{a}\left(\frac{e+1}{e-1}\right) \qquad (11.26)$$

where $r_P = a(e-1)$ is the radius (see equation (4.92)). The difference Δv_h between parabolic velocity and hyperbolic velocity is then given by

$$\Delta v_h = \sqrt{\frac{\mu}{r_P}}\left(\sqrt{e+1} - \sqrt{2}\right). \qquad (11.27)$$

Such orbits give faster transfer times than elliptic transfer orbits but are more costly in fuel, since the velocity increments required to enter and leave the transfer orbit are large.

A particular type of transfer called the *bi-elliptic transfer* may be referred to here. It follows from a comparison of the energy required to give parabolic velocity to the vehicle and the total energy for the two impulses necessary to transfer the vehicle from the orbit of radius a_1 to that of radius a_2. In the former case, by equation (11.25),

$$\Delta v_P = (\sqrt{2} - 1)v_c = (\sqrt{2} - 1)\sqrt{\frac{\mu}{a_1}}.$$

In the latter case, adding equations (11.21) and (11.22) gives

$$\Delta v_A + \Delta v_B = \left(\frac{\mu}{a_1}\right)^{1/2}\left[\left(\frac{2a_2/a_1}{1+a_2/a_1}\right)^{1/2} - 1\right] + \left(\frac{\mu}{a_2}\right)^{1/2}\left[1 - \left(\frac{2}{1+a_2/a_1}\right)^{1/2}\right].$$

Equating these two relations, a quadratic in a_2/a_1 is obtained which has as a real root $a_2/a_1 \sim 3{\cdot}4$. If a_2/a_1 is less than this value, the cotangential transfer consumes less fuel than the impulse giving the vehicle escape velocity from the orbit of radius a_1. If a_2/a_1 is greater than 3·4, the transfer energy is greater than the parabolic increment energy. This suggests that for transfer between orbits where $a_2/a_1 \gg 3{\cdot}4$, the simple cotangential ellipse may not be the most economical in fuel, but that a three-impulse transfer orbit composed of two semiellipses may be better. The procedure would be as shown in figure (11.4).

The increment $\Delta v_A'$ of velocity puts the vehicle into an elliptical orbit carrying it far outside the orbit of radius a_2 to an apocentre C. There a further increment $\Delta v_C'$ of velocity increases its energy sufficient to place it in a new elliptic orbit with pericentre B on the orbit of radius a_2, where

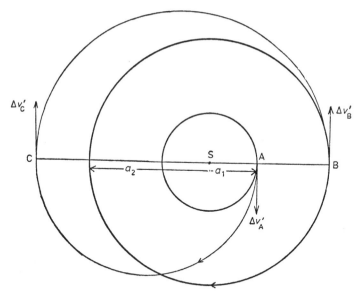

Figure 11.4

a third expenditure of fuel resulting in a velocity decrement $\Delta v'_B$ transfers the vehicle to the required circular orbit of radius a_2.

It may be shown (Ehricke 1962) that for $a_2/a_1 \geqslant 15.582$ any bi-elliptic transfer orbit of this type will result in some saving of fuel. The disadvantage of such orbits is the very large transfer time involved.

11.3.3 Changes in the orbital elements due to a small impulse

In this section the effects on the orbital elements of applying a small impulse I at an arbitrary angle to the orbit are considered. Since the radius vector does not change during the operation, all changes in the elements depend upon the velocity vector's change in magnitude and direction caused by the the application of the impulse I. Qualitatively, many of the consequences may be seen at once by remembering that the impulse's change Δv in the velocity vector v can be split into a component at right angles to the orbital plane (Δv_W) and two mutually perpendicular components lying in the plane, either along and at right angles to the radius vector (Δv_S and Δv_T), or tangential to and normal to the orbit at the vehicle's instantaneous position ($\Delta v_{T'}$ and Δv_N) (section 6.7.4). Thus

$$\Delta v = \Delta v_W + \Delta v_S + \Delta v_T = \Delta v_W + \Delta v_{T'} + \Delta v_N. \qquad (11.28)$$

It is obvious that an impulse that makes Δv_W zero will not affect the inclination or the longitude of the ascending node, since all change takes place in the plane of the old orbit.

Again, since the velocity relations for the ellipse and hyperbola are respectively

$$V^2 = \mu\left(\frac{2}{r} - \frac{1}{a}\right), \quad V^2 = \mu\left(\frac{2}{r} + \frac{1}{a}\right)$$

a change only in direction of the velocity vector will leave the element a unchanged, since r does not vary during the impulse.

An important application of the elliptic velocity relation may be mentioned at this point. Differentiating and remembering that r does not vary in this situation, we obtain

$$2V\Delta V = \frac{\mu}{a^2}\,\Delta a$$

showing that if it is desired to make the greatest change in the semimajor axis of an elliptic orbit, it is most economically obtained by applying the impulse at pericentre, where V is greatest.

Equations (6.41) may be modified to give the change $\Delta\sigma$ in any element σ of an elliptic orbit due to a small impulse I.

Writing

$$\Delta v_S = S\Delta t$$

$$\Delta v_T = T\Delta t$$

$$\Delta v_W = W\Delta t$$

the equations become

$$\Delta a = \frac{2}{n\sqrt{1-e^2}}\left(e\sin f\,\Delta v_S + \frac{p}{r}\,\Delta v_T\right)$$

$$\Delta e = \frac{\sqrt{1-e^2}}{na}\,[\Delta v_S \sin f + (\cos E + \cos f)\,\Delta v_T]$$

$$\Delta i = \frac{r\cos u}{na^2\sqrt{1-e^2}}\,\Delta v_W$$

$$\Delta\Omega = \frac{r\sin u}{na^2\sqrt{1-e^2}\,\sin i}\,\Delta v_W$$

$$\Delta\varpi = \frac{\sqrt{1-e^2}}{nae}\left[-\Delta v_S \cos f + \left(1 + \frac{r}{p}\right)\Delta v_T \sin f\right] + 2\sin^2\frac{i}{2}\,\Delta\Omega$$

$$\Delta\epsilon = \frac{e^2}{1+\sqrt{1-e^2}}\,\Delta\varpi + 2\sqrt{1-e^2}\,\sin^2\frac{i}{2}\,\Delta\Omega - \frac{2r}{na^2}\,\Delta v_S$$

(11.29)

where f and E are the true and eccentric anomalies respectively, $p = a(1-e^2)$, and $u = f + \varpi - \Omega = f + \omega$.

If e and i are very small, the transformation of section 6.7 can be used, namely the introduction of h, k, p and q given by

$$h = e \sin \varpi, \qquad k = e \cos \varpi$$

$$p = \tan i \sin \Omega, \qquad q = \tan i \cos \Omega.$$

Some of the effects exhibited by equations (11.29) are now discussed.

Apart from the consequences already mentioned above, it is seen that not only is an orthogonal component in the impulse necessary to change i and Ω, but for a given r the greatest change in i is effected if the orthogonal component is applied at a node ($u = 0°$, $180°$), while the greatest change in Ω results if the impulse is applied midway between the nodes ($u = 90°$, $270°$). The changes are maximum if $r = a(1 + e)$; that is, if the vehicle is at apocentre. The orthogonal component also affects ϖ and ϵ unless $u = 0$. If $u \neq 0$, and the orthogonal component is the only nonzero impulse component, then

$$\Delta \varpi = 2 \sin^2 \frac{i}{2} \, \Delta \Omega = \Delta \epsilon.$$

Now $\omega = \varpi - \Omega$; hence $\Delta \omega = \Delta \varpi - \Delta \Omega$, and it is found that, due to the orthogonal component,

$$\Delta \omega = -\cos i \, \Delta \Omega.$$

The right-hand side is the change in ω due to the change in the line of nodes, the origin from which it is measured. Thus if ω is measured from a fixed line in the orbital plane it is, like a, e and T (the orbital period), unaffected by the orthogonal component.

Because of the appearance of the trigonometrical functions of f and E, the magnitudes and signs of the changes in the elements a, e, ϖ and ϵ depend upon the point in the orbit at which the impulse is applied. A full discussion of the dependence of the elements upon the magnitudes of the impulse components to the velocity given by equations (11.29) is given for the ellipse and for the hyperbola by Ehricke (1962). The hyperbolic set corresponding to equation (6.41) is given by Ehricke (1961).

11.3.4 Changes in the orbital elements due to a large impulse

If a change from an ellipse with given elements to another of widely different elements is desired, or even from an ellipse to a hyperbola or vice versa, it can still be accomplished by applying one or more impulses; that is, by applying thrust for a short time. The impulses, however, must now be considered large.

In section 4.12, formulae for the rectangular components of position and velocity in terms of the orbital elements and a given time were derived; the reverse problem of obtaining the elements from the components of position, velocity and a time was also treated. In principle, the problem of transfer from an orbit of given elements (the departure orbit) to a second orbit of given elements (the destination or target orbit) may be solved by the following scheme using the two-body formulae of chapter 4.

(i) Choose a time. From the elements of the departure orbit, compute the position and velocity components of the vehicle at that time.
(ii) Compute the new velocity components at that time (the position being unchanged) required to place the vehicle into the desired transfer orbit.
(iii) Subtract the old velocity components from the new to obtain the required velocity increments, and hence the required impulse increments.
(iv) Use the elements of the transfer orbit and the time it intersects the target orbit to calculate the vehicle's position and velocity components at that time.
(v) Compute its velocity components for that time and position from the target orbit elements.
(vi) Subtract the velocity components derived in calculation (iv) from those computed in calculation (v) in order to find the velocity increments required to place the vehicle into the destination orbit.

The only constraint put on the choice of transfer orbits in the above scheme is that it should touch or intersect both departure and destination orbits. In practice, further constraints arising from the trade-off in fuel expenditure budget, transfer time, sensitivity of transfer orbit to impulse error, and relative positions of arrival and destination points (in the interplanetary case both being planets) impose further limitations on the number of possible transfer orbits.

Some general remarks on the restraints arising from such considerations are given in the following sections.

11.3.5 Variation of fuel consumption with transfer time

It was seen that a given impulse had the greatest effect on the kinetic energy of the vehicle if it was applied tangentially to the orbit. The most economical use of fuel is therefore obtained by tangential impulses. But this fuel-budgeting economy leading to cotangential transfer orbits means that they are slow transfer orbits, most of the time being spent in the true anomaly region $90° < f < 180°$, according to Kepler's second law. If we still retain the tangential impulse for changing from the departure orbit to the transfer orbit we can, by increasing the impulse, increase the semimajor axis of the transfer ellipse. Indeed, as seen in section 11.3.2, a parabolic or hyperbolic transfer orbit may be obtained. Omitting these aperiodic orbits from consideration for the moment we see that the point of intersection of transfer ellipse and destination orbit (assumed circular and coplanar with the circular departure orbit) will regress with increasing impulse as shown in figure 11.5, where the true anomalies of the points A_1, A_2 and A_3 are successively less as the impulse at P increases.

The transfer time t_T is no longer given by half the period of the transfer

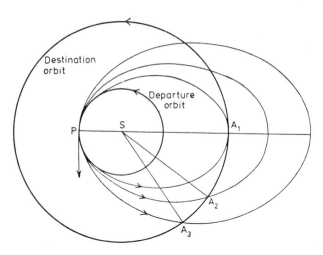

Figure 11.5

orbit, but by the time it takes the vehicle to move to a true anomaly PŜA, which we will write as f_A.

From chapter 4,

$$\tan \frac{E}{2} = \left(\frac{1-e}{1+e}\right)^{1/2} \tan \frac{f_A}{2} \qquad (11.30)$$

$$M = E - e \sin E \qquad (11.31)$$

and by

$$M = n(t-t_0), \qquad t-t_0 = M/n \qquad (11.32)$$

where

$$n^2 a^3 = \mu. \qquad (11.33)$$

The quantities μ, a and e are respectively GM, the semimajor axis and the eccentricity of the transfer orbit; t and t_0 are respectively the time the vehicle reaches the destination orbit and the time it enters the transfer orbit. Hence the transfer time t_T is $(t-t_0)$.

If the radius of the departure orbit is a_1, then

$$a(1-e) = a_1. \qquad (11.34)$$

Also, the pericentre velocity V_P in the transfer orbit is given by

$$V_P{}^2 = \frac{\mu}{a}\left(\frac{1+e}{1-e}\right) = (V_D + V_I)^2 \qquad (11.35)$$

where V_D is the velocity in the departure orbit while V_I is the velocity added by the impulse.

Hence

$$V_D = \sqrt{\frac{\mu}{a_1}}$$

and by equations (11.34) and (11.35) the quantities a and e can be found,

347

enabling equations (11.30) to (11.33) to be used to find the transfer time for a given f_A.

If the destination orbit is a circle, as in this discussion, its known radius a_2 is the radius vector of the vehicle in the transfer orbit when it reaches A, so that

$$a_2 = \frac{a(1-e^2)}{1+e \cos f_A}$$

giving f_A. If the destination orbit is an ellipse, the radius vector at any intersection point may be taken to be specified by the true anomaly, the semimajor axis and the eccentricity of the destination orbit. The velocity magnitude and direction at this point in both transfer and destination orbit can be found by using the relevant equations of chapter 4, namely

$$V^2 = \mu\left(\frac{2}{r} - \frac{1}{a}\right)$$

and

$$\sin \phi = \left[\frac{a^2(1-e^2)}{r(2a-r)}\right]^{1/2}$$

where ϕ is the angle between velocity vector and radius vector. A comparison of both velocity vectors enables the impulse necessary to convert transfer orbit to destination orbit for the vehicle to be computed in the manner shown below.

In figure 11.6, which is a generalization of figure 11.5 to the extent of making the destination orbit an ellipse, V_T and V_N are the velocities in transfer and destination orbits respectively at A, while ϕ_T and ϕ_N are the

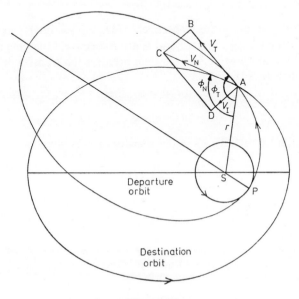

Figure 11.6

respective angles the velocity vectors make with the radius vector. A velocity vector change $\mathbf{V_I} = \mathbf{V_N} - \mathbf{V_T}$ must then be applied to convert from transfer orbit to destination orbit. It is easily seen from the parallelogram of velocities ABCD that

$$V_I = [V_N{}^2 + V_T{}^2 - 2V_N V_T \cos(\phi_T - \phi_N)]^{1/2} \qquad (11.36)$$

while

$$\tan(\phi_N - \phi_I) = \frac{V_T \sin(\phi_T - \phi_N)}{V_N - V_T \cos(\phi_T - \phi_N)} \qquad (11.37)$$

where

$$V_I = |\mathbf{V_I}|$$

and

$$\phi_I = D\hat{A}S.$$

Hence V_I and ϕ_I can be computed. For parabolic and hyperbolic transfers, the corresponding equations from chapter 4 may be used.

Figure 11.5 also shows that not only does the arrival point A regress but the angle of intersection of the transfer orbit with the destination orbit increases. This is an undesirable feature, since it leads to a larger and larger impulse being required to make the necessary orbital change if the vehicle is to enter the destination orbit at A. Thus the saving in transfer time must be balanced against the fuel expenditure in any practical case. The generalization of the problem to a transfer between two ellipses of small eccentricity, their planes inclined at a small angle to each other, does not change the main conclusion that whereas fast transfer orbits exist that intersect either one or both ellipses, such orbits involve much greater fuel expenditure than almost cotangential ones.

11.3.6 Sensitivity of transfer orbits to small errors in position and velocity at cut-off

We now consider the sensitivity of transfer orbits to errors in the velocity and radius vectors at cut-off (that is, when the impulse is ended). Such errors arise because the impulse applied is slightly different from the planned impulse required to put the vehicle into the correct transfer orbit. The transfer orbit which the vehicle enters will have elements $\sigma' = \sigma + \Delta\sigma$, where σ is the value of the planned element and $\Delta\sigma$ is the error in it due to the impulse error ΔI.

To fix our ideas we take a simple coplanar example where a vehicle is supposed at time t_0 (the cut-off time) to have a longitude l, a radius vector r and a velocity of magnitude V in a direction making an angle ϕ with the radius vector. In fact the impulse is incorrect, so that at cut-off the longitude, radius vector, velocity and velocity angle are $l + \Delta l$, $r + \Delta r$, $V + \Delta V$ and $\phi + \Delta\phi$ as shown in figure 11.7.

The elements a, e, τ (time of pericentre passage), and ω (the longitude of pericentre) of the planned elliptic orbit thus have errors Δa, Δe, $\Delta\tau$ and $\Delta\omega$, these quantities being the differences between the elements of the planned orbit and the elements of the actual orbit.

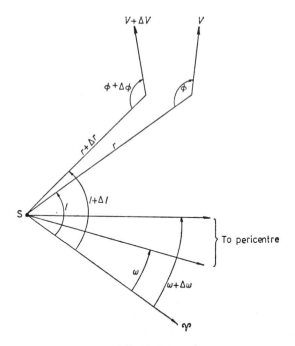

Figure 11.7

The errors may be supposed to be small so that we can obtain expressions for them by partial differentiation of the relevant equations of chapter 4. For the elliptic orbit, these are

$$V^2 = \mu\left(\frac{2}{r} - \frac{1}{a}\right) \tag{11.38}$$

$$T = 2\pi\sqrt{a^3/\mu}, \quad n = 2\pi/T \tag{11.39}$$

$$e = \left[1 - \frac{r}{a^2}(2a - r)\sin^2\phi\right]^{1/2} \tag{11.40}$$

$$r = \frac{a(1-e^2)}{1+e\cos f} = a(1-e\cos E) \tag{11.41}$$

$$E - e\sin E = n(t - \tau) \tag{11.42}$$

$$l = f + \omega \tag{11.43}$$

$$\tan\frac{f}{2} = \left(\frac{1+e}{1-e}\right)^{1/2}\tan\frac{E}{2} \tag{11.44}$$

$$\sin\phi = \left(\frac{1-e^2}{1-e^2\cos^2 E}\right)^{1/2} = \frac{1+e\cos f}{(1+e^2+2e\cos f)^{1/2}} \tag{11.45}$$

$$\cos\phi = \frac{-e\sin E}{(1-e^2\cos^2 E)^{1/2}} = \frac{-e\sin f}{(1+e^2+2e\cos f)^{1/2}}. \tag{11.46}$$

In chapter 4, the procedure was outlined for obtaining the elements a, e, ω, τ from r, V, ϕ, l and t.

Differentiating (11.38), we thus obtain

$$\frac{\Delta a}{a} = 2a \left[\left(\frac{2}{r} - \frac{1}{a} \right) \frac{\Delta V}{V} + \frac{\Delta r}{r^2} \right]. \tag{11.47}$$

Also

$$\frac{\Delta T}{T} = \frac{3}{2} \frac{\Delta a}{a}, \quad \frac{\Delta n}{n} = -\frac{3}{2} \frac{\Delta a}{a}. \tag{11.48}$$

From equations (11.40) and (11.47),

$$\frac{\Delta e}{e} = \frac{r}{a} \left(\frac{r}{a} - 2 \right) \frac{\sin \phi}{e^2} \left[\left(1 - \frac{a}{r} \right) \left(\frac{2\Delta V}{V} + \frac{\Delta r}{r} \right) \sin \phi + \Delta \phi \cos \phi \right]. \tag{11.49}$$

Using equations (11.42) and (11.43) we obtain

$$\Delta \tau = \frac{1}{n} \left[(t - \tau) \Delta n + \Delta e \sin E - \frac{r}{a} \Delta E \right] \tag{11.50}$$

and

$$\Delta \omega = \Delta l - \Delta f. \tag{11.51}$$

To obtain expressions for ΔE and Δf, we use equations (11.41), (11.45), (11.46), (11.47) and (11.49). The required expressions are

$$\Delta E = -\frac{1}{e} \left[\frac{a}{r} \left(\frac{2\Delta V}{V} + \frac{\Delta r}{r} \right) \sin E - \Delta \phi \cos E (1 - e^2)^{1/2} \right] \tag{11.52}$$

and

$$\Delta f = -\frac{1}{e} \left[\left(\frac{2\Delta V}{V} + \frac{\Delta r}{r} \right) \sin f - \left(\frac{r}{a} \cos f + 2e \right) \Delta \phi \right] \tag{11.53}$$

giving finally

$$\Delta \tau = \frac{1}{n} \left\{ (t - \tau) \Delta n + \frac{1}{e} \left[\frac{a}{r} (1 - e^3 \cos E) \left(\frac{2\Delta V}{V} + \frac{\Delta r}{r} \right) \sin E \right.\right.$$
$$\left.\left. - \sqrt{1 - e^2} \, (\cos E - e) \Delta \phi \right] \right\} \tag{11.54}$$

and

$$\Delta \omega = \Delta l + \frac{1}{e} \left[\sin f \left(\frac{2\Delta V}{V} + \frac{\Delta r}{r} \right) + \left(\frac{r}{a} \cos f + 2e \right) \Delta \phi \right]. \tag{11.55}$$

As an example of the use of the above equations in $\Delta a, \Delta e, \Delta \omega$ and $\Delta \tau$, let us suppose that the only error was in the velocity's magnitude so that

$$\Delta r = \Delta \phi = \Delta l = 0.$$

The errors in a, e, T and n are then given by

$$\frac{\Delta a}{a} = 2a \left(\frac{2}{r} - \frac{1}{a} \right) \frac{\Delta V}{V} \tag{11.56}$$

$$\frac{\Delta T}{T}=\frac{3}{2}\frac{\Delta a}{a}, \quad \frac{\Delta n}{n}=-\frac{3}{2}\frac{\Delta a}{a} \qquad (11.57)$$

$$\frac{\Delta e}{e}=\left(\frac{r}{a}-2\right)\left(\frac{r}{a}-1\right)\frac{\sin^2\phi}{e^2}\frac{2\Delta V}{V}. \qquad (11.58)$$

Suppose further that the impulse was applied at pericentre, so that

$$r=a(1-e), \quad \phi=\frac{\pi}{2}, \quad V^2=\frac{\mu}{a}\left(\frac{1+e}{1-e}\right).$$

Equations (11.56) and (11.58) become respectively

$$\frac{\Delta a}{a}=\left(\frac{1+e}{1-e}\right)\frac{2\Delta V}{V}=2\left[\frac{a}{\mu}\left(\frac{1+e}{1-e}\right)\right]^{1/2}\Delta V \qquad (11.59)$$

$$\frac{\Delta e}{e}=\frac{(1+e)}{e}\frac{2\Delta V}{V}=2\left[\frac{a}{\mu}(1-e^2)\right]^{1/2}\frac{\Delta V}{e}. \qquad (11.60)$$

It is seen that in this case the error in a is very much more sensitive for orbits in which the eccentricity is approaching unity.

An example shows just how sensitive orbits are when e is large. A transfer orbit from a circular parking orbit about 500 km above the Earth's surface to the region of the Moon's orbit requires a velocity increment of some 3·058 km s^{-1} to change the circular velocity of 7·613 km s^{-1} to the planned perigee velocity of 10·671 km s^{-1}. This is delivered by applying the appropriate impulse, a velocity error of $\Delta V/V$ occurring. The apogee distance r_A of the resulting transfer orbit will be in error by $\Delta r_A/r_A$, given by differentiating

$$r_A=a(1+e).$$

The required expression is

$$\frac{\Delta r_A}{r_A}=\frac{\Delta a}{a}+\frac{\Delta e}{1+e}. \qquad (11.61)$$

Using equations (11.59) and (11.60), this becomes

$$\frac{\Delta r_A}{r_A}=\left(\frac{4}{1-e}\right)\frac{\Delta V}{V}. \qquad (11.62)$$

The theoretical transfer orbit for the example has an eccentricity of 0·9648 and an apogee of 384400 km. Hence an error of only 30 cm s^{-1} in the cut-off velocity results in an apogee distance error Δr_A of order of 1230 km.

If the error had been solely in the length of the radius vector at cut-off, the same example shows that the error in apogee distance would be given by

$$\frac{\Delta r_A}{r_A}=\left(\frac{3-e}{1-e}\right)\frac{\Delta r}{r} \qquad (11.63)$$

resulting in an apogee error in distance of some 3231 km for an error in the radius vector of 1 km.

A similar analysis may be carried out for hyperbolic orbits. In addition, the problem of orbit sensitivity may be considered taking into account errors in inclination and longitude of the ascending node by allowing position and velocity vectors to suffer errors in all three dimensions. This more complicated problem is not different in principle from the two-dimensional case and will not be treated here.

11.3.7 Transfer between particles orbiting in a central force field

The problem of transfer from one orbit to another in a central force field is usually complicated by the consideration that the departure point and arrival point (for example, two planets in orbits about the Sun) have their own orbital motions in the departure and destination orbits. Neglecting the gravitational fields of these bodies by assuming they have infinitesimal masses, the transfer orbit between the planetary orbits must intersect the destination orbit at a point reached by the target body at that time.

Again a simple example exhibits the main features of this problem. Let two particles P_1 and P_2 revolve in coplanar circular orbits of radii a_1 and a_2 about a body of mass M. Let their longitudes, measured from some reference direction Υ be $(l_1)_0$ and $(l_2)_0$ at time t_0. The problem is to choose a transfer orbit that takes a vehicle from particle P_1 to particle P_2.

The angular velocities of the two particles P_1 and P_2 are n_1 and n_2, given by

$$n_1 = \sqrt{\frac{GM}{a_1{}^3}}, \quad n_2 = \sqrt{\frac{GM}{a_2{}^3}} \qquad (11.64)$$

so that their longitudes at time t are

$$l_1 = (l_1)_0 + n_1(t - t_0)$$

$$l_2 = (l_2)_0 + n_2(t - t_0) \qquad (11.65)$$

respectively.

The time spent by the vehicle in the transfer orbit must be the time taken by the particle P_2 to reach the point of intersection of transfer and destination orbits. This point C therefore lies ahead of the position of P_2 (namely B) when the vehicle leaves P_1 at A.

Then if $B\hat{S}C = \theta$, the transfer time t_T is given by

$$t_T = \theta/n_2. \qquad (11.66)$$

To proceed further, conditions must be laid down concerning permissible lengths of transfer time and permissible fuel expenditures. If fuel economy is the main consideration, the transfer orbit will be a cotangential ellipse between the orbits of P_1 and P_2 (unless $a_2/a_1 \geqslant 15 \cdot 582$; see section 11.3.2). Transfer time t_T is, by (11.16) and (11.24), obtained from

$$t_T = \pi \left[\frac{(a_1 + a_2)^3}{8GM} \right]^{1/2}.$$

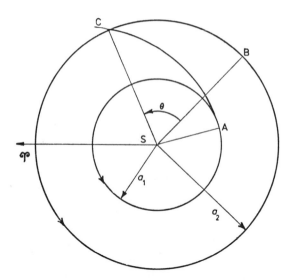

Figure 11.8

Hence by equation (11.66)

$$\theta = \pi n_2 \left[\frac{(a_1 + a_2)^3}{8GM} \right]^{1/2}. \tag{11.67}$$

The longitude of P_1 when the vehicle takes off is π radians less than the longitude of P_2 when the vehicle arrives. Thus the longitudes of the particles at vehicle departure time differ by $(\pi - \theta)$ radians or L_{12} given by

$$L_{12} = \pi \left\{ 1 - n_2 \left[\frac{(a_1 + a_2)^3}{8GM} \right]^{1/2} \right\} = \pi \left\{ 1 - \left[\frac{1 + a_1/a_2}{2} \right]^{3/2} \right\}. \tag{11.68}$$

Now by (11.65),

$$l_2 - l_1 = [(l_2)_0 - (l_1)_0] + (n_2 - n_1)(t - t_0). \tag{11.69}$$

If in (11.69), $(l_2 - l_1)$ is put equal to the right-hand side of (11.68), the resulting expression can be used to find values of t that satisfy it, giving all future epochs at which the vehicle can begin a cotangential transfer orbit from P_1 to P_2. Obviously, in the present problem such epochs are separated by a time interval S, called the *synodic period of one particle with respect to the other* and being the time that elapses between successive similar geometrical configurations of the particles and the central mass.

The synodic period is easily found from the consideration that in one synodic period the radius vector of the faster of the two bodies advances $360°$ (2π radians) on the radius vector of the slower of the two bodies. Hence

$$S(n_1 - n_2) = 2\pi$$

or, using the sidereal periods of revolution of the particles (namely T_1 and T_2, given by $n = 2\pi/T$), we have

$$\frac{1}{S} = \frac{1}{T_1} - \frac{1}{T_2}. \tag{11.70}$$

For a return of the vehicle from P_2 to P_1, the same period must elapse between successive favourable configurations for entry into a cotangential ellipse. The transfer time t_T must be the same as on the outward journey and the angle θ' between the radius vector of P_1 when the vehicle departs and that of the arrival point in P_1's orbit must be given by

$$\theta' = n_1 t_T.$$

Then for a suitable configuration of bodies, the difference in longitudes of P_2 and P_1 must be L'_{21} where

$$L'_{21} = \pi \left\{ n_1 \left[\frac{(a_1 + a_2)^3}{8GM} \right]^{1/2} - 1 \right\} = \pi \left\{ \left[\frac{1 + a_2/a_1}{2} \right]^{3/2} - 1 \right\}. \tag{11.71}$$

Also

$$l_2 - l_1 = [(l_2)_0 - (l_1)_0] + (n_2 - n_1)(t - t_0) \tag{11.72}$$

enabling (11.71) to be used with (11.72) to compute the available epochs for the return journey. The waiting interval between the arrival time at B and the first available departure time can then be found and can be added on to $2t_T$ to give the round trip time.

From symmetry considerations the minimum waiting time t_W can be readily obtained. If P_1 is α degrees 'ahead' of P_2 when a transfer from P_1 to P_2 has just ended, as in figure 11.9(a), the first available transfer back from P_2 to P_1 will begin when P_1 is α degrees 'behind' P_2.

Hence

$$t_W = \left(\frac{360° - 2\alpha}{360°} \right) S = \frac{360° - 2\alpha}{n_1 - n_2}. \tag{11.73}$$

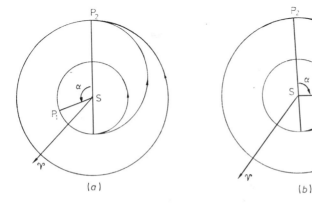

(a)　　　　　　　　(b)

Figure 11.9

Alternatively, if P_1 were α degrees 'behind' P_2 when a transfer from P_1 to P_2 had just ended, as in figure 11.9(b), the first available return from P_2 to P_1 can begin when P_1 has reached a point α degrees 'ahead' of P_2. In this case

$$t_{\mathrm{W}}=\frac{\alpha}{180} \, S=\frac{2\alpha}{n_1-n_2}. \tag{11.74}$$

To compute α, we note that by equation (11.65)

$$l_1-l_2=[(l_1)_0-(l_2)_0]+(n_1-n_2)(t-t_0).$$

Suppose $(l_1)_0$, $(l_2)_0$ were the longitudes of the particles at take-off time t_0 and l_1, l_2 were the longitudes at arrival time t. Then

$$t-t_0=t_{\mathrm{T}}, \quad (l_1)_0-(l_2)_0=\theta-\pi=n_2t_{\mathrm{T}}-\pi$$

giving

$$l_1-l_2=n_1t_{\mathrm{T}}-\pi. \tag{11.75}$$

Then α is given by

$$n_1t_{\mathrm{T}}-\pi=2\pi k+\alpha \tag{11.76}$$

where $-\pi \leqslant \alpha \leqslant \pi$, and k is a positive integer or zero. If α is positive, (11.73) gives t_{W}. If α is negative, (11.74) gives t_{W}.

The waiting time t'_{W} when the journey is from a particle in an outer orbit to one in an inner one and back again is given by

$$t'_{\mathrm{W}}=\frac{2t_{\mathrm{T}}}{n_1/n_2-1} \tag{11.77}$$

where n_1 and n_2 are the mean motions in inner and outer orbits respectively. This result arises from the consideration that during a transfer inwards the outer particle increases its longitude by an angle β less than 180°. The time when the return transfer from the inner particle can begin will therefore occur when the difference between the longitudes of outer and inner particles has increased by an angle 360° -2β. Minimum waiting time is therefore

$$t'_{\mathrm{W}}=\frac{360°-2\beta}{n_1-n_2}=\frac{2n_2t_{\mathrm{T}}}{n_1-n_2}=\frac{2t_{\mathrm{T}}}{n_1/n_2-1}.$$

If more than minimum fuel expenditure for the journey is available, not only can transfer times be cut but the waiting time at B can be shortened. The task of finding a suitable departure configuration is not very much more complicated since, once the transfer orbit has been chosen, the transfer time dictates the necessary configuration just as before. The problem becomes more complicated if the transfer is between two noncoplanar elliptic orbits of differing longitudes of pericentre, but a transfer orbit using two-body formulae can always be found describing the required configuration.

With more than minimum fuel expenditure, the flexibility of choice is greatly increased; many workers have studied the resulting problem of optimizing the transfer orbit with respect to fuel expenditure, sensitivity to error, and transfer and round trip times.

11.4 Transfer Orbits in Two or More Force Fields

In theory the gravitational field of any mass extends to infinity. At any point in space the gravitational force on a vehicle is thus contributed to by all masses in the universe. In practice, we can certainly neglect stars and other galaxies; the problems that arise due to the attractions of Sun, planets and satellites are further simplified because, in most cases, one of these bodies is dominant because of mass and proximity to the vehicle, the others providing negligible forces or merely perturbing forces. Thus the analysis of transfer orbits in a single central force field described in sections 11.3 to 11.3.7 is of practical value.

In discussing the transfer of a vehicle from the immediate neighbourhood of one mass to that of another however, the simple picture of a single force field is not adequate. From being within the first body's force field, the vehicle enters a region where both bodies' fields are comparable in intensity before proceeding onwards into the region in which the second body's field is dominant. For any high-precision study of the behaviour of the vehicle in its transfer orbit special perturbation techniques are required, at least through the two force field region. Yet reliable data regarding some general properties of such transfers may be obtained by using two-body (i.e., single force field) formulae, and in this section the mode of application of such formulae is sketched out.

11.4.1 The hyperbolic escape from the first body

Since we are dealing with the transfer of a vehicle from one force field to another, the vehicle must achieve parabolic (i.e. escape) velocity in the first field if it is going to leave it. In practice, to avoid a large time interval in effecting this manoeuvre, hyperbolic velocity is sought. Any excess of velocity over parabolic velocity dramatically cuts the time spent in the first field. The escape operation is completed when the vehicle has receded from the first mass to a distance such that the gravitational field of the first mass has no further appreciable effect on its orbit, which is now oriented with respect to the other mass.

It is assumed that the entry into the hyperbolic orbit is made from a parking orbit about the first body. This parking orbit may be elliptical or circular and may be coplanar or noncoplanar with the hyperbolic orbit. For the sake of simplicity we will consider here only a circular parking orbit coplanar with the hyperbolic orbit, and a tangential impulse. For this case the geometry of the transfer is shown in figure 11.10, where:

V_c is the circular velocity in the parking orbit of radius ρ_0,
V_e is the velocity of escape (parabolic),
V_h is the hyperbolic velocity actually achieved,
U is the point of intersection of the hyperbola's asymptotes, and
V is the velocity of the vehicle at a distance when it has just left the effective gravitational field of the central mass.

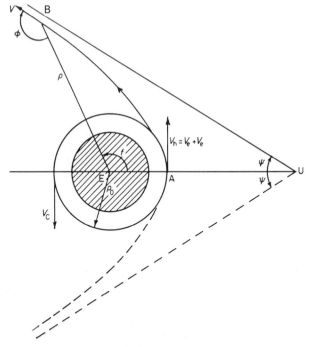

Figure 11.10

First, the velocity V is obtained in terms of the mass m of the first body, the distance ρ from its centre, the velocity of escape V_e and the additional velocity increment v_e added to give it hyperbolic velocity $V_h = V_e + v_e$ at distance ρ_0, the radius of the circular parking orbit.

Now if escape velocity at distance ρ_0 was achieved,

$$\frac{1}{2} V_e{}^2 - \frac{Gm}{\rho_0} = 0. \tag{11.78}$$

However, if the velocity at distance ρ_0 was increased from V_e to $V_e + v_e$ at 'all burnt', then the velocity V of the vehicle at distance ρ is given by

$$\frac{1}{2}(V_e + v_e)^2 - \frac{Gm}{\rho_0} = \frac{1}{2} V^2 - \frac{Gm}{\rho}$$

or, using equation (11.78), by

$$V = \left[\frac{2Gm}{\rho} + v_e(2V_e + v_e) \right]^{1/2}. \tag{11.79}$$

Expression (11.79) then gives the velocity with which a vehicle reaches a distance ρ from the body's centre when it is given, at a certain distance ρ_0, an incremental velocity v_e in addition to escape velocity V_e, where

$$V_e = \sqrt{2} V_c = \sqrt{2Gm/\rho_0}. \tag{11.80}$$

358

The distance ρ is in general many times larger than ρ_0 so that the direction of V is essentially along the asymptote UB to the hyperbola. The angle EUB therefore gives the direction of escape with respect to the direction EA where A is the point at which the motors fired.

By section 4.8, angle EÛB $(=\psi)$ is given by $\tan\psi = \pm b/a$, where $b^2 = a^2(e^2-1)$.

Now at pericentre, by equations (4.88) and (4.92), we have

$$V_h{}^2 = \frac{Gm}{a}\left(\frac{e+1}{e-1}\right), \qquad \rho_0 = a(e-1) \qquad (11.81)$$

and hence

$$e = \left(\frac{V_h}{V_c}\right)^2 - 1. \qquad (11.82)$$

Angle ψ is therefore given by

$$\tan\psi = \pm\frac{V_h}{V_c}\left[\left(\frac{V_h}{V_c}\right)^2 - 2\right]^{1/2}. \qquad (11.83)$$

If an exact direction of velocity V (i.e. the angle ϕ between radius and velocity) were required, then by (4.94)

$$\phi = 90° + \cos^{-1}\left[\frac{a^2(e^2-1)}{\rho(2a+\rho)}\right]^{1/2} \qquad (11.84)$$

where a and e come from (11.81) and (11.82), and ρ is obtained from a knowledge of the distance at which we can neglect the field due to mass m.

Since V is computed at a point just outside the effective limits of the field due to mass m, it is called the *hyperbolic excess* with which the particle escapes.

11.4.2 Entry into orbit about the second body

The vehicle will now enter an orbit about the second body of mass M. If we identify the first body E with a planet and the second with the Sun S, then in all practical cases the heliocentric orbit will be an ellipse, the elements of which are determined by the heliocentric radius vector and velocity vector of the planet and the planetocentric radius vector and velocity vector of the vehicle when it has just left the limits of the planet's effective gravitational field. The situation at this instant is shown in figure 11.11, where the problem depicted is the simple one of the hyperbolic escape orbit being coplanar with the planet's orbital plane about the Sun. (L denotes the limit of the planet's effective gravitational field.) This plane is taken to be in the ecliptic. It is also assumed that the planetocentric velocity vector is along the asymptote to the hyperbola. The vehicle at B has planetocentric radius vector ρ and velocity V given by (11.79) in a direction making an angle DB̂S with its heliocentric direction, where

$$D\hat{B}S = \psi + \theta + (l_E - l_H) \qquad (11.85)$$

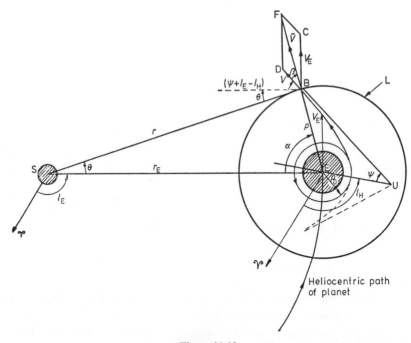

Figure 11.11

θ being the angle between the heliocentric radius vectors of planet and vehicle, l_E and l_H being the heliocentric longitude of the planet and the planetocentric longitude of the impulse point respectively.

Then ψ is known from (11.83) and l_E and l_H are given quantities. The angle θ is obtained from triangle SBE, from the equation

$$\cos\theta = \frac{1-(\rho/r_E)^2+(r/r_E)^2}{2r/r_E} \tag{11.86}$$

where r/r_E is given by

$$\left(\frac{r}{r_E}\right)^2 = 1 + \left(\frac{\rho}{r_E}\right)^2 - 2\left(\frac{\rho}{r_E}\right)\cos\alpha \tag{11.87}$$

while

$$\alpha = \pi - f + (l_E - l_H) \tag{11.88}$$

f being the hyperbolic true anomaly of the vehicle (i.e. angle UÊB).

Now the heliocentric velocity $\bar{V}$ of the vehicle is due to the planetocentric velocity V being compounded with the planet's heliocentric velocity V_E. This is done in the parallelogram of velocities BCFD.

If ϕ_E is the angle between the planet's velocity vector and its radius vector, then CB̂D is obtained from

$$\hat{CBD} = \gamma = \phi_E - \psi - (l_E - l_H) \tag{11.89}$$

which is a known quantity.

360

Hence

$$V^2 = V_E^2 + V^2 + 2V_E V \cos \gamma \qquad (11.90)$$

and

$$\tan \beta = \frac{V_E \sin \gamma}{V + V_E \cos \gamma}. \qquad (11.91)$$

Then the quantities

$$\left. \begin{aligned} l_V &= l_E + \theta \\ \phi_V &= \beta + \psi + (l_E - l_H) + \theta \\ V & \\ r & \end{aligned} \right\} \qquad (11.92)$$

enable the elements a', e', τ' and ω' (semimajor axis, eccentricity, time of perihelion passage and longitude of perihelion) to be computed from the relevant two-body formulae in the usual way.

It may be noted here in passing that for all planets, $\rho/\rho_0 \gg 1$ and $r/\rho \gg 1$, so that θ is usually about one or two degrees, while the direction of V is within a few degrees of the planetocentric radius vector.

11.4.3 The hyperbolic capture

This transfer changes an orbit about a major mass to a closed orbit about a minor mass; for example, the vehicle leaves its elliptic orbit in the heliocentric field and enters a circular or elliptic orbit in the destination planet's gravitational field.

It is theoretically possible in the three-body problem for capture to take place without an expenditure of fuel; it is probable, for instance, that the outermost retrograde natural satellites of Jupiter were once asteroids moving in heliocentric orbits. Making close encounters with the massive planet, the resultant exchange of energy and angular momentum caused each satellite to enter its present quasistable, approximately elliptic orbit about Jupiter. Calculations show however that favourable opportunities for such capture encounters are very rare and that the resultant orbits are strongly perturbed, with a strong probability that on some subsequent occasion escape will take place.

In astrodynamic practice therefore, fuel must be expended at some time during the vehicle's hyperbolic encounter with a planet in order to reduce its energy to that of a closed orbit. This process is obviously the reverse of the hyperbolic escape, the thrust acting in the same direction in which the vehicle is travelling.

In figure 11.12 the geometry of a hyperbolic capture is shown (L is again the limit of the planet's effective gravitational field). A hyperbolic encounter orbit BPJ is transformed at P by the application of a retro-impulse into a circular orbit about the planet. The retro-impulse reduces the planetocentric hyperbolic velocity V_h to circular velocity V_c. In the case illustrated (a

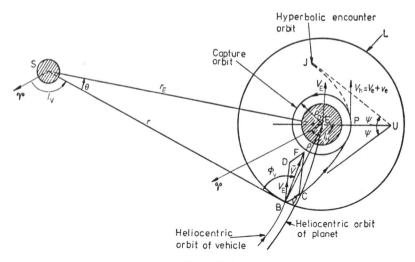

Figure 11.12

direct encounter) the impulse is applied tangentially at pericentre; it is of course possible that the encounter can be retrograde so that the resulting capture orbit is retrograde.

Once the vehicle has reached the distance ρ at which the planet's gravitational field begins to be appreciable, the heliocentric velocity $\bar{V}$, its angle ϕ_V with the heliocentric radius vector of length r, the longitude l_V of the vehicle and the corresponding quantities V_E, ϕ_E, r_E and $l_E = l_V + \theta$ for the planet enable the vehicle's planetocentric position and velocity vectors to be computed. From them and the usual hyperbolic equations, the asymptotic half-angle ψ, the pericentric longitude l_H, distance ρ_0 and velocity V_h can be found, making possible the computation of the necessary change in velocity that will convert the hyperbolic encounter to a circular orbit.

11.4.4 Accuracy of previous analysis and the effect of error

In the preceding sections, no account has been taken of the region about the lesser of the two masses in which both force fields are comparable. The concept of the sphere of influence introduced in section 6.4 is useful here. Two spheres of influence about the satellite of a primary (planet about Sun or moon about planet) were defined by the formulae

$$| \epsilon_P | \sim \frac{m'}{d^2} \left[1 - \left(\frac{d}{1+d} \right)^2 \right] \tag{11.93}$$

$$| \epsilon_S | \sim \frac{d^2}{m'} [1 - (1+d)^{-2}] \tag{11.94}$$

where m' is the satellite's mass in terms of the primary's mass; values of d (the radii of the spheres about the satellite in units of the distance separating primary and satellite) were given when values of $| \epsilon_P |$ and $| \epsilon_S |$ were adopted.

362

The latter two quantities were respectively the ratio of the satellite's perturbing acceleration on the vehicle to the primary's central force acceleration on the vehicle and the ratio of the primary's perturbing acceleration on the vehicle to the satellite's central force acceleration on the vehicle.

A figure of 0·1 for $|\epsilon_P|$ and $|\epsilon_S|$ indicates a moderately high amount of perturbation of an orbit; somewhat less, in fact, than the solar perturbations experienced on occasion by Jupiter's outermost satellites. A figure of 0·01 means a very small perturbation, especially for a vehicle that spends little time in the perturbing region (i.e. in the shell between the two spheres of influence defined by this figure and relations (11.93) and (11.94). In general therefore, we may consider the behaviour of a vehicle to be effectively a two-body problem outside and inside this shell, and a problem that requires more rigorous methods of treatment within the shell if we wish our results to have more than order-of-magnitude accuracy.

In figure 11.13, $|\epsilon_P|$ and $|\epsilon_S|$ are given for a range of values of d, and their variation with m' is also shown. Of interest is the fact that for the terrestrial planets, Mercury, Venus, Earth, Mars and Pluto, no shell exists about these bodies with both $|\epsilon_P|$ and $|\epsilon_S|$ greater than 0·1, indicating that the use of two-body formulae in two-force field feasibility studies concerning

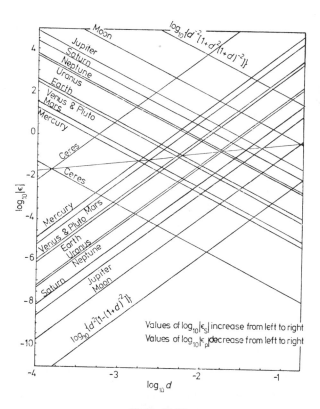

Figure 11.13

transfer between these planets should give reasonably accurate results as long as the vehicle does not linger long in the sphere of influence boundary region, a condition usually realized in practice.

Also inserted in figure 11.13 is the variation of $|\epsilon_P|$ and $|\epsilon_S|$ with d for the Earth–Moon system, giving information about the thickness of the shell around the Moon where the perturbing body is the Earth. Also in the figure is data concerning the most massive asteroid Ceres (diameter ~ 770 km, mass $\sim 1/(2\cdot46 \times 10^9)$ of the Sun's mass) showing that there is no shell about any asteroid in which both $|\epsilon_P|$ and $|\epsilon_S|$ are greater than 0·018.

The computation of the orbit through the shell in precision studies may be carried out using either Encke's method or Cowell's. About halfway inward through the shell (at the boundary of the single sphere of influence given by the formula (6.10)) the heliocentric x, y, z and $\dot{x}$, $\dot{y}$, $\dot{z}$ components of position and velocity of the vehicle are transformed by a simple change of axis to planetocentric x', y', z' and $\dot{x}'$, $\dot{y}'$, $\dot{z}'$ components of position and velocity. The method involves a knowledge of the heliocentric coordinates and velocity components of the planet at this time and is basically similar to the problem in section 2.9.2, where a transfer from heliocentric equatorial rectangular coordinates to geocentric equatorial rectangular coordinates was made. The relationships between the components of position and velocity and the orbital elements were given in section 4.12.

On entering the planet's inner sphere of influence, an unperturbed planetocentric orbit can be adopted until the vehicle exits from the sphere.

The effect of an error in the impulse that places a vehicle in a hyperbolic escape orbit is now more far reaching. In general, the position and velocity of the vehicle as it leaves the planet's effective gravitational field will be in error, being slightly different from the planned position and velocity at this time. In its turn this planetocentric error will result in a heliocentric transfer orbit so that the planned arrival point and velocity of the vehicle at the sphere of influence of the planet of destination will be changed. The hyperbolic capture orbit is now altered so that a different expenditure of energy is required to effect capture.

In section 11.3.6, an elementary analysis of the effects of impulse errors on the elements of a transfer orbit in a single-force field was made. In a similar way, relations giving the errors in the hyperbolic escape orbit of section 11.4.1 could be found in terms of errors in the impulse that transferred the vehicle from its circular parking orbit to its hyperbolic escape path. The errors in l_V, ϕ_V, V and r can then be found from the relations (11.86) to (11.92) and, by using the relevant two-body equations, the relations giving the resulting errors in the elements of the heliocentric transfer orbit may be set up. And so on.

As might be expected, the consequences of this train of error relationships are much more complicated than those for a single-force field, but one main result overshadows everything else. The extreme sensitivity to error of transfer orbits from one planet to another found by such studies makes it absolutely necessary to give any vehicle, manned or unmanned, the ability

364

to correct its orbit in flight. This involves the further necessity of adequate navigational equipment, either on the vehicle or ground-based.

A factor not explicitly mentioned before is the focusing effect of the target body's gravitational field. In order to hit the target body, it is not necessary that the approach orbit of a probe should intersect the planet but only that the pericentron of the hyperbolic encounter orbit should touch the planetary surface (figure 11.14). As long as the asymptote of the hyperbolic approach path is less than a distance OA from the centre of the planet, collision will take place. If R is the radius of the planet, the 'collision radius' OA is given by

$$OA = OU \sin\psi = (R+a) \sin\left[\pi - \cos^{-1}\left(\frac{1}{e}\right)\right]$$

or

$$OA = (R+a)\frac{\sqrt{e^2-1}}{e}$$

where $a = Gm/V^2$, and V as before is the hyperbolic excess; hence

$$OA = \left(R + \frac{Gm}{V^2}\right)\frac{\sqrt{e^2-1}}{e}. \tag{11.95}$$

The effective radius of collision can thus be much larger than the true radius of the body. This is especially true for the giant planets Jupiter and Saturn.

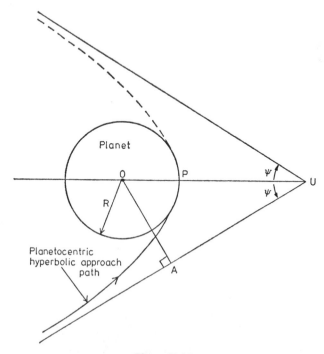

Figure 11.14

11.4.5 The fly-past as a velocity amplifier

In recent years a planetary fly-past has been used to alter the trajectory of a vehicle so that its modified heliocentric orbit takes it to some other planet. For example the fly-past of Venus by Mariner 10 took it inwards to make three subsequent fly-pasts of Mercury; the Voyager fly-pasts of Jupiter took them way out to Saturn and beyond. In this section we look at the way in which a close encounter with a planet may be used to change a space probe's heliocentric velocity, using a number of results obtained in previous sections.

Consider the simple case of a spacecraft travelling in a Hohmann cotangential ellipse between the orbits of Earth and Jupiter. The hyperbolic excess velocity V with which the spacecraft enters the sphere of influence of Jupiter is then given approximately by equation (11.22),

$$V = \left(\frac{\mu}{a_2}\right)^{1/2}\left[1 - \left(\frac{2}{1 + a_2/a_1}\right)^{1/2}\right] \tag{11.96}$$

where $\mu = GM$ (G being the constant of gravitation and M the Sun's mass) and a_1 and a_2 are the orbital radii of Earth and Jupiter respectively. We are assuming that Jupiter overtakes the spacecraft, which at this time is travelling almost tangential to Jupiter's orbit.

Now the Jovian sphere of influence radius r is given by

$$r = \left(\frac{m}{M}\right)^{2/5} a_2 \tag{11.97}$$

where m is the mass of Jupiter. Putting in the relevant values we find that $r = 0 \cdot 322$ AU. The radius of Jupiter in these units is $r_J = 0 \cdot 000477$. By equation (4.91) we can then obtain a value for a, that is, from

$$V^2 = \mu_J\left(\frac{2}{r} + \frac{1}{a}\right) \tag{11.98}$$

where $\mu_J = Gm$, by putting in the relevant values from (11.96) and (11.97).

The vehicle now performs a hyperbolic fly-past of Jupiter within its sphere of influence. Its entry into the sphere of influence may be chosen so that its perijove distance r_p is not much more than the radius of the planet r_J.

By using the relation $r_p = a(e - 1)$, we obtain

$$e = \frac{r_p}{a} + 1.$$

We also have

$$b = a(e^2 - 1)^{1/2}. \tag{11.99}$$

Now the asymptotes of the hyperbolic encounter are given by

$$\tan \psi = \pm b/a \tag{11.100}$$

and it is readily seen from figure 11.15 that the effect of the encounter is to

366

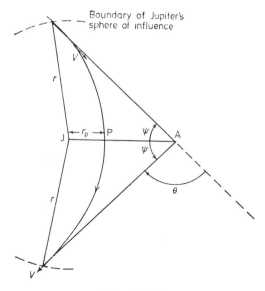

Boundary of Jupiter's
sphere of influence

Figure 11.15

rotate the direction in which the vehicle travels through an angle θ given by

$$\theta = \pi - 2\psi. \tag{11.101}$$

Then by equations (11.99), (11.100) and (11.101), we may write

$$\tan\left(\frac{\theta}{2}\right) = (e^2 - 1)^{-1/2}. \tag{11.102}$$

Substituting numerical values for a_1, a_2, Gm, GM, r and r_J, it is found approximately that

$$\left. \begin{array}{l} V = 1 \cdot 19 \text{ AU/year} \\ \theta = 160 \cdot 3° \end{array} \right\}. \tag{11.103}$$

Circular velocity at Jupiter's distance from the Sun is $V_c = 2 \cdot 76$ AU/year. The velocity of escape from the Solar System (at Jupiter's distance) is therefore $V_c\sqrt{2}$ (i.e. $3 \cdot 90$ AU/year). We see then from equation (11.103) that the effect of the encounter is to eject the spacecraft from Jupiter's sphere of influence in almost the opposite direction to which it entered and with a velocity which, added to Jupiter's orbital velocity, gives a speed greater than the velocity of escape from the Solar System.

The effect could have been further amplified by firing the vehicle's engine at perijove to increase the hyperbolic excess velocity in accordance with the principles of section 11.4.1. It is therefore seen that using a planetary mass as a velocity amplifier has practical applications.

Problems

The data in the appendices should be used where relevant.

11.1 What effect is produced in the velocity increment of a rocket operating in a gravity-free space by (i) doubling the exhaust velocity, (ii) doubling the mass ratio?

11.2 A rocket with an initial mass of 10^7 g contains 8×10^6 g of fuel. The exhaust velocity of the rocket is 2000 m s^{-1} and the fuel consumption rate is 130000 g s^{-1}. Neglecting atmospheric drag, calculate the burn-out velocity of the rocket and its height at that time when it was fired vertically upwards under gravity (take the acceleration due to gravity as a constant $=981$ cm s^{-2}).

11.3 It is proposed to put the upper stage of a two-stage rocket into a circular Earth orbit in which the velocity is 7·73 km s^{-1}. If the motor of the upper stage has an exhaust velocity of 3000 m s^{-1} (twice that of the lower-stage motor), both stages having the same mass ratio R, and the ratio of the fully fuelled upper-stage mass to that of the fully fuelled lower-stage mass is 0·15, calculate R and the initial mass of the rocket, given that the empty upper stage that goes into orbit has a mass of 10^7 g (neglect gravitational and drag losses).

11.4 Compare the velocity increment sums required to transfer a probe from a 2 AU radius heliocentric circular orbit to one of 40 AU (i) by using a single cotangential transfer orbit, (ii) by using a cotangential bi-elliptic transfer orbit with aphelion at 60 AU.

11.5 Compare the transfer times in problem 11.4.

11.6 Two circular coplanar heliocentric orbits have radii 1 AU and 3 AU. A rocket moving in the inner orbit uses its motor to provide a tangential velocity increment 1·6 times the velocity increment required to take the vehicle from the inner orbit by a co-tangential elliptic transfer orbit as far out as the outer orbit. What saving in transfer time to the outer distance is achieved?

11.7 In the preceeding problem what velocity increment is required (i) at the end of the cotangential elliptic transfer orbit, (ii) at the point of intersection of the fast transfer orbit with the outer circular orbit, to place the vehicle in the outer orbit?

11.8 Two circular heliocentric orbits have radii 1 AU and 3 AU and a mutual inclination of 5°. It is proposed to transfer a vehicle moving in the outer orbit by a single elliptic path into the inner one by applying two velocity increments. When should they be applied? Should the change in orbit inclination be made at outer or inner transfer point if a saving in fuel is to be made? Calculate the saving in the velocity increment sum if the correct decision is made.

11.9 Suppose that in the Moon-shot example of section 11, the only error was $\Delta\phi = 1'$ of arc. Find to the first order the resulting errors in the eccentricity, the size and orientation of the semimajor axis, the apogee distance and the time of perigee passage.

11.10 Two asteroids move in circular coplanar heliocentric orbits with the following elements:

Asteroid	Orbital radius (AU)	Longitude at epoch (degrees)	Epoch
A	2	139	1985 January 1·0
B	3·5	271	1985 January 1·0

An absent-minded asteroid prospector working on A decides to move his ship, with the greatest economy in fuel, to B. Find his first available take-off date. When he arrives at B he discovers that he has left his Geiger counter on A and has to go back for it. What is his minimum waiting time on B if the return journey is also made under the fuel economy condition? (Neglect the asteroids' gravitational fields.)

11.11 An interplanetary probe leaves a circular parking orbit of geocentric radius 6630 km with a tangential velocity of 12 km s^{-1}. At a distance of 1 500000 km, the direction of the geocentric velocity vector is assumed to be given by the direction of the asymptote. Find the magnitude of the error involved in this assumption.

References

Ehricke K A 1961 *Space Flight, vol 1: Environment and Celestial Mechanics* (New Jersey: Van Nostrand)
—— 1962 *Space Flight, vol 2: Dynamics* (New Jersey: Van Nostrand)

12 Interplanetary and Lunar Trajectories

12.1 Introduction

In this chapter the results obtained in previous sections will be used to examine problems arising in the transfer of space vehicles between bodies in the Solar System. We first of all consider trajectories in Earth–Moon space before discussing interplanetary operations.

12.2 Trajectories in Earth–Moon Space

The paths followed by vehicles in Earth–Moon space (i.e. within the Earth's sphere of influence of radius 900000 km) may be classified roughly as follows:

(i) Earth orbits,
(ii) Transfer orbits from the vicinity of the Earth to the vicinity of the Moon and vice versa,
(iii) Lunar orbits,
(iv) Landing on Moon or Earth.

In fact, a combination of all or some of the above four classes may describe the mission of a vehicle. Project Apollo (the landing of men on the Moon and their safe return) embodied all four classes of operation. The forces that can act on a vehicle in Earth–Moon space are due to:

(i) the vehicle's rocket motors,
(ii) the Earth's gravitational field,
(iii) the Earth's atmosphere,
(iv) the Moon's gravitational field,
(v) the Sun's gravitational field,
(vi) the planets' gravitational fields,
(vii) the Sun's radiation pressure,
(viii) electromagnetic fields and plasma streams from the Sun.

It is possible to assess immediately the relative importance of these forces.

Unless the vehicle has low-thrust motors, requiring their use for long periods of time, the motors' use will be confined to short time intervals, and without the motors the vehicle will coast under the action of the natural forces operating on it. The action of high-thrust motors can therefore be treated (as in chapter 11) as an impulse which will cause calculable changes in the vehicle's orbital osculating elements.

The effect of the Earth's atmosphere has already been considered in chapter 10 and will not be considered further, since in this chapter we will assume tacitly that any parking orbit about the Earth from which a mission begins is not occupied long enough for atmospheric drag to be appreciable.

The effect of the Sun's radiation pressure on a vehicle can certainly be important in detailed studies of many missions, especially if the probe has a high cross-sectional area-to-mass ratio, but can always be treated as a perturbation.

The effects of the planets' gravitational fields may be completely neglected, as may those due to electromagnetic fields and to plasma streams from the Sun.

The Sun's gravitational field supplies a perturbing acceleration on any body treated as moving within the spheres of influence of Earth and Moon and must be considered if much more than a feasibility study of Earth–Moon trajectories is required.

The dominant natural force acting on a vehicle in Earth–Moon space in the cases of missions (i) and (ii) and in landing on Earth is the force due to the Earth's gravitational potential. The part played by the Earth's oblateness depends upon the distance of the body from the Earth. Unless the vehicle nears or enters the Moon's sphere of influence (see below) all other forces on the vehicle may be treated as perturbations of a geocentric orbit. Within the Moon's sphere of influence the dominant force is that due to lunar gravity, giving a selenocentric orbit disturbed principally by the Earth's field.

Neglecting the solar attraction, there obviously exists on the line joining the centres of Earth and Moon a point where the gravitational forces of these two masses on a vehicle are equal in magnitude and opposite in direction. This *neutral point* is about 0·9 times the Earth–Moon distance from the Earth's centre and exhibits the relative orders of magnitude of the Earth's and the Moon's gravitational influences.

If indeed we use equation (6.11), namely

$$r'_A = \left(\frac{m'}{M'}\right)^{2/5} r_M \qquad (12.1)$$

where m', M' are the masses of Moon and Earth respectively, while r'_A and r_M are the radius of the Moon's sphere of influence and the Moon's geocentric distance respectively, we obtain

$$r'_A = 66\,190 \text{ km}$$

on substituting values for m', M' and r_M. This value is a mean one and varies with the varying Earth–Moon distance, but it indicates a distance from the Moon's centre within which it is better to use a selenocentric orbit disturbed by the Earth.

12.3 Feasibility and Precision Study Methods

The problem of predicting accurately the orbit of a vehicle in Earth–Moon space is essentially a four-body problem (vehicle, Earth, Moon and Sun) which is further complicated by consideration of any thrusts given by the vehicle's motors and possibly by the changes due to radiation pressure.

A general analytical solution is impossible and methods of general and special perturbations have to be applied. Such methods are laborious and time-consuming, special perturbations usually taking up a great deal of machine time. Therefore any approach that provides an insight into classifying orbits for a given problem into obviously unsuitable or possibly suitable ones is welcome. Such approaches are called feasibility studies, as opposed to precision studies that may be employed afterwards to further select from the class of possible suitable orbits the best one.

Feasibility studies usually depend upon setting up a model problem embodying the main features of the real problem but simplified to such an extent that deductions applicable to some degree of approximation to the real problem can be drawn from the model with a minimum of work. Some approaches that have been used by workers in recent years in this context are described below.

12.4 The Use of Jacobi's Integral

If the Sun's attraction is neglected, the orbit of the Moon about the Earth taken to be a circle, and both the Moon and the Earth assumed to be point-masses, the problem of the orbit of a vehicle within Earth–Moon space becomes the circular restricted three-body problem which was discussed in section 5.10. In this model of the Earth–Moon–vehicle system, Earth and Moon represent the two massive particles of masses $(1 - \mu)$ and μ respectively and the vehicle becomes the particle of infinitesimal mass.

The vehicle may be expected to begin any transfer manoeuvre from the Earth's vicinity to the Moon's vicinity by breaking out of a parking orbit about the Earth. For a given impulse supplied by its motors, a given increase in total energy (i.e. kinetic energy increase) will result. The vehicle's new orbit will be a geocentric ellipse, parabola or hyperbola (depending upon the size of the impulse), which will be followed faithfully by the vehicle until the Moon's attraction causes it to depart more and more from its predicted path.

Jacobi's integral and the surfaces of zero velocity derived from it (section 5.10) enable some predictions to be made concerning the flight path of the vehicle under the attractions of Earth and Moon. In the system of coordinates rotating with the Earth–Moon line (figure 5.3) the position and relative velocity of the vehicle after the break-out impulse has been applied may be readily computed. Equation (5.47), Jacobi's integral, is then used to calculate C, the constant of relative energy, by substituting these quantities into it.

Figures 5.4, 5.5 and 5.6 (in particular the first) show the surfaces for various values of C. It is seen that unless C is below a certain value C_2 (figure 5.4 (b)), it is not possible for the vehicle to reach the Moon's vicinity. This value dictates the minimum kinetic energy and therefore the minimum impulse given by the motors that is necessary if the transfer manoeuvre is to succeed. Obviously, a further decrease to C_3 (figure 5.4 (c)) is advisable (i.e. a greater impulse) in order to widen the neck through which the vehicle can pass. If the impulse is too great however, and gives rise to a small value of C such as C_6, almost all of space is available to the vehicle though it is not known what its path will be within that space. It might for instance cross Earth–Moon space and make several revolutions of the Moon as a temporary lunar satellite before, under the cumulative action of the Earth, it escapes and returns to the neighbourhood of the Earth.

12.5 The Use of the Lagrangian Solutions

These special solutions of the three-body problem, previously discussed in sections 5.7, 5.8 and 5.9, show that there exist five points in Earth–Moon space where, neglecting solar perturbations, a particle once placed there will remain with its geometrical relationship to Earth and Moon continuing unchanged. These Lagrangian points (libration points) were shown in figure 5.2. If A and B are the positions of the Earth and Moon respectively, it is found that $L_1A = 0.99$ AB, $L_2A = 0.85$ AB, $L_3A \simeq 1.17$ AB and $L_4A = L_5A = L_4B = L_5B = AB$.

It was also seen that in general the collinear points could not be considered stable positions. On the other hand, the equilateral triangle points are stable if $\mu < 0.0385$. Since for the Earth–Moon system $\mu \sim 0.01$, the points L_4 and L_5 are stable in this system. It should be remembered however that solar perturbations have been neglected.

In section 5.10.3 it was shown that the five Lagrangian points are also characterized by particular values of C (the constant of relative energy in Jacobi's integral) in that as C decreases (i.e. as the particle's initial energy is increased), the points to which the particle could be projected include in succession L_2, L_3, L_1, (L_4 and L_5) since $(1 - \mu) > \mu$ in the Earth–Moon system. Thus the circular restricted three-body problem's findings are again useful in providing some insight into the energies necessary for various types of mission in Earth–Moon space. To progress any further however, other methods must be applied.

12.6 The Use of Two-Body Solutions

Using the same model of the Earth–Moon system, valuable information about trajectories in Earth–Moon space can be obtained by using conic-section orbits to approximate to the actual trajectories. Certain feasibility studies are capable of being tackled to quite a high degree of accuracy in this way, giving data about transfer times, energies required and the shapes of orbits.

The idea of the inner and outer spheres of influence about the satellite of a primary (planet about Sun or moon about planet) introduced in section 6.4 (and used previously in section 11.4.4) can be reintroduced here.

In formulae (11.93) and (11.94), a value for $|\epsilon_P|$ and $|\epsilon_S|$ of about 0·1 allowed a moderate amount of perturbation of an orbit. If we adopt this value, putting $m' \sim 1/81·25$ (its value for the Earth–Moon system), the radii of inner and outer spheres of influence about the Moon are found from figure 11.13 to be of the order of 0·1 and 0·3 of the Earth–Moon distance.

The smaller value indicates that a probe within a distance of about 38000 km of the Moon's centre may be treated as moving in a selenocentric two-body orbit, while the larger value shows that out to some 269000 km from the Earth's centre (about 42 Earth radii) the probe moves in a geocentric two-body orbit. Since the Earth–Moon distance is about 60 Earth radii it is seen that one is able to use two-body formulae over two-thirds of the distance to the Moon for feasibility studies of moderate accuracy.

It should however be remarked that the closeness of resemblance of such orbits to the ones that would actually be pursued depends upon the length of time spent by the vehicle near the boundaries of the transition region shell. For example, a probe that moves in a geocentric ellipse of a certain major axis and eccentricity that takes it out to an apogee distance of 42 Earth radii could, because of Kepler's second law, linger within the perturbing influence of the Moon for a much longer time than one moving in an orbit of a different major axis and eccentricity. The change in the former's orbit could be expected to be larger than that in the latter's. The computation of the orbit through the shell may be carried out by Encke's or Cowell's method in the manner described before in section 11.4.4. On entering the Moon's inner sphere of influence an unperturbed selenocentric orbit can be adopted until the vehicle exits from the sphere.

Enough has been said, therefore, to indicate that feasibility studies can often use two-body conic section solutions to obtain information about trajectories in Earth–Moon space. Indeed, considering a geocentric two-body orbit alone is useful in estimating and comparing transfer times and velocities of the probe at the Moon's orbital distance from the Earth.

To fix our ideas, let us assume that the probe is in a circular parking orbit 560 km above the Earth's surface and in the plane of the lunar orbit. Its circular velocity V_c is then given by

$$V_c = \sqrt{GM/a_P}$$

where G is the constant of gravitation, M is the mass of the Earth and a_P is the radius of the Earth plus 560 km. If the probe is injected into an elliptical orbit tangential to the parking orbit with perigee a_P and apogee a_A, where a_A

is the Moon's geocentric distance, the necessary change in velocity ΔV is given by

$$\Delta V = V_P - V_c$$

where V_P is the velocity at perigee in the new orbit.
Now

$$V_P = \left[\frac{GM}{a}\left(\frac{1+e}{1-e}\right)\right]^{1/2}$$

where a and e are the transfer orbit's semimajor axis and eccentricity respectively. But

$$a(1+e) = a_A$$

$$a(1-e) = a_P$$

and hence a and e may be computed in terms of the known quantities a_P and a_A. With a and e known, ΔV may be found.

In this case the time taken to reach the lunar orbit is easily found from the period T, given by

$$T = 2\pi\sqrt{\frac{a^3}{GM}}.$$

Putting in appropriate values, T is found to be 239 h so that the transfer time is 119·5 h. This is the lunar transfer time with the least energy expenditure. In order to diminish this transfer time V must be increased. A sketch of the orbit with the probe's velocities at various times during the flight is given in figure 12.1. Since the return journey is a mirror image of the

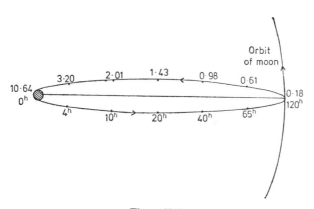

Figure 12.1

outward flight, the velocities (in km s⁻¹) are placed on the upper half of the ellipse for the sake of clarity.

It is seen how rapidly the velocity decreases as the vehicle coasts outward from Earth, exchanging kinetic energy for potential energy, and how eccentric the orbit is ($e = 0.964$), so that it bears a resemblance to a rectilinear ellipse (see section 4.8).

375

Attempts to cut down the transfer time show how highly sensitive it is to changes in perigee velocity. To diminish the time, the semimajor axis of the elliptic orbit must be increased so that the vehicle's apogee lies outside the lunar orbit. An increase of only 18·3 m s^{-1} in perigee velocity increases the apogee distance to about 70 Earth radii and cuts the transfer time to just over 80 h. If this process is continued, the elliptic orbit becomes a parabola when escape velocity V_e is reached, given by

$$V_e = V_c\sqrt{2} = 10\cdot73 \text{ km s}^{-1}$$

for the parking orbit we are considering. Transfer time to the lunar orbit is then found to be about 50 h, the velocity with which the probe crosses the lunar orbit being some 1·433 km s^{-1}. Any increase in perigee velocity beyond escape velocity turns the orbit into a hyperbola, with a further decrease in transfer time.

Any orbit beyond two-thirds of the distance from Earth to Moon will be perturbed strongly if the Moon happens to be in the vicinity of the intersection of vehicle orbit and lunar orbit when the vehicle is in that part of its orbit. In such cases the return half of the orbit (if it is an ellipse) may be transformed completely, but the general picture given above of the variation of transfer time with perigee velocity remains valid.

Some deductions may be made of the behaviour of the probe if it enters the lunar sphere of influence. Neglecting departures of the Moon from a sphere, the Moon's gravitational pull is radially symmetrical; but because of the Earth's field the effective gravitational field within the Moon's sphere of influence is distorted, the departure from radial symmetry being greatest on the Earth side of the Moon.

Any vehicle entering the lunar sphere of influence does so with some hyperbolic excess velocity so that its undisturbed selenocentric orbit will be hyperbolic. Unless its entry velocity is almost zero and a highly improbable chain of terrestrial perturbations reduces its velocity within the sphere, it will escape again along the other leg of its hyperbolic path. In any practical case therefore, an attempt to put a vehicle into an elliptic selenocentric orbit must budget for an impulse which will decrease the vehicle's velocity below escape velocity once it is well inside the lunar sphere of influence. Obviously a small transfer time which brings the probe to the Moon's vicinity with a high selenocentric velocity will require a large fuel budget for converting the hyperbolic path into a closed selenocentric orbit. On these arguments alone, if a given amount of energy for a lunar mission is available (the lunar mission being the establishment of an artificial lunar satellite), it might be better to adopt a slower transfer time. If however the object of the mission is a hard lunar landing, with no attempt at braking, a fast transfer time may be preferable. The hitherto unmentioned factor influencing such decisions is the variation in accuracy with perigee velocity.

In the discussion in section 11.3.6 on the sensitivity of transfer orbits to small errors in position and velocity at cut-off it was seen that an error of only 30 cm s^{-1} in the cut-off velocity resulted in the 384400 km apogee of a lunar transfer orbit being in error by 1230 km. If the error had been in the length of the radius vector at cut-off, the same example gave an error in apogee distance of some 3231 km for 1 km of error at cut-off. Such figures show that slow lunar transfer orbits are highly sensitive to error, requiring that the ability to make midcourse corrections be built into any vehicle as well as allowing fuel for the transformation of the hyperbolic lunar-encounter orbit into a capture orbit if desired. They also show the necessity for precision studies of lunar trajectories, taking into account the effects of the Sun's gravitational field.

12.7 Artificial Lunar Satellites

It is evident that a fast transfer orbit aiming at a lunar impact is the easiest lunar mission. The gravitational field of the Moon exercises a focusing effect in the manner described in section 11.4.4, increasing the collision cross section of the Moon. A close circumnavigation of the Moon that brings the vehicle back to the immediate vicinity of the Earth is much more difficult to achieve. To establish a vehicle in orbit round the Moon also requires a careful choice of transfer orbit, but in addition a subsequent capture manoeuvre once the vehicle has entered deeply into the Moon's sphere of influence is also required. The capture impulse must reduce the selenocentric hyperbolic velocity to elliptic or even circular velocity. A very slow transfer is too error-sensitive to be practical. In figure 12.2 the changes in circular and parabolic velocities with increase in distance from the lunar centre are given, computed from equation (4.42) after putting in the appropriate data. Also shown is the period in a circular orbit. Even if the entry into the Moon's sphere of influence is essentially parabolic, it is seen from figure 12.2 that to achieve a close circular orbit the periselenium parabolic velocity of $2 \cdot 47$ km s^{-1} has to be reduced to $1 \cdot 75$ km s^{-1}, a decrease of $0 \cdot 72$ km s^{-1}. If an elliptical orbit was allowed a smaller impulse would suffice, since any velocity below parabolic for a given distance results in an elliptic orbit. Not all elliptic orbits are suitable however, since orbits of high eccentricity would take the vehicle into the outer regions of the Moon's sphere of influence where terrestrial perturbations would render the orbit unstable, resulting in the eventual escape of the vehicle from control by the Moon or in collision of the vehicle with the Moon. Acceptable elliptic orbits are those whose aposelenia do not let the vehicle exit from the Moon's inner sphere of influence. By section 12.6, this inner sphere's radius is one tenth of the Earth–Moon distance. For satellite orbits of long life, the distance should probably be still further decreased to 20000 km.

The required periselenium velocity of the elliptic orbit of aposelenium 20000 km is then found from equation (11.21) by putting $a_1 = 1738$ km and $a_2 = 20000$ km, giving $a = 10869$ km, $e = 0 \cdot 8401$ and $V_p = 2 \cdot 37$ km s^{-1}. The required velocity of $2 \cdot 37$ km s^{-1} is only $0 \cdot 10$ km s^{-1} below parabolic velocity; thus if the lunar mission were compatible with an elliptic instead of a circular

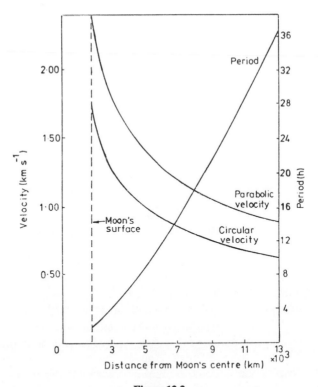

Figure 12.2

orbit about the Moon, a considerable saving in fuel could be made. The impulse need not of course be applied at periselenium or tangentially in the plane of the hyperbolic orbit; but such cases will not be dealt with here.

In the next section the perturbations suffered by an artificial lunar satellite are considered in more detail.

12.7.1 Relative sizes of lunar satellite perturbations due to different causes

The main perturbation suffered by a satellite in an elliptical orbit about the Moon will be due to the departure of the Moon's figure from a sphere, and the attractions by the Earth and the Sun. If the satellite has a large ratio of cross-sectional area to mass, then solar radiation will also produce an appreciable effect; but for most satellites this can be neglected.

It is of interest to compare the sizes of the perturbing accelerations due to the Sun, Earth, and the Moon's figure. Both the Sun and the Earth may be treated as point-masses. If m, m_E, m_S and m_V are the masses of Moon, Earth, Sun and satellite respectively, and the selenocentric radius vectors of Earth, Sun and satellite are r_E, r_S and r respectively, then by equation (6.5), if U is the potential of the Moon's field on the satellite, we may write as the equation of motion of the satellite

$$\ddot{r} = \nabla(U + R)$$

378

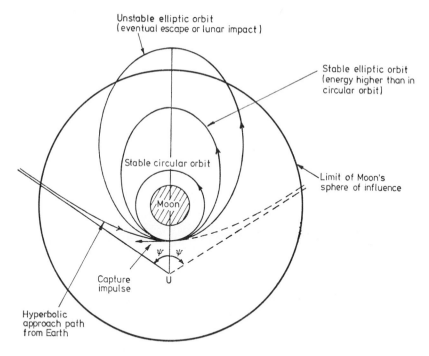

Unstable elliptic orbit
(eventual escape or lunar impact)

Stable elliptic orbit
(energy higher than in
circular orbit)

Stable circular orbit

Moon

Limit of Moon's
sphere of influence

Capture
impulse

ψ ψ

U

Hyperbolic
approach path
from Earth

Figure 12.3

where

$$U = \int \frac{dm}{\rho}$$

$$R = G\left[m_E\left(\frac{1}{r_{VE}} - \frac{\mathbf{r} \cdot \mathbf{r}_E}{r_E{}^3}\right) + m_S\left(\frac{1}{r_{VS}} - \frac{\mathbf{r} \cdot \mathbf{r}_S}{r_S{}^3}\right)\right]$$

while r_{VE} and r_{VS} are the distances between satellite and Earth and satellite and Sun respectively.

Now the main contribution to the perturbing acceleration due to the departure of the Moon from a sphere is the second harmonic. Hence U may be taken in the present problem to be given by

$$U = G\left[\frac{m}{r} + \frac{B + C - 2A}{2r^3} + \frac{3(A - B)Y^2 - 3(C - A)Z^2}{2r^5}\right]$$

where the axes X, Y and Z are fixed in the Moon. This is in fact a version of MacCullagh's formula (see section 6.5).

From the data in section 9.5, it is seen that

$$B - A \sim \frac{1}{3}(C - A)$$

and hence

$$U = G\left[\frac{m}{r} + \frac{(C - A)}{r^3}\left(\frac{2}{3} - \frac{Y^2 + 3Z^2}{2r^2}\right)\right].$$

379

For satellites near the Moon's equatorial plane, $Z \ll r$ and $Y \leqslant r$. Hence the order of magnitude of U may be found from the expression

$$U \sim G \left[\frac{m}{r} + \frac{2(C-A)}{3r^3} \right].$$

The first term gives the central force field potential due to the Moon being taken as a point-mass; the second gives the order of magnitude of the perturbing potential because of the Moon's figure.

Taking the gradient of $(U+R)$, we obtain

$$\ddot{\mathbf{r}} + \frac{Gm}{r^3} \mathbf{r} = G \left[m_E \left(\frac{\mathbf{r}_E - \mathbf{r}}{r_{VE}^3} - \frac{\mathbf{r}_E}{r_E^3} \right) + m_S \left(\frac{\mathbf{r}_S - \mathbf{r}}{r_{VS}^3} - \frac{\mathbf{r}_S}{r_S^3} \right) - \frac{2(C-A)}{r^5} \mathbf{r} \right].$$

It goes without saying that this equation is not the correct equation of motion of the satellite since the last term is only approximate, but it is formed to compare the orders of magnitude of the various perturbing accelerations on the right-hand side.

The equation is now in a suitable form to apply the argument of section 6.4. Defining $|\epsilon_E|$, $|\epsilon_S|$ and $|\epsilon_M|$ as the ratios of the perturbing accelerations of Earth, Sun and departure of the Moon's figure from a sphere to the lunar central force field acceleration, it is readily seen that

$$|\epsilon_E| \sim \frac{d_E^2}{m_E'} [1 - (1 + d_E)^{-2}]$$

$$|\epsilon_S| \sim \frac{d_S^2}{m_S'} [1 - (1 + d_S)^{-2}]$$

while

$$|\epsilon_M| \sim \frac{2(C-A)}{mr^2}$$

where $d_E = r/r_E$, $d_S = r/r_S$ and m_E', m_S' are the masses of the Moon in units of the Earth's mass and the Sun's mass in turn. The above expressions are approximately valid for d_E and d_S much less than unity, which conditions occur in practice.

Then, putting in values for the parameters involved as follows:

$$m_E' = 1/81 \cdot 25, \qquad m_S' = 1/27020000$$

and noting that

$$(C-A)/C = 0 \cdot 000627$$

while

$$C = 0 \cdot 401 \, mr_A^2$$

where r_A is the Moon's radius, we obtain

$$|\epsilon_E| \sim 81 \cdot 25 \, d_E^2 [1 - (1 + d_E)^{-2}] = 162 \cdot 5 \, d_E^3 \left[1 - \frac{3}{2} d_E + 2d_E^2 + \dots \right]$$

$$| \epsilon_S | \sim 2 \cdot 702 \times 10^7 \, d_S{}^2 [1 - (1 + d_S)^{-2}] = 5 \cdot 404 \times 10^7 \, d_S{}^3 \left[1 - \frac{3}{2} \, d_S + 2 d_S{}^2 + \; \dots \; \right]$$

and

$$| \epsilon_M | \sim 0 \cdot 000\,502 \left(\frac{r_A}{r} \right)^2 .$$

These quantities are plotted against distance from the Moon's centre in figure 12.4.

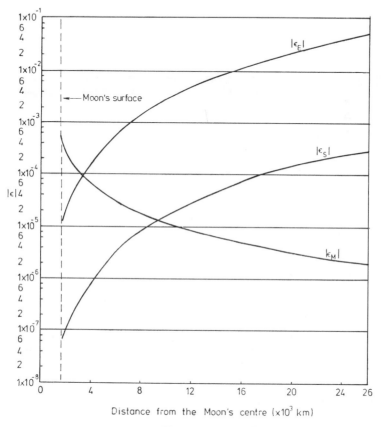

Figure 12.4

It is clear that at a distance of some 30000 km from the Moon's centre the Earth's perturbing effect is much greater than those due to the other two disturbing causes; it is about one-tenth of the central acceleration, so that a lunar satellite that reaches such a distance is probably unstable. Out to 1600 km above the lunar surface the effect of the nonspherical Moon is greater than the Earth's perturbation, with the former about ten times the latter at a height of 400 km above the lunar surface and greater than four times the latter to a height of 800 km. For the whole range, the Sun's effect is only about 0·005 times the Earth's.

381

Since perturbations due to the eccentricity and inclination of the Earth's orbit to the lunar equatorial plane will be smaller than the perturbation due to the Earth being taken to move in a circle in the plane of the Moon's equator, it is seen that out to some 1500 km from the lunar surface the major perturbing effect is due to the figure of the Moon, followed by the effect due to the Earth's circular orbit in the lunar equatorial plane. All other effects are smaller. The fact that *under this simplification* the long axis of the Moon points continuously to the Earth's centre suggests the use of Jacobi's integral in this context.

12.7.2 Jacobi's integral for a close lunar satellite

Take a set of rotating axes (Ox, Oy, Oz) with the Moon's centre as origin, Ox lying along the line joining the centres of mass of the Moon and the Earth, Oy in the lunar equatorial plane 90° ahead of Ox, and Oz perpendicular to this plane as shown in figure 12.5.

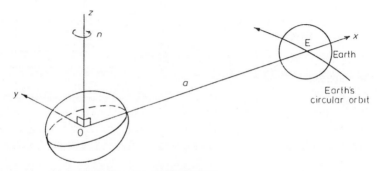

Figure 12.5

This set of axes is identical with the set fixed in the Moon along the three principal axes of inertia. The Earth's coordinates are then $(a, 0, 0)$ where a is the radius of the Earth's selenocentric orbit. If the coordinates of the satellite are (x, y, z) with respect to the rotating axes, the potential V due to the gravitational fields of Moon and Earth is given by

$$V = G \int \frac{dm}{\rho} - Gm_E \left[\frac{x}{a^2} - \frac{1}{(a-x)^2 + y^2 + z^2} \right]$$

and does not contain the time explicitly. The equations of motion of the satellite are thus

$$\ddot{x} - 2n\dot{y} - n^2 x = \frac{\partial V}{\partial x}$$

$$\ddot{y} + 2n\dot{x} - n^2 y = \frac{\partial V}{\partial y}$$

$$\ddot{z} = \frac{\partial V}{\partial z}.$$

Jacobi's integral may thereby be obtained. If they are multiplied in turn by $\dot{x}$, $\dot{y}$ and $\dot{z}$ and then added, the resulting equation may be integrated giving

$$\tfrac{1}{2}[(\dot{x}^2+\dot{y}^2+\dot{z}^2)-n^2(x^2+y^2)]=V+C.$$

It should be noted that V contains the Moon's complete potential.

Any theory of a lunar satellite must go far beyond this if information about the higher harmonics in the Moon's field is to be obtained. A suitable theory can be developed in a manner similar to Earth satellite theories, but is more complicated since the Earth's perturbing effect must be included. Not only is it far stronger than the lunar perturbation on a typical Earth satellite; the long axis of the Moon always points approximately towards the Earth's centre, raising questions of possible resonance phenomena that might cause such large-amplitude oscillations in the radius vector of the satellite that it finally crashes onto the lunar surface. Brumberg (1962), Kozai (1963), Lass and Solloway (1961), Oesterwinter (1966), and Roy (1968) are among those who have produced artificial lunar satellite theories.

12.8 Interplanetary Trajectories

Chapter 1, sections 1.1 to 1.2.5 and the tables in the appendices describe the scene of operations in travel between the planets of the Solar System. Mars and Venus are the planets most easily reached, according to energy requirements. Mars presents a much simpler landing problem than Venus since, not only is its mass less than one-seventh that of Venus, resulting in a much weaker gravitational field to overcome, but surface conditions are not nearly so rugged. Voyages to the other planets (except Mercury) are orders of magnitude more difficult to accomplish.

A number of terms frequently used in describing interplanetary configurations are illustrated in figure 12.6 in which E is the Earth and S is the Sun. The letters V and J refer respectively to an *inferior* planet (one whose orbit is inside the Earth's orbit) and to a *superior* planet (one whose orbit is outside the Earth's orbit).

A superior planet on the observer's meridian at apparent midnight is said to be in *opposition* (configuration SEJ_1). A planet whose direction is the same as that of the Sun is said to be in *conjunction* (configurations EV_1S, ESV_3, ESJ_3); an inferior planet can be in *superior conjunction* (configuration ESV_3) or in *inferior conjunction* (configuration EV_1S).

The angle the geocentric radius vector of the planet makes with the Sun's geocentric radius vector is called the planet's *elongation* (for example, configurations SEV_2 or SEJ_4). It is obvious that an inferior planet has zero elongation when it is in conjunction and maximum elongation (less than 90°) when its geocentric radius vector is tangential to its orbit (configuration SEV_2). The elongation of a superior planet can vary from zero (configuration SEJ_3) to 180° (configuration SEJ_1). When its elongation is 90° it is said to be in *quadrature* (configurations SEJ_2 and SEJ_5). These quadratures are distinguished by adding *eastern* or *western*; in the diagram the north

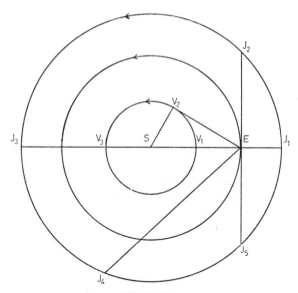

Figure 12.6

pole of the ecliptic is directed out of the plane of the paper, and so J_5 and J_2 are in eastern and western quadratures respectively.

The diagram has been drawn for coplanar circular orbits; the actual planetary orbits are ellipses of low eccentricity in planes inclined only a few degrees to each other, so that the terms defined above are obviously still applicable.

Another useful concept, the synodic period S of a planet, was defined in section 11.3.7 and may be taken in the present context to be the time between successive similar geometrical configurations of planet, Earth and Sun. If T_P and T_E are the sidereal periods of revolution of planet and Earth about the Sun respectively, then

$$\frac{1}{S} = \frac{1}{T_P} - \frac{1}{T_E}$$

for an inferior planet, while

$$\frac{1}{S} = \frac{1}{T_E} - \frac{1}{T_P}$$

for a superior planet.

These relationships are derived for circular coplanar orbits and therefore apply only approximately to the Earth and any other planet in the Solar System. The mean synodic periods for the planets are given in Appendix III.

12.9 The Solar System as a Central Force Field

The dominant gravitational field of the Sun (its mass is over one thousand times that of the most massive planet) means that in space a few million

384

kilometres away from any planet, a vehicle moves in a gravitational field closely resembling that of a simple central force field, in which the intensity falls off as the square of the distance from the Sun. The formulae and conclusions of chapter 4 and those sections in chapter 11 devoted to transfer in a single force field may therefore be used with a high degree of confidence in the study of interplanetary transfer operations.

At distances from the planets given approximately by the sphere of influence argument, there exist regions where the force fields of both planet and Sun are present in comparable intensities, and for precision studies the special perturbation methods of chapter 7 must be used; though in many feasibility studies the approximate methods sketched in chapter 11 can be applied with confidence. That this is so may be seen by studying tables 12.1 and 12.2 and also figure 11.13.

Table 12.1

Planet	Radius (r_A) of sphere of influence		
	Millions of kilometres	Fraction of planetary orbit's semimajor axis	AU
Mercury	0·112	0·00193	0·000747
Venus	0·615	0·00569	0·00411
Earth	0·925	0·00619	0·00619
Mars	0·579	0·00254	0·00387
Jupiter	48·1	0·0619	0·322
Saturn	54·6	0·0382	0·365
Uranus	52·0	0·0181	0·348
Neptune	86·9	0·0193	0·581
Pluto	3·2	0·000536	0·0211

In table 12.1 values of the radii r_A of the planetary spheres of influence are given in millions of kilometres, in astronomical units, and in fractions of the planets' mean distances from the Sun, the figures being computed by using formula (6.10):

$$r_A = \left(\frac{m}{M}\right)^{2/5} r_P$$

where m and M are the masses of planet and Sun respectively and r_P is the planet's semimajor axis. The consequence of the fall-off in intensity of the Sun's gravitational field with distance from the Sun is evident on comparing the sizes of the spheres of influence of Earth and Pluto. The latter sphere is over three times as large as the former, though the mass of Earth is about five hundred times that of Pluto.

The more flexible sphere of influence argument of section 6.4 giving an outer and inner boundary led to the graph in figure 11.13, where a shell about a planet could be defined for any accepted degree of perturbation, showing the range (i.e. the thickness of the shell) over which special or general perturbation methods had to be used. Table 12.2 gives, for two values of $|\epsilon|$, the boundaries of the shells about the planets in which such methods would be called for if perturbation ratios greater than $|\epsilon|$ were not acceptable.

The figures in tables 12.1 and 12.2 should be taken as giving merely the orders of magnitude of the spheres of influence sizes. It should be remembered too that the 'spheres' are only approximate. Nevertheless, the information embodied in the two tables and in figure 11.13 does show how the planets in the Solar System can be divided into two classes where feasibility studies are concerned.

In the first class are Mercury, Venus, Earth, Mars and Pluto (also the asteroids); in this class the use of the formulae of a central-force field (according to the methods of Chapter 11) in feasibility studies should be expected to yield fairly accurate data for interplanetary missions even when perturbation shells are neglected. For precision studies of course, special perturbation methods within the shells must be used.

In the second class are the giant planets Jupiter, Saturn, Uranus and Neptune. Feasibility studies of missions involving these planets (especially the first two) that neglect the perturbation shells about these bodies will at best provide orders-of-magnitude data about transfer times and energy budgets and cannot give real information about the actual orbits of vehicles once they have approached to within the shell boundary. Precision studies can of course always be carried out for these bodies.

12.10 Minimum-Energy Interplanetary Transfer Orbits

By assuming the planetary orbits to be coplanar and circular, the formulae of chapter 11 may be used to give information about energy requirements and transfer and waiting times that are of the right order of magnitude; more precise studies, acknowledging that in reality the orbits of the planets are ellipses of low eccentricity and low inclination to each other, do not change the picture by an order of magnitude.

A mission from the surface of a planet to the surface of another planet can be broken up into three phases:

(i) ascent from the surface of the departure planet to the boundary of its sphere of influence,
(ii) transfer in heliocentric space to the boundary of the destination planet's sphere of influence,
(iii) descent to the surface of the destination planet.

Phase (i) may involve entry into a parking orbit about the departure planet as an intermediate step for check-out purposes before an impulse puts the vehicle into the prescribed planetocentric hyperbolic escape orbit giving the required hyperbolic excess velocity at the point where it leaves

Table 12.2

Planet	$|\epsilon|=0.1$		$|\epsilon|=0.01$	
	Radius of inner boundary of shell ($\times 10^6$ km)	Radius of outer boundary of shell ($\times 10^6$ km)	Radius of inner boundary of shell ($\times 10^6$ km)	Radius of outer boundary of shell ($\times 10^6$ km)
Mercury	No shell exists with both $|\epsilon_P|$ and $|\epsilon_S|$ as large as 0·1		0·053	0·243
Venus			0·24	2·04
Earth			0·40	2·66
Mars			0·27	1·27
Jupiter	29·0	70·8	13·23	217·9
Saturn	35·4	80·5	15·7	243
Uranus	38·6	64·3	17·2	195
Neptune	64·4	107·9	30·6	342
Pluto	As for first four planets		1·9	4·9

the sphere of influence of the departure planet. For high-thrust vehicles in terrestrial planet missions (Mercury, Venus, Earth and Mars), phase (i) will last a week at most.

Phase (ii), apart from possible midcourse corrections, will consist of powerless flight under the dominant action of the Sun's gravitational field and will be described very closely by parts of ellipses (allowing for at least one midcourse correction). This phase accounts for most of the time spent in transit from one planet to another.

Phase (iii) is the reverse operation of phase (i), involving a capture operation transforming the planetocentric hyperbolic encounter orbit into a parking orbit about the planet before the final descent to the surface. Phase (iii) will last no longer than phase one in terrestrial planet missions in general.

A return mission requires the same three phases and is separated in all foreseeable practical cases from the outward mission by a waiting time whose length is specified by the orbital elements of both planets and the performance of the available vehicle. It will be remembered that this waiting time is the period that has to be spent at the destination planet before the planets and the Sun are suitably placed for the return trip to begin. Total mission time for a return trip will therefore be made up chiefly of two phase (ii) transfer times (not necessarily equal) and a waiting time.

It was seen in chapter 11 that the most economical transfer orbits between two particles in circular orbits in a single central force field consisted of cotangential ellipses (omitting the time-consuming bi-elliptic transfer). A transfer from one planet to another and back again under the consideration that a minimum of fuel is to be expended will lead to a total mission time easily obtained by the formulae of chapter 11. The first person to draw attention to such minimum-energy orbits and compute mission times for them was W Hohmann (1925). Taking the planetary orbits to be circular and coplanar, the Earth to be the departure body in all cases, and neglecting times spent in phase (i) and phase (iii) manoeuvres, the use of formulae (11.16) and (11.24) gives the transfer time t_T to be

$$t_T = \pi \left[\frac{(a_E + a_P)^3}{8\mu} \right]^{1/2}$$

where a_E and a_P are the semimajor axes of the orbits of Earth and planet respectively and μ is the product of the Sun's mass and the gravitational constant.

Now the Earth's period of revolution T_E is given by

$$T_E = 2\pi \sqrt{\frac{a_E^3}{\mu'}}$$

where $\mu' = G(M + m_E) \sim GM$, since $m_E/M \sim 1/330000$. Hence

$$t_T = (1+a)^{3/2}/5 \cdot 656 \text{ years} \tag{12.2}$$

the planetary semimajor axis a being now expressed in astronomical units.

The minimum waiting time t_W is found by using formulae (11.73) to (11.77) while the total mission time T equals $(2t_T + t_W)$. The eccentricity of the cotangential transfer orbit comes from (11.23), namely

$$e = \frac{a_2 - a_1}{a_2 + a_1}.$$

For a superior planet

$$e = \frac{a - 1}{a + 1} \tag{12.3}$$

while for an inferior planet

$$e = \frac{1 - a}{1 + a} \tag{12.4}$$

where, as in equation (12.2), the planetary semimajor axis a is in astronomical units.

Table 12.3

Planet	Transfer time t_T (years)	Minimum waiting time t_W (years)	Total mission time $T = 2t_T + t_W$ (years)	Eccentricity of transfer orbit
Mercury	0·289	0·183	0·76	0·44
Venus	0·400	1·278	2·08	0·16
Mars	0·709	1·242	2·66	0·21
Jupiter	2·731	0·588	6·05	0·68
Saturn	6·048	0·936	13·03	0·81
Uranus	16·04	0·932	33·01	0·91
Neptune	30·62	0·766	62·01	0·94
Pluto	45·47	0·061	91·00	0·95

In table 12.3 the transfer times, waiting times and total mission times for round trips to all planets are given, using minimum-energy cotangential ellipses. In addition the eccentricities of these transfer orbits are given. On examining the table, several statements may be made immediately. Manned voyages to the planets beyond Mars are rendered out of the question by the long mission times if orbits close to minimum energy have to be used. Even if unmanned probes were used, reliability of the components over such long intervals of time could not be guaranteed even if information collected by the probes' instruments could be transmitted over distances of many millions of kilometres.

The mission times for Venusian, Martian and Mercurian round trips are not impossible to contemplate for manned voyages, the interesting fact emerging that the Mercurian mission lasts only about a third and a quarter as long respectively as the Venusian and Martian missions. The important factor in these cases is the long waiting time at Mars and Venus before the return journey can be begun. It suggests that the decrease of such long waiting times by the use of different transfer orbits compatible with available energies should have a high priority in the list of factors involved in planning such voyages.

It is also illuminating to consider the actual velocity requirements for such transfer orbits. Let us calculate the velocity increments necessary to place the vehicle into particular heliocentric orbits. The first increment places the vehicle in a parking orbit about the Earth. This orbit, taken to be circular, is assumed to be at a height of 460 km so that a circular velocity of 7·635 km s^{-1} is required. To achieve parabolic or escape velocity from the Earth's field a further increment in velocity of $(\sqrt{2}-1) \times 7 \cdot 635$ km s^{-1} must be added. We suppose that this is added tangentially. In theory this would enable the vehicle to enter the heliocentric gravitational field just beyond the Earth's sphere of influence with almost zero geocentric velocity (zero hyperbolic excess) and a heliocentric velocity equal to the Earth's heliocentric velocity. In order to carry out any interplanetary mission, the actual escape should be made hyperbolically.

Expression (11.79) gives the hyperbolic excess V with which the vehicle leaves the Earth's sphere of influence (radius ρ) when it receives, at a geo-centric distance ρ_0, an incremental velocity v_e in addition to escape velocity V_e, where

$$V_e = V_c\sqrt{2} = \sqrt{2Gm/\rho_0}. \tag{12.5}$$

Rewriting (11.79) we have

$$V = \left[\frac{2Gm}{\rho} + v_e(2V_e + v_e)\right]^{1/2}. \tag{12.6}$$

In figure 12.7 for the parking orbit about the Earth of height 460 km and a radius of the outer sphere of influence ρ taken to be $2 \cdot 66 \times 10^6$ km (such that $|\epsilon_P| \leqslant 0 \cdot 01$), the hyperbolic excess V is plotted against the excess v_e to escape velocity with which the vehicle leaves the parking orbit.

For a cotangential heliocentric transfer orbit the vehicle will leave the Earth's sphere of influence either in the direction in which the Earth is travelling or in the opposite direction. If the Earth's orbital velocity is $V_\oplus$, the first case gives the vehicle a heliocentric orbital velocity of

$$V_V = V_\oplus + V \tag{12.7}$$

and in the second case the vehicle's heliocentric orbital velocity is

$$V_V = V_\oplus - V. \tag{12.8}$$

The first case places the vehicle in a transfer orbit whose perihelion distance is 1 AU; the second case gives a transfer orbit of aphelion 1 AU.

Equations (11.21) and (11.22) may be used to calculate the required velocity increment V, inserting the Earth's orbital velocity of 29·8 km s^{-1} in place of $\sqrt{\mu/a_1}$ when the transfer is to a superior planet and $\sqrt{\mu/a_2}$ when an inferior planet is the planet of destination. The second column in Table 12.4 gives the velocity increments required for cotangential transfer to the various planetary orbits. The use of figure 12.7 then allows the velocity v_e

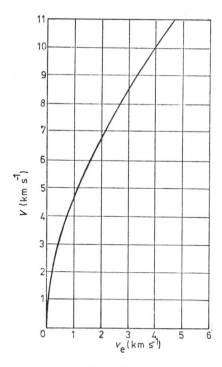

Figure 12.7

in excess of escape velocity at the parking orbit, corresponding to the required hyperbolic excess V to be found. Values of v_e appear in column three of table 12.4. Also in the table are given the hyperbolic excess V and the velocity excess v_e to achieve *heliocentric* parabolic velocity at the Earth's distance from the Sun (i.e. to achieve escape from the Solar System).

To reach any of the planets therefore, the vehicle must be capable of achieving a velocity increment of v_e km s^{-1} in excess of the escape velocity

Table 12.4

Planet	Velocity (km s^{-1}) required in addition to Earth's circular velocity $V_\oplus$	Velocity v_e (Km s^{-1}) required beyond escape velocity
Mercury	7·537	2·362
Venus	2·497	0·290
Mars	2·947	0·396
Jupiter	8·797	3·139
Saturn	10·30	4·115
Uranus	11·29	4·831
Neptune	11·66	5·075
Pluto	11·82	5·197
Interstellar space	12·34	5·608

($10\cdot80$ km s^{-1}) from the parking orbit 460 km above the Earth's surface. It may be remarked that all the planets are within the range of a rocket as powerful as a Saturn 5.

It should be pointed out that no allowance has been made in the above calculations for transformation of the resulting hyperbolic encounter with the planet of destination to an elliptic or circular capture orbit about it. Such a manoeuvre will require a considerable velocity increment in itself, since the vehicle will have to reduce its planetocentric velocity below escape velocity. The size of increment in this manoeuvre will be of the same order of magnitude as that involved in leaving the parking orbit about the planet and entering the heliocentric transfer orbit for the return journey. It should however be noted that the amount of fuel used in the escape manoeuvre from the destination planet will be less than that burned in the preceding capture operation, since the mass of the vehicle is diminished by the mass of fuel burned in the capture manoeuvre. This statement should be revalued in the light of the conclusions of section 12.16.

We can see by the above arguments that the chief obstacle to unmanned flights to the farthest reaches of the Solar System is the forbiddingly long transfer times (table 12.3). The durability of electronic equipment is not such as to guarantee that the package systems would still be operational after such lengths of time. Manned flights are obviously impractical for missions restrained to Hohmann transfers, with the possible exception of missions to Venus or Mars.

The situation, however, is not entirely gloomy. In section 11.4.5 we have seen that it is possible to use a planetary fly-by as a velocity amplifier, and the example was given where the consequence of Pioneer 10's fly-by of Jupiter was its ejection from the Solar System, its heliocentric velocity being increased from $1\cdot57$ AU/year to $3\cdot95$ AU/year. The massive planets Jupiter and Saturn can thus be used as additional power sources to boost inter-planetary probes to speeds such that they reach the outer limits of the Solar System in much shorter times. In addition, with the development of more powerful power sources, it is probable that manned exploration of the inner Solar System will become more practical. Moderately fast transfer orbits can be chosen so that the long waiting times on Mars and Venus can be slashed, especially since an added flexibility is achieved by virtue of the fact that outward and inward transfer paths need not be of the same eccentricity or have the same transfer time.

12.11 The Use of Parking Orbits in Interplanetary Missions

Considerable saving in fuel can be achieved by the use of parking orbits as storage dumps about the planets of departure and destination. The well-known analogy to this procedure is the establishment of a number of base-camps on the route to the South Pole or up the slopes of Mount Everest, in which supplies of food and fuel are left for the return journey; obviously this results in a saving of energy. In the literature of astronautics there are many studies of this use of parking orbits with application to lunar and interplanetary voyages; the Apollo Project essentially used this technique in the lunar-landing phase of the mission. We will consider the method in the following simple example of a journey conducted from the surface of planet P_1 to the surface of planet P_2 and back to the surface of planet P_1. In one case the mission is accomplished by one vehicle that uses a circum-P_1 and a circum-P_2 parking orbit only for checkout purposes ('procedure one'); in the other case, the two parking orbits are used for storing fuel tanks ('procedure two'). The mission phases are shown schematically in figure 12.8 where S is the Sun. The return journey is indicated by the dotted line and it should be remembered that, although it is shown in the diagram as a mirror image of the outward transfer orbit, a finite waiting time on P_2 is in fact necessary before take-off can occur. The orbits of P_1 and P_2 are assumed to be circular and coplanar. The sizes of the circular parking orbits are grossly exaggerated for the sake of clarity.

Then in 'procedure one', the phases of the operation are as listed in table 12.5.

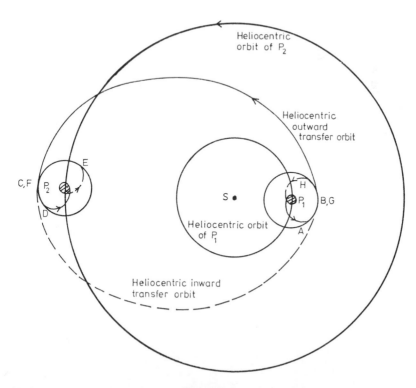

Figure 12.8

Table 12.5

Phase of mission	Characteristic velocity increment required	Remarks (all velocities are in units of the exhaust velocity of the vehicle motor, assumed constant)
$P_1 \rightarrow A$	$v_1 + v_A$	v_1 takes the vehicle to A; v_A puts it in a circular 'circum-P_1' orbit
$A \rightarrow B$	0	Coasting
$B \rightarrow C$	$v_B + v_C$	v_B puts the vehicle into a heliocentric transfer orbit; v_C puts it into a circular 'circum-P_2' orbit
$C \rightarrow D$	0	Coasting
$D \rightarrow P_2$	$v_D + v_2$	v_D takes the vehicle out of the circum-P_2 orbit; v_2 lands it
$P_2 \rightarrow E$	$v_2' + v_E$	v_2' takes the vehicle to E; v_E puts it in the circular circum-P_2 orbit
$E \rightarrow F$	0	Coasting
$F \rightarrow G$	$v_F + v_G$	v_F puts the vehicle into the heliocentric transfer orbit; v_G puts it into a circular circum-P_1 orbit
$G \rightarrow H$	0	Coasting
$H \rightarrow P_1$	$v_H + v_1'$	v_H takes the vehicle out of the circum-P_1 orbit; v_1' lands it

Since the return journey is a mirror image of the outward one, though displaced in longitude, we may assume that in magnitude

$$v_A = v_H, \qquad v_D = v_E$$

$$v_C = v_F, \qquad v_B = v_G$$

though opposite in direction. If both landing and take-off are achieved by the use of the vehicle motors only we may also set

$$v_1 \sim v_1', \qquad v_2 \sim v_2'.$$

Since we are only concerned in this section with the comparison of non-usage and usage of parking orbits for fuel storage, the staging of the vehicle will be neglected and it will be assumed that a one-stage vehicle is used. Then if m is the mass of the capsule and structure that end the flight (no fuel being left in the tanks) and M_0 is the initial mass at lift-off when the vehicle leaves P_1 at the beginning of its mission, equation (11.2) gives (neglecting for the moment any gravitational losses)

$$M_0/m = \exp V$$

where

$$V = 2(v_1 + v_A + v_B + v_C + v_D + v_2) \tag{12.9}$$

it being assumed that the unit of velocity is the value of the vehicle's motors' exhaust velocity, taken to be constant.

If we put

$$V_1 = v_1 + v_A$$

$$V_2 = v_B + v_C$$

$$V_3 = v_D + v_2$$

then

$$M_0/m = \exp (2V_1 + 2V_2 + 2V_3). \tag{12.10}$$

The second procedure is now considered. Again, a combined capsule and structure of mass m will be landed after the flight. The structure, however, is so modified that a part of it containing a store of fuel of total mass m_{AR} can be left in the circum-P_1 orbit while another part (also containing fuel) of total mass m_{CR} can be left in the circum-P_2 orbit. The schedule of phases for this vehicle of initial mass m_0 is then given in table 12.6.

Again it is assumed that

$$v_A = v_H, \qquad v_D = v_E$$

$$v_C = v_F, \qquad v_B = v_G$$

$$v_1 \sim v_1', \qquad v_2 \sim v_2'$$

each velocity increment having the same value as in 'procedure one'. In addition, if the vehicle is always empty of fuel when it regains a tank full of fuel, we may take the capsule plus structure when empty to be of mass m. In fact, since in all phases of the operation apart from the first ($P_1 \rightarrow A$)

Table 12.6

Phase of mission	Characteristic velocity increment required	Remarks (all velocities are in units of the exhaust velocity of the vehicle motor, assumed constant)	Mass of vehicle at end of phase
$P_1 \to A$	$v_1 + v_A$	v_1 takes the vehicle to A; v_A puts it into a circular circum-P_1 orbit	m_A
$A \to B$	0	Coasting; fuel-tank of mass m_{AB} detached	$m_A - m_{AB}$
$B \to C$	$v_B + v_C$	v_B puts rest of vehicle into a heliocentric transfer orbit; v_C puts it into a circular circum-P_2 orbit	m_C
$C \to D$	0	Coasting; fuel-tank of mass m_{CR} detached	$m_C - m_{CR}$
$D \to P_2$	$v_D + v_2$	v_D takes rest of vehicle out of the circum-P_2 orbit; v_2 lands it	m_2
$P_2 \to E$	$v_2' + v_E$	v_2' takes the vehicle up to E; v_E puts it into the circular circum-P_2 orbit beside fuel-tank of mass m_{CR}	m_E
$E \to F$	0	Coasting; fuel-tank m_{CR} coupled on to vehicle	$m_E + m_{CR}$
$F \to G$	$v_F + v_G$	v_F puts the vehicle into the heliocentric transfer orbit; v_G puts it into the circular circum-P_1 orbit beside fuel-tank of mass m_{AB}	m_G
$G \to H$	0	Coasting; fuel-tank m_{AB} coupled on to vehicle	$m_G + m_{AB}$
$H \to P_1$	$v_H + v_1'$	v_H takes vehicle out of circum-P_1 orbit; v_1' lands it	m_1

the structure has to contain less fuel than the procedure-one structure, it may well be less than mass m. We may then put

$$m_E \sim m_G \sim m_1 \sim m. \tag{12.11}$$

Also, by equation (11.2)

$$m_0/m_A = \exp(v_1 + v_A) = \exp V_1 \tag{12.12}$$

$$(m_A - m_{AR})/m_C = \exp(v_B + v_C) = \exp V_2 \tag{12.13}$$

$$(m_C - m_{CR})/m_2 = \exp(v_D + v_2) = \exp V_3 \tag{12.14}$$

$$m_2/m_E = \exp(v_2' + v_E) = \exp V_3 \tag{12.15}$$

$$(m_E + m_{CR})/m_G = \exp(v_F + v_G) = \exp V_2 \tag{12.16}$$

$$(m_G + m_{AR})/m_1 = \exp(v_H + v_1') = \exp V_1. \tag{12.17}$$

If the masses are all taken in units of mass m, then it is easily seen that

$$m_0 = e^{V_1}[e^{V_2}(e^{V_2} + e^{2V_3} - 1) + e^{V_1} - 1] \tag{12.18}$$

while

$$M_0 = \exp 2(V_1 + V_2 + V_3). \tag{12.19}$$

Thus $M_0 > m_0$ for all positive values of V_1, V_2 and V_3. Hence

$$\frac{M_0}{m_0} = \frac{\exp(V_1 + 2V_2 + 2V_3)}{e^{V_2}(e^{2V_3} + e^{V_2} - 1) + e^{V_1} - 1}. \tag{12.20}$$

Some numerical examples are illuminating. For modern chemical fuels, the exhaust velocity v_x is of the order of 2·5 km s^{-1}. For an Earth–Mars–Earth mission, using 460 km altitude parking orbits about both planets, $V_1 v_x$ is about 7·635 km s^{-1} (neglecting gravitational loss in ascent). For $V_2 v_x$ we remember that v_B is the velocity increment to be added to give the vehicle the required hyperbolic excess velocity to put it into the correct heliocentric transfer orbit, while v_C is the velocity increment required to transform the Mars-centred hyperbolic path of the vehicle into the circum-Mars parking orbit.

From table 12.4,

$$v_B = [(\sqrt{2} - 1) \times 7 \cdot 635 + 0 \cdot 396] \text{ km s}^{-1} = 3 \cdot 558 \text{ km s}^{-1}.$$

To obtain v_C, the hyperbolic excess V when the vehicle enters the outer Martian sphere of influence is first found from equation (11.22), namely

$$V \equiv \Delta v_B = \left(\frac{\mu}{a_2}\right)^{1/2}\left[1 - \sqrt{\frac{2}{1 + a_2/a_1}}\right]$$

where $(\mu/a_2)^{1/2}$ is the orbital velocity of Mars, and a_1 and a_2 are the Earth's and Mars' semimajor axes respectively. From the data given in the appendices

$$V = 2 \cdot 657 \text{ km s}^{-1}.$$

Equations (12.5) and (12.6) relate V to v_C thus:

$$V_e = V_c\sqrt{2} = \sqrt{2Gm/\rho_0}$$

and

$$V = \left[\frac{2Gm}{\rho} + v_e(2V_e + v_e)\right]^{1/2}.$$

Since $v_C = (\sqrt{2} - 1)V_c + v_e$, we obtain

$$V = \left[\left(\frac{2Gm}{\rho} - \frac{1}{2}V_e^2\right) + v_C^2 + 2v_C V_c\right]^{1/2}. \tag{12.21}$$

Now if escape velocity at a distance ρ is $V_{\rho e}$, equation (12.21) becomes

$$V = [V_{\rho e}^2 + (V_c + v_C)^2]^{1/2} \tag{12.22}$$

since

$$V_{\rho e}^2 = 2Gm/\rho.$$

Hence

$$v_C = [V^2 + (V_e^2 - V_{\rho e}^2)]^{1/2} - V_c. \tag{12.23}$$

For Mars, the outer sphere of influence has a radius of $1\cdot27 \times 10^6$ km (table 12.2); using this value for ρ and other relevant data from the appendices, the value of $2\cdot657$ km s^{-1} for V gives $v_C = 2\cdot073$ km s^{-1} or $v_e = 0\cdot690$ km s^{-1}. Hence

$$V_2 v_x = v_B + v_C = 5\cdot631 \text{ km s}^{-1}.$$

For ascent into the circum-Mars orbit, the equation

$$v_2' + v_E = v_D + v_2 = \sqrt{Gm/\rho_0} = 3\cdot340 \text{ km s}^{-1}$$

gives $V_3 v_x = 3\cdot340$ km s^{-1}. Then from equations (12.18), (12.19) and (12.20), we have

$$m_0 = 5060, \quad M_0 = 588\,000, \quad M_0/m_0 = 116.$$

The values obtained for the mass ratios m_0 and M_0 are of course completely impractical for one-stage rockets using chemical fuels. The ratio M_0/m_0 does however suggest that real advantages could be gained by using some form of this rendezvous technique.

As a second example, let the exhaust velocity of the vehicle be doubled to 5 km s^{-1}. Then

$$V_1 = 1\cdot527, \quad V_2 = 1\cdot126, \quad V_3 = 0\cdot668$$

and

$$m_0 = 100\cdot1, \quad M_0 = 767, \quad M_0/m_0 = 7\cdot66.$$

It may be noted how sensitive the initial mass of the vehicle is to an improvement in exhaust velocity, and also how the advantage of fuel storing in parking orbits diminishes with increase of vehicle exhaust velocity, though such storing remains very useful.

Even when step rockets are considered instead of the one-stage vehicles used in the above examples, there remains a marked advantage in using a rendezvous technique since a saving in fuel must result when mass left at an intermediate station need not be acted upon by subsequent motor thrusts. There are nevertheless certain difficulties in the rendezvous method; for example, it may not be possible to store fuel in tanks in space for an arbitrary time or couple up tanks without massive auxiliary equipment. A possible solution to this is that the fuel for the end phase $(H \rightarrow P_1)$ is not placed in orbit by the vehicle but is put into orbit by special Earth-orbit ferry rockets once the interplanetary vehicle has returned to its circum-Earth orbit. If indeed the interplanetary vehicle has a low-thrust motor with high exhaust velocity, it would probably be assembled in the circum-Earth orbit in any case since it could not ascend from surface to orbit. The end phase would therefore be conducted with powerful ferry rockets. At the other end of the interplanetary transfer orbit the vehicle would remain in orbit about Mars while another ferry rocket, carried across space by the interplanetary vehicle to the circum-Mars parking orbit, was used to carry out the planetary phases $(D \rightarrow P_2)$ and $(P_2 \rightarrow E)$. A number of ships would offer obvious advantages where the safety factor is concerned and in some studies the logistics demand that a proportion of such ships be abandoned at the end of phase $(P_2 \rightarrow E)$, together with the ferry rockets used at the planet of destination before the remaining interplanetary craft are injected into the return heliocentric transfer orbit.

Navigation problems also enter the picture, since the ships must find each other and match velocities in order to rendezvous. Such problems have however essentially been already solved in manned flights to the Moon and in rendezvous missions such as the Apollo–Soyuz flight.

Some workers have suggested that too much time and effort has been wasted in such studies, especially since interplanetary round-trip missions based on chemical fuels are almost certain never to be realized in practice; but in the author's opinion, such studies have been invaluable in emphasizing the basic principles in the field and developing a number of useful techniques that must eventually find their application (von Braun 1953).

It is clear, at the time of writing (1988), that an increasing number of new space missions must utilize some of the concepts dealt with in chapters 10, 11 and 12. The Phobos, Galileo, Magellan, MIR and Columbus missions are among the list.

Rendezvous in space of one spacecraft with another is now an everyday technique. It will be used repeatedly in the building of the space stations as indeed has already happened with the bit-by-bit construction of MIR, the USSR's station, now proceeding steadily.

The use of a planet's gravitational field to act as a velocity amplifier for a spacecraft has already found a number of applications such as the use of Venus by Mariner 10 to achieve a fly-past of Mercury, the Voyager and Pioneer uses of Jupiter to increase their heliocentric velocities to escape velocity, enabling in the case of Voyager 2 the outer planets of the Solar System to be reached in a fraction of the time a classical Hohmann transfer would have taken. The projected ESA Cassini mission to Saturn to explore the Saturnian system plans to use repeated close encounters with Titan, Saturn's largest moon, to produce orbital changes taking the spacecraft past many of the other satellites in turn.

Studies of the construction of massive solar power satellites and permanently manned space stations of large size for a multitude of scientific and technological purposes are no longer science fiction but potentially realizable, given the present state of the art. There are detailed plans to return to the Moon and establish one or more mass driver stations to deliver payloads of lunar material to low Earth orbit. Such plans seem to be technologically sensible and feasible. Energy-wise, it is more economical to ship material from Moon surface to Earth orbit than to lift it into orbit from Earth surface. It is also wiser to use the limitless supply of solar power available on the Moon, converted to electricity, to accelerate payloads on the electromagnetic launcher (the mass driver) to lunar escape velocity of 1·6 miles per second than to build a new generation of enormous rockets to lift the required massive payloads from Earth surface to Earth orbit. Logistically, for solar power satellites and large space stations we are budgeting in terms of hundreds of thousands of tons of material delivered to orbit and the Earth's satellite has more than enough to spare.

12.12 The Effect of Errors in Interplanetary Orbits

The findings of sections 11.3.6 and 11.4.4 may be applied to interplanetary orbits to obtain an idea of the sensitivity of such orbits to small errors in the position and velocity of the vehicle at a given time.

It will be remembered that the effect of an error in the impulse that places a vehicle in a hyperbolic escape orbit is far reaching. The impulse error will produce errors in the position and velocity of the vehicle as it leaves the planet's outer sphere of influence. These errors produce a slightly different heliocentric transfer orbit resulting in a changed arrival point (and time) on the sphere of influence of the planet of destination. Finally the new planetocentric hyperbolic capture orbit requires a new fuel expenditure budget to transform it into a closed planetocentric orbit.

In section 11.4.4 the way in which analytical expressions relating such error chains could be set up was indicated, and it was stated that applications of such functions showed how extremely sensitive interplanetary orbits were to initial impulse error. This sensitivity varies with the magnitude of the hyperbolic excess velocity V and also with its direction compared to the planet's orbital velocity direction; in its turn, it has been seen in section 12.10 (by figure 12.17) and equations (12.5) and (12.6) that the sensitivity of V to change in v_e, the incremental velocity in addition to escape velocity V_e from the circum-planet parking orbit, is itself a function of v_e, being most sensitive for small v_e.

A numerical example illustrates how sensitive such orbits are. In the Earth–Mars cotangential transfer, a vehicle's motors give it a velocity error Δv_e of 30 cm s^{-1} in the incremental velocity in addition to escape velocity V_e with which it leaves the circum-Earth parking orbit. What is the resulting error in its heliocentric orbit's aphelion?

By equation (12.6)

$$V = \left[\frac{2Gm}{\rho} + v_e(2V_e + v_e) \right]^{1/2}. \tag{12.24}$$

A change Δv_e of 30 cm s^{-1} gives a new hyperbolic excess velocity V_1, given by expanding equation (12.24) after substituting $(v_e + \Delta v_e)$ for v_e in it:

$$V_1 = V \left[1 + \left(\frac{v_e + V_e}{V^2} \right) \Delta v_e \right]. \tag{12.25}$$

Then from table 12.4, we have $v_e = 0.396$ km s^{-1}, $V = 2.947$ km s^{-1} while $V_e = 10.80$ km s^{-1} and $\Delta v_e = 30$ cm s^{-1}, giving $V = (V + 0.00114)$ km s^{-1}. Hence using equation (12.7) and inserting 29·8 km s^{-1} for the Earth's orbital velocity V_V, the perihelion velocity of the vehicle in its heliocentric transfer orbit is 32·7481 km s^{-1} instead of 32·7470 km s^{-1}.

By equation (11.23), namely

$$e = \frac{a_2 - a_1}{a_2 + a_1}$$

or from table 12.3, the eccentricity of the transfer orbit is 0·21. The error Δr_A in the aphelion of the orbit is thence found from equation (11.62):

$$\frac{\Delta r_A}{r_A} = \left(\frac{4}{1-e} \right) \frac{\Delta V}{V}$$

by putting

$$r_A = a_2 = 1.5237 \text{ AU} = 227.8 \times 10^6 \text{ km}$$

$$e = 0.21$$

$$\Delta V = 0.00114 \text{ km s}^{-1}$$

$$V = 32.7470 \text{ km s}^{-1}.$$

It is found to be 40200 km, or six times the diameter of Mars. A similar calculation for Jupiter gives an aphelion error in the transfer orbit for an error of 30 cm s^{-1} in v_e of 118000 km, rather less than one Jovian diameter.

. In fact, as pointed out in section 11.4.4, the effective collision cross section of a planet depends upon the body's gravitational field; thus, although the above examples indicate a high sensitivity in transfer orbit to errors in cut-off velocity, this is offset (especially in the cases of Jupiter and Saturn) by their extensive fields of influence which strongly focus trajectories in their neighbourhood. Even so, any vehicle must possess an adequate fuel supply for course-correction procedures which also involves the necessity of adequate navigational equipment, either on the vehicle or ground-based.

Problems

12.1 An astronaut on the surface of the Moon observes an artificial lunar satellite pass through his zenith with a certain angular velocity. Assuming the satellite to be in a circular orbit at a height of 400 km above the Moon's surface, calculate the observed angular velocity in degrees per second.

12.2 Calculate the selenocentric radius vector of an artificial lunar satellite moving in a circular orbit in the plane of the lunar equator that would always have the same selenographic longitude. Why is it not possible to have a satellite in such an orbit?

12.3 Find to four significant figures the distance of the so-called neutral point on the Earth–Moon line of centres from the Earth's centre as a fraction of the Earth–Moon distance (take the Moon's mass to be $1/81 \cdot 25$ that of the Earth). Find the distance from the Earth of the other point on this line at which the magnitudes of the forces of Earth and Moon on a probe are equal.

12.4 What is the order of magnitude of the ratio of the perturbing acceleration due to the Earth to the central two-body acceleration of the Moon on a probe at the neutral point?

12.5 Calculate to four significant figures the distance of L_1, L_2 and L_3 (figure 5.2) from the Earth's centre for a probe in the Earth–Moon system. (Assume the Moon's orbit about Earth is circular and the mass of the Moon to be $1/81 \cdot 25$ that of the Earth. You may take the values given for L_1A, L_2A and L_3A in section 12.5 as a first approximation.)

12.6 Compare the collision radii of Jupiter and Mars for probes approaching those planets with a hyperbolic excess velocity of 2 km s^{-1} and an eccentricity of $1 \cdot 3426$.

12.7 Calculate the two hyperbolic excess velocities at exit from the Earth's outer sphere of influence ($|\epsilon| = 0 \cdot 01$) required to place a probe on a cotangential transfer orbit that ends (i) at the perihelion, (ii) at the aphelion of Mars. (Neglect the eccentricity of the Earth's orbit, the size of its sphere of influence with respect to the size of the Earth's orbital radius, and the mass of Mars, and assume the orbits are coplanar.)

12.8 An asteroid is discovered moving in the plane of the ecliptic with a semimajor axis of $2 \cdot 0045$ AU and an eccentricity of $0 \cdot 08456$. Neglecting the eccentricity of the Earth's orbit, show that the total mission time for a reconnaisance of the asteroid to take photographs of its surface at fly-by and bring them back to within 10^6 km of the Earth for radio transmission can be as little as 2 years, even if the transfer is to be a cotangential orbit. What hyperbolic excess velocity is required from the Earth's outer sphere of influence ($|\epsilon| = 0 \cdot 01$)? (Assume that the gravitational field of the asteroid can be neglected.)

12.9 In the preceding problem calculate the duration of the flight from a circular circum-Earth parking orbit at 480 km height to the outer sphere of influence and the departure velocity from the parking orbit if the hyperbolic escape orbit is tangential to the parking orbit.

12.10 A probe is intended to sample continuously interplanetary conditions out of the plane of the ecliptic and transmit its data to Earth. To this end the probe's orbit is intended to be a circular orbit of period one year. On injection into its heliocentric orbit the probe leaves the Earth's outer sphere of influence radially with a hyperbolic velocity excess of 5 km s^{-1} and a heliocentric radius vector of 1 AU. Show that the probe's heliocentric orbit is not quite circular but that the eccentricity is less than $0 \cdot 0015$ and can consequently be neglected. Calculate (i) the inclination of the resulting orbit to the plane of the ecliptic and (ii) the probe's maximum geocentric distance. At what approximate date after the probe is injected into its heliocentric orbit does it first re-enter the Earth's sphere of influence? (Neglect the Earth's orbital eccentricity; take the radius of the Earth's outer sphere of influence to be $2 \cdot 66 \times 10^6$ km.)

References

von Braun W 1953 *The Mars Project* (University of Illinois)
Brumberg V A 1962 *Bull. Inst. Theor. Astron.* (Leningrad) **8** 705
Glaser P E 1973 *Astronautics and Aeronautics* **11** 60-8
Heppenheimer T A 1977 *Colonies in Space* (New York: Warner)
Hohmann W 1925 *Die Erreichbarkeit der Himmelskorper* (*The Accessibility of the Heavenly Bodies*) (Munich: Oldenburg)
Kozai Y 1963 *Publ. Astron. Soc. Japan* **15** 301
Lass H and Solloway C R 1961 *J. Am. Rocket Soc.* **31** 220
Oesterwinter C 1966 *Astron. J.* **71** 987
Roy A E 1968 *Icarus* **9** 82,133

Bibliography

Ehricke K A 1961 *Space Flight, vol. 1: Environment and Celestial Mechanics* (New Jersey: Van Nostrand)
—— 1962 *Space Flight, vol 2: Dynamics* (New Jersey: Van Nostrand)

13 Orbit Determination and Interplanetary Navigation

13.1 Introduction

In this chapter three closely related subjects are discussed; namely orbit determination, orbit improvement and interplanetary navigation. In *orbit determination* the elements of a body observed in the Solar System are found from the reduced observational data. The classical methods of Laplace, Gauss and others have had to be based on observations of the bodies' positions on the observer's celestial sphere (usually given in right ascension and declination). Since the orbit of the body about the Sun is a conic section (omitting perturbations from consideration) six elements have in general to be found, so that observations of the body's right ascension and declination at three different times constitute the minimum number of pieces of data required to find its orbit. This is certainly true for an elliptic or a hyperbolic orbit; in the parabolic case (since $e = 1$) only five elements are required to be found, so that in theory three right ascensions and two declinations should suffice; while for the circular case (with $e = 0$ and the longitude of perihelion meaningless), two observations of right ascension and declination should be sufficient. In practice however, various other considerations enter and it may be said that three different observations at different times are required before a satisfactory preliminary orbit can be found. To obtain an orbit that approximates to the actual orbit of the observed object is indeed the goal of orbit determination; from such an approximate or preliminary orbit an ephemeris (a table of calculated positions) that will give predictions of the body's future coordinates can be set up. These are used for tracking the object so that more observations may be collected for future orbit improvement computations, as shown below.

Observational information additional to the observed right ascensions and declinations of the object may be available in a particular astrodynamic case. Such information is usually radar-obtained and consists of range and range-rate measurements (see chapter 3). The classical orbit determination methods have therefore been modified to take advantage of such additional data.

The task of *orbit improvement*, as its name implies, is simply to obtain more accurately the elements of the body's orbit. If the preliminary orbit was reasonably close to the actual one, its orbital elements will differ from

404

the actual orbital elements by small quantities. Equations may be set up relating such quantities to the differences between the observed right ascensions and declinations of the body and its predicted position co-ordinates. The equations, which are linear, can then be solved by the method of least squares to give the corrections to the preliminary orbit's elements.

In astrodynamics, the preliminary orbital elements may well be known beforehand. For example, an interplanetary probe will have a desired pre-computed orbit; when fired, the probe may be expected to be placed in an orbit not too much different from the theoretical orbit. In such a case the orbit determination is unnecessary. In other cases, when the precom-puted orbit is not available, the preliminary orbit must be found from observations.

What is certainly new in the last few years is the possibility of observations leading to orbit determination being carried out with spaceship-based instruments. Consideration of the use of such methods is the province of *interplanetary navigation,* so called because it appears that their most extensive use will be in vehicles on lunar or interplanetary missions and not on artificial satellites. Special optical and electronic devices are involved here and the subject will be briefly discussed in the last part of this chapter. In the first part, the classical methods of orbit determination and their modern modifications will be described briefly; after that, the basic ideas used in orbit improvement will be given.

13.2 The Theory of Orbit Determination

Let the heliocentric equatorial coordinates of the Earth E and a space vehicle V at a given time be (X, Y, Z) and (x, y, z) respectively, with their heliocentric distances R and r being given by

$$R^2 = X^2 + Y^2 + Z^2, \quad r^2 = x^2 + y^2 + z^2.$$

The geocentric distance ρ of the vehicle is then related to R and r by the equation

$$r^2 = \rho^2 + R^2 - 2\rho R \cos \theta \tag{13.1}$$

where θ is the angle $S\hat{E}V$ in triangle VSE (figure 13.1) and S is the Sun. The vehicle's geocentric coordinates (x', y', z') are related to its right ascension α, declination δ, and geocentric distance ρ (we suppose the observations α and δ to have been corrected for parallax, precession, etc. according to the methods of chapter 3) by the equations

$$\left. \begin{array}{l} x' = \rho \cos \delta \cos \alpha = \rho l \\ y' = \rho \cos \delta \sin \alpha = \rho m \\ z' = \rho \sin \delta = \rho n \end{array} \right\} \tag{13.2}$$

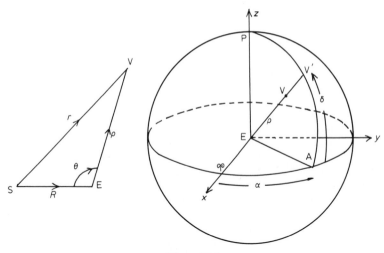

Figure 13.1

where l, m and n are the geocentric direction cosines of the vehicle. Then

$$\left.\begin{array}{l} x'=x-X=\rho l \\ y'=y-Y=\rho m \\ z'=z-Z=\rho n. \end{array}\right\} \tag{13.3}$$

Differentiating the first of equations (13.3) twice with respect to time we get

$$\dot{\rho}l+\rho\dot{l}=\dot{x}-\dot{X} \tag{13.4}$$

and

$$\ddot{\rho}l+2\dot{\rho}\dot{l}+\rho\ddot{l}=\ddot{x}-\ddot{X}. \tag{13.5}$$

But both the Earth and the vehicle, of masses m_E and m_V respectively, move in orbits around the Sun (mass M). These orbits are given by

$$\ddot{\mathbf{r}}+G(M+m_V)\frac{\mathbf{r}}{r^3}=0 \tag{13.6}$$

and

$$\ddot{\mathbf{R}}+G(M+m_E)\frac{\mathbf{R}}{R^3}=0 \tag{13.7}$$

neglecting perturbations. Then equation (13.5) becomes

$$\ddot{\rho}l+2\dot{\rho}\dot{l}+\rho\ddot{l}=-G(M+m_V)\frac{x}{r^3}+G(M+m_E)\frac{X}{R^3}. \tag{13.8}$$

Neglecting the vehicle's mass we obtain, on substitution of $(\rho l+X)$ for x in equation (13.8),

$$\left(\ddot{\rho}+\frac{GM\rho}{r^3}\right)l+2\dot{\rho}\dot{l}+\rho\ddot{l}=-GX\left(\frac{M}{r^3}-\frac{M+m_E}{R^3}\right)$$

with two similar equations in Y and Z.

406

These three equations may be solved to give $[\ddot{\rho}+(GM\rho/r^3)]$, $2\dot{\rho}$, and ρ in terms of l, m, n; $\dot{l}$, $\dot{m}$, $\dot{n}$; $\ddot{l}$, $\ddot{m}$, $\ddot{n}$; X, Y, Z; G, M, m_E and r. All of these except r are known or derived from observed quantities. This last quantity is therefore eliminated by substituting for r in the above solution for ρ, from the relation

$$r^2 = R^2 + \rho^2 + 2\rho(Xl + Ym + Zn) \tag{13.9}$$

obtained from triangle SEV since, by (13.1),

$$r^2 = R^2 + \rho^2 - 2R\rho \cos \hat{SEV}$$

and

$$\cos \hat{SEV} = -\frac{Xx' + Yy' + Zz'}{R\rho} = -\frac{1}{R}(Xl + Ym + Zn).$$

When r has been eliminated, the resulting equation is of the eighth degree in ρ. The problem of finding its roots is discussed in a number of texts, such as Moulton (1914), Danby (1962), Plummer (1918), and Herget (1948).

When r and hence ρ have been found, the vehicle's heliocentric co-ordinates (x, y, z) and velocity components $(\dot{x}, \dot{y}, \dot{z})$ may then be computed from the relations

$$x = \rho l + X$$

and

$$\dot{x} = \dot{\rho} l + \rho \dot{l} + \dot{X}$$

with the similar equations in y, z, $\dot{y}$, $\dot{z}$.

The application of the method of section 4.12 then supplies the elements of the vehicle's heliocentric orbit.

13.3 Laplace's Method

The scheme in the previous section was suggested by Laplace as a method of orbit determination. In order to use it, the first and second time derivatives of l, m and n must be found; l, m and n are directly related to the observed quantities α and δ, while $-X$, $-Y$ and $-Z$ are tabulated in the 'Astronomical Almanac' for every day of the year so that their first derivatives $\dot{X}$, $\dot{Y}$, $\dot{Z}$ are readily obtained.

If we let ρ' denote a unit vector in the line of sight from the Earth's centre to the vehicle, then

$$\rho' = \mathbf{i}l + \mathbf{j}m + \mathbf{k}n \tag{13.10}$$

where $\mathbf{i}$, $\mathbf{j}$ and $\mathbf{k}$ are unit vectors along the geocentric x, y and z axes respectively. Expanding ρ' by a Taylor series about its value ρ'_0 at time $t = 0$, we obtain

$$\rho' = \rho'_0 + \Delta t (\dot{\rho}')_0 + \tfrac{1}{2}\Delta t^2 (\ddot{\rho}')_0 + \ldots$$

where ρ' is the value of ρ' at a time interval Δt after it had a value ρ'_0, the brackets and suffix zero indicating that after the differentiation with respect to t, the values at $t = 0$ are substituted.

If Δt is sufficiently small, terms higher than Δt^2 may be neglected. Then

$$\rho' = \rho_0' + \Delta t(\dot{\rho}')_0 + \tfrac{1}{2}\Delta t^2(\ddot{\rho}')_0. \tag{13.11}$$

Three observations provide three equations in the three quantities ρ_0', $(\dot{\rho}')_0$, $(\ddot{\rho}')_0$ so that $(\dot{\rho}')_0$ and $(\ddot{\rho}')_0$ may be found, ρ_0' being already known.

Usually ρ_0 is chosen to be the middle observation. The values found for $(\dot{\rho}')_0$ and $(\ddot{\rho}')_0$ are of course approximate, but can be improved if more than three observations are available. It is then possible to write down more equations and use the set to eliminate the higher-order terms in ρ_0' first, enabling more accurate values of $(\dot{\rho}')_0$ and $(\ddot{\rho}')_0$ to be computed.

Various modifications have been made to Laplace's original method to remove practical inconveniences. One such modification by Stumpff uses the ratios of the direction cosines. Following Herget's account (1948) we let U, V, P and Q be defined by

$$\left. \begin{aligned} U &= \frac{y+Y}{x+X} = \tan \alpha, \qquad V = \frac{z+Z}{x+X} = \sec \alpha \tan \delta \\[2mm] P &= Y - UX, \qquad\qquad Q = Z - VX \end{aligned} \right\} \tag{13.12}$$

where the symbols on the right-hand sides have their previous meanings. U and V are obtained from observation; X, Y and Z are taken from the 'Astronomical Almanac'.

Then

$$y = Ux + UX - Y = Ux - P. \tag{13.13}$$

Also

$$z = Vx - Q. \tag{13.14}$$

Differentiating equations (13.13) and (13.14) twice with respect to time, we obtain

$$\left. \begin{aligned} \dot{y} &= \dot{U}x + U\dot{x} - \dot{P}, & \dot{z} &= \dot{V}x + V\dot{x} - \dot{Q} \\[1mm] \ddot{y} &= \ddot{U}x + 2\dot{U}\dot{x} + U\ddot{x} - \ddot{P}, & \ddot{z} &= \ddot{V}x + 2\dot{V}\dot{x} + V\ddot{x} - \ddot{Q}. \end{aligned} \right\} \tag{13.15}$$

Now

$$\ddot{\mathbf{r}} + \mu \frac{\mathbf{r}}{r^3} = 0 \tag{13.16}$$

where $\mu = GM$. Using the component equations of (13.16) to substitute for $\ddot{x}$, $\ddot{y}$ and $\ddot{z}$ in the last two equations of the set (13.15), we obtain

$$-\mu \frac{y}{r^3} = \ddot{U}x + 2\dot{U}\dot{x} - \mu \frac{x}{r^3} U - \ddot{P}$$

or

$$\ddot{P} - \frac{\mu}{r^3}(y - Ux) = \ddot{U}x + 2\dot{U}\dot{x}. \tag{13.17}$$

408

Using equation (13.13) we find that

$$\ddot{U}x+2\dot{U}\dot{x}=\ddot{P}+\frac{\mu P}{r^3}. \tag{13.18}$$

Similarly

$$\ddot{V}x+2\dot{V}\dot{x}=\ddot{Q}+\frac{\mu Q}{r^3}. \tag{13.19}$$

Defining D by the relation

$$D=\tfrac{1}{2}\dot{U}\dot{V}-\tfrac{1}{2}\dot{V}\dot{U} \tag{13.20}$$

and using equations (13.18) and (13.19), we find that

$$2Dx=\ddot{P}\dot{V}-\ddot{Q}\dot{U}+(P\dot{V}-Q\dot{U})/r^3 \tag{13.21}$$

and

$$4D\dot{x}=\ddot{Q}\dot{U}-\ddot{P}\dot{V}+(\dot{U}Q-\dot{V}P)/r^3. \tag{13.22}$$

Now

$$r^2=x^2+y^2+z^2$$

and hence

$$r^2=(1+U^2+V^2)x^2-2(UP+VQ)x+(P^2+Q^2). \tag{13.23}$$

By using the truncated Taylor series (13.11) and the three observations as before, the numerical values of $\dot{U}$, $\ddot{U}$, $\dot{V}$, $\ddot{V}$ may be found from the first two of equations (13.12). The last two of equations (13.12) (using the 'Astronomical Almanac' data) give values of P, Q and by differentiation, $\ddot{P}$ and $\ddot{Q}$.

The next stage consists in solving (13.21) and (13.23) by iteration to find r and x. Equation (13.22) then gives $\dot{x}$; and the first two of equations (13.15) then give $\dot{y}$ and $\dot{z}$, while (13.13) and (13.14) give y and z respectively. The elements are subsequently found as in section 4.12.

Though Stumpff's method reduces three-by-three determinants to two-by-two and is time-saving in hand computing, this benefit is achieved at the expense of having to divide the sky into regions and having special cases. If an electronic computer is available, it is better to retain the more general method.

13.4 Gauss's Method

The other basic method of orbit determination (due to Gauss) utilizes three positions and the time intervals between them; it also makes use of Kepler's second law of constant areal velocity that must be obeyed by the object in its heliocentric orbit (neglecting perturbations), and the fact that the object moves in a plane passing through the Sun's centre. In this section we do no more than sketch out the principles of the method.

The equation of a plane through the origin of a set of rectangular axes is

$$Ax+By+Cz=0 \tag{13.24}$$

where A, B and C are constants.

If the three observed positions have *heliocentric* equatorial coordinates x_i, y_i, z_i ($i = 1, 2, 3$), then we have the three equations

$$Ax_i + By_i + Cz_i = 0.$$

Eliminating the constants A, B and C, we find that

$$\begin{vmatrix} x_1 & y_1 & z_1 \\ x_2 & y_2 & z_2 \\ x_3 & y_3 & z_3 \end{vmatrix} = 0. \tag{13.25}$$

This determinantal equation may be written in the three forms

$$\left.\begin{aligned} (y_2z_3 - z_2y_3)x_1 - (y_1z_3 - z_1y_3)x_2 + (y_1z_2 - z_1y_2)x_3 = 0 \\ (x_2z_3 - z_2x_3)y_1 - (x_1z_3 - z_1x_3)y_2 + (x_1z_2 - z_1x_2)y_3 = 0 \\ (x_2y_3 - y_2x_3)z_1 - (x_1y_3 - y_1x_3)z_2 + (x_1y_2 - y_1x_2)z_3 = 0. \end{aligned}\right\} \tag{13.26}$$

Now the quantities in the brackets are the projections on the three coordinate planes of double the areas of the triangles formed by the Sun and the positions of the body taken two at a time. If we let $[i, j]$ denote the triangular area given by the Sun and the two positions at t_i and t_j then, on noting that in each equation the same plane is projected upon (for example, the yz plane in the first of equations (13.26)), we may write

$$\left.\begin{aligned} [2, 3]x_1 - [1, 3]x_2 + [1, 2]x_3 = 0 \\ [2, 3]y_1 - [1, 3]y_2 + [1, 2]y_3 = 0 \\ [2, 3]z_1 - [1, 3]z_2 + [1, 2]z_3 = 0. \end{aligned}\right\} \tag{13.27}$$

These equations may indeed be written as

$$\mathbf{r}_2 = c_1\mathbf{r}_1 + c_3\mathbf{r}_3 \tag{13.28}$$

where

$$c_1 = \frac{[2, 3]}{[1, 3]}, \quad c_3 = \frac{[1, 2]}{[1, 3]}.$$

From triangle ESV,

$$\boldsymbol{\rho} = \mathbf{r} - \mathbf{R} \tag{13.29}$$

and so

$$\boldsymbol{\rho}_2 - c_1\boldsymbol{\rho}_1 - c_3\boldsymbol{\rho}_3 = c_1\mathbf{R}_1 + c_3\mathbf{R}_3 - \mathbf{R}_2. \tag{13.30}$$

If c_1 and c_3 (the so-called 'triangle ratios') can be found, then equation (13.30) represents three linearly independent equations in the unknown geocentric distances, since the $\mathbf{R}$ are known from tables of the Sun's geocentric coordinates;

$$\mathbf{R} = -(\mathbf{i}X + \mathbf{j}Y + \mathbf{k}Z).$$

The triangle ratios c_1 and c_3 are now developed in power series in the

410

time intervals $(t_2 - t_1)$, $(t_3 - t_2)$ and $(t_3 - t_1)$. To do this, use may be made of the f and g series of section 4.12. Letting

$$(t_2 - t_1) \sqrt{GM} = \tau_3$$

$$(t_3 - t_2) \sqrt{GM} = \tau_1$$

$$(t_3 - t_1) \sqrt{GM} = \tau_2$$

and omitting all powers higher than τ^3, it is found that

$$\left. \begin{aligned} c_1 &= \frac{\tau_1}{\tau_2} \left[1 + \frac{(\tau_2{}^2 - \tau_1{}^2)}{6r_2{}^3} \right] \\ c_3 &= \frac{\tau_3}{\tau_2} \left[1 + \frac{(\tau_2{}^2 - \tau_3{}^2)}{6r_2{}^3} \right]. \end{aligned} \right\} \tag{13.31}$$

If the scalar product of equation (13.30) is taken with

$$\left(\frac{\mathbf{\rho_1}}{\rho_1} \times \frac{\mathbf{\rho_3}}{\rho_3} \right)$$

and the expressions for c_1 and c_3 in equation (13.31) substituted into the resulting equation, it is found that a solution for ρ_2 of the form

$$\rho_2 = A + B/r_2{}^3$$

is obtained. This is an equation in the two unknowns ρ_2 and r_2 because A and B are functions of the observations and the tabulated quantities. In order to find ρ_2 and r_2 we may proceed as in Laplace's method and use equation (13.9) written as

$$r_2{}^2 = R_2{}^2 + \rho_2{}^2 + 2\rho_2(X_2 l_2 + Y_2 m_2 + Z_2 n_2)$$

as a second equation in r_2 and ρ_2.

Having found r_2 and ρ_2, equations (13.30) give ρ_1 and ρ_3; hence $\mathbf{r_1}$, $\mathbf{r_2}$ and $\mathbf{r_3}$ can be found from (13.29).

The elements can then be obtained from $\mathbf{r_2}$ and $\mathbf{\dot{r}_2}$ as usual, where $\mathbf{\dot{r}_2}$ has been computed numerically from $\mathbf{r_1}$, $\mathbf{r_2}$, $\mathbf{r_3}$ and t_1, t_2, t_3.

Gauss in fact proceeded in a rather different manner. The positions $\mathbf{r_1}$ and $\mathbf{r_2}$ define the plane of the orbit. The remaining elements are obtained from two equations involving two unknowns. Gauss derived one of the equations from the ratio of the area of the triangle defined by $\mathbf{r_1}$ and $\mathbf{r_3}$ to the area of the sector formed by $\mathbf{r_1}$, $\mathbf{r_3}$ and the arc of the orbit between these points. He found the other equation by using Kepler's equation at t_1 and t_3. There is no doubt that Gauss's method is more complicated than Laplace's, though subsequent workers have devised variations that avoid a number of these complexities.

13.5 Olbers's Method for Parabolic Orbits

This method bears some resemblance to that of Gauss but differs in that it makes use of Euler's equation for parabolic motion. If s is the length of the chord between two positions r_1 and r_3 occupied at times t_1 and t_3 by a body moving about the Sun (mass M) in a parabolic orbit, it may be shown that

$$(r_1+r_3+s)^{3/2} - (r_1+r_3-s)^{3/2} = 6(t_3-t_1)\sqrt{GM}. \tag{13.32}$$

Dividing throughout by $(r_1+r_3)^{3/2}$ and defining η by

$$\eta = \frac{2(t_3-t_1)\sqrt{GM}}{(r_1+r_3)^{3/2}} \tag{13.33}$$

equation (13.32) becomes

$$\left(1+\frac{s}{r_1+r_3}\right)^{3/2} - \left(1-\frac{s}{r_1+r_3}\right)^{3/2} = 3\eta. \tag{13.34}$$

Tables of $s/(r_1+r_3)$ as a function of η exist (for example Bauschinger 1901).

Olbers assumed that if the time intervals between the observations were short, the 'triangle ratios' (the same c_1 and c_3 defined in the previous section) were proportional to the time intervals. Thus

$$c_1/c_3 = (t_3-t_2)/(t_2-t_1) = \tau_1/\tau_3. \tag{13.35}$$

Rewriting equation (13.30) in the form

$$\rho_2 - c_1\rho_1 - c_3\rho_3 = c_1\mathbf{R}_1 + c_3\mathbf{R}_3 - \mathbf{R}_2 = \mathbf{V}, \tag{13.36}$$

we introduce a vector $\mathbf{U}$ coplanar with $\mathbf{V}$ and $\boldsymbol{\rho}_2$. The scalar product of equation (13.36) and $(\boldsymbol{\rho}_2/\rho_2) \times \mathbf{U}$ is then taken so that only terms in ρ_3 and ρ_1 remain, and the resulting equation is

$$\rho_3 = M\rho_1 \tag{13.37}$$

where

$$M = -\frac{c_1\left(\dfrac{\boldsymbol{\rho}_1}{\rho_1}, \dfrac{\boldsymbol{\rho}_3}{\rho_3}, \mathbf{U}\right)}{c_3\left(\dfrac{\boldsymbol{\rho}_3}{\rho_3}, \dfrac{\boldsymbol{\rho}_2}{\rho_2}, \mathbf{U}\right)} \tag{13.38}$$

the quantities in parentheses being scalar triple products. Olbers then used Euler's relationship (13.34) with equation (13.38) along the following lines.

The chord s is given by

$$s^2 = (\mathbf{r}_3 - \mathbf{r}_1) \cdot (\mathbf{r}_3 - \mathbf{r}_1).$$

But by equation (13.7),

$$r_1^2 = R_1^2 + \rho_1^2 + 2\rho_1(X_1l_1 + Y_1m_1 + Z_1n_1)$$

or

$$r_1^2 = a_1 + b_1\rho_1 + d_1\rho_1^2. \tag{13.39}$$

Similarly,

$$r_3{}^2 = a_3 + b_3\rho_3 + d_3\rho_3{}^2. \tag{13.40}$$

Hence

$$s^2 = a_2 + b_2\rho_1 + d_2\rho_1{}^2 \tag{13.41}$$

by using equations (13.29) and (13.35) to eliminate ρ_3. If U is known, M and hence s may be found.

Now the three positions of the Earth at t_1, t_2 and t_3 are related by the equation

$$\mathbf{R}_2 = C_1\mathbf{R}_1 + C_3\mathbf{R}_3 \tag{13.42}$$

where C_1 and C_3 are the triangle ratios for the Earth's heliocentric orbit (see equation (13.28)).

Then approximately, as in equation (13.35),

$$C_1/C_3 = \tau_1/\tau_3$$

so that

$$c_1 = \gamma C_1, \qquad c_3 = \gamma C_3. \tag{13.43}$$

But by equation (13.36),

$$\mathbf{V} = c_1\mathbf{R}_1 + c_3\mathbf{R}_3 - \mathbf{R}_2$$

and hence, using equations (13.42) and (13.43)

$$\mathbf{V} = \mathbf{R}_2(\gamma - 1).$$

Thus as a first approximation for U, which has to be coplanar with V and ρ_2, we may take $\mathbf{U} = \mathbf{R}_2$. First approximations to s, r_1 and r_3 may then be found from equations (13.39), (13.40) and (13.41) by assuming a value for ρ_1. In its turn η can be computed from equation (13.33); and from the table of $s/(r_1+r_3)$ as a function of η, a value of $s/(r_1+r_3)$ corresponding to the computed η may be obtained. In general, this value of $s/(r_1+r_3)$ will not agree with that calculated from the first approximations to s, r_1 and r_3, but by a process of trial and error a value of ρ_1 that gives agreement can be found eventually. From equation (13.37), ρ_3 is computed and hence from (13.29) and (13.36), $\mathbf{r}_1$, $\mathbf{r}_2$ and $\mathbf{r}_3$ are obtained. The elements (of which the eccentricity is known to be unity) can be found in the usual way, Barker's equation being used to find the time of perihelion passage.

The various methods of improving this preliminary orbit without using more observational data will not be considered here.

13.6 Orbit Determination with Additional Observational Data

The advent of Earth satellites and lunar and interplanetary probes has necessitated modifications in the classical methods of orbit determination.

In the case of a newly injected artificial Earth satellite, a preliminary orbit may be found by using the measured position and velocity components at burn-out to compute elements by the method of section 4.12. This orbit

413

may be improved later when observations of the vehicle are collected by the tracking stations. An alternative method used by Briggs and Slowey (1959) uses an iterative method and a high-speed digital computer and is described below.

Suppose three tracking stations S_1, S_2 and S_3 (of known geocentric coordinates) observe the directions of a satellite at times t_1, t_2 and t_3 when the satellite in its geocentric orbit is at points V_1, V_2 and V_3 as in figure 13.2.

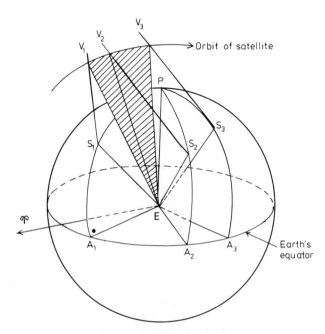

Figure 13.2

Since the orbit of the satellite lies in a plane through the Earth's centre, the three geocentric radius vectors EV_1, EV_2 and EV_3 (or r_1, r_2 and r_3) are coplanar. Let the direction cosines of the three positions as seen from S_1, S_2 and S_3 be l_i, m_i, n_i ($i = 1, 2, 3$), while the geocentric radius vectors R_i to the stations S_i are given by

$$R_i = iX_i + jY_i + kZ_i \tag{13.44}$$

where the geocentric rectangular coordinates of S_i at time t_i are X_i, Y_i, Z_i, and i, j and k are unit vectors as before.

Then the topocentric vectors ρ_i to the satellite are of the form

$$\rho_i = \rho_i(il_i + jm_i + kn_i) \tag{13.45}$$

the topocentric distances being unknown.

414

Also

$$\mathbf{r}_i = \mathbf{i}x_i + \mathbf{j}y_i + \mathbf{k}z_i \tag{13.46}$$

and

$$\boldsymbol{\rho}_i = \mathbf{r}_i - \mathbf{R}_i. \tag{13.47}$$

Omitting perturbations, the vectors $\mathbf{r}_i$ are coplanar so that

$$\mathbf{r}_1 \times \mathbf{r}_2 \cdot \mathbf{r}_3 = 0. \tag{13.48}$$

Using equation (13.47), equation (13.48) becomes

$$\begin{vmatrix} X_1 + \rho_1 l_1 & X_2 + \rho_2 l_2 & X_3 + \rho_3 l_3 \\ Y_1 + \rho_1 m_1 & Y_2 + \rho_2 m_2 & Y_3 + \rho_3 m_3 \\ Z_1 + \rho_1 n_1 & Z_2 + \rho_2 n_2 & Z_3 + \rho_3 n_3 \end{vmatrix} = 0. \tag{13.49}$$

If values for the distances ρ_1 and ρ_2 are now assumed, equation (13.49) may be used to obtain ρ_3. A convenient way of doing this is to compute $\mathbf{r}_1$ and $\mathbf{r}_2$ from equations (13.45) and (13.47), and then find quantities L, M and N from the relation

$$\mathbf{r}_1 \times \mathbf{r}_2 = \mathbf{i}L + \mathbf{j}M + \mathbf{k}N. \tag{13.50}$$

Then

$$\rho_3 = -\frac{LX_3 + MY_3 + NZ_3}{Ll_3 + Mm_3 + Nn_3} \tag{13.51}$$

enabling $\mathbf{r}_3$ to be found.

The differences in true anomaly f may now be calculated from the relations

$$\left. \begin{array}{l} \sin (f_3 - f_i) = \pm |\mathbf{r}_i \times \mathbf{r}_3|/r_3 r_i \\ \cos (f_3 - f_i) = \mathbf{r}_3 \cdot \mathbf{r}_i/r_3 r_i \end{array} \right\} \quad (i = 1, 2). \tag{13.52}$$

For direct orbits, $\sin (f_3 - f_i)$ is given the same sign as the z component of $\mathbf{r}_i \times \mathbf{r}_3$.

Now from the equation for the ellipse

$$r = \frac{a(1 - e^2)}{1 + e \cos f} \tag{13.53}$$

it may be shown that

$$\tan f_3 = -\frac{r_1(r_2 - r_3) \cos (f_3 - f_1) + r_2(r_3 - r_1) \cos (f_3 - f_2) + r_3(r_1 - r_2)}{r_1(r_2 - r_3) \sin (f_3 - f_1) + r_2(r_3 - r_1) \sin (f_3 - f_2)} \tag{13.54}$$

and that

$$e = \frac{r_i - r_j}{r_j \cos f_j - r_i \cos f_i} \tag{13.55}$$

for any i and j and $\cos f_j \neq \cos f_i$. Also

$$a = \frac{r_i(1 + e \cos f_i)}{1 - e^2}. \tag{13.56}$$

From equation (13.54) f_3 is found. Of the two possible choices for f_3, the choice is taken which makes e positive in (13.55) after (13.52) has been used to find f_1 and f_2. Equation (13.56) then gives a. The time of the perigee passage τ immediately prior to the times of observation of the satellite may now be found from the familiar relationships from chapter 4:

$$\tan \frac{E_i}{2} = \sqrt{\frac{1-e}{1+e}} \tan \frac{f_i}{2} \tag{13.57}$$

$$\bar{M}_i = E_i - e \sin E_i \tag{13.58}$$

and

$$\tau = t_i - \frac{\bar{M}_i}{n_i} \tag{13.59}$$

where $n_i = (Gm)^{1/2} a^{-3/2}$, $m=$ mass of the Earth, $G=$ constant of gravitation, and $\bar{M}_i$, E_i and f_i are the values of the mean, eccentric, and true anomalies of the satellite at the time of observation t_i.

At this stage, if the computed elements are used to provide time intervals between the observations, they will be found to disagree with the observed time intervals since estimates only of the topocentric distances ρ_1 and ρ_2 were used. By using a high-speed digital computer and an iterative procedure analogous to the Newton–Raphson method, the values for ρ_1 and ρ_2 are corrected until the predicted and observed time intervals agree.

Having found the correct orbit the remaining elements i, Ω and the argument of perigee ω can be easily computed. The inclination follows from equation (13.50) in that -

$$i = \cos^{-1} \left[\frac{|N|}{L^2 + M^2 + N^2} \right] \tag{13.60}$$

while Ω is given by

$$\Omega = \tan^{-1} \left(\frac{M}{L} \right) \pm 90° \tag{13.61}$$

where the sign chosen is that of the product LN. The argument of perigee ω is found from any observation by using

$$\omega = \pm u_i - f_i \tag{13.62}$$

where

$$u_i = \cos^{-1} \left(\frac{x_i \cos \Omega + y_i \sin \Omega}{r_i} \right) \tag{13.63}$$

the sign chosen being that of z_i.

Once the elements of a preliminary orbit are known, the theory of an artificial Earth satellite may be used to compute the secular perturbations in mean motion, right ascension of the node and argument of perigee, providing an ephemeris so that when more observations are accumulated the orbit can be improved. If range and range-rate data are available, the classical methods of orbit determination may be modified to take advantage

416

of these additional data. For example, in the case just discussed, range data would give the ρ_i, simplifying the proceedings considerably.

It is also possible to obtain the elements of a preliminary orbit from range and range-rate data alone. In principle, it may be done from three pairs of range and range-rate observations as follows. The method is a modification of Laplace's and uses truncated f and g series. Using the same notation as before, let ρ_i, $\dot{\rho}_i$ ($i = 1, 2, 3$) be the measured ranges and range rates of an interplanetary probe at times t_1, t_2 and t_3. Then

$$(x_i - X_i)^2 + (y_i - Y_i)^2 + (z_i - Z_i)^2 = \rho_i{}^2 \qquad (13.64)$$

and

$$(x_i - X_i)(\dot{x}_i - \dot{X}_i) + (y_i - Y_i)(\dot{y}_i - \dot{Y}_i) + (z_i - Z_i)(\dot{z}_i - \dot{Z}_i) = \rho_i \dot{\rho}_i \qquad (13.65)$$

where $i = 1, 2, 3$.

Now

$$\mathbf{r}_i = \mathbf{r}_2 f_i + \dot{\mathbf{r}}_2 g_i \qquad (i = 1, 3)$$

and hence

$$
\left.
\begin{aligned}
x_i &= x_2 f_i + \dot{x}_2 g_i \\
\dot{x}_i &= x_2 \dot{f}_i + \dot{x}_2 \dot{g}_i
\end{aligned}
\right\} \quad (l = 1, 3)
\qquad
\begin{aligned}
(13.66) \\
(13.67)
\end{aligned}
$$

with similar equations in y and z. Substituting these equations into (13.64) and (13.65) we obtain, after some reduction

$$
\left.
\begin{aligned}
& r_2{}^2 f_1{}^2 + v_2{}^2 g_1{}^2 + R_1{}^2 + 2 r_2 \dot{r}_2 f_1 g_1 - 2 \sum_{xyz} X_1(x_2 f_1 + \dot{x}_2 g_1) = \rho_1{}^2 \\
& \qquad\qquad\qquad r_2{}^2 + R_2{}^2 - 2 \sum_{xyz} X_2 x_2 = \rho_2{}^2 \\
& r_2{}^2 f_3{}^2 + v_2{}^2 g_3{}^2 + R_3{}^2 + 2 r_2 \dot{r}_2 f_3 g_3 - 2 \sum_{xyz} X_3(x_2 f_3 + \dot{x}_2 g_3) = \rho_3{}^2 \\
& r_2{}^2 f_1 \dot{f}_1 + v_2{}^2 g_1 \dot{g}_1 + R_1 \dot{R}_1 + r_2 \dot{r}_2 (g_1 \dot{f}_1 + f_1 \dot{g}_1) \\
& \qquad - \sum_{xyz} X_1(x_2 \dot{f}_1 + \dot{x}_2 \dot{g}_1) - \sum_{xyz} \dot{X}_1(x_2 f_1 + \dot{x}_2 g_1) = \rho_1 \dot{\rho}_1 \\
& \qquad\qquad R_2 \dot{R}_2 + r_2 \dot{r}_2 - \sum_{xyz} X_2 \dot{x}_2 - \sum_{xyz} \dot{X}_2 x_2 = \rho_2 \dot{\rho}_2 \\
& r_2{}^2 f_3 \dot{f}_3 + v_2{}^2 g_3 \dot{g}_3 + R_3 \dot{R}_3 + r_2 \dot{r}_2 (g_3 \dot{f}_3 + f_3 \dot{g}_3) \\
& \qquad - \sum_{xyz} X_3(x_2 \dot{f}_3 + \dot{x}_2 \dot{g}_3) - \sum_{xyz} \dot{X}_3(x_2 f_3 + \dot{x}_2 g_3) = \rho_3 \dot{\rho}_3
\end{aligned}
\right\} \quad (13.68)
$$

where

$$
\left.
\begin{aligned}
r_2{}^2 &= x_2{}^2 + y_2{}^2 + z_2{}^2 \\
r_2 \dot{r}_2 &= x_2 \dot{x}_2 + y_2 \dot{y}_2 + z_2 \dot{z}_2 \\
R_i{}^2 &= X_i{}^2 + Y_i{}^2 + Z_i{}^2 \\
R_i \dot{R}_i &= X_i \dot{X}_i + Y_i \dot{Y}_i + Z_i \dot{Z}_i \\
v_2{}^2 &= \dot{x}_2{}^2 + \dot{y}_2{}^2 + \dot{z}_2{}^2.
\end{aligned}
\right\} \quad (13.69)
$$

If the three pairs of observations are made within short intervals of one another, the f and g series (and also their differentials $\dot{f}$ and $\dot{g}$) may be truncated as follows:

$$f_i = 1 - \tfrac{1}{2}ut_i^2 + \tfrac{1}{2}ust_i^3, \qquad g_i = t_i - \tfrac{1}{6}ut_i^3 + \tfrac{1}{4}ust_i^4$$

$$\dot{f}_i = -ut_i + \tfrac{3}{2}ust_i^2, \qquad \dot{g}_i = 1 - \tfrac{1}{2}ut_i^2 + ust_i^3$$

where

$$u = \frac{1}{r_2{}^3}, \qquad s = \frac{\dot{r}_2}{r_2}.$$

It should be remembered that although the independent variable is written as t, it is in a time scale such that $GM = 1$.

Then taking into account equation (13.69), the equations (13.68) constitute a set of six equations in the six unknowns x_2, y_2, z_2, $\dot{x}_2$, $\dot{y}_2$, $\dot{z}_2$, which may be solved by an iterative method. A guess is made first at u and s and the set of equations (13.68) may then be solved. The values found enable new values of u and s to be computed and a new solution made. From the components of position and velocity x_2, y_2, z_2, $\dot{x}_2$, $\dot{y}_2$, $\dot{z}_2$, at time t_2, the elements may be obtained in the usual way.

Usually, if range and range-rate data are available, there is also a fair knowledge of the vehicle's direction or elongation θ (see section 13.2) so that from the equation

$$r^2 = R^2 + \rho^2 - 2R\rho \cos \theta$$

a reasonably accurate value of r_2 can be found; first approximations to x_2, y_2 and z_2 may also be computed from equations (11.3) and checked against the second equation of the set (13.68). If rough estimates of l, $\dot{m}$ and $\dot{n}$ are also available, then by equations (13.4) first approximations to $\dot{x}_2$, $\dot{y}_2$, $\dot{z}_2$ and hence to $\dot{r}_2$ and v_2 may be obtained. Then the set of equations (13.68) may be linearized in Δx_2, Δy_2, Δz_2, $\Delta \dot{x}_2$, $\Delta \dot{y}_2$, and $\Delta \dot{z}_2$, these being the corrections to the first approximations to x_2, y_2, z_2, $\dot{x}_2$, $\dot{y}_2$ and $\dot{z}_2$.

For details of a number of methods of utilizing range and range-rate data in orbit determination the reader is referred to an important paper by Baker (1960).

13.7 The Improvement of Orbits

Let the heliocentric preliminary orbit have elements σ_i $(i = 1\text{–}6)$. Then any geocentric observed quantity ϕ at time t will be given by

$$\phi = \phi(\sigma_i, p_i, t) \qquad (i = 1\text{–}6) \tag{13.70}$$

where the six p_i stand for the Earth's elements and $\phi(\sigma_i, p_i, t)$ is a function of the twelve elements and the time.

If the σ_i are changed slightly in arbitrary ways $\delta \sigma_i$, the change in ϕ will be $\delta \phi$ given by

$$\delta \phi = \sum_{i=1}^{6} \frac{\partial \phi}{\partial \sigma_i} \delta \sigma_i. \tag{13.71}$$

Now in general, the elements of the preliminary orbit are not the elements of the orbit actually followed by the vehicle, and so the predicted quantities ϕ_{cal} will be slightly different from the actual observed quantities ϕ_{obs} at a given time. Let

$$\Delta\phi = \phi_{obs} - \phi_{cal}$$

for a given time. Then if we have n observations of ϕ made at n times $t_1, t_2 \ldots t_n$, we may write

$$
\left.
\begin{aligned}
\Delta\phi_1 &\equiv (\phi_{obs} - \phi_{cal})_1 = \sum_{i=1}^{6} \left(\frac{\partial\phi}{\partial\sigma_i}\right)_1 \delta\sigma_i \\[2ex]
\Delta\phi_2 &\equiv (\phi_{obs} - \phi_{cal})_2 = \sum_{i=1}^{6} \left(\frac{\partial\phi}{\partial\sigma_i}\right)_2 \delta\sigma_i \\[2ex]
\cdot \quad & \qquad \cdot \qquad\qquad\qquad \cdot \\
\cdot \quad & \qquad \cdot \qquad\qquad\qquad \cdot \\
\cdot \quad & \qquad \cdot \qquad\qquad\qquad \cdot \\[1ex]
\Delta\phi_n &\equiv (\phi_{obs} - \phi_{cal})_n = \sum_{i=1}^{6} \left(\frac{\partial\phi}{\partial\sigma_i}\right)_n \delta\sigma_i
\end{aligned}
\right\} \tag{13.72}
$$

where the suffices $1, 2 \ldots n$ mean that the quantities within the brackets are observed at, or evaluated for, the epochs $t_1, t_2 \ldots t_n$.

If $n=6$, the n equations in $\delta\sigma_i$ may be solved for the $\delta\sigma_i$; if $n > 6$, they can be solved for the $\delta\sigma_i$ by the method of least squares. Each $\delta\sigma_i$ can then be added onto its σ_i to give improved values of the elements. These will be the most probable values of the elements and there may also be calculated values of the probable errors of the elements.

Obviously, ϕ can take more than one form. It can be right ascension α, declination δ, range ρ or any other observed quantity that can be related analytically to the six elements of the orbit of the vehicle and those of the Earth. The quantities $\partial\phi/\partial\sigma_i$ in classical celestial mechanics can then be found by analytical differentiation. A variation of this approach that may be used when a high-speed machine is available is to obtain the $\partial\phi/\partial\sigma_i$ in numerical form. The basic idea behind this approach is given below.

Let the heliocentric rectangular differential equations of motion of the vehicle be represented by

$$
\left.
\begin{aligned}
\ddot{x} &= F(x, y, z, t) \\
\ddot{y} &= G(x, y, z, t) \\
\ddot{z} &= H(x, y, z, t)
\end{aligned}
\right\} \tag{13.73}
$$

where t represents the way in which time enters the equations through perturbations (if allowed for). The forms of the functions F, G and H are known. Then a numerical integration of the set (13.73) between epochs t_0 and t_E gives sets of values for x, y and z at epoch steps between t_0 and t_E, these values depending upon the chosen initial conditions at t_0, namely

$x_0, y_0, z_0, \dot{x}_0, \dot{y}_0, \dot{z}_0$. These values are obtained from the preliminary orbital elements in the usual way. Then, formally,

$$\left. \begin{array}{l} x = x(x_0, y_0, z_0; \dot{x}_0, \dot{y}_0, \dot{z}_0; t) \\[4pt] y = y(x_0, y_0, z_0; \dot{x}_0, \dot{y}_0, \dot{z}_0; t) \\[4pt] z = z(x_0, y_0, z_0; \dot{x}_0, \dot{y}_0, \dot{z}_0; t) \end{array} \right\}. \tag{13.74}$$

Although the forms of the functions x, y and z are not known, we can now by interpolation obtain tabulated values for x, y, z for any value of t between t_0 and t_E. If we now vary *one* of $x_0, y_0, z_0, \dot{x}_0, \dot{y}_0, \dot{z}_0$ (say x_0), giving it a slightly different value but keeping all five other initial conditions the same, a new set of values for x, y, z will be obtained in a new numerical integration for the time interval between t_0 and t_E. If *at any given time* the two values of x obtained in this way are x_2 and x_1, we may write

$$x_2 - x_1 = \left(\frac{\partial x}{\partial x_0} \right) \delta x_0$$

where δx_0 is the change we made in x_0. We may do this since although in general

$$x_2 - x_1 = \sum_{i=1}^{6} \left(\frac{\partial x}{\partial \sigma_i} \right) \delta \sigma_i$$

where σ_i is any one of $x_0, y_0, z_0, \dot{x}_0, \dot{y}_0, \dot{z}_0$, all $\delta \sigma_i$ are zero except δx_0. Then

$$\left(\frac{\partial x}{\partial x_0} \right) = \frac{x_2 - x_1}{\delta x_0}$$

the right-hand side being known for any given time between t_0 and t_E from the stored tabulated solutions. In similar fashion, we have

$$\left(\frac{\partial y}{\partial x_0} \right) = \frac{y_2 - y_1}{\delta x_0}$$

$$\left(\frac{\partial z}{\partial x_0} \right) = \frac{z_2 - z_1}{\delta x_0}.$$

Five more integrations are carried out, in each case giving one of the five remaining quantities $y_0, z_0, \dot{x}_0, \dot{y}_0, \dot{z}_0$ a slightly different value and keeping the others unchanged. In this way all quantities

$$\left. \begin{array}{l} \left(\dfrac{\partial x}{\partial \sigma_i} \right) = \dfrac{x_j - x_1}{\delta \sigma_i} \\[10pt] \left(\dfrac{\partial y}{\partial \sigma_i} \right) = \dfrac{y_j - y_1}{\delta \sigma_i} \\[10pt] \left(\dfrac{\partial z}{\partial \sigma_i} \right) = \dfrac{z_j - z_1}{\delta \sigma_i} \end{array} \right\} \quad (j = 1+i, \, i = 1, 2 \ldots 6) \tag{13.75}$$

can be tabulated for times between t_0 and t_E where σ_i is any one of the six quantities $x_0, y_0, z_0, \dot{x}_0, \dot{y}_0, \dot{z}_0$.

If now observations made between the epochs t_0 and t_E furnish values of x, y, z (that is $x_{obs}, y_{obs}, z_{obs}$ for various times), we may write

$$x_{obs} - x_{cal} = \sum_{i=1}^{6} \left(\frac{\partial x}{\partial \sigma_i}\right) \delta \sigma_i \tag{13.76}$$

with similar equations in y and z. But by the set of tabulated quantities (13.75) all the $\partial x / \partial \sigma_i$ are known, and so equation (13.76) given by the observations may be solved to give the values of the six $\delta \sigma_i$. These, added to $x_0, y_0, z_0, \dot{x}_0, \dot{y}_0, \dot{z}_0$ in turn, enable improved values of the preliminary orbit's elements to be found.

13.8 Interplanetary Navigation

The main task of a space navigation system is to find out where the ship was and what velocity it had (with respect to a known coordinate system) at a particular epoch. If this task is carried out successfully, the elements of the ship's orbit may be computed and, taking known perturbations into account, its position and velocity at any future time can be found. In general, the actual orbit will differ from the desired orbit and a midcourse correction can then be planned to place the ship into a new orbit. It may be noted that the new orbit is not necessarily the old desired orbit since the present 'erroneous' position of the ship may render it more economical in fuel expenditure to make a change to a new orbit that also achieves the mission's goal than to attempt a correction that sets the ship on the old desired course.

There are a number of navigational methods available. Some are Earth-based and some are vehicle-based, and the choice depends not only upon the mission the vehicle is to carry out and the payload mass available for navigational equipment but also upon the phase of the mission. Thus a number of methods may well enter into the navigational requirements for a single mission. The most practical are based on optical tracking, radar tracking and the use of inertial equipment (comprising stabilized platforms and accelerometers). In addition, fast electronic computers are required.

In the first case, the ship itself may be tracked optically by Earth-based instruments, though at distances of more than a few million kilometres any ship of reasonable size would be invisible to the best modern equipment. For example, at a distance of $80\,000\,000$ km a sphere of 150 metres radius and 100% reflecting power would be of the 19th magnitude (see section 3.2); well beyond the capabilities of a Baker–Nunn camera. However, optical tracking methods may be used from the ship itself, only light and moderately-sized equipment being required. Such methods will be described later.

The second method, radar tracking, can be either Earth- or ship-based, though equipment of only moderate power and range can be carried on a ship. The data supplied by such methods are highly accurate ranges and range rates and (for large radar installations) directions as well. The Deep Space Instrumentation Facility stations are certainly capable of tracking vehicles equipped with transponders well outside the orbit of Pluto. Ship-based radar will be important when the interplanetary vehicle enters its final phase on the outward journey and approaches the planet of destination. It is also important in rendezvous manoeuvres.

13.8.1 Stabilized platforms and accelerometers

The *stabilized platform* provides an inertial attitude reference system by using gyroscopes, one gyroscope with a single degree of freedom being necessary for each of the three mutually perpendicular axes. The gyroscopes are mounted on the platform, allowance being made for the vehicle's angular motion with respect to the platform by mounting the platform on two gimbals (figure 13.3). The rotation of the vehicle about a gyroscope-stabilized axis causes a torque to act on the platform and makes it rotate about that axis. In its turn, the gyroscope spin axis precesses. Its angular velocity is sensed by an electric pick-up, is amplified, and is made to govern a servo-motor that opposes the disturbing torque. In so doing, it maintains the platform in its reference attitude. In many vehicles the platform is a four-gimballed one to allow tumbling of the spacecraft without having to lock the gyros and still not throw them off axis, which can happen in the case of a three-gimballed platform.

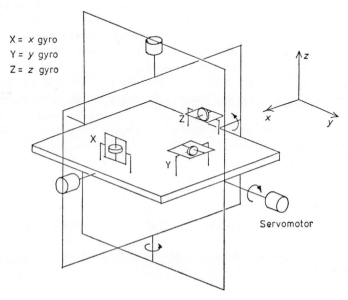

Figure 13.3

422

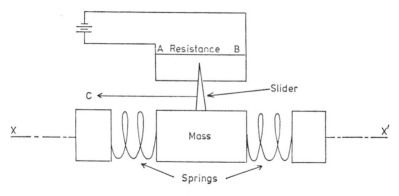

Figure 13.4

An *accelerometer* is used to measure the acceleration of the vehicle in a given direction, say the XX′ direction in figure 13.4. When the vehicle is under acceleration the mass, because of its inertia, presses back against one of the springs and carries the slider along the resistance to a point determined by the acceleration and the strength of the springs. If a voltage is applied across AB, the potentiometer output at C is proportional to the acceleration of the vehicle. Three accelerometers mounted on the stabilized platform in mutually perpendicular directions provide the necessary data for an inertial navigation system.

Before launching the vehicle, the platform is locked on to the desired reference system. During the powered phase the computer accepts the accelerometer readings, integrating them twice to obtain the components of position and velocity at any instant. In particular, at the end of the powered phase the elements of the vehicle's orbit can be computed. Comparison with the desired orbit can be made and the first midcourse correction program can be calculated. The inertial guidance system can then be used to control the manoeuvre.

13.8.2 Navigation by on-board optical equipment

Although it seems likely that an interplanetary vehicle will be in constant radio communication with Earth and that Earth-based radar installations will provide direction, range, and range-rate data, observations taken on board the vehicle can be used to navigate the craft. These observations, made optically, may be processed by an on-board computer or be radioed back to Earth for process in the larger, faster, and more versatile computers there. Wherever this is done, a method of finding position and velocity using optical observations may be developed along the following lines. In this account (Vertregt 1956), we first deal with the theory and then consider some practical difficulties before mentioning other possible sources of finding position and velocity.

The stars provide a useful reference background for space navigation, and we may take as a coordinate system the ecliptic rectangular heliocentric system using the direction of the First Point of Aries, the point on the ecliptic 90° greater in celestial longitude than Aries, and the north pole of the ecliptic as the x, y and z axes. The heliocentric celestial longitude λ, latitude β, and the radius vector r of a space vehicle are then connected to its rectangular coordinates by the relations

$$\left.\begin{aligned} x &= r \cos \beta \cos \lambda \\ y &= r \cos \beta \sin \lambda \\ z &= r \sin \beta. \end{aligned}\right\} \tag{13.77}$$

The longitude λ_P, latitude β_P, and radius vector r_P of any planet at any time will be known. If the subsequent computations are to be done on board, we may suppose the navigator has an 'Astronautical Almanac' containing such information.

In figure 13.5 the vehicle V, planet P and Sun S are shown, together with the direction of the First Point of Aries. The projections of V and P on the plane of the ecliptic, namely A and B, are also shown. The navigator at a known epoch measures:

(i) the apparent longitude of the Sun $= \lambda_S' = \Upsilon \hat{A} S$,

(ii) the apparent longitude of the planet $= \lambda_P' = \Upsilon \hat{A} B$,

(iii) the apparent latitude of the Sun $= \beta_S' = V \hat{S} A = -\beta$.

Then

$$AS = r \cos \beta, \quad BS = r_P \cos \beta_P.$$

Also, from triangle ABS

$$\frac{AS}{\sin A \hat{B} S} = \frac{BS}{\sin B \hat{A} S}.$$

But

$$A \hat{B} S = 360° + \lambda_P - \lambda_P'$$
$$B \hat{A} S = \lambda_P' - \lambda_S'$$
$$\beta = -\beta_S'.$$

Hence

$$r = r_P \frac{\cos \beta_P \sin (\lambda_P - \lambda_P')}{\cos \beta_S \sin (\lambda_P' - \lambda_S')} \tag{13.78}$$

all quantities on the right-hand side of equation (13.78) being obtained from tables of measurements. Also

$$\lambda = 180° + \lambda_S' \tag{13.79}$$

and

$$\beta = -\beta_S'. \tag{13.80}$$

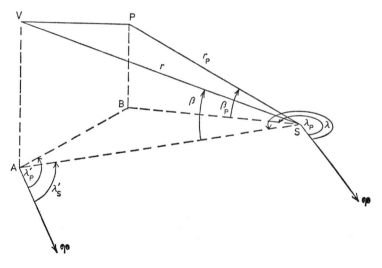

Figure 13.5

The vehicle's coordinates r, λ and β at time t are therefore known. Hence by (13.77), rectangular coordinates x, y and z at time t may be found.

A similar set of measurements taken after a suitable time interval will provide in theory enough data to obtain $\dot{x}$, $\dot{y}$ and $\dot{z}$. In practice, several sets at a number of epochs would be taken, so that for example the f and g series might be used to provide more accurate values of velocity components at one of the epochs. These, together with the coordinates of position at that epoch, could then be used to compute the orbital elements. Obviously more than one planet will usually be visible and a more accurate fix may be obtained by using all available planets and averaging.

Such a position- and velocity-finding method enables a check to be made on the stabilized platform system. Once this correction is made, the inertial navigation system can be used to control the application of the midcourse correction thrust.

13.8.3 Observational methods and probable accuracies

The navigational method of the preceding section depends upon the measurements of three angles, all referred to the ecliptic reference system: two angles involve the Sun, the third involves a planet. It might be thought that by using stars that define the plane of the ecliptic and a reference direction lying in it (not necessarily that of the First Point of Aries), an instrument based on the sextant could be used to measure the required angles. Serious difficulties arise, however, in the construction of an instrument that would be accurate enough, yet be of reasonably small mass. An accuracy of 1″ would be difficult to achieve; yet such accuracies would be required if distances are to be measured to within a few thousand kilometres.

A better method by far would be to use the whole stellar background

425

as a reference system and make differential measurements of, for example, the planet's position with respect to the positions of the nearby stars. The coordinates of the stars being known, the planet's ecliptic longitude and latitude at that instant could be calculated. The precision of an instrument required to measure the relatively small angles involved in this method would not need to be very high to achieve an accuracy of measurement of $1''$.

With respect to the measurement of the Sun's apparent longitude, the stars in the solar neighbourhood will be invisible but this difficulty may be overcome by projecting a faint image of the Sun onto the field of stars surrounding the point on the celestial sphere in opposition (see section 12.8) to the Sun. In this way, differential measurements of the position of the Sun's centre with respect to the field stars may be made; the apparent longitude λ_S'' and latitude β_S'' found in this way then yield the apparent longitude λ_S' and latitude β_S' of the Sun from the relations

$$\lambda_S' = \lambda_S'' - 180°$$

$$\beta_S' = -\beta_S''.$$

Other methods of obtaining useful position data and hence velocity data for the ship have been proposed.

The measurement of the Sun's angular diameter, which varies inversely with the ship's distance from the Sun, could be used to obtain the length of the ship's heliocentric radius vector. The intensity of solar radiation, also varying inversely as the square of the ship's distance from the Sun, would similarly provide a measurable quantity that yields the length of the radius vector. As the ship neared the sphere of influence of the planet of destination, a measurement of the planetary angular diameter would give the planetocentric distance.

It is therefore seen that, even if a ship was thrown onto its own navigation resources, on-board equipment of sufficient accuracy and of moderate size and mass can be provided, enabling all information required for navigating the vehicle to be collected.

References

Baker R M L Jr 1960 *J. Am. Rocket Soc.* Preprint No. 1220–60

Bauschinger J 1901 *Tafeln zur Theoretischen Astronomie* (*Tables on Theoretical Astronomy*) (Leipzig: Engelmann)

Briggs R E and Slowey J W 1959 *Smithsonian Institution Astrophysical Observatory Research in Space Science, Special Report* No. 27

Danby J M A 1962 *Fundamentals of Celestial Mechanics* (New York: Macmillan)

Herget P 1948 *The Computation of Orbits* (University of Cincinnati)

Moulton F R 1914 *An Introduction to Celestial Mechanics* (New York: Macmillan)

Plummer H C 1918 *An Introductory Treatise on Dynamical Astronomy* (London: Cambridge University Press)

Vertregt M 1956 *J. Br. Interplanet. Soc.* **15** 324

14 Binary and Other Few-Body Systems

14.1 Introduction

As seen in chapter 1, more than half the stars in the Galaxy are members of double, triple or greater-number systems of stars. In this chapter we will mainly consider double and triple systems, leaving many-body systems to be discussed in a later chapter. We shall first study binaries on an elementary level, beginning with the observational methods employed and the main deductions made from resulting observational data.

Binaries reveal themselves in several different ways. Firstly, the apparent closeness of some pairs of stars on the celestial sphere is statistically more frequent than might be expected from chance alignments of stars at different distances. In chapter 1 we saw that Sir William Herschel published a catalogue of the positions of many pairs of stars. The aim of this work was to make regular observations of these stars and see if the brighter of the pair exhibited parallactic motion relative to the fainter and presumably more distant component. Further observation of some of the pairs over a period of years revealed that the stars were in fact gravitationally connected and in orbit about each other. These pairs are therefore relatively close to each other in space, sufficiently close for the force of gravitation between them to be strong. They are known as *visual binaries.*

If we imagine a pair of stars brought progressively closer together, their relative mean orbital motion increasing in accordance with Kepler's third law, a situation will arise where the two stars become unresolvable to the distant observer. If the stars are also orbiting each other in a plane containing or close to the line of sight, there will be times, according to the relative positions of the stars in their orbits, when one star will eclipse the other. The eclipse would be registered by the observer as a decrease in brightness of the apparent single star. Stars of variable brightness, with a pattern of variability which can be explained on the basis of eclipses, are not uncommon. An example is the star Algol which has a regular fluctuation with a period of 2^d 20^h 49^m, this period being discovered by Goodricke in 1783. Observations of the brightness changes allow a light curve to be obtained, and from this curve orbital parameters and physical properties of the eclipsing pair may be deduced. The interpretation of the light curves of *eclipsing binary* systems therefore provides a second means of investigating such systems.

A third way is provided from the analysis of stellar spectra. Some stars, which otherwise might have been considered as being single, exhibit duplicity in their spectral lines. Each spectral line is doubled, showing that an apparent single star has two components and that the components are moving with different relative velocities with respect to the observer. Over a period of time the relative positions of the lines are seen to change, showing that the velocities of the two stars change. This can only be interpreted by considering the two components to be revolving around each other. Figure 14.1 illustrates the effect when the two stars are in orbit in a plane which contains the line of sight; typical spectra are presented for three epochs of the orbit.

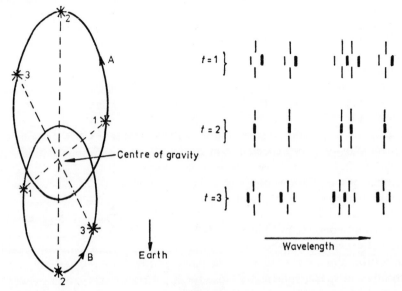

Figure 14.1

At time $t = 1$ (the top and bottom sets of lines at each epoch denote the laboratory reference), star A is receding from the Earth and star B is approaching. The spectral lines of star A (denoted by the thick lines) are thus red-shifted and those of star B blue-shifted as a consequence of the Doppler effect. At $t = 2$, both stars have no radial velocity with respect to the Earth and the spectral lines are superimposed. At $t = 3$, star A is approaching the Earth and exhibits a blue-shifted spectrum, while star B recedes and exhibits a red-shifted spectrum. Regular monitoring of the spectra shows that the stars periodically reverse their sense of radial velocity and so a period can be ascribed to their orbits.

Such a system exhibiting periodic changes of the above nature is known as a *spectroscopic binary*. By plotting how the radial velocities of each component change with time, *velocity curves* are produced. Analysis of a velocity curve allows deduction of a star's orbit about the centre of mass of the system. In some cases an apparent single star exhibits a single spectrum as expected, but it is found that the star has a radial velocity which exhibits periodic changes. This again is interpreted as the star being a component of a binary system but with the second star being too faint to contribute significantly to what would be the combined spectrum.

The three classical types of binary star were thought for many years to be the only kinds in existence but recently other kinds have been detected such as x-ray binaries and black-hole candidates. Astrophysical theory predicts that stars will end their lives as highly compact objects—white dwarfs, neutron stars or black holes. The first category consists of faint objects such as the companion of Sirius but are detectable even when invisible if they are one of a pair of stars forming a binary. The behaviour of the visible component reveals the presence of its invisible companion gravitationally bound to it. Neutron stars, even more compact and fainter than white dwarf stars, can be detected as pulsars; even after the pulsar emission has decayed, they can, like white dwarfs, be revealed if they are members of binary systems. A black hole, its name acknowledging its infinite capacity to absorb any electromagnetic radiation impinging upon it and allowing no radiation to be emitted, can reveal its presence if a 'normal' star is in orbit about the black hole.

We will discuss firstly the three classical types of binary before considering these more recently discovered forms.

14.2 Visual Binaries

The angular separation of visual binaries may either be measured by eye (with the aid of a rotatable micrometer eyepiece) or their positions may be recorded photographically for subsequent measurement in the laboratory.

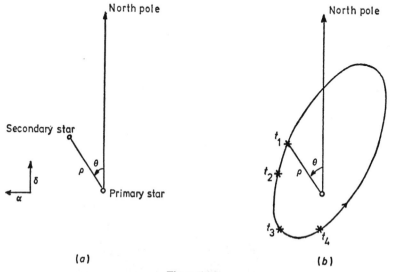

Figure 14.2

429

By making regular observations, their apparent orbits may be determined. Typical orbital periods range from a few tens to hundreds of years. Some binaries have not yet been measured over a time sufficiently long for one complete orbit to have been observed and so considerable uncertainty arises about the orbital period.

Usually one star in a binary is chosen as reference. This is conventionally the brighter of the two and it is known as the *primary star*; the other star is known as the *secondary star*. Observation is made at a chosen time t of the angular separation ρ of the stars and the position angle θ of the secondary star; the *position angle* θ is defined as the angle between the celestial north pole, the primary star and the secondary star. It is measured positively in the direction of increasing right ascension (see figure 14.2).

The elliptical orbit which is obtained directly from observations by plotting them represents what is known as the *apparent orbit*. The plane of the *true orbit* is in general tilted with respect to the tangent plane at the star perpendicular to the line of sight. What the observer sees as the apparent orbit is the projection of the true orbit on that plane. If the observer wishes to know all the parameters of the binary star orbit, he must allow for the tilt of the orbit with respect to himself. There are several standard mathematical procedures for doing this.

Any ellipse in a particular plane when projected onto another plane produces a figure which is again an ellipse, but with different characteristics. Moreover, a focus in the first ellipse when projected does not appear at the position of the focus of the projected ellipse. Thus, when the apparent orbit is examined, it is generally found that the primary star does not sit at the position of the focus of the ellipse. The necessary change in perspective required to place the primary star at the focus can be determined by one of the standard methods, so giving the inclination of the true orbit with respect to the celestial sphere. After this has been determined, all the parameters describing the true orbit may be deduced. It must be pointed out, however, that the sign of the angle of inclination is indeterminate; a positive or negative tilt of the same amount produces an identical apparent orbit. If the radial velocity of the orbiting star can be measured, the sign ambiguity can be removed. We will define the orbital elements of a binary star in a later section.

Of immediate use are the orbital period T (which is available directly from the apparent orbit) and the size of the major axis α. If the distance of the binary star is known, then we can determine the sum of the masses of the stars as follows.

If M_1 and M_2 are the masses of the primary and secondary stars, then the period of revolution T of the secondary about the primary is given by equation (4.26), viz.

$$T = 2\pi \sqrt{\frac{a^3}{G(M_1 + M_2)}} \qquad (14.1)$$

where a is the semimajor axis of the orbital ellipse and G is the universal constant of gravitation. Now the corresponding formula for the Earth's orbit about the Sun is

$$T_\oplus = 2\pi \sqrt{\frac{A^3}{G(M_\odot + M_\oplus)}}.$$

If we express the periods of revolution in years and consider that $M_\odot \gg M_\oplus$, the last expression reduces to

$$1 = 2\pi \left(\frac{A^3}{GM_\odot}\right)^{1/2}$$

and from this we see that

$$(G)^{1/2} = 2\pi \left(\frac{A^3}{M_\odot}\right)^{1/2}.$$

Substituting this into equation (14.1) gives

$$T = \left[\frac{a^3}{M_1 + M_2}\left(\frac{M_\odot}{A^3}\right)\right]^{1/2}$$

so that

$$M_1 + M_2 = \left(\frac{a}{A}\right)^3 \frac{M_\odot}{T^2}.$$

By letting the solar mass equal unity, this expression becomes

$$M_1 + M_2 = \left(\frac{a}{A}\right)^3 \frac{1}{T^2}. \tag{14.2}$$

Thus, if the period of revolution is determined and the size of the orbit is known, the sum of the masses of the two stars may be deduced in terms of the solar mass.

If d is the distance of the binary star, the apparent angular size α of the semimajor axis is given by

$$\sin \alpha = \frac{a}{d} \tag{14.3}$$

and since α is a very small angle, this may be written as

$$\alpha = \frac{a}{d}. \tag{14.4}$$

Now the parallax P of the star (section 3.7) is given by

$$\sin P = \frac{A}{d}.$$

Since P is also a very small angle, this may be written as

$$P = \frac{A}{d}. \tag{14.5}$$

431

Hence

$$\frac{\alpha}{P} = \frac{a}{A}.$$

Substituting this into equation (14.2), we have

$$M_1 + M_2 = \left(\frac{\alpha}{P}\right)^3 \frac{1}{T^2} \tag{14.6}$$

where α and P are usually measured in seconds of arc.

If it is possible to measure the stars' positions relative to the position of their centre of gravity then the ratio of the masses may be determined. This type of measurement requires very accurate positions of both stars observed against the distant star background over a long period of time. For a single star, prolonged observation over many years shows that it has a motion of its own with respect to the fainter background stars, giving it a path which is part of a great circle on the celestial sphere. If it is a binary system however, it is the centre of gravity of the system which progresses along a great circle. The two stars forming the system follow curved paths with a slow oscillation about the centre of gravity (see figure 14.3). From the positional measurements of both stars, the path of the centre of gravity and then the separate orbits may be determined.

Suppose α_1 and α_2 are the angular distances of the primary and secondary stars from the apparent centre of gravity of the system. Then we have

$$M_1 \alpha_1 = M_2 \alpha_2$$

so that

$$\frac{M_1}{M_2} = \frac{\alpha_2}{\alpha_1}. \tag{14.7}$$

If observations allow parameters to be inserted into both equations (14.6)

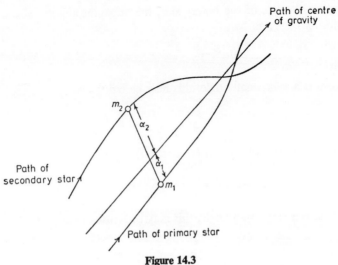

Figure 14.3

and (14.7), then the masses of the individual stars may be evaluated. Typical masses obtained from the study of visual binary stars run from 0·1 to 20 times the mass of the Sun.

14.3 The Mass–Luminosity Relation

Apart from being the source of our knowledge about stellar masses, binaries of known distance (or parallax) also provide data showing that a relationship exists between the luminosity (or intrinsic brightness) of a star and its mass. This empirical relation, known as the mass–luminosity law, can also be justified on theories of stellar structure (figure 14.4). For convenience absolute bolometric magnitude, which is directly related to the luminosity, is plotted against the logarithm (base 10) of the mass of the star, the solar mass being

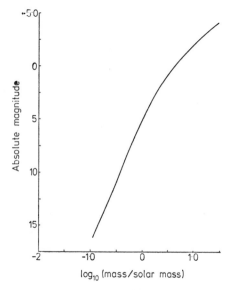

Figure 14.4

taken as unity. The Sun, with absolute bolometric magnitude +4·79 and the log of its mass as zero, thus lies on the curve.

To a good approximation, it is found that over most of the range

$$L \propto M^{3 \cdot 1}$$

where L is the luminosity of a star and M is its mass. The mass–luminosity relation can evidently be used to assign a mass to a star if its luminosity is known.

14.4 Dynamical Parallaxes

The fact that the masses of observed visual binaries do not cover a very wide range can be used to estimate the distances of those which cannot be measured by the usual parallax method. This method of distance determination is known as the method of *dynamical parallax*. The method involves a number of steps, repeated until a satisfactory answer is obtained.

(i) We assume as a first approximation that each star has solar mass. Then $M_1 + M_2 = 2$ and, by using equation (14.6) in the form

$$P = \frac{\alpha}{T^{2/3}} \frac{1}{(M_1 + M_2)^{1/3}} \tag{14.8}$$

we can obtain a first approximation to the parallax by substituting observed values for α and T and letting $M_1 + M_2 = 2$.

(ii) We now use the measured apparent magnitudes m_1 and m_2 of the binary components.

From section 3.7, we had

$$M = m + 5 + 5 \log P.$$

If $\mathcal{M}_1$ and $\mathcal{M}_2$ are the absolute magnitudes of the components, then

$$\mathcal{M}_1 = m_1 + 5 + 5 \log P$$

$$\mathcal{M}_2 = m_2 + 5 + 5 \log P.$$

Substituting the first approximation obtained in step (i) for the parallax into these equations will give first approximations for the absolute magnitudes $\mathcal{M}_1$ and $\mathcal{M}_2$ of the components.

(iii) Use is now made of the mass–luminosity relation. Using the first approximations found in step (ii) for the components' absolute magnitudes in this relation, we can read off improved values of the masses M_1 and M_2 of the components.

(iv) Use these values in equation (14.8) to derive an improved value P_2 of the parallax.

(v) Go back to step (ii) and continue *ad infinitum*.

In practice it is found that the values of P converge very quickly. The reiterative process is halted when any difference between two successive approximations is less than one in the last significant figure to which the apparent magnitudes are known. For example, if the apparent magnitudes were 0·16 and 0·85, and it was found that $P_2 = 0·15$ arc sec while $P_3 = 0·14$ arc sec, it would be meaningless to carry the process any further. It should also be noted that the quantity $\alpha T^{-2/3}$ in equation (14.8) need only be calculated once.

Even if the first guess that $M_1 + M_2 = 2$ is a poor one, the form of equation (14.8) minimizes the error, since the quantity $(M_1 + M_2)$ is raised to the power one-third. Thus, if in fact $M_1 + M_2 = 20$ (an unusually large mass for a binary) and we put as a first approximation $M_1 + M_2 = 2$, we see that $20^{1/3} = 2·714$, while $2^{1/3} = 1·260$. The factor of 10 in the sum of the masses is reduced immediately to a factor of about $2\frac{1}{2}$ in the term $(M_1 + M_2)^{1/3}$. Because of this fact, dynamical parallaxes are reliable, providing useful additions to our collection of stellar distances and masses.

434

14.5 Eclipsing Binaries

The periods of the light curves of eclipsing binaries are usually a few days, indicating that the components of this type of system are much closer together than in the cases of visual binaries. Actual shapes of light curves vary from one binary star to another, but the general characteristic of there being two falls in brightness within the period may only be interpreted by considering a system of two stars which are orbiting each other and presenting eclipses to the observer.

The basic form of an eclipsing binary light curve is depicted in figure 14.5 where, during the periods of minimum brightness, the level remains constant. This particular form would indicate that the eclipses are total. Figure 14.5(a) illustrates the configurations which produce the kind of light curve depicted in figure 14.5(b), representing the orbit that would be seen if it were possible to resolve the component stars.

By comparing figures 14.5(a) and (b) we see that, when the smaller star is in position A, each component contributes fully to the total brightness. At position B the smaller star is about to commence its passage across the disc of the larger star. In progressing from position B to C, the smaller star begins to block off light from the larger and the total light level drops smoothly. It then levels off and remains at this brightness until the smaller star arrives at position D. In moving from D to E more and more of the disc of the larger star is revealed, until at position E the light level regains its full brightness.

Full brightness is then maintained until the motion brings the smaller star to position F. At this position it commences to be eclipsed by the larger star and the light level falls. At position G the smaller star is fully eclipsed and remains so until it arrives at H. During the period from G to H the light level remains constant, but in general not at the same level as the minimum produced between positions C and D as the brightnesses of the component stars are usually different. On egress from the eclipse to position I, the light

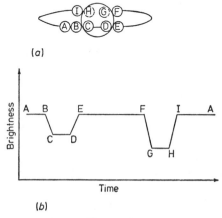

(a)

(b)

Figure 14.5

level rises until full brightness is recorded. This level is maintained until position B again when a new cycle of the light curve begins.

Let us now look at the light curve more quantitatively. Although the light curves may sometimes be expressed in terms of changes in stellar magnitude, it is more convenient here to consider them in terms of brightness changes. Suppose that the smaller star has a luminosity L_1, and the larger star a luminosity L_2. (It is generally found that $L_1 > L_2$.) Now the apparent brightness of the system is equal to the sum of the brightnesses of the two stars. They contribute to this total according to their luminosities and to the amount of their surfaces that can be seen. If the fully presented surfaces are S_1 and S_2 for the smaller and larger stars respectively and if the recorded brightness between the eclipses (i.e. full apparent brightness) is B, we may write

$$L_1 S_1 + L_2 S_2 = kB \qquad (14.9)$$

where k is a constant related to the stars' distance from the observer.

Suppose that at the first minimum (smaller star in front of larger star) the apparent brightness falls to B_1 and at the second minimum (smaller star behind larger one) the apparent brightness is B_2. It is easily seen that

$$L_1 S_1 + L_2 (S_2 - S_1) = kB_1 \qquad (14.10)$$

and also that

$$L_2 S_2 = kB_2. \qquad (14.11)$$

Let b_1, b_2 be the brightness losses at the two minima so that $b_1 = B - B_1$ and $b_2 = B - B_2$. Subtract equations (14.10) and (14.11) in turn from equation (14.9) to obtain

$$L_2 S_2 - L_2 (S_2 - S_1) = k(B - B_1)$$

or

$$L_2 S_1 = kb_1 \qquad (14.12)$$

and

$$L_1 S_1 = k(B - B_2)$$

or

$$L_1 S_1 = kb_2. \qquad (14.13)$$

Dividing equation (14.13) by equation (14.12) we have

$$\frac{L_1}{L_2} = \frac{b_2}{b_1}. \qquad (14.14)$$

This simple analysis immediately shows that the ratio of the stars' luminosities may be obtained directly from the ratio of the apparent brightness losses at the two minima.

By using equations (14.09), (14.11) and (14.14) it is easily shown that

$$\frac{S_1}{S_2} = \frac{b_1}{B - b_2}$$

and since the values of S_1 and S_2 are proportional to the square of the stellar radii R_1 and R_2, we can write

$$\left(\frac{R_1}{R_2}\right)^2 = \frac{b_1}{B - b_2}$$

436

and we can therefore write

$$\frac{R_1}{R_2} = \left(\frac{b_1}{B - b_2}\right)^{1/2}.$$ (14.15)

Thus, by measuring the maximum brightness and the brightness loss at the minima, the ratio of the radii of the stars can be deduced.

Values of the ratios of luminosities and radii of the stars helps us to compare the properties of stars which happen to be the components of an eclipsing binary system. Further analysis of the light curve can in many cases enable the radii of the stars to be related to the sizes of their orbits. The inclination of the orbit with respect to the observer may also be deduced. All this information is particularly useful if the eclipsing binary is also observed as a spectroscopic binary (section 14.6). However, the elegant methods that are applied to the light curve are beyond the scope of this text and will not be discussed here.

It may be noted though that the light curve described above represents a system which exhibits total eclipses. The fact that there are some systems which exhibit partial eclipses is clearly evident. For such systems, there is no extended period when the minima hold steady values; the light curve has two V-shaped minima, usually of different depths. Figure 14.6(a) represents such a partially eclipsing system and figure 14.6(b) illustrates the light curve.

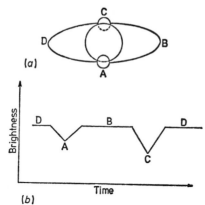

(a)

(b)

Time

Brightness

Figure 14.6

It will be seen in figure 14.6 that the maximum area of the larger star eclipsed by the smaller occurs at A. Because the eclipse is partial, the light curve immediately begins to rise again. It may easily be shown that the depths of the minima from such a light curve still allow the ratio of the luminosities to be determined. However, the ratio of the radii cannot be obtained by the simple expression (14.15). Other standard but more complicated ways are available for obtaining this information from the light curve.

The light curve can also provide knowledge of the eccentricity of the orbit of one star about another. As an example, an extreme case is illustrated

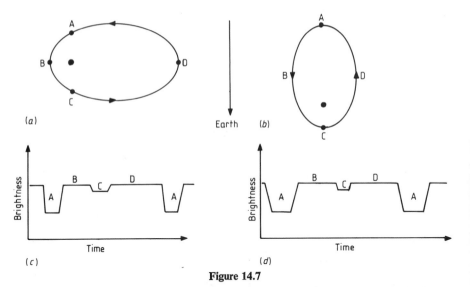

Figure 14.7

in figure 14.7(*a*) where the major axis is at right angles to the line of sight. Now both stars are subject to the law of gravitation and therefore obey Kepler's three laws. The secondary star will therefore travel at its fastest when nearest the primary star, when it is said to be at periastron. Because of this, the secondary eclipse C occurs closer to the preceding primary eclipse A (figure 14.7(*c*)) than to the following primary eclipse, and the periods of maximum brightness (B and D) are not of equal length. In contrast, the situation of figure 14.7(*b*), where the major axis is parallel to the line of sight, will produce periods of maximum brightness of equal length but minima of unequal length (figure 14.7(*d*)).

Besides providing orbital information, a detailed analysis of a light curve may provide knowledge about:

(i) departures from sphericity of the shapes of stars,
(ii) the uniformity of brightness across the stellar discs (i.e. limb darkening),
(iii) the effects of reflection (i.e. the light from one star being reflected by the other in the direction of the observer).

These are discussed briefly below:

(i) Some stars are so close together that they distort each other gravitationally, each star being elongated along the line joining their centres. Thus, as illustrated in figure 14.8, if two oblate stars revolve about each other in a plane such that eclipses occur, the light curve will contain no straight parts. It will change smoothly because the total area the stars present to the observer is never constant.

(ii) It is well known that the Sun does not have uniform brightness across its disc and that the brightness falls off towards the solar limb. This effect is known as limb darkening. From the light curves of eclipsing binaries, we know that some stars must exhibit the same effect. When the eclipse begins (see figure 14.9) the initial fall in brightness is slow, as the less bright

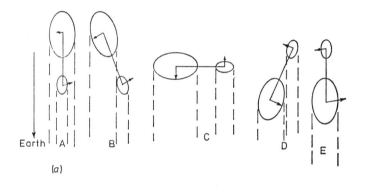

Earth

(a)

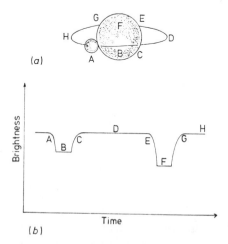

(b)

Figure 14.8

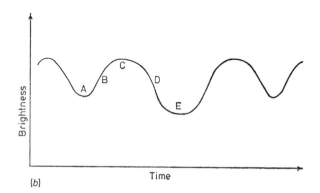

(a)

(b)

Figure 14.9

439

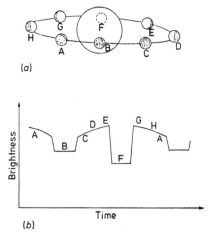

(a)

(b)

Figure 14.10

parts of the stellar disc at the limb are occulted first. The fall in brightness increases at a faster rate as the occulting star begins to cover the brighter parts of the eclipsed star. Thus the falls and rises in brightness are not linear when the stars exhibit limb darkening.

(iii) In this case the parts of the light curve between the minima are sloped and curved as shown in figure 14.10, so that although neither star is entering or emerging from eclipse, the brightness of the system is altering. What is happening is that the smaller star is showing phases analogous to those exhibited by Venus or the Moon. The side presented to the larger star appears brighter than the side turned away from it. It must be remembered however that, unlike Venus and the Moon, the smaller star is self-luminous as well.

14.6 Spectroscopic Binaries

An idea of the shapes that can be expected for a radial velocity curve can be obtained by considering three different types of orbit. For simplicity let us consider the orbit of one star about the centre of gravity and suppose the orbit to be in a plane which contains the line of sight. We shall consider the orbit as being: (a) a circle, (b) an ellipse with its major axis at right angles to the line of sight, and (c) an ellipse with its major axis along the line of sight. The orbits are illustrated in order in figures 14.11(a), (b) and (c), together with their associated radial velocity curves. It will be noted in all cases that for positions 1 and 3 the motion is transverse and the radial velocity is zero. Any measured radial velocity at these points represents the motion of the whole system with respect to the Earth.

For the circular orbit the radial velocity curve is symmetrical. The motion of the star towards and away from the observer is similar to that of simple harmonic motion, and hence the velocity curve is in the form of a sine wave.

440

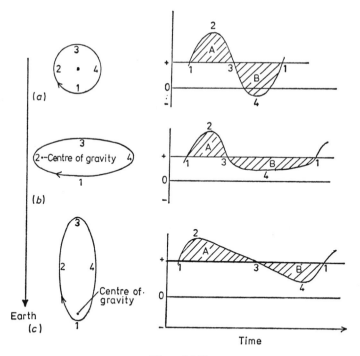

Figure 14.11

For the elliptical orbit with its major axis at right angles to the observer, Kepler's law predicts that the velocity of the star is greatest at periastron; it consequently spends a relatively short time in this part of its orbit. The velocity curve shows a sharp peak for the period through the points 1, 2 and 3. It spends a longer time with a motion which is nearly transverse. This corresponds to the orbit from point 3, through 4 and on to 1.

For the elliptical orbit with its major axis along the line of sight, the velocity changes its direction from negative to positive very quickly at point 1 near periastron. At point 3, the orbital speed is much slower than at 1. The cross-over from a positive to a negative radial velocity is consequently much slower than the opposite cross-over at point 1.

The above three examples are all special cases. When it is considered that the orbit may be set with its major axis at a different angle and the plane inclined to the observer, then the shape of the curve must reflect these facts. Since the net orbital velocity over one period is zero and since the velocity curve is a plot of velocity against time, a line of constant velocity can be drawn on the curve so that the area above the line is equal to the area below. The velocity indicated by this line represents the constant radial velocity of the binary system as a whole with respect to the Sun. When both components contribute to the spectrum two velocity curves may be obtained, corresponding to the orbits of each star about the centre of gravity of the system. It goes without saying that any determined radial velocity

441

must be corrected for the Earth's orbital motion about the Sun before the value can be plotted on the radial velocity curve.

If any binary star orbit is considered, it is possible to derive the expressions for the value of the radial velocity of each component at any particular time. Appearing in the radial velocity expression for the primary star is the product $a_1 \sin i$, and in the expression for the secondary star is the product $a_2 \sin i$, where a_1 and a_2 are the semimajor axes of the orbits about the centre of gravity; the two products are the projections of these axes onto the plane at right angles to the line of sight (i.e. i is the inclination of the plane of the orbit relative to the tangent plane on the celestial sphere). From the analysis of the two radial velocity curves, these products may be determined. The parameters a_1 and a_2, however, cannot be separated from $\sin i$ using the radial velocity data alone.

From the definition of the centre of gravity we have the relation

$$M_1 a_1 = M_2 a_2. \tag{14.16}$$

Multiply both sides of this equation by $\sin i$ to give

$$M_1 a_1 \sin i = M_2 a_2 \sin i$$

so that

$$\frac{M_1}{M_2} = \frac{a_2 \sin i}{a_1 \sin i}. \tag{14.17}$$

The numerator and denominator of the right-hand side of equation (14.16) are the very quantities which may be determined from analysis of radial velocity curves.

When both curves are obtained, it is seen that one curve is a reflection of the other about the zero-velocity line, though perhaps with a different amplitude. The ratio of the amplitudes of the two velocity curves is inversely proportional to the ratio of the masses of the stars. Thus if both curves are available, the ratio of the component masses can in fact be determined directly from the curves.

In equation (14.2), we have already shown the relationship between the sum of the masses of two stars, the size of the major axis of the orbit of one star about the other and the period of revolution. By expressing distances in terms of the astronomical unit, this equation reduces to

$$M_1 + M_2 = \frac{a^3}{T^2}. \tag{14.18}$$

By substituting the value of M_2 obtained from equation (14.16), the above equation may be written as

$$M_1 = \frac{a^3}{T^2 \left(1 + \dfrac{a_1}{a_2} \right)}. \tag{14.19}$$

In relating the two orbits about the centre of gravity to the one referred to the primary star, we have the relation $a = a_1 + a_2$ or, multiplying by $\sin i$, the relation $a \sin i = a_1 \sin i + a_2 \sin i$.

Now as we have seen, the analysis of the radial velocity curves allows $a_1 \sin i$ and $a_2 \sin i$ (and hence $a \sin i$) to be deduced. By expressing the right-hand side of equation (14.19) in terms of quantities which can be deduced we have

$$M_1 \sin^3 i = \frac{(a \sin i)^3}{T^2 \left(1 + \dfrac{a_1 \sin i}{a_2 \sin i} \right)} \tag{14.20}$$

thus showing that a value for $M_1 \sin^3 i$ may be determined. In a similar manner a value for $M_2 \sin^3 i$ may also be determined.

If only one curve is available then a quantity known as the *mass function* can be obtained. Suppose that it is the primary star which provides the spectrum for measurement. We are therefore able to determine $a_1 \sin i$ but not $a_2 \sin i$. By adding $M_2 a_1$ to both sides of equation (14.16), we have

$$M_1 a_1 + M_2 a_1 = M_2 a_1 + M_2 a_2$$

so that

$$\frac{a_1}{M_2} = \frac{a_1 + a_2}{M_1 + M_2}.$$

Since $a_1 + a_2 = a$, this may be rewritten as

$$\frac{a_1}{M_2} = \frac{a}{M_1 + M_2}.$$

Eliminating a from this equation by means of equation (14.18), we obtain

$$\frac{a_1{}^3}{T^2} = \frac{M_2{}^3}{(M_1 + M_2)^2}.$$

Multiplication of both sides of this equation by $\sin^3 i$ allows the left-hand side to be expressed in terms of measured and deduced quantities. Thus,

$$\frac{(a_1 \sin i)^3}{T^2} = \frac{(M_2 \sin i)^3}{(M_1 + M_2)^2}. \tag{14.21}$$

The right-hand side of equation (14.19) is known as the mass function of the spectroscopic binary.

14.7 Combination of Deduced Data

A summary of the information about the physical nature of binary stars which can be deduced from observations is given in table 14.1.

Table 14.1

Visual binary	Eclipsing binary	Spectroscopic binary
Angular size of the major axis Eccentricity of the ellipse Modulus of the inclination of the orbital plane If the parallax is known: the linear size of the major axis the sum of the masses If the centre of gravity is known: the ratio of the masses and hence the mass of each star.	Ratio of luminosities Ratio of radii to the radius of the relative orbit, considered as being a circle Inclination of the orbital plane Shape of the stars Eccentricity of the orbit Limb darkening	The product of the size of the major axis and the sine of the inclination of the orbital plane Eccentricity of the ellipse From knowledge of both velocity curves, the product of each mass and the cube of the sine of the inclination of the orbital plane

If the binary is both eclipsing and spectroscopic, the masses of the components and absolute values for the radii can be deduced.

14.8 Binary Orbital Elements

In the remainder of this chapter we describe certain aspects of binary systems that demonstrate how complex such systems can be and how far the majority of binaries depart from the simple two-body problem. As a preliminary, we define what is meant by the orbital elements of a binary system. These correspond to the orbital elements of a planet or satellite; because of the nature of the problem however, certain modifications must be made.

In figure 14.12, the tangent plane at the binary to the observer's celestial sphere is shown. A second sphere may be drawn about P, the primary

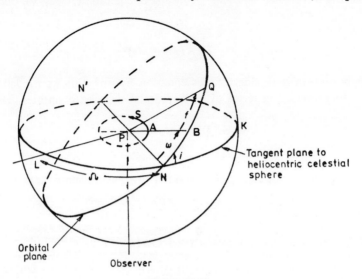

Figure 14.12

component of the binary, and the tangent plane taken to be the fixed plane of reference for measurements in this sphere. In this plane it will be possible to define a direction PL from the binary towards the north celestial pole L. This direction can then be used as a fixed reference direction in the tangent plane.

We can now define the elements of the orbit of the secondary star S about the primary P. Let the orbital plane cut the tangent plane in the nodes N and N′. Then:

$\Omega = LN =$ the position angle (measured in an easterly direction) of the ascending node,

$i = B\hat{N}K =$ the inclination of the orbital plane to the tangent plane,

$\omega = A\hat{P}N =$ argument (or longitude) of periastron (the point of closest approach of the secondary star to the primary),

$a =$ the orbital semimajor axis,

$e =$ the eccentricity (since we are dealing with a bound orbit, $0 \leqslant e \leqslant 1$),

$\tau =$ the time of periastron passage, and

$T =$ the orbital period (measured in years for visual binaries or days for eclipsing or spectroscopic binaries).

Some explanatory remarks may be made here. Although both a and T are treated as elements, and are related through Newton's form of Kepler's third law

$$\frac{4\pi^2 a^3}{T^2} = G(m_1 + m_2)$$

the masses m_1 and m_2 are themselves unknown quantities to be determined. To do this, values of T and a must be found. A binary system therefore has the seven elements Ω, i, ω, a, e, τ and T.

We have seen that unless radial velocity measurements of the components of a visual binary are available, there remains an ambiguity of 180° in the determination of the ascending and descending nodes. Without these measurements it is the custom to take $0° \leqslant i \leqslant 90°$ if the apparent motion is direct and to assume that the node for which $\Omega \leqslant 180°$ is the ascending node. Spectroscopic and eclipsing binaries provide their own problems in orbital determination and improvement. There is a lengthy literature on these matters, constantly being added to.

We now consider in more detail two of the seven elements, namely the period of revolution T and the argument (or longitude) of periastron ω.

14.9 The Period of a Binary

The period of a binary is one of the most important elements to be determined. It can usually be measured to a higher accuracy than that of any other element. In principle any phenomenon that is periodic and measurable can have its period measured to greater and greater precision if measurements are many, unambiguous and made throughout time intervals many multiples of the period in length. An eclipsing binary which has well defined primary and secondary minima and a period that is not nearly an integral number of sidereal days is ideal. An accuracy of one part in 10^9 is attainable. For a visual binary, most of which have periods greater than 10 years, the accuracy is probably one part in 10^4 or less (it is to be remembered that reliable observations for most visuals lie within the past century or less). The accuracy of spectroscopic binary periods lies between those for eclipsing and visual binaries.

Once the period has been measured accurately for (say) an eclipsing binary, predictions of times of beginnings and ends of eclipses can be made. This ephemeris can then be compared with observations of such phenomena and any change in period detected. Such changes in period are observed in many binaries. They may be sudden or periodic, and have been attributed to a number of causes. We will consider such changes in a later section. It may be remarked here, however, that corrections have to be applied to the measured period because of the radial velocity of the binary's centre of mass relative to the Sun and the Earth's orbital motion about the Sun. Such corrections are analogous to those required when observations of transits, eclipses and occultations of Jupiter's Galilean moons are compared with orbital theory and are found to be 'late' or 'early' in a systematic way, depending upon the finite velocity of light and the varying distances between Jovian satellite and Earth (it was a study of such bad satellite timekeeping that enabled Romer to measure the velocity of light in 1675).

14.10 Apsidal Motion

Consider again the simple case of an eclipsing binary, the orbital plane of which contains the line of sight. Let the eccentricity be moderate and let the major axis be at right angles to the line of sight (figure 14.7(a)). Because of Kepler's second law the secondary star will then travel fastest at periastron, so that the secondary minimum will be closer to the preceding primary minimum than to the following. If however the major axis lies in the line of sight, as in figure 14.7(b), the secondary minimum will be equidistant from the preceding and following primary minima. Later, if the major axis is again at right angles to the line of sight, but the longitude of periastron is 180° ahead of the longitude of periastron, the secondary minimum will be nearer the following primary than the preceding one. If we now consider that the orbit is rotating in its own plane (i.e. there is a secular advance of periastron) it is clear that if the eccentricity is even moderate it should be possible in the course of time to see that the secondary minima oscillate about the midpoints between the primary minima. The period of oscillation is the period of rotation of the line of apsides.

446

Examples of eclipsing binaries for which apsidal motion has been measured are γ Cygni (apsidal period 54 years), CO Lacertae (apsidal period 45 years), GL Carinae (apsidal period 27 years), AG Persei (apsidal period 83 years). Their orbital periods in days are respectively 3·00, 1·54, 2·42 and 2·03, showing that the ratio of apsidal period to orbital period is usually thousands to one.

14.11 Forces Acting on a Binary System

If the components of a binary system are point-masses, and no other forces act on them apart from gravitation, then the binary is an example of the two-body problem and the elliptic solution will completely describe the orbital motion of one component about the other. The orbital elements are therefore constant. By section 6.5 it is seen that this is the case even if the two stars are not point-masses but spherical and of finite size, with an internal density distribution that is radially symmetrical within them. It is rare, however, that this simple picture describes any particular case. There are a number of other factors that can operate to distort the basic picture. The most important of these are:

(i) presence of one or more stars gravitationally connected with the binary,
(ii) the inadequacy of Newton's law of gravitation,
(iii) departure of the components from effective point-masses, and
(iv) exchange of matter between the components or loss of mass from the system.

These factors will cause changes to occur in the binary orbital elements. Of particular interest are the changes in the period of revolution T and the argument or longitude of periastron ω. In seeking information about the structure of a binary system and its components, it therefore becomes important to assess the contributions such factors may make to the measured changes in these elements. We now consider them in turn.

14.12 Triple Systems

It was remarked in chapter 1 that between one-quarter and one-third of all binaries, on closer and prolonged examination, are found to be triple systems. It also appears that in practice the vast majority of triple systems consist of a close binary with a third star at a distance many times (in a number of cases hundreds of times) that of the close binary separation. There is in fact a dearth of systems in which all the mutual separations are of the same order.

In chapter 5 we saw that numerical experiments in the general three-body problem enabled a classification of types of orbital motion to be made; these were summarized in table 5.1. Among these classes, *interplay* was only of transient duration leading to *escape* or *ejection*, while a quasistable

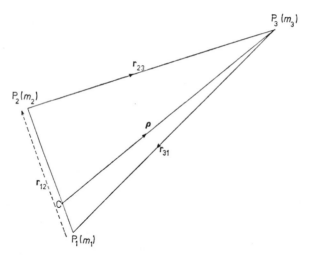

Figure 14.13

mode for a three-body system was found in *revolution* where a close binary was formed and the third body revolved about the binary at an average distance much greater than that separating the binary components. We can see quite clearly the reason for this quasistability when we set up the three-body problem in the Jacobi coordinate form, as was done in section 5.11.1.

It will be remembered that C is the centre of mass of P_1 and P_2; $\mathbf{r}$ is the vector from P_1 to P_2 while $\boldsymbol{\rho}$ is the vector from C to P_3. Then it was found (from equations (5.90) and (5.91)) that

$$\ddot{\mathbf{r}} + \mu \mathbf{f}(\mathbf{r}) = (M - \mu)|\mathbf{f}(\boldsymbol{\rho} - \nu\mathbf{r}) - \mathbf{f}(\boldsymbol{\rho} + \nu^*\mathbf{r})| \qquad (14.22)$$

and

$$\ddot{\boldsymbol{\rho}} = -M|\nu^*\mathbf{f}(\boldsymbol{\rho} - \nu\mathbf{r}) + \nu\mathbf{f}(\boldsymbol{\rho} + \nu^*\mathbf{r})| \qquad (14.23)$$

where

$$M = m_1 + m_2 + m_3, \qquad \mu = m_1 + m_2$$

$$\nu = m_1/\mu, \qquad \qquad \nu^* = m_2/\mu$$

and

$$\mathbf{f}(\mathbf{x}) = G\mathbf{x}/x^3.$$

Now in a triple system the case almost always found in practice is that in which $(r/\rho) = \epsilon$, where $\epsilon \ll 1$. Also $\nu < 1$ and $\nu^* < 1$. Hence we may expand $|\boldsymbol{\rho} - \nu\mathbf{r}|^{-3}$ and $|\boldsymbol{\rho} + \nu^*\mathbf{r}|^{-3}$ by the binomial theorem in the usual way. After a little reduction, and remembering that $\nu + \nu^* = 1$, we find that equation (14.22) (to the order of ϵ^2) becomes

$$\ddot{\mathbf{r}} + G\mu \frac{\mathbf{r}}{r^3} = Gm_3 \left\{ \frac{3\boldsymbol{\rho}}{\rho^3} \left[\epsilon u - \tfrac{1}{2}\epsilon^2(\nu - \nu^*)(1 - 5u^2) \right] \right.$$

$$\left. - \frac{\mathbf{r}}{\rho^3} \left[1 + 3\epsilon u(\nu - \nu^*) - \tfrac{3}{2}\epsilon^2(\nu^2 - \nu\nu^* + \nu\nu^{*2})(1 - 5u^2) \right] \right\} \qquad (14.24)$$

448

where

$$u = \frac{\boldsymbol{\rho} \cdot \mathbf{r}}{\rho r} \leqslant 1.$$

Similarly equation (14.23) becomes, to the order of ϵ^2,

$$\ddot{\boldsymbol{\rho}} + GM \frac{\boldsymbol{\rho}}{\rho^3} = 3GM \left\{ \frac{\boldsymbol{\rho}}{\rho^3} \left[\frac{\nu \nu^* \epsilon^2}{2} (1 - 5u^2) \right] + \mathbf{r} \frac{\nu \nu^*}{\rho^3} \left[\epsilon u - \tfrac{1}{2} (\nu - \nu^*)(1 - 5u^2) \right] \right\}.$$

$$(14.25)$$

It is then seen that the ratio of the largest term on the right-hand side of (14.24) to the central two-body term on the left-hand side is $m_3 \epsilon^3 / \mu$. Stellar masses are usually not widely different from each other and so $m_3 \sim \mu$. Hence the perturbing acceleration of the third mass on the binary is of the order of ϵ^3. For most triple systems $\epsilon < 10^{-2}$ so that $\epsilon^3 < 10^{-6}$. The perturbation is therefore small; much smaller than, for example, that of Jupiter on Saturn.

In equation (14.25) the ratio of the largest term on the right-hand side to the two-body term on the left-hand side is of the order of $\nu \nu^* \epsilon^2$. Now $\nu \nu^* \leqslant (1/4)$ and $\epsilon^2 < 10^{-4}$, so the perturbation is again small. Hence in both cases, namely the orbital motion of the binary system and the orbital motion of the third mass about the centre of mass of the binary, they are slightly perturbed elliptic motions. Over an astronomically long time however, the three-body computer experiments tell us that most triple systems end in escape leaving a binary and a field star, so that it is perhaps not surprising that the fraction of triple to binary systems is as low as one-quarter to one-third.

It is obvious from the above arguments that the most common form of quadruple stellar system, where two close binaries are gravitationally bound but the separation between the pairs is much greater than the separation of the components in each binary, must also be quasistable. Indeed the star Castor (α Geminorum) illustrates this principle in spectacularly convincing form. It consists of six component stars in three spectroscopic binaries, which we shall refer to as A, B and C. Their periods of revolution are respectively 9, 2 and 0·8 days. Binary B revolves about binary A with a period of several hundred years; binary C on the other hand revolves about A and B with a period of several thousand years.

Going back to the case of a close binary attended by a distant third star, it is readily seen that the orbital elements of the orbit of the secondary component about the primary will change. Because the disturbing function of the problem is small, Lagrange's planetary equations may be used to produce a general perturbation theory giving the changes (short- and long-period and secular) in the orbital elements. A lunar-type development is usually favoured, which is understandable if we recall how useful the Jacobi coordinate system is in both lunar and triple star problems.

In particular the longitude of periastron ω will change. In the special case of a coplanar triple star problem, with the third star's orbit circular and of period T', the apsidal period U' is given in terms of the close binary period T and the masses by an expression of the form

$$T'/U' = \frac{0.75 \, m_1}{m_1 + m_2 + m_3} \left(\frac{T}{T'}\right) + \text{higher-order terms.}$$

In practice, $m_1/(m_1 + m_2 + m_3) \sim 1/3$ and $T/T' < 10^{-2}$ so that

$$T/U' = \frac{TT'}{T'U'} < 10^{-4}$$

which is not negligible in comparison with measured values of T/U.

Lyttleton (1934), Brown (1936, 1937) and Kopal (1959) are among those who have studied the much more difficult triple-star problem where the third body's orbit is elliptic, its orbital plane being inclined to the close binary orbital plane with both planes also inclined to the tangent plane to the observer's celestial sphere. It goes without saying that the close binary orbital period is also modified by the presence of the third body.

14.13 The Inadequacy of Newton's Law of Gravitation

Newton's law of gravitation is sufficiently accurate in celestial mechanics and astrodynamics for almost every case yet encountered. One notable exception was the residual 43 arcsec per century advance of the perihelion of Mercury unaccounted for by Newtonian gravitational law perturbations by the other planets but accounted for beautifully by Einstein's law of gravitation. Within the Solar System the advance of perihelion is much smaller for planets other than Mercury, for the change of perihelion per orbital period is inversely proportional to the planet's semimajor axis. The larger semimajor axes and longer periods of the other planets therefore produce perihelion changes according to Einstein's law of gravitation too small to be detected.

It is perhaps appropriate that binary systems, discovered in the late eighteenth century, which verified for the first time that the Newtonian gravitation law operated far outside the Solar System, should also provide additional convincing proof that Einstein's theory holds.

In a close binary system, even if the component stars are point-masses gravitationally, a relativistic advance of periastron should take place. According to Kopal, the ratio of the relativistic apsidal motion period U'' to the orbital period T is given by

$$U''/T = 1.57 \times 10^5 A(1 - e^2)/(m_1 + m_2)$$

where the masses m_1 and m_2 are in units of the solar mass and A is the semimajor axis of the binary orbit in units of the solar radius. Hence for a close binary with massive stars $T/U'' \sim 10^{-5}$, showing that the relativistic apsidal advance rate could be of the same order as that due to the presence of a third body.

450

In fact the discovery in 1974 by Taylor and Hulse that the pulsar PSR 1913 + 16 is a member of a binary system provided a conclusive test not only of Einstein's law of gravitation but also of another of Einstein's predictions, the existence of gravitational radiation. The measured parameters of the binary pulsar are given in table 14.2.

Table 14.2

Parameter	Measured value†
1 Pulse period	$0 \cdot 059\ 029\ 995\ 271 \pm 2$ s
2 Rate of change of pulse period	$8 \cdot 63 \pm 0 \cdot 02 \times 10^{-18}$ s s^{-1}
3 Orbital eccentricity	$0 \cdot 617\ 193 \pm 3$
4 Orbital period	$27\ 906 \cdot 981\ 61 \pm 3$ s
5 Periastron advance rate	$4 \cdot 2261° \pm 7$ per year
6 Rate of change of orbital period	$-2 \cdot 30 \pm 0 \cdot 22 \times 10^{-12}$ s s^{-1}

† Errors are given either explicitly (parameters 2, 5) or as errors in the last figure.

The fortunate provision of two point-masses, one incorporating a highly accurate clock, orbiting each other every 7·75 h in a strong gravitational field, has enabled not only Einstein's theory of relativity to be tested but also other, more modern, theories. Taylor and Weisberg, from a six-year study of the binary, show that Einstein's general theory of relativity is the best description we yet have of gravity.

The measured rate of advance of periastron of 4·2261° per year is in excellent agreement with Einstein's theory.

It is of interest also that Einstein's theory predicts that the emission of gravitational radiation, the detection of which is the goal of experimental physicists today, from the binary pulsar PSR 1913 + 16 should cause the orbital period to decrease at a rate of $2 \cdot 40 \times 10^{-12}$ s s^{-1}, a prediction beautifully confirmed by observation. This result must strengthen the faith of those searching for gravitational radiation that their search will ultimately be rewarded.

14.14 The Figures of Stars in Binary Systems

If we again make the 'thought experiment' of setting up a widely separated binary system with the two non-rotating stars moving in ellipses about their centre of mass, they will be spherical and act as point-masses. If we decrease the separation, the period will of course decrease according to Kepler's third law, and there will come a time when the gravitational interaction between them will raise perceptible tides upon them, each star being elongated along the lines joining their centres. If the stars are also rotating their figures will be flattened as well, just as the Earth's figure is by its rotation. Kopal has suggested

451

that the stars in a close binary would rotate at angular velocities given by the maximum orbital angular velocity. The light curve of such an eclipsing binary will contain no straight parts (see figure 14.8).

Just as the Earth's gravitational potential could be described by a series expression, the harmonic constants of which could be evaluated by observing the changes in the orbits of artificial Earth satellites, so the external gravitational potential of a rotating and tidally distorted star can be expressed by a suitable harmonic series. Likewise, a series giving the total gravitational potential due to both stars can be found. This series, minus the point-mass gravitational potential of the system, then becomes the disturbing function to be used in the Lagrange planetary equations that will give the perturbations in the orbital elements. In particular the line of apses advances with a specific secular rate modified by periodic vibrations of small amplitude. Under certain simplifying assumptions the secular rate of apsidal advance per orbital revolution is $\Delta\omega$, given by

$$\Delta\omega = 2\pi\left\{k_{12}r_1^5\left[\frac{m_2}{m_1}h_2(e)+g_2(e)\right]+k_{22}r_2^5\left[\frac{m_1}{m_2}h_2(e)+g_2(e)\right]\right\} \quad (14.26)$$

where

$$h_2(e)=15f_2(e)+g_2(e)$$

$$f_2(e)=(1-e^2)^{-5}(1+3e^2/2+e^4/8)$$

$$g_2(e)=(1-e^2)^{-2}$$

$$k_{i2}=[3-y_2(r_i)]/[4+2y_2(r_i)] \quad (i=1,2)$$

and $y_2(r_i)$ satisfies the differential equation

$$r\frac{dy_2}{dr}+y_2^2-y_2-6+6\frac{\rho}{\bar{\rho}}(y_2+1)=0.$$

The star's density ρ is a function of r, while $\bar{\rho}$ is the mean density of the star. The quantity r varies from zero to r_i (the star's fractional radius) and y is zero at $r=0$.

The parameters k_{12}, k_{22} therefore have values that are dependent on the internal structure of the stars, being zero if the stars are point-masses and 0·75 if the stars are homogeneous. The terms $f_2(e)$ and $g_2(e)$ appear respectively from the tidal and rotational distortions.

We define $\bar{k}_2$ by

$$\bar{k}_2=\frac{c_1k_{12}+c_2k_{22}}{c_1+c_2}$$

where c_1 and c_2 are the coefficients of k_{12} and k_{22} in equation (14.26) and $\bar{k}_2$ is the quantity that is actually found in practice from observations of the rate of apsidal advance. It will give information about the internal structure of the binary components and, by comparing it with values computed for various stellar models, will yield information about stellar evolution in binary systems.

If we let the period of apsidal rotation due to the figures of the stars be U'', then $T/U'' = \Delta\omega/2\pi$, so that

$$k_2 = \frac{T/U''}{c_1 + c_2}.$$

A typical value of k_2 from astrophysical theory is 10^{-2} while $(c_1 + c_2)$ is usually between 10^{-2} and 10^{-3}. Hence $T/U'' \sim 10^{-4}$ to 10^{-5}. It is therefore seen that the presence of a third body, the relativistic gravitational effect and the departure of the binary components from possessing point-mass gravitational potentials all contribute a fair share to the apsidal advance.

14.15 The Roche Limits

Let us now consider a close binary system in the light of the restricted three-body problem. Referring back to section 5.9.3 we recall that the surface of zero velocity depends upon the value of the Jacobi constant, which in turn depends upon the initial position and velocity of the infinitesimal particle. For various values of C the surface consisted in part of two lobes, each surrounding a massive body and the larger lobe surrounding the larger mass. For a particular value of C the lobes became joined at the Lagrangian libration point L_2 between the finite masses. If the two massive bodies are now taken to be the two components in a binary system, this particular surface of zero velocity about the two components, often called the *Roche limit*, enables a number of deductions to be made. Following Kopal (1955) we see that it implies an upper limit to the size of a component. If particles in the outer layers of a binary component have energies in excess of this C value and cross the zero velocity surface, they may enter the other star's lobe or become part of a cloud of material about both stars, or even leave the system altogether. If the star within a lobe extends as far as the lobe surface, then particles forming the outermost layers of the star need have very little kinetic energy to escape. Kopal has therefore divided all binaries into three classes:

(i) systems in which neither star fills its lobe,
(ii) systems in which one component fills its lobe, and
(iii) systems in which both fill their lobes.

Although magnetohydrodynamic forces act within the outer layers of a star, and although strictly speaking the Jacobi integral holds only in the circular restricted problem with both finite masses acting as point-masses, the model seems to correspond closely to reality. It is known, for example, that in Algol-type binary systems the secondary components fill their lobes; according to Batten (1973) no well observed system is known in which either component exceeds its lobe size to any extent.

The outer layers or atmosphere of a star will then tend to be stripped off if the particles are close to the surface of zero velocity, or if the surface alternatively expands and contracts because of the binary's orbital eccentricity, or if explosive outbursts take place from time to time as is believed to occur with some stars. Thus in a binary with an eccentric orbit, a large secondary component could be just inside its Roche lobe at apiastron but overflow its zero velocity surface at periastron, material from its atmosphere streaming through the tubular neck of the surface opened up at the Lagrangian point L_2.

Again, according to the standard theory of stellar structure and evolution, stars will swell in size as they exhaust their supply of hydrogen in their cores, their radii increasing by a factor between 10 and 10^2. A star in a binary system undergoing this part of its evolution may fill and overflow its Roche limit, its partner then falling heir to much of the excess material. Such processes show that close binaries of this proximity cannot be treated as isolated stars either gravitationally or astrophysically. Not only do they distort each other's figure and exchange gas but they also affect each other's evolution.

14.16 Circumstellar Matter

From the above arguments it is clear that in association with the two members of a close binary there should be material surrounding the binary. This circumstellar matter has been detected in the study of many binaries. It can take the form of gas streams, discs and envelopes or clouds about both components. It makes its presence known by superimposing additional emission and absorption lines in the binary spectra, by distorting the radial velocity measurements giving the velocity curves and by modifying the light curve.

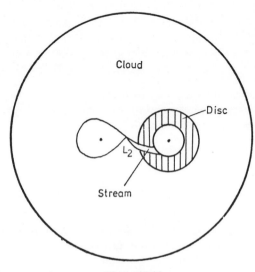

Figure 14.14

Batten (1970) has suggested a general model that can apply to any system containing circumstellar matter, though individual details will acquire more or less importance from system to system. He suggests defining a *characteristic volume* for a binary system, a cylinder of radius twice the semimajor axis of the orbit centred on the system's centre of mass, extending above and below the orbit by an amount equal to the radius of the smaller component. The cylindrical shape was chosen in recognition that in many systems, for example in Algol-type binaries, it would appear that circumstellar matter would be concentrated in or near the orbital plane. This is not invariably the case and in other systems it is probable that mass may be being shed isotropically, so that a spheroid may more accurately define the shape of the surrounding cloud.

Three features, apart from the stars themselves, may be found inside the characteristic volume. There may be *streams* running from one star to the other. A stream of gas may be ejected from a component through the Lagrangian point L_2, the gas particles following trajectories dictated by their energies and the gravitational attractions of the stars, ending on the other star or contributing to the *discs*. Either or both stars may possess a disc. The disc lies in the orbital plane and is gravitationally bound to the star it surrounds, moving with the star as it pursues its orbit about the centre of mass of the system. The third feature is the *cloud* of gas surrounding both components and confined roughly within the characteristic volume of the system. More tenuous than the average disc or stream, it will have its own rotation in that the gas particles it consists of will have their own complicated orbital motions under the binary's gravitational force. The binary components will plough their way through the cloud.

Circumstellar matter in its several manifestations of stream, disc and cloud must have effects on the binary's orbital elements. We have seen in chapter 10 that atmospheric drag on an artificial Earth satellite orbit will secularly decrease the eccentricity and semimajor axis and therefore the orbital period. The first-order effect on the longitude of perigee is small and periodic. On analysis, similar effects are found to take place in the corresponding orbital elements of a binary orbit due to drag by a circumstellar gas cloud on the stellar components.

The transfer of mass between the components, or the loss of mass from the star system altogether, cause changes in the orbital elements. This is a far more complicated dynamic problem than the circumstellar cloud-drag problem. The orbital period can increase or decrease secularly depending upon the mass-flow conditions. The simplest case is where mass is lost isotropically from the system.

By Kepler's third law,

$$\frac{4\pi^2 a^3}{T^2} = G(m_1 + m_2) = GM$$

giving, for a constant semimajor axis a, the following relation between the change in period ΔT for a loss of mass ΔM:

$$\frac{\Delta T}{T} = -\frac{\Delta M}{2M}.$$

Wood (1950) suggested that an abrupt change of period could be caused by one component losing mass in an eruptive prominence outburst, the material being ejected at a high speed. Something of the order of 10^{-7} of a solar mass lost would be required to change the period by about one second. Even if the mass is only transferred from one component to the other, changes of period should result. Since it is now believed that in many close binaries at least 10^{-7} to 10^{-6} of a solar mass are transferred each year, some abrupt period changes may indeed be due to eruptive behaviour.

If a continuous stream of material goes on, it may be shown that if the total mass and angular momentum of the system are conserved,

$$\frac{\Delta T}{T} = \frac{3(2\mu - 1)}{\mu(1-\mu)}\frac{\Delta M}{M}$$

where $\mu = m_2/M$ and m_2 is the mass of the component gaining mass. Since ΔM is assumed positive, the period increases or decreases according to whether the mass transfer is from the less to the more massive component or vice versa.

The transfer of material from one component in a binary to the other can reveal the existence of neutron stars or black holes. Both can be sources of very energetic radiation, in particular x-rays, caused by the violent accretion of matter from one component or the circumstellar disc onto the massive compact component. Such a system is known as an x-ray binary. As has earlier been remarked, only the observation of such binaries gives the chance of deducing the existence of a black hole.

14.17 The Origin of Binary Systems

It is still not at all clear how binary systems are formed. This is unsatisfactory when we recall that over half the stars are members of binary systems. At least three theories have much to recommend them and it is probable that not all binaries have had the same mode of origin.

One theory, backed up by many computer studies of simulated few-body star clusters, suggests that stars form out of the interstellar medium in small groups. It will be seen in the following chapter that such groups tend to be unstable. Some stars escape from the group and one or more binary systems form. Triple and higher-number sub-systems may also come into being. In addition, it has been suggested that when the original set of stars condenses out of the interstellar cloud, some pairs are so close together that they become gravitationally bound almost immediately. The theory is plausible but does not explain the existence of so many very close binaries.

The second theory is the fission hypothesis, again the subject of many studies. On this scenario a rapidly rotating star becomes unstable and splits, forming a close binary system. There is no space to go into the many arguments for and against this theory of origin. What seems clear is that not all binary systems can have originated by fission, even if some mechanism is suggested to separate the originally very close components resulting from the fission process.

The third theory is probably the least likely of the three. It suggests that in the general field of stars, two stars can enter into orbit about each other by a close encounter. We have already seen that when a spacecraft approaching a planet along a hyperbolic trajectory makes a planetary fly-by it will (having positive energy) recede along the other arm of the hyperbola. For it to be captured by the planet, its excess kinetic energy must be removed. The spacecraft uses its rocket motor to do this; for a pair of stars a third body must remove the excess energy. Schmidt showed that the central bulge of the Galaxy could fill this role; other workers have suggested a third star or the local interstellar medium. The probability of such processes occurring, however, is very low indeed. It would certainly not explain the number of binaries in existence or their ranges of separations and eccentricities and it would be even more improbable that such processes would give rise to the estimated number of triple and higher-number systems.

Problems

14.1 When seen through a telescope a star is observed to be a close double with components with magnitudes $8 \cdot 3^m$ and $7 \cdot 6^m$. What is the magnitude of the star when unresolved?

14.2 An eclipsing binary has a constant apparent magnitude $4 \cdot 35^m$ between minima and apparent magnitude $6 \cdot 82^m$ at primary minimum. Assuming that the eclipse is total at primary minimum, calculate the magnitudes and the relative brightness of the components.

14.3 The following data refer to the binary system ρ Her: orbital period 34·4 years; parallax 0·10''; angular semimajor axis of the relative orbit 1·35''; angular semimajor axis of the orbit of the primary relative to the centre of mass of the system 0·57''. Calculate the masses of the two components in solar mass units.

14.4 The binary star Capella has a total magnitude of $0 \cdot 21^m$ and the two components differ in magnitude by $0 \cdot 5^m$. The parallax of Capella is 0·063'': calculate the absolute magnitudes of the two components.

14.5 The two components of a binary star are of approximately equal brightness. Their maximum separation is 1·3'' and the period is 50·2 years. The composite spectrum shows double lines with a maximum separation of 0·18 Å at 5000 Å. Assuming that the plane of the orbit contains the line of sight, calculate (i) the total mass of the system in terms of the solar mass, and (ii) the parallax of the system.

14.6 The true period of an eclipsing binary is 3·12 days and its velocity in the line of sight (away from the Sun) is 30 km s⁻¹. Show that its apparent period is greater than the true one by 27 seconds.

14.7 The centre of mass of a spectroscopic binary has no radial velocity relative to the Sun. Show that the heliocentric radial velocity R of one component of the star is given by

$$R = \frac{na \sin i}{(1-e^2)^{1/2}} \left[\cos(\omega + f) + e \cos \omega \right]$$

where n, a, e, i and ω are the mean motion, semimajor axis, eccentricity, inclination and argument of periastron, and f is the true anomaly, the orbital elements being defined for a barycentric orbit (i.e. with respect to the centre of mass).

14.8 Calculate the dynamic parallax of a visual binary star, given that the period of revolution of the components is 67·4 years, the angular semimajor axis of the orbit is 3·14 seconds of arc, and that the components have apparent magnitudes of 4·15^m and 6·35^m.

References

Batten A H 1970 *Publ. Astron. Soc. Pacific* **82** 574
—— 1973 *Binary and Multiple Systems of Stars* (Oxford: Pergamon)
Brown E W 1936 *Mon. Not. R. Astron. Soc.* **97** 56, 62
—— 1937 *Mon. Not. R. Astron. Soc.* **97** 116, 388
Hulse R A and Taylor J H 1975 *Astrophys. J. Lett.* **195** L51
Kopal Z 1955 *Ann. Astrophys.* **18** 379
Lyttleton R A 1934 *Mon. Not. R. Astron. Soc.* **95** 42
Taylor J H and Weisberg J M 1982 *Astrophys. J.* **253** 908
Wood F B 1950 *Astrophys. J.* **112** 196

Bibliography

Aitken R G 1935 *The Binary Stars* (New York: McGraw-Hill)
Cowling T G 1938 *Mon. Not. R. Astron. Soc.* **98** 734
Gyldenkerne K and West R M (eds) 1970 *Mass Loss and Evolution in Close Binaries* (*IAU Colloquium No. 6*) (Copenhagen University Observatory)
Kopal Z 1950 *The Computation of Elements of Eclipsing Binary Systems* (Harvard Observatory Monograph: Cambridge, Mass.) No. 8
—— *Dynamics of Close Binary Systems* (Dordrecht: Reidel)
Russel H N and Moore C E 1939 *The Masses of The Stars* (Princeton University Press)
Smart W M 1956 *Textbook on Spherical Astronomy* (London: Cambridge University Press)
Sterne T E 1939 *Mon. Not. R. Astron. Soc.* **99** 451

15 Many-Body Stellar Systems

15.1 Introduction

In this chapter we leave those dynamical problems where the number n of gravitating bodies is few and enter a field where n is many (between 10^2 and 10^3 in the case of an open or moving cluster of stars, between 10^4 and 10^6 for a globular cluster and between 10^7 and 10^{11} for a galaxy). Apart from the case of small open clusters, we are concerned with problems where statistical methods are now applicable, so many are the particles involved. The methods of statistical mechanics may therefore be employed; in addition, the system of gravitating bodies may be shown to operate under conditions similar to those in a fluid, so that a hydrodynamical approach is also possible.

In these respects the analogy of a gas is illuminating. One classical approach treats the gas as an assembly of molecules, whose properties are described by the kinetic theory of gases with the molecular motions obeying a Maxwellian distribution of velocities. A second approach forgets that the gas is made of many discrete particles moving and colliding with each other, and considers it to be a continuous medium exhibiting density, pressure and viscosity, with its properties described by hydrodynamical theory.

It is still possible, however, to adopt celestial mechanics and consider the orbits of individual stars in the stellar system concerned. This approach also sheds light on the structure, evolution and stability of such stellar systems. In the present text, lack of space presents a full discussion of the vast field covered by stellar kinematics and dynamics. All we can do is discuss certain fundamental properties and theorems of many-body dynamics which highlight some important results.

15.2 The Sphere of Influence

In the case of an interplanetary probe it was seen how useful was the concept of the sphere of influence, a volume of space about a planet within which a probe was effectively on a planetocentric orbit but disturbed by the Sun, and outside which the probe had an essentially interplanetary orbit. We can apply this concept to the case of a star in a stellar system. If the stellar system is roughly spherical its integrated gravitational field is approximately equivalent to that of a point-mass at its centre equal to the sum of all the stellar masses. Let it be M and let a star on the outskirts of the stellar system have mass m and be distant R from the centre. Then by relation 5.70, the radius r of the sphere of influence is given by

$$r = R\left(\frac{m}{M}\right)^{2/5}.$$

We consider two examples:

(i) For a globular cluster, $m/M \sim 10^{-6}$ and $R \sim 10$ pc. Hence $r \sim 0.04$ pc ~ 8000 AU.

(ii) For the central bulge of the Galaxy and the Sun, $m/M \sim 10^{-11}$ and $R = 10^4$ pc. Hence $r \sim 0.4$ pc ~ 80000 AU.

On both the outskirts of a globular cluster and in the solar neighbourhood, the average separation d of the stars (omitting binaries) is of the order of 4 pc. It is therefore seen that the cluster force field on the one hand and the central galactic bulge on the other is always dominant unless two stars make a close approach to one another. In other words, apart from close encounters, a star's galactic or cluster orbit is not appreciably disturbed by the gravitational attraction of individual stars. It is easy to see that the argument still holds for stars within the cluster or the galactic bulge.

How often does such a close encounter take place? If we define $q = \frac{4}{3}\pi r^3$ as the volume of the encounter sphere, d as the mean distance between stars and ν the star density given by d^{-3}, the probability of a k-fold close encounter is p_k, given by Poisson's formula as

$$p_k = \frac{1}{k!}(\nu q)^k e^{-\nu q}.$$

Now

$$\nu q = \frac{4}{3}\pi\left(\frac{r}{d}\right)^3 \sim 4 \times 10^{-6}$$

so we see that apart from close binary encounters, themselves very highly improbable, multiple encounters in a stellar system hardly ever occur and can be neglected.

Small perturbations will occur continually on a random basis as the star follows its orbit. These perturbations are due to distant encounters with other stars, the perturbation being the smaller the more distant the encounter. Being small, each has very little effect on the star but there does exist a statistical chance that the star's velocity could be changed appreciably by repeated distant encounters. We consider the effects of such encounters in the next section.

15.3 The Binary Encounter

Suppose two stars S_1 and S_2 of masses m_1 and m_2 approach each other to the extent that they perturb each other's orbit. By the results given in chapter 4, it is then clear that we can treat this case as one where the star S_2 makes a hyperbolic encounter with S_1. The star S_2 comes effectively from an infinite distance with an initially unperturbed velocity $\mathbf{V}$ along the hyperbola APB (figure 15.1), approaches pericentre P at distance ρ_P from S_1 and departs along the arm PB.

Drawing the asymptotes DOD' and FOF' of the hyperbola, it is seen that the effect of the stellar encounter is to convert the relative velocity vector $\mathbf{V}$ along DOD' to a velocity vector $\mathbf{V}'$ along FOF', where $|\mathbf{V}| = |\mathbf{V}'|$ and where the original direction DOD' has been turned through an angle D'OF', which we shall denote by θ.

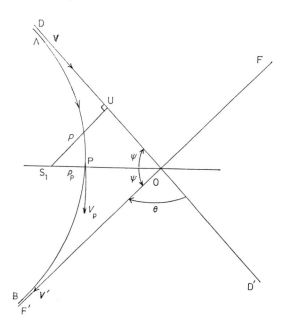

Figure 15.1

Then

$$\tan \frac{\theta}{2} = \tan \left(\frac{\pi}{2} - \psi\right) = \cot \psi = a/b = (e^2 - 1)^{-1/2} \tag{15.1}$$

using the results given in section 4.8. Also, if ρ_P and V_P are the pericentre distance and velocity respectively,

$$\rho_P = a(e-1), \quad V_P^2 = \frac{\mu}{a}\left(\frac{e+1}{e-1}\right)$$

where $\mu = G(m_1 + m_2)$, giving

$$V_P^2 = \mu(e+1)/\rho_P. \tag{15.2}$$

By the energy integral,

$$\tfrac{1}{2} V^2 = \tfrac{1}{2} V_P^2 - \frac{\mu}{\rho_P} = C \tag{15.3}$$

461

where C is the energy constant, so that on substituting for $V_P{}^2$ from equation (15.2) in (15.3), we obtain

$$e-1=V^2\rho_P/\mu. \tag{15.4}$$

But by the conservation of angular momentum

$$\rho_P V_P = pV$$

where $p = S_1 U$ is the closest approach the stars would have made if they had not attracted each other. Hence by equation (15.2),

$$\rho_P = p^2 V^2/[\mu(1+e)]$$

so that on substitution into (15.4)

$$e^2 - 1 = V^4 p^2/\mu^2$$

giving, from (15.1)

$$\tan\frac{\theta}{2} = \mu/V^2 p. \tag{15.5}$$

This formula (Jeans 1928), together with the formula giving the magnitude ΔV of the velocity increment vector, contains all the required information. The velocity increment vector $\Delta \mathbf{V}$ is obtained from the following considerations. In triangle ABC (figure 15.2), AB and AC are the initial and final relative velocity vectors $\mathbf{V}$ and $\mathbf{V}'$, separated by angle θ. Then $\Delta \mathbf{V}$ is vector BC and it is easily seen that in magnitude

$$\Delta V/V = 2\sin(\theta/2). \tag{15.6}$$

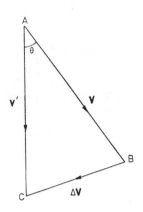

Figure 15.2

Consider now the number rate of such encounters. Let a star S move with velocity V through a volume occupied by other stars (assumed to be at rest) in which ν is the star density. Then the velocity vector $\mathbf{V}$ and all distances p perpendicular to the velocity vector will define a cylindrical volume within which all the stars will experience an encounter with S with undisturbed passing distance less than p. In unit time therefore, the number of such encounters is

$$\pi p^2 \nu V \qquad (15.7)$$

and the reciprocal of this gives τ, the average time interval between encounters. Hence

$$\tau = 1/\pi p^2 \nu V. \qquad (15.8)$$

The quantity $l = \tau V = 1/\pi p^2 \nu$ is defined as the *mean free path* of the star for a given value of p and is the mean distance the star will travel between encounters.

Let us now put in some numerical values. Following Jeans, we consider an encounter to be very close if the deflection θ exceeds $90°$. Taking as representative values $V = 20$ km s^{-1} and $\nu = 10^{-1}$ stars per cubic parsec (with each star roughly one solar mass), we put $\theta = 90°$ in equations (15.5) and (15.6). From the first, we have

$$p = \mu/V^2.$$

From the second,

$$\Delta V = (2V^2)^{1/2}.$$

Then for a very close encounter, it is found that $p = 4.5$ AU, $l = 6.7 \times 10^9$ pc and $\tau = 2.9 \times 10^{14}$ years. Since the diameter and age of the Galaxy are of the order of 3×10^4 pc and 10^{10} years respectively, it is seen that very close encounters effectively never happen. In table 15.1, these and other results are displayed.

Table 15.1

	Very close	Close	Moderately close	Normal
p (AU)	4·4	40·0	508·1	7000
θ	90°	12·7°	1°	0·036°
τ (years)	$3 \cdot 37 \times 10^{14}$	$4 \cdot 14 \times 10^{12}$	$2 \cdot 56 \times 10^{10}$	$1 \cdot 35 \times 10^8$
l (pc)	$6 \cdot 88 \times 10^9$	$8 \cdot 46 \times 10^7$	$5 \cdot 24 \times 10^5$	$2 \cdot 77 \times 10^3$
$\Delta V/V$	1·41	0·22	0·017	0·0013

The radius of the Solar System is 40 AU. A star that approached the Sun to that distance would strongly perturb the planetary orbits. Nevertheless it is seen that the probability of such an occurrence is very low indeed. Even normal encounters bringing two stars as close to each other as 7000 AU do not occur for any one star more than twice in each rotation of the Galaxy. The perturbations from such an encounter are small.

It will be remembered that we could take the effect of firing a rocket motor to be impulsive (i.e. it produced a change of velocity but no change in position during the burn), so short a time did it take compared with the orbital period. In the same way the effect of a stellar encounter may be taken to be impulsive, producing a velocity change but no alteration in the stellar coordinates. That this is so may be seen by noting that the duration of the encounter is roughly given by the time it takes a body travelling at 20 km $s^{-1}=4\cdot22$ AU/year to travel a distance of the order of 7000 AU. This figure of 1659 years is small compared with the Galaxy's rotation period, which is of the order of 250×10^6 years.

15.4 The Cumulative Effect of Small Encounters

We consider now the cumulative effect of many such feeble encounters on the path of a star. Since the encounters are distant, θ is small and we may take $\sin(\theta/2)=\tan(\theta/2)$. Hence by (15.5) and (15.6) we have

$$\Delta V/V = 2\mu/V^2 p \qquad (15.9)$$

where as before $\mu = G(m_1+m_2)$.

The interesting result follows that if the star passed by a star cluster, the effect of the cluster stars would be additive. Thus if V and p are the average relative velocity and encounter distance of the star with respect to the cluster stars, the overall effect of the encounter with the cluster is N times the effect of a single cluster star of average mass, where N is the number of stars in the cluster. Clouds of interstellar dust, which can amount to masses of the order of 10^5 solar masses, can also act as perturbing objects.

As far as non-cluster stars' cumulative effects are concerned, their effect on a star may be found by the following argument, due to Jeans. In unit time the number of encounters producing deflections greater than θ was, by equation (15.7),

$$\pi p^2 \nu V. \qquad (15.10)$$

But by equation (15.5)

$$\cot\left(\frac{\theta}{2}\right) = V^2 p/\mu \qquad (15.11)$$

so that the number of encounters in unit time is, by equations (15.10) and (15.11),

$$\frac{\pi \nu \mu^2}{V^3} \cot^2\left(\frac{\theta}{2}\right).$$

If we differentiate this expression we obtain the number of encounters in unit time producing a deflection between θ and $\theta+d\theta$. This is

$$\frac{\pi \nu \mu^2}{V^3} \frac{\cos(\theta/2)}{\sin^3(\theta/2)} d\theta.$$

Again, since θ is small, we may write $\cos(\theta/2)=1$ and $\sin(\theta/2)=\theta/2$, giving $(8\pi\nu\mu^2/V^3\theta^3)\,d\theta$. By the theory of errors, since the small deflections

464

$\theta_1, \theta_2, \theta_3 \ldots$ are random, they must be added according to the law of errors. Hence the total probable deflection Θ is given by

$$\Theta^2 = \theta_1{}^2 + \theta_2{}^2 + \ldots$$

Let $\theta_1, \theta_2 \ldots$ be the deflections between two limits α and β occurring within a time t. By integrating, we then have

$$\Theta^2 = \frac{8\pi\nu\mu^2 t}{V^3} \int_\alpha^\beta \frac{d\theta}{\theta}$$

or

$$\Theta^2 = \frac{8\pi\nu\mu^2 t}{V^3} \ln\left(\frac{\beta}{\alpha}\right). \tag{15.12}$$

The upper limit β may be taken to be $\pi/2$. The value of the lower limit α is dictated by the consideration that equation (15.12) is accurate only if the deflections $\theta_1, \theta_2 \ldots$ are independent. But if the minimum value of θ is very small, the corresponding distance of closest approach must be large. If this is so it is likely that several stars lie within this distance; their tiny deflections will tend to cancel each other out. We must therefore choose the lower limit such that this corresponds to a distance which is comparable with the average distance between neighbouring stars. We may take this to be given by $\nu^{-1/3}$.

Now by equation (15.5), since θ is very small, $\theta = 2\mu/V^2 p$. Hence we can write

$$\alpha = 2\mu\nu^{1/3}/V^2.$$

Inserting the representative values $V = 20$ km s^{-1}, $\nu = 10^{-1}$ stars/pc^3, and taking each star to be of solar mass, we find that

$$\alpha \sim 2 \times 10^{-5} \sim 4 \text{ arc seconds}$$

and

$$\Theta^2 \sim 3\cdot14 \times 10^{-14} \, t \ln\left(\frac{\beta}{\alpha}\right). \tag{15.13}$$

If these tiny deflections eventually produce a resultant deflection equal to $\pi/2$ in a time T, the value of T is obtained from (15.13) by putting $\Theta = \pi/2$, giving

$$T = \frac{10^{14}\,\pi}{4\ln\left(\dfrac{\beta}{\alpha}\right)} \text{ years.}$$

Putting $\alpha \sim 2 \times 10^{-5}$ and $\beta = \pi/2$ we obtain a minimum value of T, so that $T \geqslant 7 \times 10^{12}$ years.

It is therefore seen that the effects of stellar encounters are negligible in that the vast majority of stars will follow orbits essentially undisturbed by their immediate neighbours.

15.5 Some Fundamental Concepts

We now consider some fundamental concepts in the light of the above discussions.

We want to know how a stellar system evolves. We assume that the process of evolutionary change is orderly, or at least assume that most of the stellar systems observed in the universe (star clusters and galaxies) behave in this fashion. At any instant therefore, the state of a system is almost in equilibrium. Such a state is called a *quasi-steady state*. Thus, just as in astrophysics where the evolution of a star has been studied by considering an orderly sequence of stellar models, each of which is taken to be in equilibrium, we can consider stellar system models each of which is in a quasi-steady state (i.e. in an equilibrium which is only very slowly changing).

At any point in the stellar system, the stars in its vicinity have velocities that are statistically distributed about a mean velocity (which may be zero). The difference between a star's velocity and the mean velocity is its *residual velocity*. The reference point within an element of volume containing a number of stars and travelling with the mean velocity of the stars is called the *centroid*.

Thus the few thousand stars in the neighbourhood of the Sun (and including the Sun) form a local group of stars with its centroid travelling at a speed of about 250 km s^{-1}, the orbital velocity of objects at a distance of almost 10^4 pc from the galactic centre. The members of the group, however, have their own residual velocities within the group (of the order of 20 km s^{-1}) with respect to the centroid. The so-called *solar motion* is defined with respect to this centroid and is therefore the Sun's residual velocity.

Returning to the analogy of a gas, we recall that at each point in a flow of gas there will be a systematic velocity, the individual molecules near that point having residual velocities in a Maxwellian distribution according to the kinetic theory of gases. In stellar kinematics the usual distribution law is the Schwarzschild ellipsoidal law, of which the Maxwellian distribution is a special case.

Let us now introduce some other basic terms. If we neglect the differing masses of the stars (not too drastic a step in practice) and take the stars to be particles, then the state of any star is given by its coordinates x, y, z and its velocity components u, v, w with respect to a fixed set of rectangular axes. We can in fact define a state vector s, in a six-dimensional phase space, whose components are x, y, z, u, v, w. This vector defines a point in that phase space describing the state of the star at that moment. If we know the distribution of such points in the phase space, then we know the state of the stellar system. The function describing such a distribution is called the *phase density function*. If it can be determined, then the other quantities describing the stellar system can be derived from it.

Consider a six-dimensional element of volume dQ of sides dx, dy, dz, du, dv, dw defined in the following way. All the points defined by those state vectors whose components lie between x and $x+dx$, y and $y+dy$ ...,

w and $w+dw$ will define and lie within such an element of volume dQ. Let the number be dN. Then the number of points (stars) per unit volume in that small region is dN/dQ. This phase density f will change from point to point in the phase space. It is therefore a function of x, y, z, u, v, w. If the stellar system is evolving, it will also be a function of time. Hence $dN/dQ=f(x, y, z; u, v, w; t)$ or

$$dN = f(x, y, z; u, v, w; t)\, dQ. \tag{15.14}$$

We may define two other functions related to the phase density function, namely the star density function and the velocity distribution function. The *star density function* v is the number of stars per unit volume in space at the point considered, namely the point with coordinates x, y, z. It is therefore given by $dn=(x, y, z; t)\, dx\, dy\, dz$, or

$$dn = v(x, y, z; t)\, dq \tag{15.15}$$

where $dq = dx\, dy\, dz$. Clearly the relationship between the star density function and the phase density function is

$$v = \int\int\int f\, du\, dv\, dw \tag{15.16}$$

the integration being taken over all the velocity space.

The *velocity distribution function* gives the distribution of velocities within a volume element centred at a given point (x, y, z), where the values of x, y, z are treated now as parameters, that is to say they are constant for a given position. If dp is the velocity volume element, given by $dp = du\, dv\, dw$, then dn/dp is the density of points at the position (x, y, z) within the velocity volume element, where

$$dn/dp = f(x, y, z; u, v, w; t). \tag{15.17}$$

We will return to consider the velocity distribution function later.

15.6 The Fundamental Theorems of Stellar Dynamics

Let U be the gravitational potential at a point of radius vector $\mathbf{r}$ in a stellar system. Then the force per unit mass at the point has components given by $\ddot{\mathbf{r}} = \nabla U$ or, referring to the rectangular axes x, y, z,

$$\ddot{x}=\frac{\partial u}{\partial x}; \quad \ddot{y}=\frac{\partial u}{\partial y}, \quad \ddot{z}=\frac{\partial u}{\partial z}. \tag{15.18}$$

These equations may also be written in the form

$$\dot{u}=X, \quad \dot{v}=Y, \quad \dot{w}=Z \tag{15.19}$$

where (x, y, z) and (u, v, w) may be taken to be the coordinates and velocity components of a star.

After a small time interval dt let the coordinates and velocity components of the star be (x_1, y_1, z_1) and (u_1, v_1, w_1) respectively; for the x component and corresponding velocity component we may then write

$$x_1 = x + u \, dt, \qquad u_1 = u + \dot{u} \, dt = u + X \, dt \tag{15.20}$$

with similar relations for the y and z components and velocity components.

Let the stars (dN in number) that occupied the phase space volume element dQ now occupy a volume element dQ_1, where $dQ = dx \, dy \, dz \, du \, dv \, dw$ and $dQ_1 = dx_1 \, dy_1 \, dz_1 \, du_1 \, dv_1 \, dw_1$. Now

$$dQ_1 = \frac{\partial(x_1, y_1, z_1; u_1, v_1, w_1)}{\partial(x, y, z; u, v, w)} \, dQ = J \, dQ.$$

Using equation (15.20) the Jacobian J may be written out as

$$
\begin{vmatrix}
1 & 0 & 0 & dt & 0 & 0 \\[2mm]
0 & 1 & 0 & 0 & dt & 0 \\[2mm]
0 & 0 & 1 & 0 & 0 & dt \\[2mm]
\dfrac{\partial X}{\partial x} dt & \dfrac{\partial X}{\partial y} dt & \dfrac{\partial X}{\partial z} dt & 1 & 0 & 0 \\[3mm]
\dfrac{\partial Y}{\partial x} dt & \dfrac{\partial Y}{\partial y} dt & \dfrac{\partial Y}{\partial z} dt & 0 & 1 & 0 \\[3mm]
\dfrac{\partial Z}{\partial x} dt & \dfrac{\partial Z}{\partial y} dt & \dfrac{\partial Z}{\partial z} dt & 0 & 0 & 1
\end{vmatrix} .
$$

To the first order in dt, this reduces to unity. Hence

$$dQ_1 = dQ. \tag{15.21}$$

Now by (15.14) we have

$$dN = f(x, y, z; u, v, w; t) \, dQ \tag{15.22}$$

and, taking $t_1 = t + dt$, we also have

$$dN = f(x_1, y_1, z_1; u_1, v_1, w_1; t_1) \, dQ_1. \tag{15.23}$$

Expanding equation (15.23) by Taylor's theorem to the first order and equating (15.22) and (15.23), we have

$$\frac{\partial f}{\partial t} + u \frac{\partial f}{\partial x} + v \frac{\partial f}{\partial y} + w \frac{\partial f}{\partial z} + X \frac{\partial f}{\partial u} + Y \frac{\partial f}{\partial v} + Z \frac{\partial f}{\partial w} = 0 \tag{15.24}$$

or

$$\frac{\partial f}{\partial t} + u \frac{\partial f}{\partial x} + v \frac{\partial f}{\partial y} + w \frac{\partial f}{\partial z} + \frac{\partial U}{\partial x} \frac{\partial f}{\partial u} + \frac{\partial U}{\partial y} \frac{\partial f}{\partial v} + \frac{\partial U}{\partial z} \frac{\partial f}{\partial w} = 0 \tag{15.25}$$

both (15.24) and (15.25) being forms of Boltzmann's equation. In deriving the formulae we have tacitly assumed that the effect of encounters is negligible compared with the effect of the potential U produced by the system as a whole. This would be equivalent to neglecting molecular collisions in the kinetic theory of gases.

Now let the operator D/Dt be defined by the relation

$$\frac{D}{Dt} = \frac{\partial}{\partial t} + \sum_{i=1}^{6} \frac{dx_i}{dt} \frac{\partial}{\partial x_i}$$

where x_i stands for x, y, z, u, v, w in turn and D/Dt is the Stokes derivative (i.e. the total time derivative) of a function in six-dimensional phase space. By equations (15.19) and (15.24) we see that

$$\frac{Df}{Dt} = 0. \tag{15.26}$$

Now the number of points dN does not vary with time, and so

$$\frac{D}{Dt}(dN) = 0. \tag{15.27}$$

But $dN = f \, dQ$, so that by equations (15.26) and (15.27) it is seen that $(D/Dt)(dQ) = 0$, which is a restatement of the relation (15.21);

$$dQ_1 = dQ.$$

This is Liouville's theorem, which we have already encountered (chapter 5). It states that in the motion of a dynamical system any volume of phase space remains constant.

15.6.1 Jeans's theorem

Equation (15.24) is a partial differential equation of the first order in the variables x, y, z, u, v, w and t. The equations of motion (15.18) may be written in the form

$$\frac{dx}{u} = \frac{dy}{v} = \frac{dz}{w} = \frac{du}{X} = \frac{dv}{Y} = \frac{dw}{Z} = dt. \tag{15.28}$$

The standard method of solving equation (15.24) is Lagrange's method. Equations (15.28) form six independent equations and so the integrals of (15.28) are six in number. In general they are of the form

$$I_k(x, y, z; u, v, w; t) = C_k \quad (k = 1, 2 \ldots 6) \tag{15.29}$$

where the C_k are constants. Then the general solution of the partial differential equation (15.24) is any function of the six integrals, that is

$$f = F(I_1, I_2 \ldots I_6) \tag{15.30}$$

where F is any function of $I_1, I_2 \ldots I_6$.

Hence it is seen that the phase density f is constant along the path of a star in phase space; it is also a function of the six quantities that remain constant along the star's path. This is Jeans's theorem. By equation (15.30) it is also seen that the coordinates and velocity components appear in the phase density only in combinations that are integrals of the motion.

There is a further restriction on the phase density f. In a gravitating system with potential U caused by the system, Poisson's equation must be satisfied at all points in the system. If ρ is the mass per unit volume at the point (x, y, z), Poisson's equation may be written as

$$\nabla^2 U = \frac{\partial^2 U}{\partial x^2} + \frac{\partial^2 U}{\partial y^2} + \frac{\partial^2 U}{\partial z^2} = -4\pi\rho.$$

Again assuming that all the stars are of the same mass m, we have by equation (15.16) the result that the number of stars per unit space volume is the star density function ν given by $\nu = \iiint f\, du\, dv\, dw$. Then $\rho = m\nu$, giving

$$\nabla^2 U = -4\pi m \iiint f\, du\, dv\, dw. \tag{15.31}$$

15.7 Some Special Cases for a Stellar System in a Steady State

If a stellar system is in a steady state, neither the phase density f nor the potential U are explicit functions of time t. Thus

$$\frac{\partial U}{\partial t} = \frac{\partial f}{\partial t} = 0.$$

It is also seen by equation (15.16) that if $\partial f/\partial t = 0$, then $\partial \nu/\partial t = 0$ (i.e. the star density function ν at any point is independent of time). Then equations (15.24) and (15.28) are now reduced to

$$u\frac{\partial f}{\partial x} + v\frac{\partial f}{\partial y} + w\frac{\partial f}{\partial z} + X\frac{\partial f}{\partial u} + Y\frac{\partial f}{\partial v} + Z\frac{\partial f}{\partial w} = 0 \tag{15.32}$$

and

$$\frac{dx}{u} = \frac{dy}{v} = \frac{dz}{w} = \frac{du}{X} = \frac{dv}{Y} = \frac{dw}{Z} \tag{15.33}$$

respectively. There are only five independent integrals now, so that $f = F(I_1, I_2, I_3, I_4, I_5)$, where

$$I_k\,(x, y, z, u, v, w) = C_k \quad (k = 1, 2 \ldots 5).$$

When values are attached to the constants C_k, these integrals define the phase path of the star.

470

The energy integral may be formed from equation (15.33). We have

$$u \, du = X \, dx = \frac{\partial U}{\partial x} \, dx$$

$$v \, dv = Y \, dy = \frac{\partial U}{\partial y} \, dy$$

$$w \, dw = Z \, dz = \frac{\partial U}{\partial z} \, dz.$$

Adding and integrating, we obtain

$$I_1 \equiv u^2 + v^2 + w^2 - 2U = C_1 \tag{15.34}$$

or

$$I_1 \equiv V^2 - 2U = C_1 \tag{15.35}$$

where V is the velocity.

Most galaxies have rotational symmetry. For such stellar systems, U is a function of z and $\rho = (x^2 + y^2)^{1/2}$, where the z axis is taken to be the rotation axis and ρ is the cylindrical radius. Thus $U = U(\rho, z)$ and consequently

$$x \frac{\partial U}{\partial y} = y \frac{\partial U}{\partial x}$$

or

$$xY = yX.$$

From equation (15.33) we have

$$Y \, du = X \, dv$$

so that

$$y \, du = x \, dv$$

giving

$$I_2 \equiv yu - xv = C_2$$

on integration.

Hence in the case of a stellar system with rotational symmetry and in a steady state we have the relation $f = F(I_1, I_2)$, where the two integrals are the energy and angular momentum integrals.

One of the most important classes of stellar systems is the one which includes all systems whose mass distribution is spherically symmetric, such as the globular clusters and those elliptical galaxies of Hubble type EO which show no ellipticity. Here the potential U is evidently a function only of the distance r from the system centre, so that

$$U = U(r) = U[(x^2 + y^2 + z^2)^{1/2}].$$

In addition to the energy integral I_1, we now have three angular momentum integrals:

$$\left. \begin{array}{l} I_2 \equiv yw - zv = C_3 \\ I_3 \equiv zu - xw = C_4 \\ I_4 \equiv xv - yu = C_5 \end{array} \right\} \tag{15.36}$$

which is a consequence of the fact that we may substitute

$$X = \frac{x}{r}\frac{\partial U}{\partial r}, \qquad Y = \frac{y}{r}\frac{\partial U}{\partial r}, \qquad Z = \frac{z}{r}\frac{\partial U}{\partial r}$$

into the equations (15.33). Then

$$f = F(I_1, I_2, I_3, I_4). \tag{15.37}$$

15.8 Galactic Rotation

The shape of the Galaxy (a flat disc with central bulge and spherical halo of globular clusters) suggested that it was a rotating system. Observations of the neighbouring galaxy M31 in Andromeda revealed its rotation and it is now believed that most stellar systems are rotating.

Let us consider what we mean by rotation of a system made up of individual stars and dust and gas clouds. Even if the system has no angular velocity, the stars will still follow their own orbits. For example, it is conceptually possible to have two concentric systems, each of which is the exact mirror image of the other in that, although each system consists of the same number of stars all revolving about the common centre in the same direction, that direction is direct for one system and retrograde for the other. At any point in the common system we would therefore find that in a volume element centred at that point, half the stars would be moving in one direction, the other half in the opposite direction. The mean velocity (or centroid velocity) would be zero and we would say that the whole system showed no trace of rotation because the centroid velocities throughout the system were all zero.

In considering the rotation of a stellar system we are therefore concerned with the angular velocities of the centroids. In particular we are concerned with the distribution of centroid angular velocities throughout the system. If they are all the same, the system rotates as a solid body. If not, we want to know how the angular velocity varies with distance from the centre.

15.8.1 Oort's constants

One particularly fruitful line of investigation of galactic rotation was carried out by the Dutch astronomer Oort (1927a,b, 1928). In what follows we consider only the first-order theory. Let S and X be the positions of the Sun and a star in the Galaxy, C being the galactic centre. Let both lie in the equatorial plane of the Galaxy at distances R and R_1 for Sun and star respectively. (We should however note that, strictly speaking, S and X should refer to the centroids of the groups of stars about the points S and X.) In addition the velocities V and V_1 of S and X are the centroid velocities, both velocity vectors lying in the galactic plane.

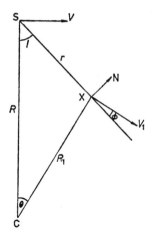

Figure 15.3

Then

$$V = \omega R, \qquad V_1 = \omega_1 R_1$$

where ω and ω_1 are the angular velocities of S and X about the galactic centre C.

We consider S and X, distance r from each other, to be so close that $r/R \ll 1$. Then the observed radial velocity of X relative to S due to galactic rotation is

$$\rho = V_1 \cos \phi - V \sin l \qquad (15.38)$$

where ϕ is the angle between the line SX produced and the vector V_1. Similarly the transverse velocity u of X relative to S is given by

$$u = V_1 \sin\phi - V \cos l. \qquad (15.39)$$

From triangle SXC, we have

$$90° - \phi = l + \theta. \qquad (15.40)$$

Also

$$\frac{\sin \theta}{r} = \frac{\sin l}{R_1} \qquad (15.41)$$

and

$$R_1{}^2 = R^2 + r^2 - 2Rr \cos l$$

so that

$$R_1 = R\left[1 + \left(\frac{r}{R}\right)^2 - 2\frac{r}{R} \cos l\right]^{1/2}$$

or, to the first order in r/R,

$$\Delta R = R_1 - R = -r \cos l. \qquad (15.42)$$

473

Inserting $90° - l - \theta$ for ϕ in equations (15.38) and (15.39) and expanding, we obtain

$$p = V_1(\sin l \cos \theta + \cos l \sin \theta) - V \sin l \qquad (15.43)$$

$$u = V_1(\cos l \cos \theta - \sin l \sin \theta) - V \cos l. \qquad (15.44)$$

We may write

$$V_1 = V + \frac{dV}{dR} \Delta R$$

to the first order, so that using (15.42) we have

$$V_1 = V - r \frac{dV}{dR} \cos l.$$

Now θ is small and so $\cos \theta \simeq 1$. Also in equation (15.41) we may replace R_1 by R. Hence (15.43) and (15.44) become, on using the amended form of (15.14) and neglecting second-order terms factored by $(r/R)(dV/dR)$,

$$p = rA \sin 2l \qquad (15.45)$$

and

$$u = rA \cos 2l + rB \qquad (15.46)$$

where

$$A = \frac{1}{2}\left(\frac{V}{R} - \frac{dV}{dR}\right), \qquad B = -\frac{1}{2}\left(\frac{V}{R} + \frac{dV}{dR}\right) \qquad (15.47)$$

or

$$B = A - \frac{V}{R} = A - \omega \qquad (15.48)$$

since $V = \omega R$.

Differentiating, we have

$$\frac{dV}{dR} = \omega + R \frac{d\omega}{dR}.$$

Hence A and B can take the alternative forms

$$A = -\tfrac{1}{2} R\left(\frac{d\omega}{dR}\right), \qquad B = -\omega - \tfrac{1}{2} R\left(\frac{d\omega}{dR}\right). \qquad (15.49)$$

If we had taken X to be outside the galactic equatorial plane in galactic latitude b as measured from S, the first-order analysis would have given the following expressions for the radial velocity p, the proper motion in longitude μ_l and latitude μ_b due to galactic rotation:

$$\left.\begin{array}{l} p = rA \sin 2l \cos^2 b \\[4pt] \kappa\mu_l = (A \cos 2l + B) \cos b \\[4pt] \kappa\mu_b = -\tfrac{1}{2} A \sin 2l \sin 2b \end{array}\right\} \qquad (15.50)$$

where $\kappa = 4 \cdot 74$ and μ_l and μ_b are expressed in seconds of arc per year.

Equations (15.50) are the first-order equations giving the radial velocity p, proper motion components μ_l and μ_b in galactic longitude and latitude, at a centroid distant r from the Sun, caused by galactic rotation. The constants A and B are called Oort's constants.

Values of A and B can be found from measurements of ρ, μ_l and μ_b for groups of stars in many directions from the Sun. For each group of stars mean values of ρ, μ_l and μ_b are determined; one hopes that by this method the random or residual motions of the group members will largely cancel out, leaving only the effect of the group centroid velocity. For each group, the mean values of l and b are of course known. A reasonably accurate value for the distance r can be found if the stars are taken from a narrow magnitude range and are of approximately the same spectral type. Since r appears only in the expression for the radial velocity ρ (this being proportional to r), the most distant stars are chosen. In practice the very luminous O- and B-type stars are used. Recent values of A and B are:

$$\left. \begin{array}{l} A = +0 \cdot 020 \text{ km/s/pc} \\ B = -0 \cdot 013 \text{ km/s/pc.} \end{array} \right\} \tag{15.51}$$

Since $R \sim 10^4$ pc and $r \ll R$ has been assumed to obtain the expressions (15.50), r should be kept below 10^3 pc. Second-order expressions have been developed, enabling larger values of r to be taken.

15.8.2 The period of rotation and angular velocity of the galaxy

Substituting the values of A and B given in equation (15.51) into the expression (15.48), we obtain a value for ω of $0 \cdot 033$ km/s/pc. If we require the value of ω in seconds of arc per year, we put

$$\omega = A - B = \frac{0 \cdot 033 \times 31 \cdot 56 \times 10^6 \times 206\,265}{206\,265 \times 149 \cdot 6 \times 10^6}$$

since 1 pc = 206 265 AU, 1 AU = $149 \cdot 6 \times 10^6$ km, the number of seconds of time in one year is $31 \cdot 56 \times 10^6$ and one radian = 206 265 seconds of arc. Hence $\omega = 0 \cdot 0070$ seconds of arc per year. It may be remarked that the rotation is in the direction of decreasing galactic longitude.

This value of ω corresponds to a period of rotation T of $1 \cdot 86 \times 10^8$ years. These values for ω and T refer of course to the neighbourhood of the Sun. Since $\omega = V/R$ and ω is now found, a knowledge of V or R will provide R or V. The centroid velocity V has been determined with reference to the system of globular clusters which have very small speeds about the galactic centre. The centroid distance R has also been measured from studies of RR Lyrae stars near the galactic centre. Considering all the available data, it appears that $R = 8500$ pc and $V = 250$ km s^{-1}.

The angular velocity ω about the galactic centre with respect to an inertial frame is not in fact readily available. T given by ω is $1 \cdot 86 \times 10^8$ years so that any inertial frame must be known to within a small fraction of one revolution in 2×10^8 years.

A laboratory gyroscope is in principle essentially a possibility but in practice hopelessly imprecise. If the Earth was completely isolated and spherical its axis of rotation would be fixed in direction but precession of that axis occurs with a period of 26 000 years and with unknown errors in its determination. If we go to planetary orbital precession, even in Jupiter's case, its semimajor axis rotates once in 10^6 years and the value is not known to better than four or five figures.

The system of globular clusters is therefore a much better candidate to supply an inertial system of the required accuracy. Even better is the use of distant galaxies.

15.8.3 The mass of the Galaxy

Let F be the force per unit mass due to the Galaxy's gravitational field operating at the Sun's distance R from the galactic centre. Then if V denotes the circular velocity as before, equating gravitational force to centrifugal force gives $V^2 = RF$.

Then

$$2V\frac{\mathrm{d}V}{\mathrm{d}R} = F + R\frac{\mathrm{d}F}{\mathrm{d}R}.$$

Using equation (15.47) we have

$$A = \frac{1}{4}\frac{V}{R}\left[1 - \frac{R}{F}\left(\frac{\mathrm{d}F}{\mathrm{d}R}\right)\right] \qquad (15.52)$$

and

$$B = -\frac{1}{4}\frac{V}{R}\left[3 + \frac{R}{F}\left(\frac{\mathrm{d}F}{\mathrm{d}R}\right)\right]. \qquad (15.53)$$

To get any further we must make some hypotheses concerning the distribution of mass within the Galaxy. Oort suggested that the gravitational field was largely due to a spherical central mass M_1, and to a spheroidal and uniform distribution of matter of mass M_2 concentric with the spherical central mass. The Sun could be taken to lie outside the spherical mass but inside the spheroid. This model, admittedly crude, must bear some resemblance to the truth and so results from its adoption should be of the right order of magnitude.

Then the force F per unit mass is given by $F = F_1 + F_2$, where F_1 and F_2 are due to the central mass and spheroid respectively. Now the attraction of a spherical mass is proportional to the inverse square of the distance from its centre, so that

$$F_1 = CR^{-2} \qquad (15.54)$$

where C is a constant. At a point inside a spheroid, we have seen in chapter 6 that the attractive force is proportional to the distance to the centre, and thus

$$F_2 = ER \qquad (15.55)$$

where E is a constant. Hence

$$F = CR^{-2} + ER. \qquad (15.56)$$

It should be remarked that this expression can only hold within a limited range of R. Obviously it leads to absurd values of F as $R \to 0$ or as $R \to \infty$. But we have already stated that the Sun lies outside the central mass and within the spheroidal distribution, thus restricting the range of values R can take.

Differentiating (15.56) with respect to R, we obtain

$$\frac{R}{F}\frac{dF}{dR} = \frac{F_2 - 2F_1}{F}.$$

Substitution of this expression in the relations (15.52) and (15.53) gives

$$A = \frac{3}{4}\left(\frac{VF_1}{RF}\right)$$

and

$$B = -\frac{V(F_1 + 4F_2)}{4RF}.$$

Eliminating V/R between these expressions gives

$$\frac{F_1}{F} = \frac{4A}{3(A-B)}.$$

Using the values for A and B from equation (15.51), we find that $(F_1/F) = 0.8$ and $(F_2/F) = 0.2$, showing that the attraction of the central mass is dominant.

To obtain the actual masses M_1 and M_2 we note firstly that the force of attraction per unit mass due to the central mass M_1 is

$$F_1 = GM_1/R^2.$$

The force F_2 is that experienced at a point inside a homogeneous spheroid distant R from the spheroid centre. Let the spheroid have mass M_2, let it be of uniform density ρ and let it have semiaxes a, b and c. If we have rotational symmetry, $a = b$. Let $a > c$. Then the components (X, Y, Z) of the force per unit mass at the point with coordinates x, y, z, within the spheroid are defined by

$$X = -G\alpha\rho x, \quad Y = -G\alpha\rho y, \quad Z = -G\gamma\rho z$$

where

$$\frac{\alpha}{2\pi a^2 c} = \int_0^\infty \frac{du}{(a^2+u)^2(c^2+u)^{1/2}} \qquad (15.57)$$

and

$$\frac{\gamma}{2\pi a^2 c} = \int_0^\infty \frac{du}{(a^2+u)(c^2+u)^{3/2}}. \qquad (15.58)$$

Let the x axis pass through the Sun. Then $x = R$ and $y = z = 0$. Hence

$$F_2 = -X = G\alpha\rho R.$$

Now

$$2\alpha + \gamma = 4\pi \qquad (15.59)$$

and also

$$2\alpha a^2 + \gamma c^2 = \frac{4\pi a^2 c}{(a^2 - c^2)^{1/2}} \tan^{-1}\left[\frac{(a^2 - c^2)^{1/2}}{c}\right].$$

In the Galaxy, $c/a \sim 0.1$ so that, neglecting $(c/a)^2$ and higher orders, we may write

$$2\alpha a^2 + \gamma c^2 = 2\pi^2 ac. \qquad (15.60)$$

From equations (15.59) and (15.60) we obtain

$$\alpha = \frac{\pi c(\pi a - 2c)}{a^2 - c^2} \sim \frac{\pi^2 c}{a}.$$

Now the mass M_2 is given by

$$M_2 = \frac{4\pi a^2 c\rho}{3}$$

so that

$$\alpha\rho = \frac{3\pi M_2}{4a^3}.$$

Hence

$$F_2 = \frac{3\pi G M_2 R}{4a^3} \qquad (15.61)$$

where a is the equatorial radius of the Galaxy (of the order of 1.5×10^4 pc). Now $R = 8500$ pc, so that $R/a \sim 0.57$.

Now $F = F_1 + F_2$, so that

$$F = \frac{G M_1}{R^2} + \frac{3\pi G M_2 R}{4a^3}. \qquad (15.62)$$

Hence

$$\frac{dF}{dR} = \frac{F}{R} - \frac{3G M_1}{R^3}. \qquad (15.63)$$

By equation (15.52), we had

$$A = \frac{V}{4R}\left(1 - \frac{R}{F}\frac{dF}{dR}\right). \qquad (15.64)$$

Subsituting from equation (15.62) in (15.64), we obtain

$$A = \frac{3G M_1}{4\omega R^3}$$

478

giving

$$M_1 = \frac{4A\omega R^3}{3G}.$$ (15.65)

Substituting for M_1 from (15.65) into (15.62) and replacing F by $\omega^2 R$, we obtain

$$M_2 = \frac{4a^3\omega}{3\pi G}\left(\omega - \frac{4}{3}A\right)$$ (15.66)

so that

$$\frac{M_2}{M_1} = \frac{1}{\pi}\left(\frac{a}{R}\right)^3\left(\frac{\omega}{A} - \frac{4}{3}\right).$$ (15.67)

All the quantities on the right-hand sides of equations (15.65), (15.66) and (15.67) can have values assigned to them. Thus $A = +0\cdot020$ km/s pc, $\omega = 0\cdot033$ km/s/pc, $R = 8500$ pc, $a = 1\cdot5 \times 10^4$ pc and $G = 6\cdot667 \times 10^{-8}$ in cgs units.

It is found that in solar mass units $M_1 = 1\cdot2 \times 10^{11}$ and $M_2 = 0\cdot67 \times 10^{11}$, giving $M = M_1 + M_2 = 1\cdot9 \times 10^{11}$ times the mass of the Sun. More recent studies, adopting more sophisticated models of the Galaxy, do not alter the order of magnitude of this value.

15.8.4 The mode of rotation of the Galaxy

One topic of interest in the dynamics of stellar systems such as the Galaxy is their mode of rotation. Does the system rotate like a solid body, or (like Saturn's ring system) does each particle within it obey Kepler's laws, with the angular velocity decreasing with increasing distance from the centre? Observational evidence provides a partial answer.

Considering only stars in the galactic equatorial plane (i.e. $b=0$), we had from equation (15.50)

$$\rho = rA \sin 2l$$

$$\kappa\mu_l = A\cos 2l + B$$

$$= A(\cos 2l + 1) - \omega$$

where κ is a constant, A and B are Oort's constants and r and l are the radius vector and galactic longitude of a star as seen from the Sun's position. The angular velocity of the Sun is given by ω. From (15.47) it is seen that neither A nor B depends upon the star's coordinates.

The behaviour of ρ and $\kappa\mu_l$ with changing l should, for a given value of r, behave in the systematic way shown in the graphs of figure 15.4.

If the Galaxy rotates in the Sun's neighbourhood as a solid body then there will be no radial velocities. In fact it is found that the radial velocities behave as in figure 15.4(a). This does not necessarily imply galactic rotation by itself. For example, if stars in the Sun's vicinity moved in straight lines but with velocities decreasing linearly with increasing distance from the

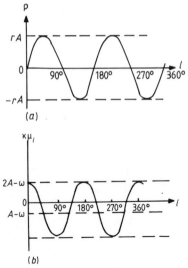

(a)

(b)

Figure 15.4

galactic centre (as in figure 15.5(*a*)), then we would obtain the relative field of stellar velocities shown in figure 15.5(*b*), which is obtained by subtracting the Sun's velocity from all stellar velocities. This relative field would in its turn give rise to the systematic distribution of radial velocities with longitude sketched in figure 15.5(*c*). This distribution agrees with graph (*a*) in figure 15.4.

The distribution of observed proper motion components seen in figure 15.5(*d*) does not agree with that sketched in figure 15.4(*b*). Proper motion in all longitudes would be positive or zero, whereas observation shows them to be positive, zero or negative, depending upon the longitude.

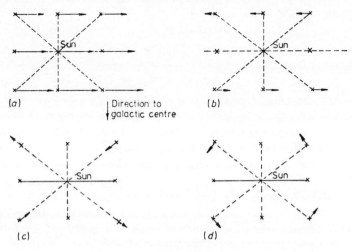

(a)   Direction to galactic centre   (b)

(c)   (d)

Figure 15.5

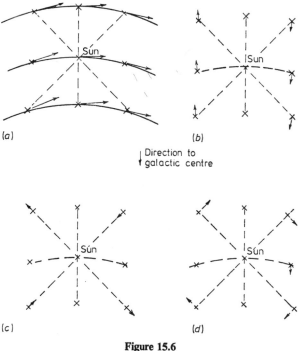

Direction to
galactic centre

Figure 15.6

If however the stars are in orbit about the galactic centre, so that the curvature of the orbits are taken into account when the velocity vectors are drawn, we have the situation as sketched in figure 15.6 where, in figure 15.6(a), constant speeds have been assumed to simplify the picture. While the radial velocity pattern is essentially unchanged, the proper motion pattern is seen to be reversed. Taking into account both curvature of orbit and decreasing speed with increasing distance from the galactic centre, we can obtain a a proper motion pattern that agrees with the one in figure 15.4(b) given by observation.

In recent years additional observational information from radioastronomy measurements has augmented our knowledge of galactic rotation as well as enabling maps of the distribution of interstellar material to be drawn. Neutral hydrogen emits radiation with a wavelength of 21 cm, which can be detected by a radio telescope. Each cloud of neutral hydrogen is in orbit about the galactic centre and therefore has a radial velocity relative to the Sun. The wavelength of the radiation it emits is therefore altered by the Doppler effect. The difference $\Delta\lambda$ between the theoretical value λ and the measured value gives the radial velocity v through the Doppler formula $\Delta\lambda/\lambda = v/c$, where c is the velocity of light. There may be many clouds intersected by any line of constant galactic longitude drawn from the Sun across the Galaxy and they will lie at different distances, having different densities, so that the observed profile of intensity with wavelength around the 21 cm wavelength will be complex for any given galactic longitude. Careful collation

481

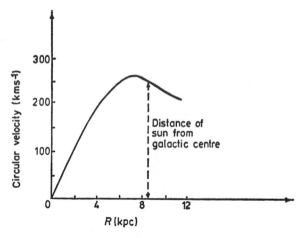

Figure 15.7

of all the data has however enabled detailed deductions to be made about galactic rotation and spiral structure.

In figure 15.7 the circular velocity V at various distances R from the galactic centre is given, based on optical data from O- and B-type stars and radio observations of the 21cm radiation.

Out to a distance of about 7·5 kpc the Galaxy rotates like a solid body with constant angular velocity (i.e. $\omega = V/R = \text{constant}$). Beyond that distance a maximum velocity is reached and thereafter the circular velocity diminishes with distance. Empirically, and from theoretical studies of rotationally symmetric stellar systems in a steady state with an ellipsoidal velocity distribution law, it appears that V is given by a law of the form

$$V = k_1 R/(1 + k_2 R^2) \tag{15.68}$$

where k_1 and k_2 are constants. The maximum value of V is reached when $R = k_2^{-1/2} = R_0$. The Sun is in the region just beyond R_0, where circular velocity is diminishing with distance from the galactic centre.

We can obtain values for k_1 and k_2 as follows. From equation (15.68), we obtain on differentiation

$$\frac{dV}{dR} = \frac{k_1(1 - k_2 R^2)}{(1 + k_2 R^2)^2}. \tag{15.69}$$

But by equation (15.47) it is seen that

$$A + B = -\frac{dV}{dR}, \qquad A - B = \frac{V}{R}. \tag{15.70}$$

Hence, by eliminating dV/dR and V between relations (15.68), (15.69) and (15.70), we find that

$$k_1 = -\frac{(A - B)^2}{B}, \qquad k_2 = -\frac{A}{BR^2}. \tag{15.71}$$

482

Note that in these expressions R is the Sun's distance from the galactic centre, since A and B are Oort's constants measured from the Sun's position in the Galaxy.

The velocity law (15.68) is easily explained in semiquantitative terms. Stars far from the almost spherical galactic nucleus exist in regions where quasi-Keplerian orbits are traced because the massive galactic nucleus acts approximately as a point-mass. In such a region the angular momentum is constant (i.e. $R^2\omega = h$). Hence $RV = h$. For stars close to the nucleus or within it, the force is proportional to the mass M contained within a sphere of radius R. But $M = (4/3)\, m\pi\nu R^3$, where ν is the star number density in the nucleus and m is the average mass of a star. Hence if we equate centrifugal force per unit mass to gravitational force per unit mass, we have $V^2 = GM/R$; that is

$$V^2 = \tfrac{4}{3}\, \pi m G \nu R^2.$$

Hence in this inner region, $(V/R) = \omega = $ constant.

Within this region all stars in circular orbits have the same orbital period, of the order of 10^8 years. Other galaxies such as M31 in Andromeda and M33 in Triangulum show the same rotation patterns, an inner region rotating as a solid body plus an outer region where velocity diminishes with increasing distance from the galactic centre.

15.8.5 The gravitational potential of the Galaxy

It is reasonable to assume that the gravitational potential U experienced by a star in the Galaxy is, to a close approximation, that due to a stellar system which is symmetrical about an axis of rotation and about a plane perpendicular to that axis. The potential U is therefore a function of R and z only, where R is measured in the plane of symmetry from the galactic centre while z is measured from that plane, parallel to the axis of rotation.

Now the expression (15.68), due to Paranago, gives the circular velocity V at a distance R from the galactic centre in the equatorial plane, k_1 and k_2 being constants. The form of this expression is confirmed by studies of galactic long-period cepheids and from studies of nearby galaxies such as M31 and M33.

In the galactic plane, equating centrifugal and gravitational forces, we have

$$V^2 = -R\frac{dU_p}{dR}$$

where U_p is the gravitational potential at a distance R from the galactic centre.

Then using (15.68) we have

$$\frac{dU_p}{dR} = -\frac{k_1^2 R}{(1 + k_2 R^2)^2}.$$

This can be integrated to give

$$U_p = \frac{k_1^2}{2k_2(1+k_2R^2)} + C.$$

Now, outside the Galaxy, the potential must be given approximately by GM/R. If we let R_1 be the approximate equatorial radius of the Galaxy, it follows that

$$\frac{k_1^2}{2k_2(1+k_2R_1^2)} + C \doteqdot \frac{GM}{R_1}.$$

By (15.71) and a knowledge of the values of A and B we can compute k_1 and k_2. They are respectively of order $72 \text{ km s}^{-1} \text{ (kpc)}^{-1}$ and 0.024 (kpc)^{-2}. The mass M of the Galaxy is of order 1.3×10^{11} solar masses, the mass of the Sun is 1.99×10^{30} kg and R_1 is approximately 13 kpc. Hence it is found that C is much less than $(k_1^2/k_2)/(1+k_2R_0^2)$ where R_0 is approximately the distance of the Sun from the galactic centre. In fact, within the Galaxy, C may be neglected. Hence

$$U_p = \frac{k_1^2}{2k_2(1+k_2R^2)} = \frac{U_c}{1+k_2R^2}$$

where $U_c = k_1^2/2k_2 \sim 1.1 \times 10^{-25} \text{ km}^2 \text{ s}^{-2}$.

Outside the galactic plane, U_p is multiplied by a function ϕ of z which is chosen so that it decreases as z changes from $z = 0$ both positively and negatively, and gives $\phi(0) = 1$, $\phi(\pm\infty) = 0$, $(d\phi/dz)_0 = 0$, $(d\phi/dz)_{\pm\infty} = 0$ and $(d^2\phi/dz^2)_0 < 0$.

Paranago chose, from theoretical and empirical considerations,

$$\phi(z) = e^{-\lambda z^2}$$

where λ has the value $5.9 \times 10^{-35} \text{ km}^{-2}$.

Hence the gravitational potential of the Galaxy U is finally given by

$$U(R, z) = \frac{U_c e^{-\lambda z^2}}{1+k_2R^2}.$$

15.8.6 Galactic stellar orbits

It has been seen that for the local group of stars a centroid may be defined, possessing a centroid velocity. The stars (including the Sun) have velocities dispersed about the centroid velocity with residual velocities (the difference between the star's galactic velocity and the centroid galactic velocity) of the order of 20 km s^{-1}. It is found that the velocity distribution function ϕ is given by an expression of the form

$$\phi(u, v, w) = C \exp\left\{-\frac{u^2}{2\sigma_x{}^2} - \frac{v^2}{2\sigma_y{}^2} - \frac{w^2}{2\sigma_z{}^2}\right\}$$

where ϕ is the number of stars within the group with residual velocities between u and $u+du$, v and $v+dv$, and w and $w+dw$, measured with respect to axes x, y and z respectively; C is a constant and σ_x, σ_y, σ_z are the standard deviations also defined with respect to those axes. The x axis is drawn towards the galactic centre, the y axis lies tangential to the galactic rotation and the z axis is perpendicular to the galactic plane. Values of σ_x, σ_y and σ_z are about ± 28, ± 20 and ± 15 km s^{-1} respectively. In contrast, the centroid velocity of the local group is about 250 km s^{-1}.

This Schwarzschild three-axis ellipsoidal velocity distribution may be explained as a consequence of the fact that the stars in the local group, although temporarily in the same volume element, have slightly different galactic orbits. Some are circular; most are elliptical with small but differing eccentricities and differing inclinations to the galactic equatorial plane. Lindblad (1933) showed in fact that observed movements of stars in such orbits would show an ellipsoidal distribution with stars streaming away from or towards the galactic centre as viewed from the Sun.

It is easy to show that such almost circular and low inclination orbits are possible for stars within a stellar system such as the Galaxy. For most stars the residual velocities are an order of magnitude smaller than the centroid rotation velocity, so that the orbits do not depart much from circular coplanar orbits. Following a development by Lindblad we let r, θ and z be the cylindrical coordinates of a star X in a stellar system with rotational symmetry, the z axis being the axis of symmetry (figure 15.8). We also assume the stellar system to have a plane of symmetry perpendicular to the axis of symmetry, the reference direction CD lying in this plane so that θ is the azimuthal angle measured to the projection CH of the radius vector CX of the star. We also let CH be r.

The equations of motion of X are then

$$
\left.
\begin{aligned}
\ddot{r} - r\dot{\theta}^2 &= \frac{\partial U}{\partial r} \\[2mm]
\frac{1}{r}\frac{d}{dt}(r^2\dot{\theta}) &= \frac{1}{r}\frac{\partial U}{\partial \theta} \\[2mm]
\ddot{z} &= \frac{\partial U}{\partial z}
\end{aligned}
\right\}
\qquad (15.72)
$$

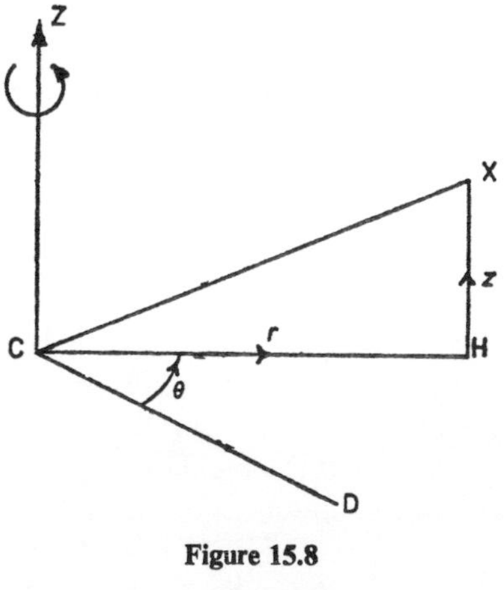

Figure 15.8

where U is the gravitational potential acting at X due to the stellar system.

Now by symmetry,

$$U = U(r, z) = U(r, -z) \tag{15.73}$$

so that $(\partial U/\partial \theta) = 0$ giving, from the second of equations (15.72), the relation

$$r^2 \dot\theta = h. \tag{15.74}$$

Now let the star be moving in the plane of symmetry of the stellar system with no component of velocity in the z direction. Hence $z = \dot z = 0$, and by symmetry $(\partial U/\partial z)_{z=0} = 0$. Also $(\partial^2 U/\partial r \partial z)_0 = 0$.

Hence (15.72) becomes

$$\left. \begin{aligned} \ddot r - r\dot\theta^2 &= \frac{\partial U}{\partial r} \\ r^2 \dot\theta &= h. \end{aligned} \right\} \tag{15.75}$$

Now try for a solution with $r = r_0 = $ constant and $\omega = (d\theta/dt) = \omega_0 = $ constant. Then by the second of equations (15.75) we have

$$r_0^2 \omega_0 = h.$$

By the first of (15.75),

$$\ddot r - r_0 \omega_0^2 = \left(\frac{\partial U}{\partial r} \right)_0.$$

If $\ddot r = 0$, we have

$$r_0 \omega_0^2 + \left(\frac{\partial U}{\partial r} \right)_0 = 0 \tag{15.76}$$

which, putting $V_0 = \omega_0 r_0$, gives

$$\frac{V_0^2}{r_0} + \left(\frac{\partial U}{\partial r} \right)_0 = 0.$$

But this is the equation for a particle moving in a gravitational field due to a potential U. Hence a circular orbit is possible, the star pursuing such an orbit with constant angular velocity ω_0 given by $r_0^2 \omega_0 = h$.

Let us now disturb the star slightly from its circular motion so that its coordinates are

$$r = r_0 + \rho$$

$$\theta = \theta_0 + \phi$$

$$z = z_0 + \zeta$$

$$= \zeta \text{ (since } z_0 = 0)$$

where ρ, ϕ and ζ are small variable quantities. Note that θ_0 is not constant, whereas r_0 is. We substitute these new variables into equations (15.72) and linearize the resulting equations to obtain the differential equations for ρ, ϕ and ζ in much the same way that we did in chapter 5 when we considered the stability of the Lagrange solutions of the circular restricted three-body problem.

486

First we expand $U(r, z)$ to the second order in the small quantities $\rho = r - r_0$ and $\zeta = z - z_0 = z$. Hence, remembering that

$$\left(\frac{\partial U}{\partial z}\right)_0 = \left(\frac{\partial^2 U}{\partial r \partial z}\right)_0 = 0$$

we have

$$U = U_0 + \left(\frac{\partial U}{\partial r}\right)_0 (r - r_0) + \frac{1}{2}\left(\frac{\partial^2 U}{\partial r^2}\right)_0 (r - r_0)^2 + \frac{1}{2}\left(\frac{\partial^2 U}{\partial z^2}\right)_0 z^2.$$

Partially differentiating this expression with respect to r and z, we obtain

$$\frac{\partial U}{\partial r} = \left(\frac{\partial U}{\partial r}\right)_0 + \left(\frac{\partial^2 U}{\partial r^2}\right)_0 \rho, \qquad \frac{\partial U}{\partial z} = \left(\frac{\partial^2 U}{\partial z^2}\right)_0 \zeta. \qquad (15.77)$$

Substituting $r = r_0 + \rho$ and $\theta = \theta_0 + \phi$ in the relation $r^2\dot\theta = h$, and retaining only first-order terms, we obtain

$$\dot\phi = -2\left(\frac{\omega_0}{r_0}\right)\rho. \qquad (15.78)$$

The third equation in (15.72) gives

$$\ddot\zeta = \left(\frac{\partial^2 U}{\partial z^2}\right)_0 \xi. \qquad (15.79)$$

Using the first of (15.77), (15.76), and the first equation in (15.72), we obtain after a little reduction

$$\ddot\rho = \left\{\left(\frac{\partial^2 U}{\partial r^2}\right)_0 + \frac{3}{r_0}\left(\frac{\partial U}{\partial r}\right)_0\right\}\rho. \qquad (15.80)$$

Equations (15.78), (15.79) and (15.80) comprise the required set of differential equations. In these equations the coefficients of ρ and ζ are constant. The behaviour of ρ, ζ and ϕ depend upon the signs of these coefficients.

When the star crosses the equatorial plane it is moving in the direction of increasing z. The z component of the force on it is negative so that $\partial U/\partial z < 0$ for $z > 0$; likewise $\partial U/\partial z > 0$ for $z < 0$. Hence at $z = 0$, $\partial U/\partial z$ is a decreasing function of z, that is $\partial^2 U/\partial z^2 < 0$.

Hence equation (15.79) is the equation for simple harmonic motion, its solution being

$$\zeta = \alpha_1 \sin(n_1 t + \beta_1)$$

where $n_1 = [-(\partial^2 U/\partial z^2)_0]^{1/2}$ and α_1 and β_1 are constants of integration.

In equation (15.80) the coefficient of ρ may be written as

$$\left[\frac{1}{r^3}\frac{\partial}{\partial r}\left(r^3\frac{\partial U}{\partial r}\right)\right]_0.$$

Now the magnitude of the gravitational force at distance r is $F(r) = -(\partial U/\partial r)$. Even if all the mass were concentrated at the galactic centre, F would decrease no faster with increasing r than r^{-2}. Hence $-r^3 (\partial U/\partial r) = r^3 F(r)$ is an increasing function of r. It is then clear that the expression

$$\frac{1}{r^3} \frac{\partial}{\partial r} \left(r^3 \frac{\partial U}{\partial r} \right)$$

must be negative. Equation (15.80) is therefore also the equation of simple harmonic motion, giving the solution

$$\rho = \alpha_2 \sin (n_2 t + \beta_2)$$

where

$$n_2 = \left\{ -\left[\frac{1}{r^3} \frac{\partial}{\partial r} \left(r^3 \frac{\partial U}{\partial r} \right) \right]_0 \right\}^{1/2}$$

and α_2 and β_2 are constants of integration. Finally, by substituting this solution into equation (15.78), we obtain the solution for ϕ

$$\phi = \alpha_3 + \alpha_3' \cos (n_2 t + \beta_2)$$

where $\alpha_3' = 2\omega_0 \alpha_2 / n_2 r_0$.

The interpretation of these results is that the star performs an elliptical motion about the circular orbit's reference point in a period $T_1 = 2\pi/n_1$ while oscillating to and fro through the galactic plane in a vibration of period $T_2 = 2\pi/n_2$. Calculations show that for a star in the solar neighbourhood the values of T_1 and T_2 are approximately 150×10^6 and 80×10^6 years respectively. For comparison we remember that the period of revolution about the galactic centre at the Sun's distance is about 186×10^6 years.

15.8.6 The high-velocity stars

It has been stated that the residual velocities of most of the members of the local group of stars are of the order of 20 km s^{-1} and are orientated according to Schwarzschild's ellipsoidal velocity distribution. If however we select out those stars with residual velocities greater than 100 km s^{-1}, we find a marked asymmetry in their distribution. None of these high-velocity stars is moving in the direction in which the Sun revolves round the Galaxy's centre. Most have velocity vectors lying in the semicircle bisected by the opposite direction.

This asymmetry is explainable if we remember that the local group centroid's rotational velocity is of the order of 250 km s^{-1}, which is essentially circular velocity for the Sun's distance from the galactic centre. The velocity of escape, if we consider as a rough approximation that the material inside the Sun's galactic orbit acts as a point-mass, is $\sqrt{2} \, V_{circ} \sim 350$ km s^{-1}. Thus any stars having velocities greater than 100 km s^{-1} and proceeding in the same direction as that of the Sun's velocity could exceed the velocity of escape and presumably would be in the process of departing from the Galaxy altogether. The many stars that show high velocities relative to the Sun are therefore moving with speeds much less than circular velocity at the Sun's distance. They are still revolving about the galactic centre in the same direction as the Sun but their orbits must be markedly elliptical. It can be calculated that many of them in the Sun's vicinity must be near the apocentres of their orbits and that the pericentres must lie deep in the galactic nucleus. Such stars also show a higher z component in their velocities, showing that their orbits are also more highly inclined to the galactic plane.

One such group of high-velocity stars are the RR Lyrae variables. These are Population II stars, much older than the Population I stars of the galactic disc. The globular clusters are also, according to this viewpoint, high-velocity 'stars' or objects, moving even more slowly than the RR Lyrae stars with respect to the local group and forming an almost spherical distribution about the galactic centre. They also consist of the older Population II stars. The implication is that the Galaxy is composed of a set of sub-systems; the older the sub-system is, the more spherical it is. Even the galactic nucleus falls into this scheme, being an oblate spheroid composed of Population II stars.

15.9 Spherical Stellar Systems

We now consider briefly the dynamics of spherical stellar systems such as the open clusters and globular clusters that are observed to exist in the Galaxy. As a preliminary we will apply the 'sphere of influence' criterion

$$r = R \left(\frac{m}{M}\right)^{2/5}$$

to the sphere of influence of (i) an open cluster and (ii) a globular cluster against the attraction of the galactic bulge.

Case (i); Representative values for an open cluster in the solar neighbourhood are: $m = 10^2$, $M = 10^{11}$, $R = 10^4$ (masses in solar mass units, distance R in pc). It is then found that $r \sim 2 \cdot 5$ pc.
Case (ii); For a globular cluster we may put $m = 5 \times 10^5$, $M = 10^{11}$ and $R = 10^4$, giving $r \approx 76$ pc.

Now the measured radii of open clusters lie between 1 and 10 pc, the majority being less than 3 pc. For globular clusters, radii are found to lie between 10 and 75 pc, with an average around 25 pc. The agreement in both cases is therefore good, suggesting that while tidal effects on clusters by the attraction of the galactic bulge will not be negligible, the cluster sizes have adjusted themselves to withstand the disruptive effects of such tides. Indeed there is observational evidence that the sizes of globular clusters are proportional to their distance from the galactic centre; there

489

is also evidence that their outer parts are extended along an axis passing through the galactic centre.

Other disruptive mechanisms exist. For example, any massive interstellar cloud passing by a cluster will tend to expand the cluster, increasing the speeds of the cluster stars. The cumulative effects of such encounters will in time cause the stars to escape, ultimately leading to the destruction of the open cluster. For a small open cluster the characteristic time to disruption is of the order of 10^8 years; for denser open clusters it may be as long as 5×10^9 years.

For a small dense open cluster with only a few members, individual encounters with other members of the cluster may boost a star's speed to near the velocity of escape from the cluster. It therefore leaves the cluster and wanders away, robbing the cluster of some of its kinetic energy. The cluster in consequence shrinks. After repeated escapes, the cluster dwindles to perhaps a binary or triple stellar system.

In a globular cluster, the rate of escape of its members is low. The strong general gravitational field of the cluster holds them in bound orbits so strongly that the probability of them building up the necessary escape velocity by a succession of random encounters is small. A globular cluster is therefore stable and will survive for at least 10^9 to 10^{10} years.

15.9.1 Application of the virial theorem to a spherical system

We can make these ideas a little more precise by considering the relevance of the virial theorem. In chapter 5, section 5, it was found that for a system of n gravitating particles of masses m_i ($i = 1, 2 \ldots n$), we had the relation $T - U = C$ and also

$$\ddot{I} = 4T - 2U = 2U + 4C = 2T + 2C$$

where

$I = \sum_{i=1}^{n} m_i \mathbf{R}_i{}^2 =$ moment of inertia of the system about its centre of mass,

$T = \frac{1}{2} \sum_{i=1}^{n} m_i \dot{\mathbf{R}}_i \cdot \dot{\mathbf{R}}_i =$ kinetic energy of the system,

$U = \frac{1}{2} G \sum_{i=1}^{n} \sum_{j=1}^{n} \dfrac{m_i m_j}{r_{ij}}, i \neq j = -$ potential energy of the system,

$C =$ total energy of the system $=$ constant,

$\dot{\mathbf{R}}_i, \dot{\mathbf{R}}_i$ are the radius and velocity vectors of the ith particle,

$r_{ij} = |\mathbf{R}_j - \mathbf{R}_i|$, the origin being the centroid of the system.

Then since both U and T are positive, if C is positive I will be positive and I will increase indefinitely, leading to the escape of at least one of the masses.

490

Now if the star cluster is in a steady state, I is not a function of time and so

$$T = \tfrac{1}{2} U = -C \qquad (15.81)$$

i.e. the sum of the potential energy and twice the kinetic energy is zero. If $\overline{V^2} = \dot{\mathbf{R}} \cdot \dot{\mathbf{R}}$ is the root mean square velocity and M is the total mass of the system, we may write

$$T = \frac{1}{2} \sum_{i=1}^{n} m_i \dot{\mathbf{R}}_i \cdot \dot{\mathbf{R}}_i = \tfrac{1}{2} M \overline{V^2}. \qquad (15.82)$$

Also, if the system is a homogeneous sphere of radius R, the potential energy is obviously given approximately by

$$U = -GM^2/R \qquad (15.83)$$

since the average separation of any two stars is the radius of the sphere, the average value of $m_i m_j$ is m^2 where m is the average mass of a star, given by $m = M/n$, and it should be remembered that in the double summation every term is counted twice. Hence by equations (15.81), (15.82) and (15.83), we have

$$\overline{V^2} \simeq \frac{GM}{R}.$$

Now for a star at the edge of the cluster, the velocity of escape V_e is given by $V_e = (2GM/R)^{1/2}$. It is then seen that if there is a Maxwellian distribution of the velocities, some stars will be able to have velocities greater than escape; such stars will therefore leave the cluster.

A large number of studies have been made to develop these ideas, giving rise to the concept of the *relaxation time* for a stellar system. If one or more stars leave the cluster, the time it takes for the cluster to set up a new equilibrium distribution of velocities is the relaxation time. This time is closely related to the disintegration time of the system. A value of the relaxation time may be found from the formula

$$T = \frac{1}{16} \left(\frac{3\pi}{2} \right)^{1/2} \left(\frac{nR^3}{Gm} \right)^{1/2} \bigg/ \ln \left(\frac{n}{2^{3/2}} \right)$$

where n is the number of stars in the system, R is the radius of the system and m is the average mass of a star. If we use solar mass units and R is measured in parsecs, this formula reduces to

$$T = 8 \times 10^5 \left(\frac{nR^3}{m} \right)^{1/2} \bigg/ \log_{10} (n - 0 \cdot 5) \text{ years.}$$

The disintegration half-life of the system, which is the time it takes for half the stars to escape, is $133T$. For the Pleiades open cluster $T \sim 5 \times 10^7$ years, so that $133\,T \sim 6 \times 10^9$ years. For most globular clusters, $T \sim 10^{10}$ years.

15.9.2 Stellar orbits in a spherical system

Certain statements can be made about the orbits of stars in a stellar system possessing spherical symmetry. In section (15.7) we saw that, the gravitational potential U being a function only of the distance r from the centre of the system, we had four integrals I_1, I_2, I_3, I_4. The first is the energy integral

$$I_1 \equiv V^2 - 2U = C_1$$

and the others are the angular momentum integrals, which can be summarized in vector form as

$$\mathbf{I} \equiv \mathbf{r} \times \dot{\mathbf{r}} = \mathbf{r} \times \mathbf{V} = \mathbf{C}$$

where $\mathbf{r}$ and $\mathbf{V}$ are the radius and velocity vectors of a star in the system. Thus the plane of a stellar orbit does not change its orientation in a spherically symmetric system (apart from the rare occasion when a close encounter between the star in question and another star in the system takes place). We may then write the equations of motion of the star in the plane polar coordinate form

$$\ddot{r} - r\dot{\theta}^2 = \frac{dU}{dr} = -\frac{GM(r)}{r^2} \tag{15.84}$$

$$r^2\dot{\theta} = h \tag{15.85}$$

where h is the angular momentum constant and

$$U(r) = G \int_r^\infty \frac{M(r)}{r^2}\, dr. \tag{15.86}$$

As the star is within the spherically symmetric system at a distance r from the centre, the force acting on it is due to the mass $M(r)$ within a sphere of radius r.

Eliminating θ between equations (15.84) and (15.85), we obtain

$$\ddot{r} - \frac{h^2}{r^3} = \frac{dU}{dr}.$$

Multiplication by $\dot{r}$ and integration of the resulting equation gives

$$\frac{\dot{r}^2}{2} + \frac{h^2}{2r^2} = U + C' \tag{15.87}$$

where C' is a constant. This is the energy relation. Let V_R and V_T be the radial and transverse velocity components, so that

$$V_R = \dot{r}, \qquad V_T = r\dot{\theta}.$$

Then, using equations (15.85) and (15.87), we have

$$rV_T = h \tag{15.88}$$

and

$$V_R{}^2 + V_T{}^2 - 2U = C \qquad (15.89)$$

or

$$V^2 - 2U = C \qquad (15.90)$$

where C is a constant. The relations (15.88) and (15.89) are all that are required to determine the orbit properties.

Circular orbits are possible. If so $V_R = 0$, $V_T = $ constant, $h \neq 0$ and $r = r_0 = $ constant. Hence $U = U(r_0) = $ constant. All of these are consistent with equations (15.87), (15.88) and (15.89). Rectilinear orbits through the system centre are also possible. If so $V_T = 0$, $h = 0$, $V_R = \dot{r}$ is variable as is also U, their relations being given by equations (15.87) and (15.89). It is also readily seen that if the orbit is neither circular nor rectilinear it must lie between two concentric circles whose radii give the apocentre and pericentre distances.

By equation (15.87), we have

$$\dot{r}^2 + \frac{h^2}{r^2} - 2U = C. \qquad (15.91)$$

If the star reaches pericentre or apocentre $\dot{r}$ becomes zero, and therefore under these conditions we have

$$\frac{h^2}{r^2} - 2U = C. \qquad (15.92)$$

The roots of this equation thus give the pericentre and apocentre distances. Such an orbit will be an oval of some kind which may precess in its orbital plane. In particular, if the orbit lies far out in the spherical system then the motion and orbit will be approximately Keplerian, since the vast bulk of the stellar system will behave as a point-mass at the centre. On the other hand, a star whose orbit lies deep within the core of the system will suffer an attractive force proportional to the distance from the centre. Hence U will be of the form $U = -cr^2$, where c is a positive constant. From (15.92) we then obtain

$$h^2 + 2\,cr^4 = Cr^2$$

a biquadratic equation whose roots give the major and minor axes of the approximately elliptic orbit performed by a star under this law of force. Unlike the Keplerian orbit, the centre of the ellipse is at the centre of the system and the angular velocity is $\sqrt{c}$, a constant which is the same for every orbit in this central region.

15.9.3 The distribution of orbits within a spherical system

If there were no escape of stars from a spherical system, it would in time tend towards an equilibrium state. There would be a Maxwellian velocity distribution, the star density becoming that of an isothermal polytrope.

A stellar system behaving in this way acts as a spherical mass of gas, with stars replacing the molecules or atoms. There is a vast literature on polytropic gas spheres, which are described by solutions of Emden's equation, providing relations among the pressure, density and kinetic temperature of the particles. Plummer, von Zeipel and Eddington were among those who sought to apply the theory of polytropic gas spheres to spherical systems such as globular clusters. In fact the application can only be approximate since the continual escape of stars will finally lead to the total disintegration of the system.

Various other approaches are possible. In the previous section we have seen that in a spherically symmetric stellar system equations (15.88) and (15.89) determine a star's orbital properties, and that in general there are pericentre and apocentre distances. Let r_a be the apocentre distance. When the star is at apocentre $V_r = 0$. Hence r_a and V_{Ta} (the latter being the transverse velocity at apocentre) will define the orbit. The distribution function ϕ for these 'orbital elements' can then be set up. Thus $d\nu = \phi(r_a, V_{Ta}) \, dr_a \, dV_{Ta}$ will be the number of stars with apocentres at distances between r_a and $r_a + dr_a$ and apocentric transverse velocities between V_{Ta} and $V_{Ta} + dV_{Ta}$.

By assuming that Schwarzschild's velocity distribution law is obeyed in the system, it is possible to show that

$$\nu = \frac{\pi^{3/2} A \exp (2p^2 U)}{p^3 (1 + kr^2)} \qquad (15.93)$$

where the Schwarzschild function f is given by

$$f = A \exp [p^2 (2U - V_R^2 - V_T^2 - kr^2 V_T^2)].$$

In these expressions A, p and k are constant parameters, while U is the gravitational potential, V_R is the radial velocity $\dot{r}$ and V_T is the transverse linear velocity $r\dot{\theta}$. Eddington carried out this investigation and arrived at equation (15.93) for the density ν, which is a particular case of the general solution derived from Jeans's theorem (Eddington 1913, 1915).

From investigations of this nature it is possible to show that the fraction of circular orbits and rectilinear orbits is small, and that few stars remain near the cluster centre or pass close to it.

15.10 Computer Experiments

With the advent of computers in recent years and spectacular progress in the development of data-handling techniques, it is only natural to apply such methods to stellar dynamics, not only to check the main results arrived at by more or less rigorous analytical procedures but also to tackle problems outwith the scope of analysis. It is obviously possible to use a 'brute-force' technique, by getting a modern high-speed, large capacity computer to integrate numerically the equations of motion of the stars in the stellar system, i.e. to integrate

$$\ddot{\mathbf{r}}_i = -G \sum_{j=1}^{N} m_j \frac{\mathbf{r}_i - \mathbf{r}_j}{|\mathbf{r}_i - \mathbf{r}_j|^3}, \qquad (j \neq i, i = 1, 2, \ldots, N)$$

where $\ddot{\mathbf{r}}_i$ is the acceleration of the ith star of mass m_i, its radius vector $\mathbf{r}_i$ being measured with respect to some inertial frame. Unfortunately, in all such calculations, as we found in celestial mechanics when N is rarely as high as 10, various inbuilt limitations exist that must be considered and taken care of before even the simplest problem ($N \geqslant 3$) can be tackled and brought to yield useful and understandable results. These include the growth of round-off error, the types of orbit encountered, the available computing facilities, the numerical integration procedure and the possibility of close approaches of two or more bodies. When, as in the stellar dynamical case, N can be between 10^5 and 10^6 for a typical globular cluster, or as high as 10^{11} for a galaxy, the problems are enhanced by orders of magnitude.

Nevertheless, various strategies can be adopted to circumvent such limitations and carry out investigations that show that much of the earlier analytical work was valid in the conclusions it drew and speculations it made about stellar systems. Numerical studies show that the virial theorem holds; the relaxation time formula agrees closely in its results with relaxation time numerical computations. Stars escape and the star cluster adjusts itself, within a time roughly of the order of the relaxation time, to a Maxwellian distribution. Close binaries form and play a large part in the escape of stars and the further evolution of the cluster. The strategies available to us, however, still do not permit us to study in detail systems containing more than a few thousand bodies, for example a small globular cluster.

A concept known as the *crossing time*, t_{cr}, of the system is a useful timescale unit in cluster affairs. As its name implies

$$t_{cr} = 2R/v$$

where $2R$ is a measure of the diameter of the cluster and v^2 is the mean square speed of the cluster stars, the virial theorem being supposed to be satisfied by the cluster. For a typical cluster $t_{cr} \sim 10^6$ years.

Then it is quickly ascertained that for a numerical integration of the cluster stars' equations of motion on a small microcomputer, no sophisticated strategies being adopted, the computation progress would proceed at much the same speed as the real cluster, it taking of order 10^6 years of computing time to follow a star across the cluster!

A first step to improve this dire state of affairs is taken by choosing a numerical integration procedure that allows time steps for the outer stars in the cluster, whose accelerations change more slowly and with less amplitude than those of the stars in the dense cluster core, to be much longer than those for the centre stars.

A further procedure is the application of a *tree code*. This makes use of the fact that groups of stars far from the star in question tend to cancel out

495

the fluctuations in their net force-field. Thus in applying a tree code, the cluster stars are placed on branches, sub-branches, sub-sub-branches, and so on according to the force they apply. Groups of more distant stars, whose contributions on the star in question can be summed in a barycentric approximation, occupy a 'coarse, thick' branch; only the star's nearest and strongest disturbers need to be placed each one on one of the ultimate twigs in the sub-division.

This procedure would appear at first glance to be cumbersome and time-wasting but in practice it pays off handsomely especially if a way can be found to ensure the tree design alters as slowly as possible. For more information the reader is referred to Heggie (1988) and to Hut and McMillan (1986).

A common test of the inevitable accumulation of error as a computation proceeds is to reverse the computation at a suitable time and try to recapture the initial conditions at the original time. This is never achieved exactly. At best the reverse computation creates its own burden of round-off error so that the stars arrive back at positions only to some extent resembling those they started out from. At worst, and this is far more probable, close encounters have taken place. During such close encounters, the forces between the participating stars are large; they are also extremely sensitive in their effects on the future trajectories of the stars to the exact distance apart of the stars during the encounter. Unless some form of regularization is adopted during the encounter, error is maximized; in any case, on the 'return trip' the inevitable accumulation of round-off error makes it certain that the encounter will never be precisely retraced—it may even be missed! For a discussion of error in stellar dynamic calculations and the reliability or otherwise of statistical results, such as the mean rate of escape of stars from the cluster, see Heggie (1988).

Heggie has pointed out that in a recent simulation on computer of a small cluster of 3151 stars, approximately 15 CPU days were taken to compute the evolution of the cluster over 10^7 years. This is a considerable improvement over the figure given at the beginning of this section, where the computation proceeded only as fast as the real cluster. Nevertheless, it is still inadequate for even an average-sized globular cluster and further progress awaits not just the arrival of a new generation of computers but more likely the creation of fundamentally new ways of tackling the problem. As far as a stellar system of the magnitude of a galaxy is concerned, computers remain as before merely the handmaidens of analysis.

Problems

15.1 Given that the Sun is $8\cdot5$ kpc from the centre of the Galaxy and has a period of revolution about the centre of 200 million years, calculate the approximate mass of the Galaxy within the Sun's orbit in solar mass units. Assume a circular orbit and a spherical distribution of material within it, and neglect the material outside the Sun's orbit.

15.2 Observations of the 21 cm line of neutral hydrogen reveal that, after correction for local solar motion and Earth's orbital velocity, the maximum line-of-sight velocity in a direction making an angle of 30° with the direction to the centre of the Galaxy is 210 km s^{-1}. Calculate the mass of the Galaxy in solar mass units on the assumption that its mass is concentrated at its centre and that the Sun is 8·5 kpc from the centre.

15.3 For an angular distance θ from the galactic centre the observed maximum Doppler shift in the 21 cm line for material in the galactic plane is l cm. Assuming that those parts of the Galaxy which lie at angular distances not less than θ from the centre rotate as if the whole mass of the galaxy were concentrated at the centre, prove that the rotation velocity V_0 at the Sun's distance from the centre is given by

$$V_0 = \frac{cl}{21[(\operatorname{cosec}\theta)^{1/2} - \sin\theta]} \text{ km s}^{-1}$$

where c is the velocity of light in kilometres per second.

15.4 A star at a distance of 10 pc has an apparent magnitude of $0·0^m$ whilst the globular cluster 47 Tuc, at a distance of 4·6 kpc, has an apparent magnitude of $4·0^m$. Assuming that the single star is representative of those in the globular cluster, estimate the number of stars in the cluster.

15.5 Two stars lying in the galactic plane have longitudes l and $(90-l)$, their proper motions in galactic longitude being μ_1 and μ_2 respectively. Assuming circular orbits and that the Galaxy acts as a point-mass, show that

$$l = \frac{1}{2}\cos^{-1}\left[\frac{1}{3}\left(\frac{\mu_2 - \mu_1}{\mu_2 + \mu_1}\right)\right].$$

15.6 If the gravitational attraction at the Sun's distance from the galactic centre were due two-thirds to a central point-mass and one-third to mass distributed uniformly throughout a spheroid (the Sun being within the spheroid), prove that $A + B = 0$, where A and B are Oort's constants.

15.7 In the case of a spherically symmetric stellar system in a steady state, show that a solution of Boltzmann's equation is

$$f = f(V_R^2 + V_T^2 - 2U, rV_T)$$

where V_R and V_T are the radial and transverse velocities respectively ($V_R = \dot{r}$, $V_R^2 + V_T^2 = u^2 + v^2 + w^2$) and U is the gravitational potential.

15.8 An observer on a planet in orbit about a star moving in a circular orbit of radius r about the centre of a spherical star cluster of uniform density ρ and radius R finds that asymmetry of stellar motions for high-velocity stars sets in at a speed v relative to the observer's star. Prove that the observer's star's orbital speed v_c is given by

$$v_c = v\{[(3R^2/r^2) - 1]^{1/2} - 1\}.$$

References

Eddington A S 1913 *Mon. Not. R. Astron. Soc.* **74** 5
—— 1915 *Mon. Not. R. Astron. Soc.* **75** 366
Heggie D C 1988 *Long-Term Behaviour of N-body Dynamical Systems* ed A E Roy (Dordrecht: Reidel)
Hut P and McMillan S L W (ed) 1986 *The Use of Supercomputers in Stellar Dynamics* (Berlin: Springer-Verlag)
Jeans J H 1928 *Astronomy and Cosmogony* (Cambridge: CUP)
Lindblad B 1933 *Handbuch der Astrophysik Vol 2* (Berlin: Springer-Verlag) p 1033
Oort J H 1927a *Bull. Astron. Inst. Netherlands* **3** 275
—— 1927b *Bull. Astron. Inst. Netherlands* **4** 79
—— 1928 *Bull. Astron. Inst. Netherlands* **4** 269

Bibliography

Becker W and Contopoulos G (ed) 1970 *IAU Symposium No. 38* (Dordrecht: Reidel)

Binney J and Tremaine S 1988 *Galactic Dynamics* (Princeton: Princeton University Press)

Blaauw A and Schmidt M (ed) 1965 *Galactic Structure* (University of Chicago Press)

Chandrasekhar S 1960 *Principles of Stellar Dynamics* (New York: Dover)

Contopoulos G (ed) 1966 *IAU Symposium No. 25* (New York: Academic)

Eddington A S 1914 *Stellar Movements and The Structure of The Universe* (London: Macmillan)

Hayli A (ed) 1975 *IAU Symposium No. 69* (Dordrecht: Reidel)

Kozai T (ed) 1974 *IAU Symposium No. 62* (Dordrecht: Reidel)

Lecar M (ed) 1970 *IAU Colloquium No. 10* (Dordrecht: Reidel)

Ogorodnikov K F 1965 *Dynamics of Stellar Systems* (Oxford: Pergamon)

Oort J H 1977 *Ann. Rev. Astron. Astrophys.* **15** 295

Smart W M 1938 *Stellar Dynamics* (London: Cambridge University Press)

Tapley B D and Szebehely V (ed) 1973 *Recent Advances in Dynamical Astronomy* (Dordrecht: Reidel)

Toomre A 1977 *Ann. Rev. Astron. Astrophys.* **15** 437

Trumpler R J and Weaver H F 1953 *Statistical Astronomy* (Berkeley and Los Angeles: University of California Press)

Answers to Problems

Chapter 2

2.1 (i) 638 nautical miles; (ii) 1444 nautical miles.

2.2 3 h 3·4 min.

2.3 56° 00′; 36·7 min after leaving Prestwick (Hint: at its highest northerly latitude, the aircraft's course is 270° E of N).

2.4 0ʰ, 0° 00′; 6ʰ, 23° 27′ N; 12ʰ, 0° 00′; 18ʰ, 23° 27′ S (Hint: the Sun's path against the stellar background is the ecliptic. One revolution of this is performed by the Sun in one year, the Sun being at ♈ on or about March 21).

2.5

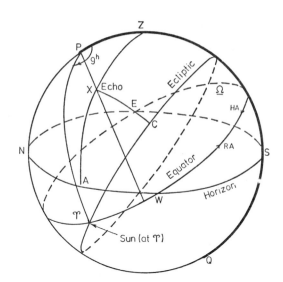

Figure A.1

(i) RA ~ 2ʰ, DEC ~ 60° N; (ii) Longitude ~ 50°, Latitude ~ 30° N.

2.6 RA = 2ʰ 24ᵐ, DEC = 59° 35′ N.

2.7 8 PM.

2.8 6° 33′.

2.10 23ʰ 05ᵐ 54ˢ.

2.11 The LST = HA ☉, DEC ☉ = LST − HA ☉; therefore ♈ and also the Sun can be placed in the diagram and hence the ecliptic can be inserted.

2.12 *B* is (44° 30′ S, 14° 08′ W); 11·00ʰ, January 8 (Hint: in such problems, especially those that involve crossing the International Dateline, it is convenient to use the GMT with date, known as the *Greenwich Date*, as an intermediary).

2.13 09ʰ 23ᵐ 41·6ˢ.

2.14   Clock errors (LST−clock reading) are −13·4^s and −12·3^s. Clock rate is +1·1^s/24 sidereal hours.

2.15   17^h 07^m 29^s. (Hint: obtain the GMT of observation, then the interval of mean solar time between this GMT and 0^h GMT December 13. Convert this interval to sidereal time to obtain the GST of the observation. Then apply the longitude.)

2.16   22^h 21^m 31^s.

2.17   13^h 44^m 32^s of sidereal time; 02^h 40^m 22^s LST; 285° 40′ E of N.

Chapter 3

3.1   6·67 times.

3.2 $M=1·26$ for Sirius, $M=2·95$ for Procyon.

3.3   158·5 to 1.

3.4   5·1 km s^{-1} receding.

3.5   28° 00′ 31″.

3.6 $\phi=66°$ 46′ 39″ N, $\delta=56°$ 29′ 34″ N.

3.7   (i) RA$=0^h$, DEC$=0°$; (ii) RA$=6^h$, DEC$=+23°$ 27′. (Hint: no calculation is involved. Draw a diagram and insert the appropriate equinoxes and equators.)

3.8 $d\alpha/dt=4·4274^s$, $d\delta/dt=4·0069^s$ (use five-figure logarithms).

3.10   Topocentric RA$=2^h$ 11^m 40$^s=32°$ 55′. Topocentric DEC$=32°$ 56′ N. Geocentric RA$=5^h$ 44^m 48$^s=86°$ 12′. Geocentric DEC$=54°$ 35′ N. (Hint: an accurately drawn diagram is often useful in checking decisions about the quadrants into which angles fall.)

Chapter 4

4.2   (Velocity at perihelion)/(velocity at aphelion)$=60·16$. Corresponding ratio for angular velocities$=3619$.

4.4   One-half the Earth's distance from the Sun. (Hint: use equation (4.82) and $r=(p/2)[1+\tan^2 (f/2)]$. Differentiate $(t-\tau)$ for a maximum after setting $r=1$.)

4.7 $E=116°$ 31′.

4.8   Vertically upwards; 145800 km; 54·40 h. (Hint: the angle is irrelevant except that it shows that no part of the orbit lies within the Earth. The second orbit is a covelocity one with respect to the 'vertically upwards' shot (a rectilinear ellipse).)

4.9   Mass of Sun/mass of Earth$=331700$.

4.10   The orbit is an ellipse with elements $a=0·7227\times$ (Earth's orbital semimajor axis), $e=0·5762$, $i=11°$ 57′, $\Omega=300°$ 19′, $\omega=315°$ 53′, $\tau=$November 20·72, 1962.

Chapter 5

5.1   Examination of equations (5.2) shows that if R_i, r_{ij} constitute a solution to the n-body problem, then DR_i, Dr_{ij} are also solutions if all times are multiplied by $D^{3/2}$.

5.2   Spheres centred on the 'fixed' body as shown by equation (4.16). A rectilinear ellipse.

5.3   This n-body problem is completely soluble, reducing to simple harmonic motions. For any mutual radius vector r_{ij}, we thus have

$$r_{ij}=a_{ij}\cos Nt+b_{ij}\sin Nt$$

where a_{ij} and b_{ij} are constants, $N=(k\Sigma_i m_i)^{1/2}$, where k is the gravitational constant for this law. The radius vector R_i from the centre of mass to the body of mass m_i is therefore given by

$$R_i=A_i\cos Nt+B_i\sin Nt$$

where A_i and B_i are constants.

5.4 $T=2\pi/N$; the centre of mass moves with constant velocity.

Chapter 6

6.1 Hint; $d^2x/dt^2 = g$ gives $x = \frac{1}{2}gt^2 + at + b$. Varying a and b and using the condition described in equation (6.26), we get $\dot{a}t + \dot{b} = 0$; subsequently we obtain $\dot{a} = \epsilon gx$.

6.2

$$e = e_0 + \frac{Gm_1\sqrt{1-e_0^2}}{na_0^2 e_0}\left\{\left(\Sigma\, kP \sin Q\right)t - \Sigma\frac{P'\cos Q'}{hn + h_1 n_1}\,[k + h(1 - \sqrt{1 - e_0^2})]\right\}.$$

6.3 The disturbing function is $\mu\lambda\rho^2/3r^3$ in this problem; use the $\dot{\varpi}$ equation of the set (6.30).

6.4 If M_0 is the mass at $t = 0$, then $M = M_0 - \alpha t$ where α is a small positive constant. The disturbing function R is then given by

$$R = -\frac{G\alpha t}{r} \simeq -\frac{G\alpha t}{a}\,[1 + e\cos(nt + \epsilon - \varpi)]$$

for small e. Hence by examining the relevant equations in the set (6.30), it is seen that $\dot{\Omega} = di/dt = 0$. The other four equations can be immediately integrated after substitution of $\partial R/\partial e$ etc. into them to give the first-order perturbations.

6.5 Hint: $D = -T'$; the use of equation (6.42), the first equation of the set (6.41) and the relevant two-body equations of chapter 4 produce the required expression.

Chapter 10

10.5 $46° 23'$; $106° 51'$.
10.6 $\dot{\omega} = 5\cdot28°/\text{day}$; $\dot{\Omega} = -3\cdot51°/\text{day}$.

Chapter 11

11.1 (i) Velocity increment is doubled; (ii) new increment = old increment + $0\cdot6932v_e$.
11.2 $2\cdot6153$ km s^{-1}; $54\cdot99$ km.
11.3 (Hint: the expression in R as unknown can be solved by a method of successive approximations.) $R = 6\cdot705$, $M_0 = 5\cdot141 \times 10^8$ g.
11.4 (i) $\Delta v_1 = \Delta v_A + \Delta v_B = (0\cdot5376 + 0\cdot2186)\pi = 0\cdot7562\pi$ AU/year.

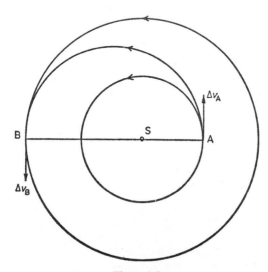

Figure A.2

(ii) $\Delta v_2 = \Delta v'_A + \Delta v'_C + \Delta v'_B = (0.5532 + 0.1654 + 0.0302)\pi = 0.7488\pi$ AU/year.

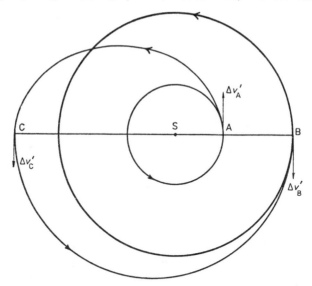

Figure A.3

11.5 Ratio of bi-elliptic transfer time to single transfer time = 5·468.

11.6 Saving in transfer time = (1·4142 − 0·6012) years = 0·8130 years.

11.7 (i) 0.3382π AU/year added tangentially; (ii) 1.139π AU/year in a direction making an angle of 12° 31' with the radius vector.

8.8 At the ascending and descending nodes of the inner orbital plane upon the outer orbital plane. The change in inclination should be made at outer transfer point. Saving in velocity increment = $(0.8273 − 0.7980)\pi = 0.0293\pi$ AU/year.

11.9 $\Delta e = \Delta a = 0$; $\Delta \omega = 2.036'$; $\Delta \tau = -0.374$ s.

11.10 1986 January 26·8; 1·930 year.

11.11 About 1'.

Chapter 12

12.1 0·216°/s.

12.2 88 260 km. (Hint: the Moon's angular velocity of rotation is 0·009 58 rad h⁻¹, which is the satellite's angular velocity of revolution.) This distance is greater than the radius of the Moon's sphere of influence (66 190 km).

12.3 $d_1 = 0.9002$, $d_2 = 1.125$.

12.4 0·14.

12.5 $AL_1 = 0.9929$, $AL_2 = 0.8490$, $AL_3 = 1.1678$. (Hint: the angular velocity n of AB about the centre of mass of the system gives $G(m_E + m_M)$ from $n^2 a^3 = G(m_E + m_M)$, since $a = 1$. Equating centrifugal force on the probe to the net gravitational force gives a quintic equation in the unknown distance which can be solved by a method of successive approximations.)

12.6 21 193 000 km for Jupiter; 9450 km for Mars.

12.7 (i) 2·30 km s⁻¹; (ii) 3·51 km s⁻¹.

12.8 The period of transfer to asteroid when it is at aphelion is exactly 1 year; therefore, on probe's return along the other half of the transfer orbit to Earth's orbit, it meets Earth again. Required hyperbolic excess velocity is 0.3408π AU/year$=5.078$ km s^{-1}.

12.9   Using equations (4.91), (4.88) (with $f=0$), (4.92), (4.97), (4.100) and (4.101) and the relevant data in the appendices, flight time is 143 h and departure velocity is 11.9 km s^{-1}.

12.10   Let the periods of Earth and probe be T_E and T. Then $T=T_E=1$. Also $r=r_E=1$ AU. Hence $V=V_E$. Also $\gamma=90°-\theta/2$. If $S\hat{P}B\neq90°$, the probe's orbit is not circular. In figure A.4,

$$FD=DP \sin D\hat{P}F=V \sin(S\hat{P}D-90°)=V \sin(90°-\gamma).$$

Hence

$$FD=V \sin \frac{\theta}{2}.$$

Also

$$S\hat{P}B=90°+C\hat{P}B=90°+\frac{V}{\bar{V}}\sin\frac{\theta}{2}.$$

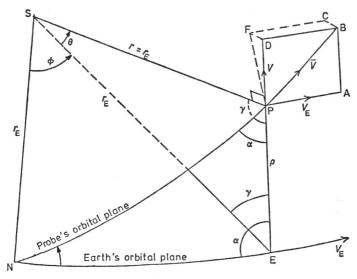

Figure A.4

Hence the eccentricity e of the probe's orbit is given by

$$e^2=1-\frac{r}{a}\left(2-\frac{r}{a}\right)\sin^2 S\hat{P}B\simeq\sin^2\left(\frac{V}{\bar{V}}\sin\frac{\theta}{2}\right).$$

But $\theta\leqslant1°$ and $(V/\bar{V})\sim0.168$. Hence $e\leqslant0.0015$.

If orbits of probe and Earth intersect at N and $N\hat{E}P=\alpha$, then $APE=PAB=\alpha$. Also

$$\sin\frac{\theta}{2}=\frac{\rho}{2r_E}$$

and since $\bar{V}=V_E$,

$$V=2V_E \cos \alpha.$$

Now if $N\hat{S}E=\phi$, the spherical triangle NPE gives

$$\sin\frac{i}{2}=\sin\frac{\theta}{2}\operatorname{cosec}\phi$$

503

giving

$$\sin \frac{i}{2} = \frac{\rho}{2r_E} \operatorname{cosec} \phi.$$

But

$$\cos \alpha = \frac{1 - \cos \theta}{\sin \theta} \cot \phi = \tan \frac{\theta}{2} \cot \phi$$

and

$$\cos \alpha = \frac{V}{2V_E} = 0.08389$$

giving $\alpha = 85° \, 11'$. Hence θ is found to be $1° \, 1 \cdot 1'$ and $\phi = 6° \, 03'$. Putting in values for ρ and r_E, i is found to be $9° \, 40'$.

The probe keeps in step with Earth (apart from a libration) and thus the maximum geocentric distance occurs about three months after injection into orbit when

$$\rho_E = 2r_E \sin \frac{i}{2} = 25 \cdot 24 \times 10^6 \text{ km.}$$

The probe re-enters the Earth's sphere of influence about six months after launch.

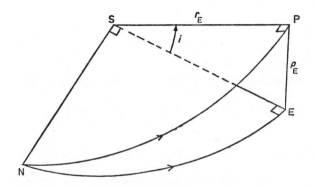

Figure A.5

Chapter 14

14.1   Magnitudes cannot be added but brightnesses can. If B, B_1 and B_2 are the brightnesses of the binary and its components, then $B = B_1 + B_2$. This gives us (see section 3.2) the relation $10^{-0 \cdot 4m} = 10^{-0 \cdot 4 \times 8 \cdot 3} + 10^{-0 \cdot 4 \times 7 \cdot 6}$ from which we obtain $m = 7 \cdot 14$.

14.2 $m_1 = 6 \cdot 82$, $m_2 = 4 \cdot 47$, $B_2/B_1 = 8 \cdot 71$.

14.3   The components' masses are $0 \cdot 878$ and $1 \cdot 201$.

14.4 $0 \cdot 0238$ and $-0 \cdot 4762$.

14.5   Use the Doppler formula to obtain the orbital velocity and hence semimajor axis in kilometres. Use equation (14.2) to obtain the sum of masses $= 2 \cdot 388$ solar masses. Hence by equation (14.6), the parallax is found to be $0 \cdot 0715''$.

14.6   Each 'signal' of the eclipse has to travel a distance $30 \times 3 \cdot 12 \times 24 \times 3600$ km further than the previous one. Light takes 27 s to cover that distance.

14.7   The celestial sphere for a spectroscopic binary is shown below. The component C is orbiting about the binary centre of mass at G. The radial component of velocity along GZ is required. Thus $R = \dot{z}$, where (x, y, z) are the coordinates of the component C. Now $z = r \sin BD$. From triangle NBD, by the sine formula, we have

$$z = r \sin BD = r \sin (\omega + f) \sin i.$$

504

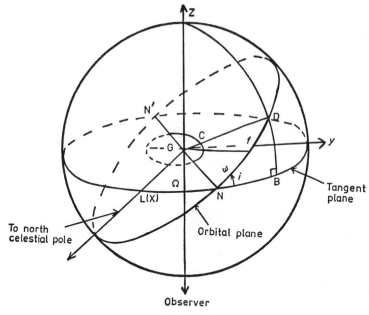

Figure A.6

We have also the two-body elliptic formulae

$$r=\frac{a(1-e^2)}{1+e\cos f}, \quad r^2\frac{df}{dt}=h=\sqrt{\mu a(1-e^2)}, \quad n^2a^3=\mu.$$

Differentiating the relation for z and using these formulae, we obtain the required expression.

14.8 $P_0=0\cdot151$, using $M_1+M_2=2$ in equation (14.6). First approximations to the absolute magnitudes (using the apparent magnitudes) give $5\cdot04$ and $7\cdot24$. From the mass–luminosity relation (figure 14.4), we get second approximations to the masses of $0\cdot851$ and $0\cdot550$; $M_1+M_2=1\cdot401$ then gives $P_1=0\cdot169$. A new cycle does not change P in the last significant figure.

Chapter 15

15.1 $1\cdot347\times10^{11}$. (Use Newton's method of calculating the mass of a planet in terms of the Sun's mass.)

15.2 $1\cdot044\times10^{11}$.

15.4 5315 (Use the formulae $M=m+5-5\log_{10}d$ and $B_1/B_2=10^{-0\cdot4(m_1-m_2)}$).

15.5 Hint: use the second equation of equations (15.50) with $b=0$.

15.8 Hint: asymmetry sets in when $v_c+v=$ velocity of escape $=v_{esc}$.

$$v_c^2=\frac{GM(r)}{r}=\tfrac{4}{3}\pi G\rho r^2.$$

For escape, we have

$$\tfrac{1}{2}v_{esc}^2-U(r)=0.$$

But

$$\frac{-dU}{dr}=\frac{GM(r)}{r^2}$$

giving

$$U(r) = A - \tfrac{2}{3}\pi G\rho r^2.$$

For $r = R$, we have

$$U(R) = \frac{GM}{R} = A - \tfrac{2}{3}\pi G\rho R^2$$

so that

$$A = 2\pi G\rho R^2.$$

Hence

$$\tfrac{1}{2}v_{esc}^2 = 2\pi G\rho(R^2 - \tfrac{1}{3}r^2).$$

Substitution in

$$v_c + v = v_{esc}$$

gives

$$v_c = v\{[(3R^2/r^2) - 1]^{1/2} - 1\}^{-1}.$$

Appendix I: Astronomical and Related Constants

1. The Sun

Radius $= 696\,000$ km $= 432\,000$ mi
Mass $= 1\cdot99 \times 10^{30}$ kg
Mean density $= 1\cdot41 \times 10^{3}$ kg m^{-3}
Surface gravity $= 2\cdot74 \times 10^{2}$ m s$^{-2} = 27\cdot9\ g_{\oplus}$

2. The Earth

1 astronomical unit (AU) $= 149\,596\,000$ km $= 92\,955\,000$ mi (measured by radar)
Equatorial radius $(a) = 6378\cdot12$ km $= 3963\cdot2$ mi
Flattening $(f) = 1/298\cdot2 = 0\cdot003\,352$
Polar radius $(b) = 6356\cdot74$ km $= 3949\cdot9$ mi
Eccentricity $(e) = 1/12\cdot22 = 0\cdot081\,83$
Mass $= 5\cdot98 \times 10^{24}$ kg
Mean density $= 5\cdot52 \times 10^{3}$ kg m^{-3}
Normal gravity $(g_{\oplus}) = (9\cdot8064 - 0\cdot0259 \cos 2\phi)$ m s^{-2}, where ϕ is the geodetic latitude
$b = a(1-f)$, $\quad b^2 = a^2(1-e^2)$, $\quad e^2 = 2f - f^2$

3. The Moon

Mean radius $= 1738$ km $= 1080$ mi
Mass $= 7\cdot35 \times 10^{22}$ kg
Escape velocity from surface $= 2\cdot37$ km s^{-1}
Surface gravity $= 1\cdot62$ m s$^{-2} = 0\cdot16\ g_{\oplus}$

4. Physical Constants

Speed of light $= 299\,791$ km s^{-1}
Constant of gravitation $= 6\cdot668 \times 10^{-11}$ N m^2 kg^{-2}

Solar parallax $P_\odot = 8 \cdot 794\,05''$
Obliquity of ecliptic $= 23° \; 27' \; 08 \cdot 26'' - 46 \cdot 84''\; T†$
Constant of general precession $= 50 \cdot 2564'' + 0 \cdot 0222''\; T†$
Constant of aberration $= 20 \cdot 49''$
Constant of nutation $= 9 \cdot 207''$

5. Time

Length of the second

1 ephemeris second $= 1/31\,556\,925 \cdot 975$ of length of tropical year at $1900 \cdot 0$
1 mean solar second (smoothed) $= 1$ ephemeris second (to about 1 part in 10^8)
1 mean solar second $= 1 \cdot 002\,737\,909\,3$ mean sidereal seconds

Length of the day

1 mean solar day $= 1 \cdot 002\,737\,909\,3$ sidereal days $= 24^h\;03^m\;56 \cdot 5554^s$ mean sidereal time $= 86\,636 \cdot 5554$ mean sidereal seconds
1 sidereal day $= 0 \cdot 997\,269\,5664$ mean solar days $= 23^h56^m\;04 \cdot 0905^s$ mean solar time $= 86\,164 \cdot 0905$ mean solar seconds

Length of the month

Synodic	$29 \cdot 530\,59^d$	or $29^d\;12^h\;44^m\;03^s$
Sidereal	$27 \cdot 321\,66$	or 27 07 43 12
Anomalistic	$27 \cdot 554\,55$	or 27 13 18 33
Nodical	$27 \cdot 212\,22$	or 27 05 05 36
Tropical	$27 \cdot 321\,58$	or 27 07 43 05

Length of the year

Tropical (Υ to Υ)	$365 \cdot 242\,20^d$	or $365^d\;05^h\;48^m\;46^s$
Sidereal (fixed star to fixed star)	$365 \cdot 256\,36$	or 365 06 09 10
Anomalistic (perigee to perigee)	$365 \cdot 259\,64$	or 365 06 13 53
Julian	$365 \cdot 25$	or 365 06 00 00

6. Mathematical Constants and Conversion Factors

1 radian $= 57 \cdot 295\,78° = 3437 \cdot 747' = 206\,264 \cdot 8''$
$e = 2 \cdot 718\,281\,83$
$\pi = 3 \cdot 141\,592\,65$
$\log_{10} x = 0 \cdot 434\,294\,48 \ln x$
$\ln x = 2 \cdot 302\,585\,09 \log_{10} x$
1 km s$^{-1} = 2236 \cdot 9$ mi/h
1 km $= 0 \cdot 621\,37$ mi $= 3280 \cdot 8$ ft
1 mi $= 1 \cdot 6093$ km

† T is measured in Julian centuries from 1900.

1 International Nautical Mile = 1·8520 km = 1·1508 statute mile = 6076·11 ft
1 Astronomical Unit (AU) = 149 600 000 km = 92 960 000 mi
1 parsec (pc) = 206 264·8 AU = 3·085 721 × 10^{13} km = 1·917 431 × 10^{13} mi
1 light year = 63 238·4 AU = 9·460 47 × 10^{12} km = 5·878 62 × 10^{12} mi
1 light year = 0·306 59 pc; 1 parsec = 3·2617 light year
1 ft/s = 0·304 80 m s^{-1}
1 AU/year = 2·9456 mi/s = 4·7404 km s^{-1}
1 kg = 2·204 622 lb = 0·984 206 × 10^{-3} ton
1 lb = 0·453 592 kg

7. The Galaxy

Diameter = 30 000 pc
Mean thickness = 2000 pc
Total mass = 1·4 × 10^{11} solar masses
Sun's distance from galactic centre = 8500 pc
Rotational velocity in solar neighbourhood = 250 km s^{-1}
Oort's constants of galactic rotation are:

$$\left.\begin{array}{l} A = +0\cdot020 \text{ km/s/pc} \\ B = -0\cdot013 \text{ km/s/pc} \end{array}\right\} \pm 0\cdot002.$$

8. SI Units

Basic SI units

Physical quantity	Name of unit	Symbol
length	metre	m
mass	kilogram	kg
time	second	s
electric current	ampere	A
thermodynamic temperature	degree Kelvin	K

Derived SI units for special quantities

Physical quantity	Name of unit	Symbol	Dimensions
energy	joule	J	$kg \, m^2 \, s^{-2}$
power	watt	W	$kg \, m^2 \, s^{-3}$
force	newton	N	$kg \, m \, s^{-2}$
electric charge	coulomb	C	$A \, s$
magnetic flux density	tesla	T	$kg \, s^{-2} \, A^{-1}$
frequency	hertz	Hz	s^{-1}

Conversion of cgs units to SI units

$1 \text{ ångström} = 10^{-10} \text{ m}$

$1 \text{ dyne} \quad = 10^{-5} \text{ N}$

$1 \text{ erg} \quad = 10^{-7} \text{ J}$

$1 \text{ millibar} \quad = 10^{2} \text{ GN m}^{-2}$

$1 \text{ gauss} \quad = 10^{4} \text{ T}$

$1 \text{ erg s}^{-1} \quad = 10^{-7} \text{ W}$

Appendix II: The Earth's Gravitational Field

The gravitational potential U of the Earth at an exterior point at a distance r from the centre of mass is given by

$$U = \frac{GM}{r} \left[1 - \sum_{n=2}^{\infty} J_n \left(\frac{R}{r} \right)^n P_n (\sin \phi) \right]$$

where ϕ is the geocentric latitude of the point, P_n is the Legendre polynomial of order n, G is the gravitational constant, M is the mass of the Earth, R is the Earth's equatorial radius and the J_n are constants.

Recently determined values of the constants J_n and of GM and R are given below (King-Hele *et al* 1967, 1969, Kozai 1964, 1969).

$$10^6 J_2 = 1082 \cdot 63 \pm 0 \cdot 01 \qquad 10^6 J_3 = -2 \cdot 51 \pm 0 \cdot 01$$
$$10^6 J_4 = -1 \cdot 60 \pm 0 \cdot 01 \qquad 10^6 J_5 = -0 \cdot 13 \pm 0 \cdot 01$$
$$10^6 J_6 = 0 \cdot 50 \pm 0 \cdot 01 \qquad 10^6 J_7 = -0 \cdot 36 \pm 0 \cdot 01$$
$$10^6 J_8 = -0 \cdot 12 \pm 0 \cdot 02 \qquad 10^6 J_9 = -0 \cdot 10 \pm 0 \cdot 02$$
$$10^6 J_{10} = -0 \cdot 35 \pm 0 \cdot 02 \qquad 10^6 J_{11} = 0 \cdot 20 \pm 0 \cdot 04$$
$$10^6 J_{12} = -0 \cdot 04 \pm 0 \cdot 03 \qquad 10^6 J_{13} = -0 \cdot 12 \pm 0 \cdot 05$$

$$GM = 398\,603 \cdot 2 \text{ km}^3 \text{ s}^{-2}$$

$$R = 6378 \cdot 165 \text{ km}$$

Legendre polynomials

If $P_n(x)$ is the Legendre polynomial of degree n in x, then

$$P_n(x) = \frac{1}{2^n n!} \frac{d^n}{dx^n} [(x^2 - 1)^n]$$

and hence

$$P_2(x) = \tfrac{1}{2} (3x^2 - 1)$$
$$P_3(x) = \tfrac{1}{2} (5x^3 - 3x)$$
$$P_4(x) = \tfrac{1}{8} (35x^4 - 30x^2 + 3)$$
$$P_5(x) = \tfrac{1}{8} (63x^5 - 70x^3 + 15x)$$
$$P_6(x) = \tfrac{1}{16} (231x^6 - 315x^4 + 105x^2 - 5)$$
etc.

It should be noted that in the above expression for U, tesseral and sectorial harmonics are not included.

References

King-Hele D G, Cook G E and Scott D W 1967 *Planet. and Space Sci.* **15** 741
—— 1969 *Planet. and Space Sci.* **17** 629
Kozai Y 1964 *Publ. Astron. Soc. Japan* **16** 263
—— 1969 *Smithsonian Institution Astrophysical Observatory, Special Report* No. 295

Bibliography

Allen C W 1973 *Astrophysical Quantities* 3rd edn (London: Athlone)

Appendix III

Mean Elements of the Planetary Orbits† (Epoch 1960 January 1·5, Ephemeris time)

Planet	Mean distance from Sun (AU)	Mean distance from Sun (10^6 km)	Sidereal period (tropical years)	Synodic period (days)	Mean daily motion n (°/day)	Orbital velocity (km s^{-1})
Mercury	0·387099	57·9	0·24085	115·88	4·092339	47·8
Venus	0·723332	108·1	0·61521	583·92	1·602131	35·0
Earth	1·000000	149·5	1·00004	—	0·985609	29·8
Mars	1·523691	227·8	1·88089	779·94	0·524033	24·2
Jupiter	5·202803	778	11·86223	398·88	0·083091	13·1
Saturn	9·538843	1426	29·45772	378·09	0·033460	9·7
Uranus	19·181951	2868	84·01331	369·66	0·011732	6·8
Neptune	30·057779	4494	164·79345	367·48	0·005981	5·4
Pluto‡	39·43871	5896	247·686	366·72	0·003979	4·7

Planet	Inclination	Eccentricity	Mean longitude (deg) of node	Mean longitude (deg) of perihelion	Mean longitude (deg) at epoch
Mercury	7·00399	0·205627	47·85714	76·83309	222·62165
Venus	3·39423	0·006793	76·31972	131·00831	174·29431
Earth	0·0	0·016726	0·0	102·25253	100·15815
Mars	1·84991	0·093368	49·24903	335·32269	258·76729
Jupiter	1·30536	0·048435	100·04444	13·67823	259·83112
Saturn	2·48991	0·055682	113·30747	92·26447	280·67135
Uranus	0·77306	0·047209	73·79630	170·01083	141·30496
Neptune	1·77375	0·008575	131·33980	44·27395	216·94090
Pluto‡	17·1699	0·250236	109·88562	224·16024	181·64632

† From *Explanatory Supplement to the Astronomical Ephemeris and the American Ephemeris and Nautical Almanac*. Reprinted by permission of Her Majesty's Stationery Office, Atlantic House, Holborn Viaduct, London E.C.1.

‡ The elements for Pluto are osculating values for epoch 1960 September 23.0 ET.

Approximate Elements of the Four Major Asteroids† (Epoch 1957, June 11, 0ʰ ET, Referred to Mean Equinox and Ecliptic of 1950·0)

Asteroid	Mean distance from Sun (AU)	Mean daily motion (°/day)	Inclination (deg)	Eccentricity
Ceres	2·7675	0·21408	10·607	0·07590
Pallas	2·7718	0·21358	34·798	0·23402
Juno	2·6683	0·22612	12·993	0·25848
Vesta	2·3617	0·27157	7·132	0·08888

Asteroid	Ω (deg)	ϖ (deg)	Mean anomaly at epoch (deg)	Diameter of asteroid (km)
Ceres	80·514	152·367	279·880	770
Pallas	172·975	122·734	271·815	490
Juno	170·438	56·571	329·336	190
Vesta	104·102	253·236	79·667	390

† From *Explanatory Supplement to the Astronomical Ephemeris and the American Ephemeris and Nautical Almanac.* Reprinted by permission of Her Majesty's Stationery Office, Atlantic House, Holborn Viaduct, London, E.C.1.

Appendix IV

Physical Elements of the Planets†

Planet	Equatorial radius (km)	Reciprocal of flattening	Mass (Earth=1)	Mean density (g cm⁻³)
Mercury	2420	∞	0·056	5·13
Venus	6200	∞	0·817	4·97
Earth	6378	297	1·000	5·52
Mars	3400	192	0·108	3·94
Jupiter	71370	16·1	318·0	1·33
Saturn	60400	10·4	95·2	0·69
Uranus	23530	16	14·6	1·56
Neptune	22300	50	17·3	2·27
Pluto	1500?	?	0·002?	1?

Planet	Surface gravity (Earth=1)	Velocity of escape (km s⁻¹)	Rotation period	Inclination of equator to orbit
Mercury	0·36	4·2	58^d 16^h	47°
Venus	0·87	10·3	243^d (retro)	~179°
Earth	1·00	11·2	23^h 56^m 04^s	23° 27'
Mars	0·38	5·0	24^h 37^m 23^s	23° 59'
Jupiter	2·64	61	9^h 50^m 30^s‡	3° 04'
Saturn	1·13	37	10^h 14^m‡	26° 44'
Uranus	1·07	22	10^h 49^m	97° 53'
Neptune	1·41	25	14^h?	28° 48'
Pluto	?	1?	6·39^d	?

† From *Explanatory Supplement to the Astronomical Ephemeris and the American Ephemeris and Nautical Almanac.* Reprinted by permission of Her Majesty's Stationery Office, Atlantic House, Holborn Viaduct, London E.C.1.

‡ For these planets, the period of rotation depends upon the latitude, increasing towards the poles. In the case of Pluto, the values given are very uncertain.

Saturn's Rings†

Ring			Radius (10^3 km)
A Outer	{	Outer edge	136·0
		Inner edge	119·7
B Inner	{	Outer edge	117·0
		Inner edge	90·5
C Crape		Inner edge	74·6

† The thickness of the rings is less than 20 km.

Each ring consists of hundreds of ringlets, some elliptical. The recently discovered additional rings are:

Ring	Radius (10^3 km)
D	76·6
E	210 (inner)
	294·6 (outer)
F	139·2
G	168

The nine narrow rings of Uranus lie between 42 000 and 51 000 km from the planet's centre. From inner to outer boundary they are designated 6, 5, 4, α, β, η, γ, δ and ϵ.

Satellite Elements and Dimensions

Planet and satellite	Mean distance (10³ km)	Sidereal period (days)	Inclination of orbit to planet's equator or planet's orbit§	Eccentricity of mean orbit	Radius of satellite (km)	Reciprocal mass (1/planet)
Earth						
Moon	384·4	27·321 66	5°08'§	0·054 90	1738	81·3
Mars						
I Phobos	9·4	0·318 91	0°57'	0·0210	11	?
II Deimos	23·5	1·262 44	1°18'	0·0028	6	?
Jupiter						
XVI 1979J3	128	0·295	~0	<0·004	20	?
XV 1979J1	128	0·297	~0	~0	12	?
V Amalthea	181	0·498 18	0°24'	0·003	135×78	?
XIV 1979J2	222	0·678	0°48'	0·015	40	?
I Io	422	1·769 14	0	0	1826	21 300
II Europa	671	3·551 18	0	0	1560	39 000
III Ganymede	1071	7·154 55	0	0	2500	12 700
IV Callisto	1884	16·689 02	0	0	2450	17 800
VI Himalia	11 480	250·57	27°38'	0·157 98	85	?
VII Elara	11 740	259·65	24°46'	0·207 19	40	?
X Lisithea	11 860	263·55	29°01'	0·130 29	10	?
XIII Leda	11 100	239	26°42'§	?	4	?
XII Ananke	21 200	631·1	147°	0·168 70	10	?
XI Carme	22 600	692·5	164°	0·206 78	10	?
VIII Pasiphae	23 500	738·9	145°	0·378	10	?
IX Sinope	23 700	758	153°	0·275	10	?

Satellite Elements and Dimensions (*cont.*)

Planet and satellite	Mean distance (10³ km)	Sidereal period (days)	Inclination of orbit to planet's equator (or orbit)*	Eccentricity of mean orbit	Radius of satellite (km)	Reciprocal mass (1/planet)
Saturn						
1980S28	137·7	0·60192	0° ?	~0	20 × 10 × ?	$2 \cdot 2 \times 10^{-11}$†
1980S27	139·4	0·61300	0° ?	~0·0024	70 × 50 × 40	$1 \cdot 0 \times 10^{-9}$†
1980S26	141·7	0·62854	0° ?	~0·0042	55 × 45 × 35	$6 \cdot 4 \times 10^{-10}$†
1980S1	151·4	0·69433	0° 8'	~0·007	110 × 90 × 80	$6 \cdot 5 \times 10^{-9}$†
1980S3	151·5	0·69467	0° 20'	~0·009	70 × 60 × 50	$1 \cdot 5 \times 10^{-9}$†
Mimas	185·6	0·94221	1° 31'	0·0201	195	15 000 000
Enceladus	238·1	1·36908	0° 01'	0·00444	255	7 000 000
Tethys	294·7	1·88521	1° 06'	0	525	910 000
1980S25	294·7	1·88521	~0°	~0	17 × 11 × 11	$1 \cdot 5 \times 10^{-11}$†
1980S13	294·7	1·88521	~0°	~0	17 × 14 × 13	$2 \cdot 2 \times 10^{-11}$†
Dione	377·5	2·73313	0° 01'	0·00221	560	490 000
1980S6	377·4	2·73921	0° 12'	0·005	18 × 16 × 15	$3 \cdot 1 \times 10^{-11}$†
Rhea	527·2	4·51075	0° 21'	0·00098	765	250 000
Titan	1 221·6	15·91029	0° 20'	0·0289	2575	4 150
Hyperion	1 483·0	21·28121	0° 26'	0·104	205 × 130 × 110	5 000 000
Iapetus	3 560·1	79·15492	14° 43'	0·02828	720	300 000
Phoebe	12 950·0	549·14775	150°*	0·16326	100	?

† Assuming mean densities of 1 g cm^{-3}.

518

Satellite Elements and Dimensions (*cont.*)

Planet and satellite		Mean distance (10³ km)	Sidereal period (days)	Inclination of orbit to planet's equator (or orbit)*	Eccentricity of mean orbit	Radius of satellite (km)	Reciprocal mass (1/planet)
Uranus†							
V	Miranda	124	1·414	0	<0·01	150?	1 000 000
I	Ariel	192	2·520 38	0	0·002 8	400	67 000
II	Umbriel	267	4·144 18	0	0·003 5	300	170 000
III	Titania	438	8·705 88	0	0·002 4	600	20 000
IV	Oberon	587	13·463 26	0	0·000 7	500	34 000
Neptune							
I	Triton	354	5·876 83	159° 57'	0	1850	750
II	Nereid	5570	359·4	27° 27'*	0·76	150?	
Pluto							
	Charon	20	6·39	0?	0?	600?	10?

† The Voyager fly-by of Uranus discovered a number of small satellites in addition to the five major Uranian satellites listed here.

Index

Cook A 296, 297
Cook G E 512
Coordinates
—, barycentric 29
—, ecliptic 25-6
—, equatorial 23-5
—, galactic 40-1
—, generalized 195-8
—, geographic 19-21
—, horizontal 22-3
—, ignorable 196
—, Jacobian 153-61, 289-91, 448-9
—, measured 60
—, orbital plane 29-30
—, rectangular 29, 34-9
—, standard 60
—, topocentric 25, 63
Copenhagen problem 144-7
Covelocity orbits 80-1
Cowell P H 208, 237
Cowell's method 208-9, 211, 213, 263, 374
Cowling T G 458
Critical
— argument 261, 263, 265-6
— inclination 319, 321
Crommelin A D 208, 237
Crossing time 495

Damoiseau M C 294
Danby J M A 120, 163, 205, 407, 426
Darwin G H 146, 147, 208
Data observational 50
—, reduction of 50-69
Day
—, apparent solar 41, 42
—, mean solar 45, 508
—, sidereal 41, 42, 45, 508
De Pontécoulant P G 294
De Vaucouleurs G 18
Declination 24, 25, 33, 34
Delaunay C 280, 281, 294, 295
Delaunay's lunar theory 280, 281, 294, 295
Departure 31
Deprit A 141, 163, 294, 295, 296, 298
Dermott S F 248, 257, 278
Descartes R 125
Differentiation, numerical 232
Digital orrery 266
Dione 246, 261, 262
Disturbing function 167-8, 181-2, 183-4, 186, 270, 452
— for artificial satellite 314-6
—, resolution of 191-4
Dollfus A 245
Doppler C 52
Doppler
— effect 52, 245, 428, 481
— radar 52, 53, 286, 296, 416, 422
Dormand J R 263, 278
Douglas M R 277

Drag
—, atmospheric 308, 311, 323-8
— coefficient 324
Dynamical
— ellipticity 289, 305
— parallax 433-4

Earth 2, 3, 19-21, 31-2, 167, 301-11, 313, 370, 375-7, 507, 511-3
— as a planet 301-11
—, atmosphere of 307-9
—, gravitational potential of 176-7, 311, 313, 370, 511-3
—, interior of 306
—, magnetic field of 306-7, 309-11
—, shape of 165, 176, 301-6, 511-2
—, size of sphere of influence 374-5
Earth-Moon system 125, 139, 141, 147, 165, 167, 168, 240, 280-97, 299, 371
Earth-Moon trajectories 352, 370-83
Eccentric anomaly 82-5
Eccentricity 28, 76
Eckels A 165, 331
Eckert W J 263, 278, 295, 296
Eckstein M C 331
Eclipsing binary 13, 427, 435-40, 445
Ecliptic 25-6
— coordinate system 25-6
— latitude 26
— longitude 26
—, obliquity of 25
—, pole of 25
Eddington A S 494
Ehricke K A 338, 343, 345, 369, 403
Einstein A 450
Einstein's law of gravitation 450-1
— theory of relativity 450-1
Electronic methods of observation 51, 52, 53
Elements
— of the asteroid orbits 514
— of the orbit 26-9
— of the planetary orbits 513
— of the satellite orbits 517-9
—, osculating 164
Elias J H 278
Elliot J 245, 278
Ellipse 21, 27-8, 76
—, osculating 164
Ellipsoid
—, Clarke 302
—, Hayford 21, 302
—, International 302, 305
—, potential of 176, 179
—, reference 21
Elliptic orbit 26-9, 76-89
Ellipticity 21
—, mechanical (dynamical) 289, 305
— of star 438-9
Elongation 383
Empirical stability criteria 271-6
Emslie A G 108, 110, 230, 237, 279